PROCEEDINGS OF THE CONFERENCE

Condensed Matter Physics Aspects of Electrochemistry

27 August – 9 September 1990
ICTP, TRIESTE, ITALY

Editors

M. P. Tosi
(ICTP)

A. A. Kornyshev
(Acad. of Sci., USSR)

World Scientific
Singapore • New Jersey • London • Hong Kong

Published by

World Scientific Publishing Co Pte Ltd
5 Toh Tuck Link, Singapore 596224
USA office: 27 Warren Street, Suite 401-402, Hackensack, NJ 07601
UK office: 57 Shelton Street, Covent Garden, London WC2H 9HE

British Library Cataloguing-in-Publication Data
A catalogue record for this book is available from the British Library.

CONDENSED MATTER PHYSICS ASPECTS OF ELECTROCHEMISTRY

ISBN-13 978-981-02-0560-7
ISBN-10 981-02-0560-0

PREFACE

This volume collects the proceedings of the Working Party on "Electrochemistry: Condensed Matter, Atomic and Molecular Physics Aspects", held for two weeks in the summer of 1990 at the International Centre for Theoretical Physics (ICTP) in Trieste. The goal of the meeting was to discuss those areas of electrochemistry that are accessible to the modern methods of theoretical condensed matter, atomic and molecular physics, in order to stimulate insight and deeper involvement by theoretical physicists into the field.

What is particularly interesting for a physicist in Electrochemistry ? The answer one may find by reading some earlier reviews and conference proceedings[1-8]. The correct answer should start with the importance of the field in itself. A beautiful answer has been given by Frumkin in his scientific testament, " a word in favor of electrochemistry"[9]. Here, we wish to stress two main points.

Electrochemistry is a science with important technological applications. These are, in particular:

- Energy conversion and power sources
- Transducers of information signals
- Electroanalysis
- Materials processing
- Electrolysis, electrocatalysis and electrosynthesis
- Inhibition of corrosion
- Biomedical engineering (heart stimulators, artificial kidney, acceleration of bone mending etc.).

The roots in the fundamentals of other sciences are no less important. Electrochemistry underlies many phenomena in inorganic chemistry, colloid science, biophysics and biochemistry. It is a powerful partner in a number of problems in the physics of liquids and solutions, solid state physics, surface science.

Electrochemistry and "electrophysics" developed hand in hand from the first steps in the study of electricity. The names of Galvani, Volta and Faraday are usually cited in this

---

1. H. Gerischer, D. M. Kolb and J. Sass, Adv. Phys. **27** (1978) 437.
2. Nontraditional Approaches to the Study of the Solid-Electrolyte Interface, eds. T. E. Furtak, K. L. Kliewer and D. W. Lynch, Surf. Sci. **101** (1980).
3. Physical Electrochemistry, eds. W. Schmickler and J. Ulstrup, Chem. Phys. **141**(1) (1980).
4. Studies in Physical and Theoretical Chemistry, eds. W. H. Hansen, D. M. Kolb and D. W. Lynch, vol. **29** (Elsevier, Amsterdam 1983).
5. Electrodynamics and Quantum Phenomena at Interfaces (Mezniereba, Tbilisi 1986).
6. Structure and Dynamics of Solid/Electrolyte Interfaces, eds. H. Gerischer and D. Kolb, Ber. Bunsenges. Phys. Chem. **91**(4) (1987).
7. Chemistry and Physics of Electrified Interfaces, ed. S. Trasatti, Electrochim. Acta **34**(12) (1989).
8. Interdisciplinary Conference on Electrified Interfaces, eds. D. Henderson and O. Melroy, Electrochim. Acta (1991) in press.
9. A. N. Frumkin, Selected Works: Electrode Processes (Moscow, Nauka 1987) p. 24.

context. The best example from the present century is the Debye theory of electrolytes, which has had an enormous impact on many-body physics. The fundamental equation of electroanalysis, i.e. the Lippmann-Gibbs equation, is the basic formula of electrocapillarity (as discussed in the Landau-Lifshitz book on Electrodynamics of Continuous Media). This list could be continued.

The next round of impetuous compenetration of these sciences, which began in the seventies, was induced by the "surface science boom". It was recognized that interfacial electrochemistry provides a number of possibilities for the study of metal or semiconductor surfaces *in situ*, i.e. in contact with the electrolyte. The latter plays a twofold role. The experience in "clean electrochemistry" allows one to deal with a solution instead of vacuum almost at an equivalent degree of purity. Secondly, and most importantly, the metal in contact with the electrolyte can be strongly charged by applying a moderate voltage, owing to the very large interfacial capacitance. A new and well controlled variable thus appears in the system. Moreover, the electric field produced by this charge is localized - due to screening by the electrolyte - within a thin interfacial region. Hence, the response to charging probes just the properties of the interfacial region.

The core of the ICTP Working Party was a set of topically grouped plenary lectures, accompanied by contributed seminars and by the formulation of joint research projects. In the tradition of the ICTP, it was not a meeting of pure theoreticians: about half of the lecturers were professional experimentalists - experts in electrochemistry, physical chemistry, surface science, technical applications. A finite set of topics was chosen for discussion at the meeting:

- Liquids, solvation, solutions
- The interface (structure, characterization, electric properties, adsorption)
- Electrodynamics, optics, photo-emission
- Charge transfer kinetics (homogeneous and heterogeneous reactions and processes)
- Superconducting electrodes
- Fractal electrodes
- Applied research (energy conversion and power sources, electrocatalysis, electroanalysis of turbulent flows).

A number of important directions in modern electrochemistry are still missing from this list. These are various aspects of hydrodynamics and mass transfer, diffusion and migration processes, multi-stage reactions, chaos and dissipative structures, superionics, bioelectrochemistry.

It was a common view of the participants in the meeting that its main fruit, which is a coherent picture of a large part of theoretical electrochemistry as a "condensed matter science", was worthy of record. In order to achieve a thorough analysis of the state of the art, it was also necessary to address some long-standing questions, which are often passed by because of their complexity and lack of sufficient knowledge. To create a background for an approach to such problems, some of the lecturers were welcomed to speak about matters that they had already reviewed in their publications. Such contributions are present in this volume as extended abstracts or brief notes, with updated references. In fact, the special character of the ICTP Working Party makes this book closer to a collective monograph than to typical conference proceedings.

In conclusion, we wish to acknowledge the support of Professor Abdus Salam and of the ICTP Solid State Advisory Committee, especially Professor Stig Lundqvist and Professor Norman March. We thank all who contributed to the scientific outcome of the Working Party: the lecturers (whose talks were deep, comprehensive and exciting) and the participants (whose questions, seminars and enthusiasm stimulated the work). Special thanks are due to the ICTP secretarial staff, particularly to Ms Deisa Buranello and Ms Marina De Comelli. One of us (A. A. K.) acknowledges the support of his regular visits to the ICTP given by the A. N. Frumkin Institute. His contribution to the final editorial process was facilitated by his stay at the Technical University of Munich as a Humboldt Award Guest Professor (Humboldt Foundation, FRG).

Alexei Kornyshev and Mario Tosi
Moscow-München and Trieste

CONTENTS

ADSORPTION

ELECTRODYNAMICS, SPECTROELECTROCHEMISTRY, SCANNING TUNNELING MICROSCOPY

INTRODUCTORY NOTES

ELECTROCHEMICAL PHYSICS

HEINZ GERISCHER
Fritz-Haber-Institut der Max-Planck-Gesellschaft,
Faradayweg 4-6, D-W-1000 Berlin 33, F. R. G.

Abstract

In this contribution, areas are emphasized, where the concepts of physics are fundamental for the interpretation of electrochemical processes or new techniques, developed in physics, have promoted the understanding of such processes. The short text and the references may be used as a guide to more detailed information.

1. Introduction

Electrochemistry is an interdisciplinary science. Its understanding combines concepts of chemistry, particularly structural and kinetic chemistry and physics, especially electrostatics, electrodynamics and solid state physics. It also has close links to biology. In this lecture the interrelation with physics have been exemplified with emphasis on some recent developments which have stimulated progress in both fields. Three general topics have been chosen, for which a selection of references is given in the following.

2. Surface Properties of Electrodes and Adsorption from Electrolytes

The chemical composition of the interface can now be analyzed by new spectroscopic techniques which have been developed in surface physics under ultrahigh-vacuum (UHV) conditions. Some of them can only be applied ex-situ after transfer of the electrode into UHV. Other techniques in which the signal can pass the electrolyte can be used in-situ.

Ex-situ methods are particularly photoelectron emission and electron energy loss spectroscopy (Auger PS, UPS, XPS, EELS) [1-6] and electron diffraction techniques (LEED, RHEED) [2,3,7-10].

In-situ techniques do not include the risk that the surface composition or structure is changed during the transfer into the vacuum. They can also be applied for dynamic processes at electrodes. Particularly instructive for the adsorption of organic molecules is IR reflection spectroscopy [11,12], in some special cases Raman reflection spectroscopy [13,14]. Information on electronic properties of electrode surfaces and their variation with surface composition or electrode potential can be obtained from reflection spectroscopy with visible or UV light [15-17]. Finally, X-rays can also be used in-situ for monolayer and surface studies where a high intensity source from synchrotron radiation is avialable. With grazing incidence of the X-rays and small variations of the angle of incidence one can obtain structural information [18,19]. The analysis of the interference with neighboring atoms reveils distances in adsorbates or in the electric double layer (EXAFS) [20,21]. A most recent technique giving structural information about the electrode surface in-situ up to atomic resolution is scanning tunneling microscopy (STM) [22-24].

3. Influence of Solid State Properties on Electrode Behavior

The electronic properties of the electrode control the distribution of excess electric charge in the electrical double layer and can also have a decisive influence on the charge transfer reaction at the interface. If the conductivity is metallic, the double layer properties depend more on the crystallographic surface orientation than on the electronic band structure. Semimetals [25] and particularly semiconductors [26,27] have a diffuse space charge layer which is reflected in a very different and asymmetric dependence of the differential capacity on the bias as compared to metallic electrodes.

The difference in band structure is also seen in the charge transfer kinetics. While simple redox reactions on metals depend little on the electron density at the Fermi level [28], one can distinguish at semiconductors between reactions where electrons of the conduction band are predominantly involved or holes of the valence band [29,30]. At metals, the bias controls the electric field strength in the Helmholtz double layer and affects the height of the activation barrier; at semiconductors the bias controls the concentration of electrons or holes at the surface, and this changes the reaction rate in parallel like the concentration of one of the reactants in a homogeneous reaction [31]. This corresponds to a variation of the entropy factor in kinetics. Surface orientation has a more subtle influence on the charge transfer reaction rates. This is particularly important in multistep charge-transfer reactions at metals in which intermediates being strongly adsorbed on the surface are generated [32]. Such processes are closely related to heterogeneous catalysis in the gas phase (electrocatalysis), and many questions are left open.

4. Photoeffects at Electrodes

Light absorption in electrodes generates excited electronic states. If their lifetime is long enough and they can reach the interface, they can induce charge transfer reactions in which the excitation energy diminishes the activation barrier very much or makes the reaction even activationless. In metals the lifetime is very short. The effects are therefore very small. They are equivalent to photoelectron emission into vacuum with three important differences: 1. The barrier height depends on the solvent and can be varied by the bias [33,34]. 2. The reaction yield depends on the presence of scavengers in the electrolyte that capture the charge carriers irreversibly [35]. 3. Not only excited electrons can be emitted and react with suitable scavengers but also excited holes, i.e., deep-lying vacant electronic states in the conduction band of metals [36]. Such differences should be most attractive for physicists to investigate solid state properties which are not accessible in vacuum [37].

The reactivity of excited electronic states is much more pronounced at semiconductor elcetrodes due to their long lifetime therein. This will be dealt with in another contribution to this workshop (see chapter on "The Semiconductor-Electrolyte Interface"). With semiconductor electrodes also the redox reactions of excited molecules of the electrolyte, if they are adsorbed on the surface, can be studied [38,39]. This is connected with the mechanism of spectral sensitization in photography [40]. The result of electronic excitation in solid electrodes can also be luminescence. This has been found at metal electrodes [41] and with much higher quantum yield at semiconductor electrodes [42,43].

5. References

1. P. M. A. Sherwood, *Surface Sci.* **101** (1980) 619.
2. A. Hubbard, *Acc. Chem. Res.* **13** (1980) 177.
3. P. N. Ross and F. T. Wagner, in *Advances in Electrochemistry and Electrochemical Engineering,* Vol. 13, eds. H. Gerischer and C. W. Tobias, (Wiley, New York, 1984), p. 69.
4. W. N. Hansen, *J. Electroanal. Chem.* **150** (1983) 133.
5. T. Solomun, R. McIntyre, W. Richtering and H. Gerischer, *Surface Sci.* **169** (1986) 414.
6. P. M. A. Sherwood, *Chem. Soc. Rev.* **14** (1985) 1.
7. A. Hubbard, R. Ishikawa and J. Kutakaru, *J. Electroanal. Chem.* **86** (1978) 271.
8. A. S. Homa, E. Yeager and B. D. Cahan, *J. Electroanal. Chem.* **150** (1983) 181.
9. M. S. Zei, Y. Kakai, G. Lehmpfuhl and D. M. Kolb, *J. Electroanal. Chem.* **150** (1983) 201.
10. D. M. Kolb, *Z. Phys. Chem. (Frankfurt) N.F.* **154** (1987) 179.
11. A. Bewick and S. Pons, in *Advances in Infrared and Raman Spectroscopy,* Vol. 12, eds. J. H. Clark and R. E. Hesters, (Wiley, New York, 1985), p. 1.
12. B. Beden and C. Lamy, in *Spectroelectrochemistry,* ed. R. J. Gale, (Plenum Press, New York, 1988), p. 189.
13. R. K. Chang and T. E. Furtak, eds., *Surface Enhanced Raman Scattering,* (Plenum Press, New York, 1982).
14. R. L. Birke and J. R. Lombardi, in *Spectroelectrochemistry,* ed. R. J. Gale, (Plenum Press, New York, 1988), p. 263-348.
15. J. D. E. McIntyre, in *Advances in Electrochemistry and Electrochemical Engineering,* Vol. 9, ed. R. M. Müller, (Wiley, New York, 1973), p. 61.
16. A. Tadjeddine, D. M. Kolb and R. Kötz, *Surface Sci.* **101** (1980) 277.

17. D. M. Kolb, in *Spectroelectrochemistry,* ed. R. J. Gale, (Plenum Press, New York, 1988), p. 87-188.
18. M. G. Samant, M. F. Toney, G. L. Borges, L. Blum and O. R. Melroy, *J. Phys. Chem.* **92** (1988) 220; *Surface Sci.* **193** (1988) L29.
19. G. Materlik, M. Schmäh, J. Zegenhagen and W. Uelhoff, *Ber. Bunsenges. Phys. Chem.* **91** (1987) 292.
20. L. Blum, in *Advances in Chemical Physics,* Vol. 78, eds. I. Prigogine and S. A. Rice, (Wiley, NewYork, 1990), p. 171-222.
21. L. R. Sharp, W. R. Heineman and R. C. Elder, *Chem. Rev.* **90** (1990) 705.
22. R. Sonnefeld and P. K. Hansma, *Science* **232** (1986) 211.
23. R. Christoph, H. Siegenthaler, H. Rohrer and H. Wiese, *Electrochim. Acta* **34** (1989) 1011.
24. J. Wiechers, T. Twomey, D. M. Kolb and R. J. Behm, *J. Electroanal. Chem.* **243** (1988) 225; **248** (1988) 451.
25. H. Gerischer, R. McIntyre, D. Scherson and W. Storck, *J. Phys. Chem.* **91** (1987) 1930.
26. S. R. Morrison, *Electrochemistry of Semiconductor and Oxidized Metal Electrodes,* (Plenum Press, New York, 1980).
27. Yu. V. Pleskov, in *Progress in Surface and Membrane Science,* Vol. 7, eds. J. F. Danielli, M. D. Rosenberg and D. A. Cadenhead, (Academic Press, NewYork 1983), p. 57.
28. T. Iwasita, W. Schmickler and J. W. Schultze, *Ber. Bunsenges. Phys. Chem.* **89** (1985) 138.
29. H. Gerischer, in *Advances in Electrochemistry and Electrochemical Engineering,* Vol. 1, eds. P. Delahey and C. W. Tobias, (Wiley, New York, 1961), p. 139.
30. H. Gerischer, *Electrochim. Acta* **35** (1990) 1677.
31. H. Gerischer, *J. Phys. Chem.* **95** (1991), in press.
32. J. Clavilier, C. Lamy and J. M. Leger, *J. Electroanal. Chem.* **125** (1981) 249.
33. A. M. Brodskij and Yu. V. Pleskov, in *Progress in Surface Sciences,* Vol. 2, part 1, ed. S. G. Davison, (Pergamon Press, New York, 1982), p. 1.
34. J. K. Sass and H. Gerischer, in *Photoemission and the Electronic Properties of Surfaces,* eds. B. Feuerbacher, B. Fitton and R. F. Willis, (Wiley, New York, 1978), p. 469.
35. Yu. Ya. Gurevich, Yu. V. Pleskov and Z. A. Rotenberg, *Photoelectrochemistry,* (Plenum Publ. Co., New York, 1980).
36. T. Watanabe and H. Gerischer *J. Electroanal. Chem.* **122** (1981) 73.
37. N. M. Alpatova, L. I. Krishtalik and Yu. V. Pleskov, in *Topics in Current Chemistry,* Vol. 138, (Springer, Berlin, 1986), p. 149-219.
38. H. Gerischer and F. Willig, in *Topics in Current Chemistry,* Vol. 61, (Springer, Berlin, 1976), p. 31.
39. B. A. Parkinson, *Langmuir* **4** (1988) 967.
40. B. Bressel and H. Gerischer, *Ber. Bunsenges. Phys. Chem.* **89** (1985) 1083.
41. R. McIntyre and J. K. Sass, *Phys. Rev. Lett.* **56** (1986) 651.
42. W. S. Hobson and A. B. Ellis, *J. Appl. Phys.* **54** (1983) 5956.
43. G. Chmiel and H. Gerischer, *J. Phys. Chem.* **94** (1990) 1612.

CONDENSED MATTER PHYSICS IN ELECTROCHEMISTRY

A.A.KORNYSHEV

The A.N.Frumkin Institute of Electrochemistry
of the Academy of Sciences of the USSR
Leninskii prosp.31, Moscow 117071, USSR *

AND

International Center for Theoretical Physics (IAEA - UNESCO)
P.O.Box 586, Miramare, 34100 Trieste, Italy

AND

Medical Biophysics Research Laboratory, University Clinics r.d.I.
Technical University of Munich
*Ismaningerstr.,22, D-8000 Munchen 80, F.R.G.***

ABSTRACT

Some topics in electrochemistry are considered from the condensed matter physics view point in relation to the problems discussed in this book. Examples of successful application of condensed matter physics to electrochemistry are discussed together with prospective problems and pressing questions.

1. PREAMBLE

The "invisible college" of scientists who work in application of condensed matter, atomic and molecular physics to electrochemistry is not too large; a part of it was represented at this meeting. Members of this community are basically aware of various aspects of the field of research and the progress achieved by their colleagues. So, this short overview is oriented not on experts of this kind, but rather on physicists who wish to learn about the capabilities of their powerful science in electrochemistry. Some more practical electrochemists could find a summary of what theoretical physics is doing or could do for electrochemistry relevant. Last but not

---

* Permanent address.
** Present address.

Fig.1. A flight over "condensed matter, atomic and molecular physics in electrochemistry" (as imagined in Miramare).

least, it is oriented on students to share with them the beauty of electrochemistry as a physical science and focus on the prospects of further research. In so doing, I suggest to use a "sea-gull principle" (Fig.1): to make a flight high above this region, diving from time to time to catch a particularly alluring "fish".

After this "fishy" preamble I feel enough defended to meet the discontent of an experienced reader that the diving does not go too deep.

2. Solvent, Solvation, Solutions

Solvent is an ingenious invention of nature:

* It carries the reactants and allows them to approach and meet each other by a stochastic diffusion-migration process.
* It provides the energy bath to allow and facilitate the reactants transformations.

Electrochemistry deals, primarily, with charged or electrically polarized reactants. Therefore, it is particularly interested in the ion solvent interaction and solvation of ions.

In order to understand how the ion interacts with the surrounding liquid one has to know something about the liquid. Simple (one component) liquids are of minor interests for electrochemistry which is concerned with multi-component liquids associated, in many cases, by hydrogen bonds, like, e.g. water.

The two complimentary branches of the molecular theory of liquids are the analytical statistical-mechanical theory and the molecular dynamics or Monte Carlo computer simulation. Based on the intermolecular pair potentials they calculate (in their own way) the structure factors, thermodynamic characteristics, functions of dynamic response. Advances in this field could not be possible without a remarkable progress in experimental determination of partial structure factors for different groups of polar molecules obtained by a combination of different diffraction methods.[1] Similar information for solvation complexes of ions and electrolyte solution has also been obtained.[2] Dynamic properties were a subject of an immense amount of works.[2]

More complex liquids are typical of more involved pair potentials which are easier tractable in a computer simulation than in analytical theory. Still the latter makes gradual progress towards inclusion of more complicated potentials. Indeed, from a simple dipolar fluid (which is "composed" of molecules pictured by an arrowed circle: a hard ball with a point dipole in the centre) to site-site interaction models [4-6]. Work is in progress to treat the system of charges which represents water molecule in a way which provokes hydrogen bonding and gives reasonable structural and thermodynamic properties.[7,8] The kind of approach called central force model, popular in computer simulation[8], treats water as an electroneutral system of "oxygen atoms" and "protons". Effective charges are attributed to them with their interactions written keen enough to provide self-organization in water molecules and reasonable hydrogen bonded ensembles. This approach allows to calculate also the dynamic properties: intra- and intermolecular vibration spectra and lower frequency micro- dynamic properties.[8]

Progress in the theory of liquids is used without delay in description of solute solvent ensembles.[9-12] However, if there is a "chemical interaction" between the solvent and the solute, a description without quantum chemistry[13] is incomplete (if it is very strong some problems can be simplified by considering solvation complexes as separate entities). Both the statistical mechanics and quantum chemistry of solvation complexes has given interesting solvation patterns. Molecular dynamics simulations [14](which account for chemical effects via the *ab initio* pair potentials involved in simulation) has elucidated important dynamic properties of solvation complexes: ion structure making and structure breaking effects[15].

In this book, modern status of the theory of liquids and solutions, relevant for electrochemistry, is in some way reflected in a paper by BOPP, devoted to computer simulation of ionic solvation, and in a paper by BADIALI, concerned with the structure of semi-infinite systems.

One may ask about the rating of older phenomenological approaches to solvation. Elegance and simplicity of the Born formula[16] for the solvation free energy (calculated as a free energy gain due to charging a sphere immersed in a continuum with a dielectric constant ε) still attract scientists in applications of ideas of physical electrochemistry to inorganic chemistry, biophysics, colloid and separation science. Being used in "provinces", it is out of fashion in the "capital". Born model applied to hydration of classical "test" ions (alkali halides) is known to give a proper order of magnitude with about 50% excess over the experimental values. The excess has been explained by the breakdown of classical *macroscopic* electrostatics when applied to calculate the electric fields of so tiny objects as ions. Semi- phenomenological modification of the Born theory on the basis of *nonlocal electrostatics*[17] provided the reduction of the calculated free energies. The reduction was due to "solvent" structure effects which enter into the calculation through the *nonlocal static dielectric function of the solvent*[17] (see the first abstract by KORNYSHEV).

Krishtalik et al[18] have measured the electrostatic counterparts of the free energy of solvation of large compact ions of size about 4 - 5 Å in a series of aprotic solvents. The data fitted perfectly the Born equation ! The puzzle is where are the nonlocal effects?

In view of recent attacks on the elementary form of the nonlocal dielectric function which has been used in a number of applications of nonlocal electrostatics it is possible that we are to modify the question: what is the true form of the nonlocal dielectric function of water and other polar solvents to be used in nonlocal electrostatic calculations of solvation free energies? On the background of this most pressing question, reservations about the fundamental drawbacks of nonlocal electrostatics (linear response approximation, semi-macroscopic character, etc.[17]) or additional "nonelectrostatic" contributions to solvation energy become (tentatively!) of secondary importance. Two disputing seminar contributions (FONSECA & LADANY versus KORNYSHEV, KOSSAKOWSKI & VOROTYNTSEV) are devoted to this fascinating problem. While FONSECA & LADANY believe that the singular behavior of the dielectric function on the real axis of wave-vectors k together with a range of negative values is practically an inevitable feature of all the statistical mechanical models of molecular fluids, KORNYSHEV et al focus on detailed analysis where the singularity comes from and show the essential instability of the results, primarily in the small k region, to the parameters of the model. The both groups agree that the singular dielectric functions give rise to the free energies of solvation greater than the Born energies which is in contradiction with the experiments. FONSECA & LADANY attribute it to the drawbacks of nonlocal electrostatics, while KORNYSHEV et al draw arguments and examples suggesting to be careful with possible artifacts in determination of nonlocal dielectric functions from microscopic models of molecular fluids. The results of the latter group are rather negative than positive: it is shown that in order to be sure in obtained values of dielectric functions in the region of small k one should know with enormous precision all the charged molecular groups partial structure factors up to the distances >100 Å ! Only an extreme interest to this topic can justify optimism in expectations that we, nevertheless, soon get some reliable results in this field.

Electrolyte solutions, a classical branch of electrochemistry, was for years a proving ground for statistical mechanics of systems of charged particles. For a long time, this was a statistical theory of cold electroneutral ionic plasma immersed in a continuum of dielectric constant ε (the "primitive model") in the bulk or near interfaces . The Debye theory in the bulk or Gouy-Chapman theory near a hard charged wall were rigorously reproduced in the limit of low ionic concentrations.[19,20] Efforts were made to understand deviations from these limits with the increase of electrolyte concentration and the charge of the wall[19,21].

Advances in the molecular theory of liquids changed the fashions. Bulk or semi-infinite ion-molecular mixture theories (in which ions and solvent molecules are considered on the same footing) replaced the primitive model.[4,9,10,22] The proximity of such an approach to reality is limited, first of all, by a model of the solvent and of ion-solvent interaction. As for the double layer near the metallic electrodes, the semi-infinite ion-molecular mixture models have not yet accounted for image forces which might be quite important there. The lack of realism in the model of a metal may be of minor importance for a description of the peripheral part of the double layer, far from the electrode surface, i.e. in the "diffuse layer" problem.

Activity on computer simulation based on the *ab initio* interaction potentials between metal *atoms* and the particles of the solution has been started with impressive success.[23] Note, however, that they yet also left screening by the free electrons of the metal beyond consideration.

Properties of the "ion-molecular" double layer are summarized in a paper by BADIALI.

Molten salts - *the electrolytes without a solvent* - are important electrochemical systems, primarily due to their use in fuel cells. On the other hand they are of extreme interest as a pure model system of statistical mechanics: cold dense two-component ionic plasma. The paper by TOSI is devoted to these systems with a bias on screening properties and the structure of nonlocal dielectric function.

3. Electrode/Electrolyte Interface

A prominent part of physical electrochemistry is concerned with theoretical and experimental studies of the structure, electric and optical properties of the metal/electrolyte interface. A profile of the interface, shown schematically in Fig.2, displays system components figuring in this problem. Depending on the type of bonding between the metal and the particles of solution one may distinguish three cases:

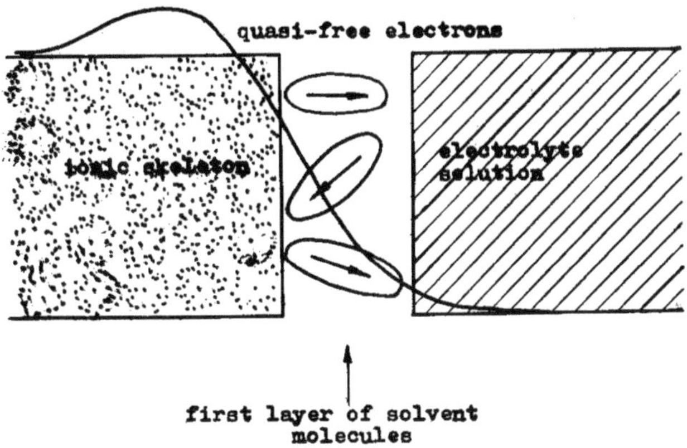

Fig.2. Schematic profile of the metal electrolyte interface: main system components.

* Ions and molecules of the solution are coupled to the metal by covalent bonds.
* Metal-solution interaction is purely "physical".
* Intermediate case.

In the first case, theory cannot succeed without quantum chemistry.

It benefits from high vacuum *ex situ* measurements[24] and electrochemical voltammograms, as employed for the study of adsorption.[25] Abstract by PIRUG and a paper by HUBBARD and PING GAO reflect the state of art in this field. "Physical" theory of chemisorption can also be helpful in this case. We shall make our first "dive" in it a bit later.

Systems of the second class and partly of the third one were described during the last decade by a combination of the density functional theory of metal surfaces with *some* description of the bounding solvent and the metal solvent interaction. Progress has been achieved not in the microscopic description of the interface but rather in a theory of its response to charging. Advances in this field have been reviewed in detail[26,27]. Second abstract by KORNYSHEV and a paper by BADIALI and AMOKRANE present some updated information on this problem. The latter focuses, in particular, on the solution of the inverse problem. Here, the information about the response of the solvent at the interface to charging is extracted from experimental capacitance curves under acceptance of the density functional results for the behavior of the free electron subsystem of the metal.

Convincing results were obtained on the basis of relatively rough description of electrolyte.[27] Activity towards a more sophisticated fully selfconsistent electron-ion-molecular double layer theory should be quoted[28] (for discussion see also[27]).

Progress in the *electronic double layer theory* was due to application of a *jellium* model in which the ionic skeleton of the metal is represented by a semi-infinite homogeneous background of positive charge. Account for lattice discreteness is making its first steps in electrochemistry.

Most adequate approach to single crystal electrodes would be an application of Kohn-Sham scheme to periodically repeated slab geometry.[29,30] (Each slab is a "slice" of the given crystal parallel to the surface crystal plane. It has a number of layers large enough to provide the bulk-like behavior in the middle of each slab. The gaps between the slabs are taken broad enough to eliminate their mutual influence. Surface of each slab, thus, simulates the surface of single crystal. The periodicity is used to work with effective Bloch set of electronic wave- functions.). Note, that there is no electrolyte in Ref.29 but a formal possibility for the slabs to be charged.

The structure of the ionic lattice of a semi-infinite metal has been accounted for via the field which acts from the skeleton on the free electrons with no discreteness in the variational trial functions representing the distribution of free electrons near the interface. Some reasonable effects of the crystallographic orientation of single crystal electrodes were obtained along these lines[31,32].

Before the beginning of the "electronic double layer boom", the description of the interface without specifically adsorbing ions was dominated by the "compact layer"-type models. In this kind of models the first layer of solvent molecules bounding the metal was treated as a quasi 2D-phase localized between the plates which mimic the "metal" and the "bulk" electrolyte (Fig.3)[26,33]. Though inadequate literally, this model gave an experience about a possible contribution to the overall response to charging of the molecular reorientations[26,33].

However, even within these idealized framework the problem has never

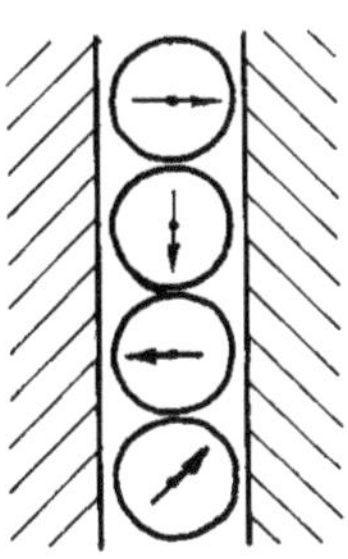

Fig.3. Compact layer model of the metal/electrolyte interface

been solved beyond the mean field approximation. Indeed, the simplest (and hardly realistic) variant of the problem which we are to deal here is the statistical mechanics of a 2D-system of hard dipolar spheres in an external electric field which interacts at distances larger than the hard sphere radius via the screened pair potential $V(R) \sim \exp[-R/(L/2\pi)]$ where L is the distance between plates and R is the distance between the centers of the dipoles. Due to the factor of 2π the range of pair potential is localized to nearest neighbors. The problem thus can be mapped on a 2D XY model in external field with nearest neighbors interaction. However, the lateral distribution of the spheres would not necessarily be regular and rigid. It can be changed with the variation of electric field.

Of course, the molecules are not spheres and the metal and electrolyte does not form ideal mirror planes. This picture is, thus, far from being a typical cartoon of the electrochemical interface. However it will be a luck for condensed matter physics if we could find among the electrochemical systems those which can be described by this picture. Studies of the capacitance dependence upon charge can then contribute to the statistical physics and theory of the critical phenomena in 2D-systems. It should be noted that the mercury water interface has been recently studied by the method of a laser induced temperature jump.[34] Phase transitions related to reorientations of water dipoles were claimed to be observed and rationalized by Landau-type mean field theory[35]. Similar studies on solid electrodes[36] gave no evidences of critical phenomena.

More realistic approach to the role of the water molecules interacting with a jellium metal deals with an appropriate Monte Carlo simulation subject to the model interaction potentials. Progress in this direction is discussed in a paper by GUIDELLI and ALOISI.

Reading the contributions by KORNYSHEV, by BADIALI and AMOKRANE, and by GUIDELLI and ALOISI one can acknowledge an unusual consensus between the three independent groups in opinion about the nature of interfacial relaxation induced by variation of voltage. This phenomenon, viz. the deformation of surface electronic profile accompanied by corresponding adjustment of the distance of closest approach of solvent molecules to the electrode, predicted in DFF simulations[27], has striking regularities of a deep fundamental nature. They explain the basic shape of the dependence of the metal/electrolyte interfacial capacitance upon the electrode charge. However, the *field-induced interfacial relaxation has never been directly observed* (except for, may be, s-polarized electroreflectance[27]; reservations are due to other possible interpretations of electroreflectance signals[27]).

4. Chemisorption from solutions

The situation without chemisorption of either ions or solvent molecules is, in fact, very rare. It takes place for a small group of metal/electrolyte combinations and in a certain range of electrode polarizations. Does it mean that it is hopeless to describe most of the contacts by the methods of condensed matter physics? To answer this question we should recall the theory of chemisorption at the metal/vacuum interface. Next step will be to discuss how much these concepts could be extended to the case of chemisorption from solutions. We preclude it by a trivial comment. The chemisorption theory is needed if we are interested in the electronic state of the adsorbate, coupling energies versus position on a single crystal plane and other fine details. For years, electrochemistry was satisfied by much rougher concepts dealing with phenomenological adsorption isotherms,[37] Grahame-Parsons analysis,[38] etc., doing well without sophisticated theory. With the development of quantum theory of electrode processes it became clear that a more detailed picture of adsorbate states and the double layer structure is required for adequate description of such kinetic characteristics like current-voltage plots, cyclic voltammogrammes, etc.

4.1. *Theory of chemisorption on free metal surfaces.*

In seventies, the two main approaches have been developed for a description of the metal - adsorbate systems: density functional formalism (DFF)[39] and application of the Anderson model[40].

The idea of DFF was to describe metal/adsorbate system as an electron gas in the external field of ionic skeleton and of the adsorbate core. Both the skeleton and the core can be described with different degree of sophistication. The skeleton may be considered as a discrete lattice to account for crystal structure effects. Bound electrons of the inner shells of the skeleton and the core can be incorporated by means of a trick usual in solid state physics: replacement of the external electrostatic potential produced by the skeleton and the core by corresponding pseudopotentials. The subsystem of free electrons is described by a functional of electronic density which has proved to describe fairly well the properties of the free electrons in the bulk. Account for the fact that the external field is highly inhomogeneous near the skeleton edge and the adsorbate core (and all the relevant "events" takes place right in this region!) is taken, as usually in DFF, by a mysteriously efficient local density approximation (LDA).[41] It accounts for the inhomogeneity only via the first gradient expansion term in the kinetic part of the energy functional, while the exchange-correlation terms are retained the same as in the bulk.

This approach, expected to work for s-p electrons, is sometimes applied (with the help of artificial tricks) to d-metals.

Instead of pseudopotential description of core electrons, some authors assume a given distribution for inner electrons in the skeleton and adsorbate core, add these densities to the free electron density, write the energy as a functional of the overall density (of the form "standard" for the free electron gas but, alas, unjustified for strongly localized electrons!) and then minimize it with respect to the free electron density[42].

There are two realizations of DFF.[39] One is the Kohn - Sham scheme which reduces the minimization procedure to a system of effective Schrödinger equations in a self-consistent potential which depends on electronic density. Another one is the direct variational approach which employs trial functions for electronic density. The latter reduces the minimization problem to a transcendent equation on the parameters involved in the trial functions. Since the resulting equation is, depending on the form of the employed density functional, a generalization of the Thomas - Fermi equation, this approach is sometimes called the extended Thomas - Fermi method. Both realizations of DFF have been used in chemisorption theory.[43,44-46]

Application of the Anderson model to chemisorption, was initially elaborated by Newns[40] and is now known as the Newns - Anderson model. Its main idea is to write the system Hamiltonian in the second quantization form including the effective single electron states in the metal, the localized state(s) on the adsorbate, and the transfer term responsible for the electron exchange between the metal and the adsorbate. The second term generally incorporates a contribution due to the interaction of two electrons on the adsorbate orbital with oppositely oriented spins. Thus, this approach is a logical extension of the LCAO idea in which one of the molecules is replaced by a system with continuum electron spectrum (the metal). The prescriptions of the Newns - Anderson model, when solved, are qualitatively clear. The exchange with the electron sea in the metal makes the electron level on adatom virtual. Its half-width is the broader, the larger the matrix element of transfer between the metal and the adsorbate. The occupation of the level is greater, the larger the difference between the Fermi energy and the input energy level of the localized orbital. If the latter is much higher than the Fermi energy the level is practically empty.

The term responsible for the interaction of two electrons with opposite spins, proportional to the product of electron density operators for these two states, makes the problem a many body one. In chemisorption, some approximation are employed to simplify the situation. The most popular is the "Hartree-Fock" which replaces the product of the two operators by a product of a density operator and its mean value. This "act of violence" leads to a simple self-consistent equation on the mean occupation number of electrons on the localized orbital. Leading to qualitatively reasonable results, Hartree-Fock became the most popular way to hide the many-body dust under the carpet and to make the Newn-Anderson model practically useful. Another way to overcome the many body difficulties was a bit more honest: to get rid of the unpleasant term under assumption that only one electron can occupy the orbital. This is possible when the Coulomb interaction of the two electrons on the localized orbital is the largest parameter of the problem. The solution then is even more simple. Since the electron spin on the orbital disappears from the problem, this approximation is called the "spinless". More accurate variational approach to the Newns - Anderson model has also been developed[47]. The complicated exact solution of the Anderson model, obtained later by Wiegmann[48], to my knowledge has not been utilized in the chemisorption context.

4.2. *Theory of chemisorption from solutions.*

How much these old ideas have been utilized in the theory of chemisorption from solutions?

Practically nothing has been done within DFF since nobody knew how to describe the solvent on a level compatible with the description of free electrons. The trial function version, initially suggested by Appelbaum and Hammann[44] and elaborated by Dunaevskii et al[45,46] seems to be promising, at least for a start. In particular, Dunaevskii and Potekhina studied adsorption of Na^+ on different crystal faces of Al for different positions of an adatom above the unit cell (atop, bridge and site positions). The way to do it was to vary system energy with regards to variational parameters which determine the perturbation of the electron density of the metal due to adsorbate (the background electron density distribution in absence of the adsorbate taken from the DFF simulation of the free metal surface).

It is a big problem how to incorporate solvent in the DFF simulation of the metal/electrolyte interface without ionic adsorption[27]; the case with adsorption is even more complicated. To introduce dielectric screening in the half-space bounding the skeleton is obviously not enough. This is not only because of the local inhomogeneity of the solvent around the adsorbate (which could be taken into account by the spatial dispersion of the solvent permittivity), but also due to deeper questions. Is it the static dielectric constant that characterizes the solvent response or some frequency dependent quantity? To get an answer it might be very instructive to study some appropriate exactly solvable models. Furthermore, we must take into account the interaction of electron shells of solvent molecules with the electrons on the adsorbate. Interaction with closed shells can be accounted for by the repulsive pseudopotentials of the Esbjerg - Norskov type.[26]

Impressive results have been obtained in the Kohn - Sham calculation of one water molecule on single crystal metal surfaces[49]. (New information about it in relation to various experimental data one can find in the abstract by PIRUG). From these calculations one may feel how hard could it be to extend them to larger amount of molecules.

Anyway, a big field for thought and activity is open here.

A parametrization type approach, the Newns - Anderson model conforms easier to a desired extension. It is possible to add to the Newns - Anderson Hamiltonian terms responsible for some boson field plus the term responsible for coupling of the electron on the localized orbital with the bosons. This approach has been used by Hewson and Newns[50] to incorporate interactions with plasmons at the metal surface (image forces). Later, bosons have been used to account for vibrations of the adsorbate along the metal - adsorbate bond[51]. (Similar approach was earlier used by Khomskii[52] in the context of a bulk problem: stabilization of mixed valence by electron-phonon interaction; the boson field was due to optical phonons in the crystal lattice). No matter, what is the nature of the boson field, the problem itself is formally stated; it later got a name: *local plaron problem*. The particular nature of the boson field is important in approximations determining the spectrum of frequencies. Various kind of adiabatic approximations can be used depending on the typical frequencies and the energy parameters of the Anderson model.[50] A variety of papers were devoted to different aspects of local polaron problem; updated list of references can be found in Refs.53,54.

Schmickler was first to apply local polaron model for a description of electrosorption (adsorption of ions from solution)[55]. The boson field represented polarization fluctuations in the solvent (polar phonons). He described the solvent polarization fluctuations by a set of effective oscillators (the quasi-continuous variety of oscillation branches can mimic the continuous absorption frequency spectrum of the system which may correspond to dissipative processes though the effective oscillators themselves are not supposed to have damping[56]). Using this approach and assuming that the characteristic frequency of charge transfer between the metal and the adsorbate is faster than the frequencies of the effective oscillators , Schmickler has shown that the interaction with the solvent retains valid the Anderson - Newns solution in the Hartree - Fock approximation but with the corresponding renormalization of the basic energy parameters. Broader class of systems is considered in the extension of this approach on the case when there are polarization modes faster than the charge transfer frequencies and image forces are involved[57]. From a consideration of a single adatom, a step has been done towards the ensembles of adsorbates to study the coverage dependence[58]. The effect of the mutual depolarization of adatoms due to electrostatic interactions was investigated in a spirit of Muscat and Newns treatment of submonolayers adsorbed at a metal/vacuum interface.[59] New results were obtained following an entirely different picture of electrostatic interactions at the metal/electrolyte interface.[58]

Single adatom chemisorptrion with one oscillation mode was considered by Kravtsov and Mal'shukov[51]. They have found that when the interaction of the electron at the localized level with the oscillation mode is strong and with the metal weak enough, two solutions of the Newns - Anderson model exist which correspond to two different equilibrium values of the normal coordinates of the oscillation. The two solutions have generally different energies. Under certain conditions with the variation of the strength of the adsorbate-metal and the adsorbate-oscillation coupling, a transition from one adsorbate state to another may take place. Kravtsov and Mal'shukov obtained this fundamental result within the framework of spinless approximation. A possibility to meet this effect also in the Hartree-Fock has been demonstrated by Schmickler[60]. Kravtsov and Malshukov[59] have shown that the splitting of the solution in the single adatom system can be manifested in adsorbate layers in the form of "phase transitions" to an intermediate phase: superlayers with doubling of the period and then, finally, to phase incommensurate with the structure of the substrate lattice. This is the result of electrostatic interaction of adatoms which can make under certain conditions inhomogeneous distributions of the two allowed orbital occupation numbers energetically more profitable than the homogeneous one. This might lead to the structures, which are more complicated than those which one might expect if all adatoms had the same occupation number. Kravtsov and Mal'shukov[61] have built a full picture of phase transitions in such systems.

Schmickler[60] suggested that a kind of Kravtsov - Malshukov transition takes place in electrosorption in the so called underpotential deposition (u.p.d.) systems.[62] Indeed the cyclic voltammograms for the u.p.d. of Tl on Ag (111), Pb on Ag (111) or Au (111) show up sharp current spikes[63-65]. Schmickler refers to Adzic et al[63] who have argued that the current spikes are due to changes in the charge state of the adsorbate. Schmickler made one step more having suggested that these changes are of the Kravtsov and Malshukov type. He rationalized this idea within the extension of their model to electrosorption. His key point was that just in electrochemistry on of the most important parameters of the Newns - Anderson model can be controllably varied: the energy level of the adatom orbital varies with the electrode potential. The more negative it is, the higher the energy of the electron on the adatom orbital, the lower the orbital occupation number. Estimates made for typical systems helped Schmickler to show how the adatom orbital changes its occupancy with the variation of electrode potential: from continuous to abrupt change at the points where the Kravtsov - Malshukov transitions takes place.

Schmickler made some speculations on the substrate crystal structure effect. He assumed that on terraces the interaction of the adsorbate with the solution is fairly strong, while the interaction with the metal is relatively weak - in contract to the case of adsorption on steps. Then, according to the Kravtsov and Malshukov theory, the case of splitted solution will be more typical for adsorption on terraces, while a single solution will be more typical for the adsorption on steps. According to Schmickler, this is the reason why sharp spikes in voltammograms are seen for flat (111) surfaces while on the stepped (110) surfaces only broad Langmuir type voltammograms are observed.

Schmickler's explanation is beautiful but not many electrochemists share the enthusiasm about it. The opponents argue that there can be other sound reasons for the spikes: adsorption-desorption of other ions or field induced surface reconstruction. Furthermore, Schmickler's analysis is incomplete. He made qualitative conclusions on the basis of single adatom behavior, while measuring voltammograms one may scan a considerable range of coverages. Schmickler's work does not exploit the ideas about the adatom superstructures[51,61] and this seems to be an exciting topic for further research.

Recently, Volokitin[66] obtained an exact solution for a *two level* local polaron model. In this model a transfer between two localized states is considered instead of one localized state ("adsorbate") and a continuum of states ("metal"). This solution does not show up any abrupt transitions. This does not mean that such a conclusion is of importance for chemisorption. Somehow, the results of electrochemical experiments (UPD voltammograms[63-65]) are quoted in discussions about an existence of abrupt transitions in the local polaron problem.[66]

A local polaron theory of chemisorption from solutions has been developed by Mishra and Rangarajan on the basis of "super" operators technique[67]. Aimed to go beyond the adiabatic approximations, this theory was elaborated in rather an involved form. In their last paper,[67] the authors have studied few examples of application of their theory to electrosorption, but I am not aware of any further progress in this direction.

The ideas of Refs.58-61 can be, possibly, used in a theory of field induced surface reconstructions. Such a phenomenon was claimed to be discovered in Brookhaven National Laboratory[68]. When a negative voltage is applied to the gold (001) and (111) single crystal surfaces the top layer reconstructs to form a hexagonal surface layer that contains more atoms than the bulk layer. The reconstruction disappears above certain positive potentials, but fully recovers when one returns back to negative electrode polarizations. This is an intriguing phenomenon in view of novel surface structures of free surfaces of Au(001) which have been observed with the variation of temperatures from more than one thousand K to room temperatures[69]:disordered, distorted-hexagonal, rotated distorted hexagonal. The transition between the first two phases goes with the formation of incommensurate overlayers.

Reservations about the interpretation of BNL observations are due to the possibility of formation of adsorbate superlayers : do we see the crystal surface transformations or rearrangements in adsorbate layers induced by voltage variation?[69] Adsorption of cations at negative potentials can definitely facilitate surface reconstruction, so that we possibly see some combined effect of adsorption induced surface reconstruction. Theory of this phenomena is a promising field for interplay of condensed matter physics and electrochemistry.

4.3. *Nonlocal electrostatics*

Nonlocal electrostatics is an efficient method to calculate the electric fields at the microscopic scales by taking into account the spatial dispersion of the dielectric constant of the medium.[17,70] In the theory of chemisorption from solutions, nonlocal electrostatics is used to calculate those quantities which are determined by the electric fields of adsorbates[58,71,70]. This direction of research is represented in this book by VOROTYNTSEV's first paper. Being primarily determined by the screening properties of the metal, these results are less affected by variance in the large wave-vector behavior of the dielectric properties of the solvent.

Note that Vorotyntsev theory of ionic adsorption isotherms assumes the fixed and identical electron occupancy on each adion. Electrostatic interactions in the ensemble of the Newns - Anderson adsorbates at the metal/vacuum interface (with no boson modes involved) were found under certain conditions to favor inhomogeneous distributions of charge occupancy on adsorbates along the surface.[72] Estimates[58]show that this case is unrealistic for electrosorption. This does not exclude the existence of inhomogeneous distributions of Kravtsov and Malshukov type due to interaction with the solvent "polar phonons". Adsorption isotherms with account for Kravtsov - Mal'shukov transitions is an interesting subject for further work.

4.4. *Chemisorption chemistry*

These speculations were about the problems in electrosorption which are well formulated in a physical language. In the "adsorption" section of this book we have almost no contributions of this kind. VOROTYNTSEV's chapter is an exception pleasant for a theorist: it demonstrates beautiful physics! His results, related to adsorption of ions at a weak electron exchange with the metal, allow to estimate the value of this exchange from the experimental adsorption isotherms and capacitance measurements. Frequently, life happens to be more complicated and the progress is hardly possible without real *chemistry*. Paper of HUBBARD and PING GAO represents the "chemical" view on adsorption at electrodes. A seminar contribution by BORKOWSKA and JARZABEK is focused on experimental studies of the role of the nature of the metal in the adsorption of the solvent. LIPKOWSKI et al summarizes *in situ* electrochemical and *ex situ* vacuum studies of adsorption of organic molecules on gold electrodes in relation to their crystal structure. During the last two decades, direct powerful methods for the study of adsorption with the help of isotope technique have been developed. Experimental competence is required here to avoid artifacts or misinterpretation; subtle points are discussed in a lecture by KAZARINOV.

5. SEMICONDUCTOR/ELECTROLYTE INTERFACE AND OTHER INTERFACES

Theory of space charge at the semiconductor/electrolyte interface has been developed many years ago on a level compatible to the Gouy-Chapman description of the double layer in the electrolyte.[73] Based on the electronic theory of semiconductors, the electrochemistry of semiconductors played an important role in the investigation of semiconductor surfaces and contributed fundamentals for solar energy conversion. The two last papers by GERISCHER give the reader a chance to get an updated information on the state of art in this field.

In contrast to phenomenological theory, the microscopic aspects of the solvent effect on the electronic structure of semiconductor surfaces have not been developed on a level reached in surface science. This is a promising subject of research, albeit not without help of quantum chemistry.

Studies of other interfaces were not discussed at this meeting. To complete the picture I just mention the activity on few of them.

The interface of immiscible liquids (usually, water/oil interface)[75] is relevant for phase transfer catalysis[76] and as a model for photosynthetic systems[77] and ion transfer membranes[78]. This interface is chargeable but the properties of the electric double layer are not yet completely clear.

Various dielectric/dielectric interfaces are of great importance in colloid science and biological physics. They are now being studied theoretically by the same methods which proved to be efficient for a description of the metal/electrolyte interface.[79-81]

The interfaces including solid electrolytes (*superionic conductors*) are traditionally discussed at separate meetings. The studies of the bulk properties of superionic conductors[82] are a part of modern solid state physics. As for the interfacial properties, usually some extension of ideas of phenomenological double layer theory are employed[83] with a lack of detailed analysis of microscopic structure of the surface of the solid electrolyte. Some new trends here are concerned with the study of fractal structures both in the bulk and at the surfaces (LIU, this issue).

6. Electrodynamics of the Electrochemical Interface

The "electrochemical renaissance"[84] of 80th was much due to understanding by condensed matter physicists that the solid/ electrolyte interface gives a complimentary and in some aspects better possibilities for the study of solid surfaces than the high vacuum systems. This is primarily because one can probe the subtle interfacial properties by the voltage modulation. Actually, this was known and used for years in electromodulation spectroscopy. The potential dependent quantities are only those which belong to the microscopically narrow interfacial region. Thus, the potential modulation cuts off the "irrelevant" background information about the bulk and outlines the interfacial properties.

Phenomenological electrodynamics of ideally smooth metal surfaces and metal electrolyte interfaces[85] provides a parameterization for various electromodulated signals in visible. However, in order to interpret the parameter dependence on the potential, electronic structure of the surface and other things one needs a microscopic theory of the interface. This is a challenging field, though developing with big difficulties. The progress in this direction is reviewed in Ref.27 (in the book updated in the second abstract by KORNYSHEV).

Nonlinear optics was found recently to be of essential interest for electrochemistry of smooth metal surfaces. This is the electromodulated optical second harmonic generation from the metal/electrolyte interface. The SHG signal is enormously sensitive to the properties of the boundary layer few Å thick. This method is now mostly used to identify the adsorption submonolayer structures of a certain symmetry by studying the azimuthal dependence of the signal. However, the background signal from a bare electrode surface is not less important. This is particularly because the "jellium theory" of the interface can give very instructive results for the interpretation of experimental data. A paper by LIEBSCH represents the latter direction of research. The SHG studies of adsorption superlayers via their influence on the non-linear in-plane and perpendicular polarizability of metal electrons at the interface and the direct SHG monitoring of the formation or lifting the surface reconstruction are discussed by PETTINGER & FRIEDRICH. Note that LIEBSCH benefits from his Kohn-Sham picture of the electronic response of the metal surface, while the electrolyte figures in his description only through a possibility to charge the electrode. Steps towards a more consistent description of the solvent effect in the *in situ* SHG are quoted in the abstract by KORNYSHEV.

Note that some striking phenomena related to optical properties of the adsorbates may be understood without a detailed information about the metal/electrolyte interface. One may concentrate here more on the singularities of the optical response of the metal adsorbate complexes, adsorption layers and surface states.[85,27]

This is, probably, not the case of *surface enhanced Raman scattering* (SERS). Raman scattering from adsorbates at metal surfaces was first observed by Fleischmann et al[86] for piridine molecules adsorbed at a silver electrode. Later Jeanmaire and Van Duyne[87] and Albrecht and Creighton[88] discovered that the effective Raman scattering cross-section of the adsorbate is about five orders of magnitude greater than one could have expected for the same molecule in the bulk of the solution. This opened a flood of works.[89] Understanding the nature of SERS was a challenge which, even now, is not completely relaxed.[89,90] It became an important source of interest of condensed matter physicists in interfacial electrochemistry.

It became soon clear that SERS is a result of an interplay of local molecular modes with surface collective modes on the adsorbate. Furthermore, since the huge enhancement is observed for sufficiently rough[90] surfaces, the interest to electrodynamics of rough surfaces got an additional impulse.

Electrodynamics of rough surfaces in contact with electrolyte has not yet been developed in any respect that essentially differs the theory from the electrodynamics of the same surfaces exposed to vacuum. One can imagine difficulties on this way related to unusual double layer effects and adsorption not yet accounted for in model calculations. The experiment, however, differs very much. In the electrochemical confinement, all the benefits of a possibility to charge the electrode are employed. Moreover, in electrochemical cells the surface roughness can be controllably created or eliminated by oxidation - reduction cycling.

In the electrodynamics of microscopically rough surfaces all the "roughness size" parameters are much smaller than the incident light wave-length. The theory deals usually with two extremes. The first one is when the characteristic roughness correlation length is much greater than the height fluctuation amplitudes. This is a case of *slightly rough surfaces*. Opposite extreme deals with a single isolated bump (usually modelled by a sphere or ellipsoid) of size much smaller than the wave-length of light . With a finite surface concentration of bumps increasing - so that the distances between bumps become smaller than the wave-length - severe problems arise. No exact solution has yet been obtained for a system of bumps but some progress has been achieved within the framework of some *mean field* approximate schemes, like random phase approximation, etc. Electrodynamics of rough surfaces is being feeded by a good level, modern theoretical physics.[91] News in this field are presented by URBAKH.

Modern aspects of SERS spectroscopy of electrodes are discussed by PETTINGER & FRIEDRICH in the interplay with SHG spectroscopy. FUNTIKOV summarizes the investigations of the nature of electromagnetic field enhancement by comparative studies of SERS, in situ photoemission and electroscattering from electrochemically pretreated surfaces.

The interest to the impedance of electrodes (a quasi-static part of electrodynamics) was recently renewed in connection with attempts to understand the nature of the so called *constant phase element* (CPE) of the equivalent circuit which is attributed to the roughness of the electrode.[92] Fractal models of roughness have been considered[93-96] in relation to the theory of CPE. Different models suggested different relationships between the CPE exponent and the surface fractal dimension of the electrode. Discussion about the existence of a universal relationship has been closed by experimental studies and counter-examples[97]. They both have shown the lack of universality. Updated review of this studies together with new results on the bulk fractal impedance are presented by LIU.

Old Gouy - Chapman theory has been revisited, now for rough metal surfaces. The space charge in the electrolyte (*diffuse double layer*) near a curved metal surface is a very complicated electrostatic problem even in the linearized Poisson-Boltzmann approximation. Goldstein et al[98] have recently suggested an original hybrid of a theory and computer simulation to solve this problem. They mapped it on a problem of a random walk near the curved surface with a special kind of boundary condition. The random walk process was then simulated in a computer experiment. They have studied the cases of surfaces build via the algorithm of certain deterministic fractals. Within a single class of fractals it was easy to vary the fractal dimension and to follow, thereby, corresponding changes in the system properties, as obtained by the random walk simulation. It was found that the diffuse layer inverse capacitance is proportional to the Debye length in the power of one half of the surface fractal dimension. (When the latter is equal to 2, the result reduces to the Gouy - Chapman one.) It is not clear now, how much universal is this result since it was obtained only for a certain class of fractals.

Electrostatics of fractal objects is a complicated subject. The first exact analytical results have been obtained for, presumably, the simplest model: charged Cantor bar[99].

7. *In situ* SCANNING TUNNELING MICROSCOPY

The work to adjust STM to the measurements at the interface metal/ liquid electrolyte is in progress in a number of electrochemical laboratories. The visual image of the surface is now not the only goal of investigations. The new round of studies addresses obtaining the density of states and its dependence on the potential, correlation functions in dynamic systems, etc. In addition to the known problems of the resolution and adequacy of the image at the metal/vacuum interface, one might expect new adventures at the metal/electrolyte interface. Typical questions to avoid artifacts are:

* What is a perturbing influence of the tip?
  Can it brush off weakly adsorbed atoms?
* What is the effect of electrolyte concentration on the screening of the electric field between the tip and the metal?
  What is the effect of specific adsorption onto the tip?
  Does the latter depend on the profile of the metal surface?
* What is the role of solvent local fluctuations near the metal surface in the formation of the tunneling barrier which calibrates the tunneling current and thus controls the image?

Experts know a number of tricks to control the situations and avoid, at least partially, some problems of this kind (*atoms should be round,* etc.). A survey of the basic principles, the basic tricks and first results in the *in situ* STM is given by GIMZEWSKI and SASS with a thorough discussion of the problem of tunneling trajectories subject to much more complex electron - molecule interaction mechanisms, as compared to tunneling in vacuum.

When the euphoria after the invention of STM have passed, it became clear that not always what you see in STM is what you actually have. Exaggerated landscapes or, alternatively, some times strange indifference to the nature of the metal - all suggested that a proper quantum mechanical theory of the tunneling process is needed. The approach to the problem can be simplified by an assumption - seemingly, reasonable - that one atom or a small cluster of atoms at the tip actually take part in the single electron process of tunneling from the tip to the electrode. That may save enormous efforts in attempts to take into account a complicated surface geometry of the tip. News in the theory of STM and experimental studies are expected to be presented at the STM-91 Conference in Interlaken, Switzerland, the Proceedings of which would appear in May 1992 in the journal *Ultramicroscopy.*

8. CHARGE TRANSFER KINETICS. ELEMENTARY ACT

Theory of charge transfer processes both in the bulk of the solution and at interfaces is a subject of long standing investigations.[100-105] This field unites the scientists from the chemical physics, physical chemistry, inorganic chemistry and electrochemistry communities, being based on quantum physics, statistical mechanics and the theory of transport processes.

Fundamental principles of the processes in the bulk and at interfaces have much in common. Each process is conventionally separated into stages:

* Transport stage (which provides the reactants to come together or a reactant to approach the electrode).
* The elementary act of reaction.

Types of problems which one meets in the theory of elementary act is most easily elucidated in the case of electron transfer reactions. The reaction is a result of a transition of electron from a donor to acceptor (the electrode may stand for one of them). The energy level of the electron in a donor or acceptor state is "breathing" in response to the fluctuation of coordinates of those degrees of freedom of the medium which interact strongly with the kinetic electron. Due to Franck - Condon principle the transition may happen only when the levels of the initial and final states are eventually equalized. This equalization occurs at some intermediate configuration of the medium coordinates and at this *reaction configuration* (or near it) the electron transfer may take place. The activation energy of the elementary act of electron transition is determined by the work needed for a medium to fluctuate to the intermediate configuration in a classical fashion.

Reactions of *nonadiabatic* and *adiabatic* types are distinguished in the theory. In nonadiabatic reactions, the probability of electron hop when the system have reached the intermediate configuration is low. The medium modes then passes the intermediate configuration many times before the act of transfer occurs. In adiabatic reactions the hop takes place with a high probability when the intermediate state is reached.

The theory of nonadiabatic reactions is built on a concept of *channel diabatic terms* of the initial and final states. These are the potential energy surfaces (as a function of slow medium coordinates) corresponding to the initial and final states of the reactants. When the matrix element of transfer (the perturbation is due to the interaction of the electron at the donor in the initial state with the acceptor) is large this concept breaks down. The *adiabatic splitting* then takes place and one should operate with the adiabatic terms (Fig.4). The phenomenon is then reduced to the so called *Landau - Zener* problem.

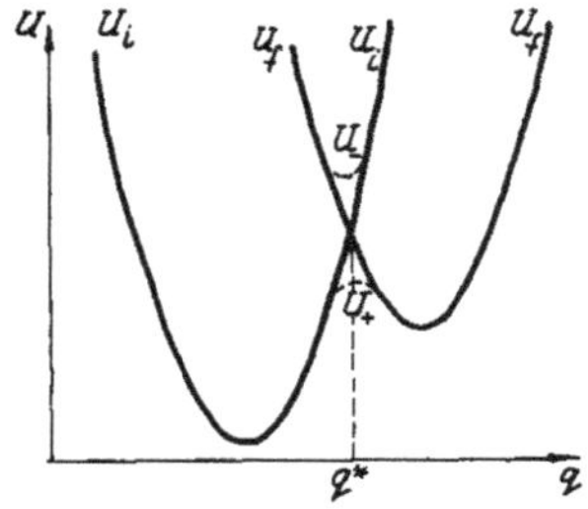

Fig.4. One dimensional diabatic and adiabatic potential energy surfaces

In the case of nonadiabatic reactions, the preexponential factor in the probability of transfer is determined by the the Landau - Zener factor. In adiabatic reactions, it is determined by the characteristic frequency of fluctuations of the medium modes in question.

The particular feature of the reactions in liquid state is the multi-dimensional character of the potential energy surfaces, be they nonadiabatic or adiabatic. A way to describe dissipative fluctuations of solvent polarization or other modes strongly interacting with the reactants is to expand them in a set of normal modes with a quasi-continuous dependence of normal frequencies on the branch number. With this trick one may reproduce the continuous frequency spectrum of the response functions which characterize the solvent. Expansion coefficients (oscillator strengths) are the unknown parameters but the sum or summation rules can be derived to relate the relevant combination of this parameters through the observable response functions.[56]

Thus the theory of electrochemical charge transfer reactions in the bulk and at interfaces is a "descendant" of the theory of radiationless[106] or radiative charge transfer processes in solids[107], albeit,developed quite far from its origin. The multi-dimensional energy surfaces,[108] continuous frequency spectrum and tunneling along the coordinates of the slow degrees of freedom with the frequencies greater than $kT/4\hbar$ [109], rearrangement of heavy particles[110], continuous electron spectrum in the case of electrode reactions[111], bridge assisted processes (superexchange) [112], various abnormal regimes (low temperature kinetics[113], highly exothermic reactions[114], etc.) - all this could be mentioned to give a flavor of a variety of new problems which electrochemistry has provoked to solve.

In electron transfer process, Condon approximation helped for years to separate quantum chemistry from kinetics; the electron matrix elements were considered independent of the medium coordinates (dependent of the reactants orbital structure and interreactant separation and mutual orientation). This was realized recently and a proper formalism has been suggested in the context of a number of processes typical of labile electron distributions in the initial or/and final states. These are, e.g. the reactions of solvated electrons[116], lousy bound electrons[117] and, possibly, some electrode reactions. "*Lability principle*" in electrochemistry has been formulated: the more labile is the electronic distribution in the reactants, the lower are the activation barriers[118]. There is hardly a universal proof of this statement (as an exercise one may try to look for counter examples) but it does hold in the variety of cases. One of the counter examples has been raised in a study of the role played by delocalization of electron distribution when the solvent coordinates pass the transition point[119]; the delocalization was concluded to give a contribution to the activation barrier. The validity of these arguments was severely questioned.[120,104]

As we have already noted, the theory distinguished the transport of reactants to meet each other and the elementary act of the reaction. The elementary act was considered to take place when the mutual arrangement of the reactants corresponded to being in the reaction zone. How to get in that zone was regarded to be a problem of diffusion kinetics (also a good but yet poorly developed field for application of condensed matter physics!). Such a separation of phenomena is not always true. Indeed fluctuation of the solvent interacting with the dissolved charged reactants affects both the microdynamics of ion motion[121] in the solution and the elementary act.

Attempts to go beyond this separation are making first steps.[122] The majority of works yet concentrated on the each state separately.

Finally a huge field of activity is concerned with solvent dynamics. It was argued that the diabatic and adiabatic potential energy surfaces are a poor language to describe the movements of the solvent modes, generally stochastic and dissipative. Instead of the effective harmonic oscillators approach, an alternative more "microscopic" approach has been suggested in which the classical random motion of the solvent polarization was described by the Langevine type equations coupled to quantum electron transitions.[123] Good old Kramers theory[124] was thus revisited[125] which opened a flood of works.[126] [A paper by HYNES et al represents this direction of research: it gives an overview on recent solvation dynamics studies and summarizes *the point of view of the authors* on the role of the electronic subsystem of the solvent (c.f.Ref.104)]. The relationship between the novel Kramers approach and the dynamic potential energy terms treatment is now more or less clear and there is a certain class of results to which one can come starting from any of these approaches.[104]

The fundamentals of the charge transfer reactions are considered in this book in a broad overview by ULSTRUP et al (the dark and light induced processes in physics, chemistry and biology). A summary of recent theoretical ideas (with a bias to electrode kinetics) is given by KUZNETSOV followed by an updated essay of KRISHTALIK about the critical experiments in quantum electrochemistry .

A little more about recent advances in the theory of electrode processes. After the heroic period in the study of isotope effects in proton transfer and the experimental proof of theoretically predicted quantum nature of proton discharge in the oxygen evolution reactions[127] one should acknowledge some relaxation period. Of the most interesting results obtained in this period one should probably mention the theory of reactions going via intermediate adsorbate states[128] where the Newns - Anderson model for an adsorbate was successfully employed.[129]

New boom is coming in the theory of electrode processes related to new types of reactions and electrodes. Leaving aside such exotic fruits as size quantized reactions at superfine electrodes[130],etc., we are to mention the phenomena which does not seem to be exotic any more after the discovery of high T_c superconductors.

Reactions at superconducting electrodes (in contact with solid electrolytes or low freezing temperature liquid electrolytes) draw great interest. First experimental results have been published[131] and a theory was proposed[132]. The theory is based on the standard principles of the usual low temperature superconductivity. It suggests a number of novel features of the electrode process with respect to reactions on normal metals. They are related to the "abnormal" spectrum of the density of states of normal electrons and a contribution of the second order reactions with participation of Cooper pairs. Attempts to explain the observed hump in the temperature dependence of the current at a given voltage have also been made. Extension of the theory towards account for new ideas on the nature of high T_c superconductivity is welcome. It should be noted, however, that the existing theory is based on the most general features of superconductivity (no matter what is the nature of them). They are accounted for by phenomenological assumptions: an existence of electron pairing, the energy gap (with a specific temperature dependence), a threshold law for the normal electrons density of states, etc. Changing the kind of superconductor would presumably, affect the values of the typical parameters involved, like, e.g. the radius of a Cooper pair, etc. The progress in the experiment and the theory is represented here by the papers of LORENTZ, SALMANN-ISCHENKO & BREITER and by KUZNETSOV's second paper, respectively).

Studies of the low temperature electrochemistry (down to 120°K) at a normal conductor/frozen electrolyte interface (hydrates of perchloric acid) have been started long before the high T_c boom by STIMMING's group (briefly reviewed in the book). The key point of these studies is the evidence that the contact of electrolyte with Au,Ag or Cu seems to remain perfect in the whole temperature range - below and above the freezing point(judging from smooth temperature dependence of the capacity) and thus the fundamentals of the low temperature charge transfer kinetics can be safely studied. The latter includes some interesting features in the current - voltage curves which awaits a proper theory. Double layer effects in kinetics may be essentially interesting in this system: the frozen electrolyte reminds more a superionic conductor with one mobile charge carrier specie. The role of the double layer effects in such systems (exhibiting the lattice saturation phenomena) has been investigated in the context of solid electrolytes[83] ;it could be of importance in Stimming's observations, as well.

Big congresses are being devoted to electrochemistry of modified electrodes.[133] The interest is especially high to polymer film covered electrodes. Some theoretical aspects of these systems have been recently studied[134]. Future progress would require understanding the electronic structure of doped polymer films, aspects of electron and ion transport through film (sometimes, porous) with possible importance of percolation phenomena. All the hits of the modern condensed matter physics are in, so that this could be a good subject for the next meeting.

Another hot topic is the charge transfer accross the fractal electrodes[135].

9. APPLIED ELECTROCHEMISTRY

Editors efforts to outline at least a few promising aspects of the use of condensed matter, atomic and molecular physics in the fundamentals of applied electrochemistry have been met with much less enthusiasm. The nice exception is the electrochemistry of semiconductors (which never fails in this context), and we are happy to present an overview by GERISHER on solar energy conversion.

A top expert summary of modern principles and engineering ideas in power sources is given by BREITER. There is no direct relation of this paper to condensed matter physics, but it would definitely help a condensed matter physicist to understand the future of his car battery.

A short review of the theory of turbulent mass transport in electrochemical cells is given by VOROTYNTSEV. He summarizes suggestions and the first probes in the use of electrochemical confinement for the study of velocity correlation functions (so much needed in the theory of turbulence!). In the case of success a top mark to electrochemistry will be put by the physics community.

In connection with some more ambitious marks, a reader might ask whether he could find in this book something about *cold fusion* or *nuclear electrochemistry*? Indeed, the success of the ideas of Fleischmann and Pons[136] could have enormously increased the rating of electrochemistry in the eyes of humanity. The failure, not speaking of a fraud, could tentatively lower down its shares.

There are still many adepts of cold fusion in the United States (not only in Utah National Cold Fusion Institute or Texas A&M University), Japan and India. An updated list of the research groups involved and the claims on the evidences in favor of this phenomenon have been recently compiled[137]. In addition to speculations about possible mechanisms of cold fusion[138], there are rumors about existence of yet unpublished quite involved theories. To my knowledge the most serious and objective theoretical analysis of the problem has been published by Leggett[139].

Among the experimentalists at this meeting, there was none who did not try to reproduce or extend Fleischmann and Pons experiments. Alas, all failed to observe any expected fusion products beyond the background (Bockris raised his contra-arguments why this is so[137]). In Germany, the popularity of nuclear electrochemistry among the researchers is not more than that of the atomic power among the Greens: Atomkraeft? Nein, danke! As for theorists, they were not keen enough to see any convincing mechanisms allowing for cold fusion. However, more frankly, I think that they simply compensated the lack of fantasy by an excess of laziness to which they fully trusted following the slogan: *wait to make a theory of a phenomenon prior a concensus among experimentalists about its existence* (if you have not predicted it earlier by your own!). Such a position is disputable[140] and could lead to a loss of a big game. In the cold fusion case, however, the future will show what a position was wise.

10. Acknowledgements

A big part of these notes has been written during my 1990 and 1991 visits to ICTP, for the support of which I wish to acknowledge Professor Abdus Salam, IAEA and UNESCO. The work has been accomplished during my stay at the Technical University of Munich as a Humboldt Research Award Guest Professor; I am thankful to Humboldt Foundation for financial support and to Professor Gregor Cevc and his colleages for generous hospitality which I met at the TUM Laboratory of Medical Biophysics.

11. References

1. J.E.Enderby and G.W.Nielsen, *Rep.Prog.Phys.*, **44** (1981) 593; in *Water. A Comprehensive Treatise*, vol.6, ed.F.Franks (Plenum, New York, 1979) Chapter 1.; P.A.Egelstaff, *Adv.Chem. Phys.* 53 (1983) 1.
2. G.Kabish and E.Kalman, in Ref.3,p.1; E.Kalman and G.Pálimkás, in Ref.3, p.501.
3. *The Chemical Physics of Solvation*, eds. R.R.Dogonadze, E.Kalman, A.A.Kornyshev and J.Ulstrup, Part B (Elsevier, Amsterdam, 1986).
4. I.R.Yukhnovskij and M.F.Golovko, *Statistical Theory of Classical Equilibrium Systems* (Naukova Dumka, Kiev, 1980) in Russian.
5. G.Stell, *Statistical Mechanics*, Modern Theoretical Chemistry, vol.5A, ed. B.Berne (Plenum, New York, 1977).
6. D.Chandler, *Fluids Simple and complex*, The Liquid State of Matter, eds. E.W.Montroll and J.E.Lebowitz (North Holland, Amsterdam, 1982).
7. G.G.Malenkov, *The Structure of Aqueous Systems* (Nauka, Moscow, 1991) in Russian.
8. K.Heinzinger, *Physica* B 131 (1985) 196.
9. L.Blum and F.Vericat, in *The Chemical Physics of Solvation*, eds. R.R.Dogonadze, E.Kalman, A.A.Kornyshev and J.Ulstrup, Part A (Elsevier, Amsterdam, 1985).p.143.
10. M.F.Golovko and I.R.Yukhnovskij, *ibid.* p.207.
11. K.Heinzinger, *ibid.* p.313.
12. G.G.Malenkov, *ibid.* p.355.
13. A.Karpfen and P.Schuster, *ibid.* p.265.
14. A.Geiger, *Ber.Berichte.Bunsenges.Phys.Chem.* 85 (1981) 52.
15. M.N.Buslaeva and O.Ya.Samoilov, in *The Chemical Physics of Solvation*, eds. R.R.Dogonadze, E.Kalman, A.A.Kornyshev and J.Ulstrup, Part A (Elsevier, Amsterdam, 1985), p.391.
16. M.Born, *Z.Phys.*,1 (1920) 45.
17. A.A.Kornyshev, in *The Chemical Physics of Solvation*, eds. R.R.Dogonadze, E.Kalman, A.A.Kornyshev and J.Ulstrup, Part A (Elsevier, Amsterdam, 1985), p.77.
18. M.V.Alpatova, L.I.Krishtalik and E.V.Ovsyannikova, *Electrochim.Acta*, in Press (1991).
19. H.L.Friedman, *Ionic Solution Theory* (Wiley-Interscience, New York,1962).

20. D.Henderson, L.Blum and M.Lozada-Cassou, *Prog.Surf.Sci.* 13(1983) 4615.

21. M.Plischke and D.Henderson, *Electrochim.Acta* **34** (1989) 1863.

22. D.Henderson, *Chem.Phys.* **141** (1990) 79.

23. K.Heinzinger and E.Spohr, *Electrochim.Acta* **34** (1989) 1849; E.Spohr, *Chem.Phys.* **141** (1990) 87.

24. D.M.Kolb, *Z.Phys.Chem.(N.F.)*154 (1986) 179.

25. J.Clavilier, K.El Achi and A.Rodes, *Chem.Phys.* **141** (1990) 1.

26. A.A.Kornyshev in *The Chemical Physics of Solvation*, eds. R.R.Dogonadze, E.Kalman, A.A.Kornyshev and J.Ulstrup, Part C (Elsevier,Amsterdam, 1985) p.355.

27. A.A.Kornyshev, *Electrochim.Acta* **34** (1989) 1829.

28. W.Schmickler and D.Henderson, *Prog.Surf.Sci.* **22** (1986) 323.

29. K.-M.Ho, B.N.Harmon and S.H.Liu, *Phys.Rev.Lett.*44 (1980) 1531; D.M.Kolb, W.Boeck, K.-M.Ho and S.-H.Liu, *ibid.* **47** (1981) 1921; K.-M.Ho, C.L.Fu, S.H.Liu, D.M.Kolb and G.Piazza, *J.Electroanal. Chem.* **150** (1983) 1235.

30. E.V.Chulkov, V.M.Silkin and I.Ya.Sklyadneva, *Surf.Sci.* **231** (1990) 9; E.V.Chulkov, V.M.Silkin and E.N.Shirkalov, *Fiz.Met.Metalloved.* **64** (1987) 213; E.V.Chulkov, V.M.Silkin, A.G.Lipnitskii and V.E.Panin, *Electrochim. Acta* **34** (1989) 19.

31. S.Amokrane and J.-P.Badiali, *Electrochim.Acta.* **34** (1989) 39; *J.Electro-anal. Chem.* **266** (1989) 21.

32. W.Schmickler, *Chem.Phys.* **141** (1990) 95.

33. W.R.Fawcett, *Israel J.Chem.* **18** (1979) 3.

34. V.A.Benderskij and G.I.Velichko, *J.Electroanal.Chem.* **140** (1982) 1; *Elek-trokhimiya* **23** (1987) 490.

35. A.M.Brodsky and L.I.Daikhin, *Elektrokhimiya* **25** (1989) 435.

36. J.F.Smalley, C.V.Krishnan, M.Goldman, S.W.Feldberg and I.Ruzic, *J.Electroanal.Chem.* **248** (1986) 323; S.Feldberg, private communication.

37. B.B.Damaskin and R.V.Ivanova, *Usp.Khim.* **48** (1979) 1747.

38. D.C.Grahame and R.Parsons, *J.Am.Chem.Soc.* **83** (1961) 1291.

39. N.D.Lang, *Solid State Phys.*,**28** (1973) 255.

40. J.P.Muscat and D.M.Newns, *Prog.Surf.Sci.* **9** (1978) 1.

41. *Local Density Approximation in Quantum Chemistry and Solid State Physics*, J.P.Dahl and J.Avery (Plenum, New York, 1984).

42. A.I.Gubanov and S.M.Dunaevskii, *Fiz.tverd.Tela* **19** (1977) 1369.

43. N.D.Lang, in *Theory of Inhomogeneous Electron Gas*, eds. S.Lundqvist and

N.S.March (Plenum, New York, 1983); M.Manninen and J.K.Norskov , in *The Chemical Physics of Solvation*, eds. R.R.Dogonadze,E.Kalman,A.A.Kornyshev and J.Ulstrup, Part A (Elsevier, Amsterdam, 1985), p.475.
44. J.R.Appelbaum and D.R.Hamann, *Phys.Rev.* B6 (1972) 1122.
45. S.M.Dunaevskii and N.D.Potekhina, *Fiz.tverd.Tela* 21 (1979) 2304.
46. O.P.Burmistrova, G.G.Vladimirov and S.M.Dunaevskii, *ibid.* 22 (1980) 836.
47. K.Shönhammer, *Z.Phys.* B 21 (1975) 389; *Phys.Rev.* B 13 (1976) 4336.
48. P.Wiegmann and A.M.Tsvelik, *Adv.Phys.* 32 (1983) 453.
49. J.E.Müller and J.Harris, *Phys.Rev.Lett.* 26 (1984) 2493; H.P.Bonzel, G.Pirug and J.E.Müller, *ibid.* 58 (1987) 2138; J.Müller, *Surf.Sci.* 178 (1986) 589; J.Harris and J.E.Müller in Ref.41.
50. A.C.Hewson and D.M.Newns, *Jap.J.Appl.Phys.Suppl.* 2 (1974) 2121.
51. V.E.Kravtsov and A.G.Mal'shukov, *Zh.exper.teor.Fiz.* 75 (1978) 691 (English translation: *Soviet Phys. JETP* 48 (1978) 248.)
52. D.I.Khomskii, *Solid State Commun.* 27 (1978) 775.
53. O.M.Braun and A.I.Volokitin, *Surf.Sci.* 131 (1983) 148.
54. O.M.Braun, A.I.Volokitin and V.P.Zhdanov, *Usp.fiz.Nauk* 158 (1989) 421.
55. W.Schmickler, *J.Electroanal.Chem.* 100 (1979) 253.
56. R.R.Dogonadze, A.A.Kornyshev and A.M.Kuznetsov, *Teor.Mat.Phys.* 15 (1973) 127 (English translation: *Theor.Math.Phys.* (USSR) 15 (1973) 407).
57. A.A.Kornyshev and W.Schmickler, *J.Electroanal.Chem.* 185 (1985) 253.
58 A.A.Kornyshev and W.Schmickler, *J.Electroanal.Chem.* 202 (1986) 1.
59. J.P.Muscat and D.M.Newns, *J.Phys.C: Solid State Phys.* 7 (1974) 2630.
60. W.Schmickler, *Chem.Phys.Lett.* 115 (1985) 216.
61. V.E.Kravtsov and A.G.Ma'lshukov, *Zh.exp.teor.Fiz.* 77 (1979) 180. (English translation: *Soviet Phys. JETP*, 77 (1979) 91).
62. D.M.Kolb, in *Advances in Electrochemistry and Electrochemical Engineering*, vol.11, eds. H.Gerischer and C.W.Tobias (Wiley, New York, 1978).
63. G.Staikov, K.Jüttner, W.J.Lorentz and E.Budevski, *Electrochim. Acta* 23 (1978) 319.
64. A.Hamelin and J.Lipkowski, *J.Electroanal.Chem.*171 (1984) 317.
65. R.Adzic, E.Yeager and B.Cahan, *J.Electrochem.Soc.* 12 (1974) 474.
66. A.I.Volokitin, *Teor.Mat.Fiz.* 80 (1990) 399.
67. A.K.Mishra, S.K.Rangarajan, *J.Electroanal.Chem.* 229 (1987) 51, 67; 230 (1987) 1, 23.
68. B.M.Ocko, J.Wang, A.Davenport and H.Isaacs, *Phys.Rev.Lett.* 65 (1990)

1466.

69. S.G.J.Mochrie, D.M.Zehner, B.M.Ocko and D.Gibbs, *Phys.Rev.Lett.* 64 (1990) 2925; D.Gibbs, B.M.Ocko, D.M.Zehner and S.G.J.Mochrie, *Phys.Rev.* A (1990) in Press.
70. M.A.Vorotyntsev, in *The Chemical Physics of Solvation*, eds. R.R.Dogonadze, E.Kalman, A.A.Kornyshev and J.Ulstrup, Part C (Elsevier, Amsterdam, 1986), p.401.
71. M.A.Vorotyntsev, *J.Res.Inst.Catalysis (Hokkaido University)* 30 (1982) 167; M.A.Vorotyntsev and S.Ivanov, *Zh.exp.teor.Fiz.* 88 (1985) 1729; M.A.Vorotyntsev, *Elektrokhimiya* 21 (1985) 351; A.V.Tsheglov, M.A.Vorotyntsev, B.B.Damaskin, and R.V.Ivanova, *Elektrokhimiya* 21 (1985) 1126.
72. L.A.Bolshov and A.P.Napartovich, *Zh.exp.teor.Fiz.* 64 (1973) 1404.
73. V.V.Pleskov and Yu.Ya.Gurevich, *Semiconductor Photoelectrochemistry* (Consultants Bureau, New York, 1986).
74. Yu.V.Pleskov, *Solar Energy Conversion. A photoelectrochemical Approach* (Springer, Berlin, 1990).
75. *The Interface Structure and Electrochemical Processes at the Boundary between two Immiscible Liquids*, ed.V.E.Kazarinov (Springer, Berlin,1987).
76. W.P.Weber and G.W.Gokel, *Phase Transfer Catalysis in Organic Synthesis* (Springer, Berlin, 1977).
77. F.A.Sidditti and H.Ti Tien in *Topics in Bioelectrochemistry and Bioenergetics*, vol.5, ed. G.Milazzo (Wiley, New York, 1983), p.157.
78. J.Koryta, *Anal.Chim.Acta* 61 (1972) 329; 91 (1977) 329; 91 (1977) 1; 111 (1979) 1; 139 (1982) 1.
79. M.A.Vorotyntsev and S.N.Ivanov, *Elektrokhimiya* 23 (1987) 215; 774.
80. M.A.Vorotyntsev and S.N.Ivanov, *ibid.* 24(1988) 805; 25 (1989) 554.
81. A.A.Kornyshev and S.Leikin, *Phys.Rev.* A 40 (1989) 6431; S.Leikin and A.A.Kornyshev, *J.Chem.Phys.* 92 (1980) 6890.
82. *Physics of Superionic Conductors, Topics in Current Physics*, vol.15 ed. B.Salomon (Springer, Berlin, 1979).
83. A.A.Kornyshev and M.A.Vorotyntsev, *Electrochim.Acta* 26 (1981) 303.
84. E.Yeager, *Surf.Sci.* 101 (1980) 1.
85. A.M.Brodsky and M.I.Urbakh, *Prog.Surf.Sci.* 15 (1984) 121; *Usp.fiz.Nauk* 138 (1982) 413; A.M.Brodsky, L.I.Daikhin and M.I.Urbakh, *Usp.Khim.* 54 (1985) 3.
86. M.Fleischmann, P.I.Hendra and A.J.McQillain, *Chem.Phys.Lett.* 26 (1974)

163; A.J.McQillain, P.I.Hendra and M.Fleischmann, *J.Electroanal.Chem.* **65** (1975) 933; M.Fleischmann, P.I.Hendra, A.J.McQillain, and R.L.Paul and E.S.Reid, *J.Raman Spectrosc.* **4** (1976) 269.

87. D.J.Jeanmaire and R.P.Van Duyne, *J.Electroanal.Chem* **84** (1977) 1; R.P.Van Duyne, in *Chemical and Biochemical Applications of Lasers*, ed. C.B.Moor (Academic press, New York, 1979), vol.**4**, Chapter 5; J.de Physique (Paris) **38** (1977) C5.
88. M.G.Albrecht and J.A.Creighton, *J.Am.Chem.Soc.* **99** (1977) 5215.
89. R.K.Chang and T.E.Furtak, *Surface Enhanced Raman Scattering* (Plenum, New York, 1982).
90. A.G.Mal'shukov, in *The Chemical Physics of Solvation*, eds.R.R.Dogonadze, E.Kalman, A.A.Kornyshev and J.Ulstrup, Part C (Elsevier, Amsterdam, 1988) p.433.
91. M.I.Urbakh, *Electrochim.Acta* **34** (1989) 1777.
92. R.de Levie, *Electrochim.Acta* **8** (1963) 751; **10** (1965) 113.
93. A.Le Méhauté and G.Crépy, *Solid State Ionics* **9/10** (1983) 17.
94. S.H.Liu, *Phys.Rev.Lett.* **55** (1985) 529; S.H.Liu, T.Kaplan and L.J.Gray, *Solid State Ionics* **18/19** (1966) 65; T.Kaplan and L.J.Gray, *Phys.Rev.* B **32** (1985) 7360.
95. L.Nyikos and T.Paikossy, *Electrochim.Acta* **30** (1985) 1533.
96. M.Keddam and T.Takenouti, *C.R. Acad.Sci.(Paris) Ser II* **302** (1986) 281.
97. J.B.Bates, J.C.Wang and Y.T.Chu, *Solid State Ionics* **18/19** (1986) 1045; B.Sapoval, *ibid.* **23** (1987) 253.
98. B.Duplantier, R.E.Goldstein, V.Romero-Rochin and A.I.Pesci, *Phys.Rev. Lett.* **65** (1990) 508; R.E.Goldstein, A.I.Pesci and V.Romero-Rochin, *Phys.Rev.* A **41** (1990) 5504.
99. A.A.Kornyshev and M.A.Vorotyntsev, *Physica* A **191** (1991) 98.
100. R.Marcus and N.Sutin, *Biochim.Biophys.Acta* **811** (1985) 265.
101. J.Ulstrup, *Charge Transfer Processes in Condensed Media , Lecture Notes in Chemistry* **10** (Springer, Berlin, 1979).
102. R.R.Dogonadze and A.M.Kuznetsov, in *Comprehensive Treatise of Electrochemistry*, vol. **7** (Plenum, New York, 1983), p.1.
103. A.M.Kuznetsov, J.Ulstrup and M.A.Vorotyntsev in *The Chemical Physics of Solvation*, eds.R.R.Dogonadze, E.Kalman, A.A.Kornyshev and J.Ulstrup, Part C (Elsevier, Amsterdam, 1988) p.163.
104. A.M.Kuznetsov, *Modern Aspects of Electrochemistry* **20** (1089) 95.

105.A.M.Berezhkovskii, A.N.Drosdov and V.Yu.Zitserman, *Russ.J.Phys.Chem.* 62 (1988) 1353.

106.A.M.Stoneham, *Theory of defects in Solids* (Clarendon, Oxford, 1975).

107.H.H.Kristofel', *Theory of Impurity Centers of Small Radii in Ionic Crystals* (Nauka, Moscow, 1974) in Russian.

108.E.M.Itskovich and A.M.Kuznetsov, *Elektrokhimiya* 16 (1980) 755.

109. M.A.Vorotyntsev, R.R.Dogonadze and A.M.Kuznetsov, *Dokl.Akad.Nauk SSSR*, 209 (1973) 1135.

110.R.R.Dogonadze and A.M.Kuznetsov, *Elektrokhimiya* 3 (1967) 380.

111.R.R.Dogonadze, A.M.Kuznetsov and M.A.Vorotyntsev, *J.Electroanal. Chem.* 25 (1970). app.17; *Croatica Chem.Acta* 44 (1977) 1498.

112.A.M.Kuznetsov and Yu.I.Khatkats, *Elektrokhimiya* 13 (1977) 1498.

113.R.R.Dogonadze, A.M.Kuznetsov, M.A.Vorotyntsev and M.G.Zakharaya, *J.Electroanal. Chem.* 75 (1977) 315.

114.R.R.Dogonadze, A.M.Kuznetsov and M.A.Vorotyntsev, *Z.Phys.Chem.(NF)* 100 (1976) 1.

115.A.M.Kuznetsov, *Elektrokhimiya* 18 (1982) 594, 598, 736; *Nuovo J.de Chimie* 5 (1981) 427.

116.E.M.Itskovich,A.M.Kuznetsov and J.Ulstrup, in *The Chemical Physics of Solvation*, eds.R.R.Dogonadze, E.Kalman, A.A.Kornyshev and J.Ulstrup, Part C (Elsevier, Amsterdam, 1988) p.275.

117.A.M.Kuznetsov and J.Ulstrup, *Phys.Stat.Sol.*(b) 114 (1982) 673.,

118.A.M.Kuznetsov, *Elektrokhimiya* 19 (1983) 1596.

119.H.J.Kim and J.T.Hynes, *J.Phys.Chem.* 94 (1990) 2736; *J.Chem.Phys.* (1991) in Press; *Int.J.Quantum Chem.(Symphosium)* (1991) in Press.

120.A.M.Kuznetsov, *J.Chem.Phys.*(1991) submitted.

121.A.M.Kuznetsov, *Faraday Discussion, Chem.Soc.*,74 (1982) 49.

122.A.M.Kuznetsov, *Elektrokhimiya* 22 (1986) 291.

123.L.D.Zusman, *Chem.Phys.*49 (1980) 295; L.D.Zusman, A.B.Gelman, B.Ya.Shapiro, *Chem.Phys.Lett.* 167 (1990) 457.

124.H.A.Kramers, *Physica* 7 (1940) 284.

125.R.F.Grote and J.T.Hynes, *J.Chem.Phys.* 73 (1980) 2715.

126.P.Hänggi, P.Talkner and M.Borkowetc, *Rev.Modern.Phys.* 62 (1990) 251; J.T.Hynes, in *Theory of Chemical Reactions*, vol 4, ed.M.Bayer (CRC, Boca Raton, 1985), p.171.

127.L.I.Krishtalik, *Charge Transfer Reactions in Electrochemical and Chemi-*

LIQUIDS, SOLUTIONS, SOLVATION, DIELECTRIC PROPERTIES

COMPUTER SIMULATIONS OF IONIC SOLVATION

PHILIPPE BOPP

Institut für Physikalische Chemie
Rheinisch-Westfälische Technische Hochschule
Templergraben 59, D-5100 Aachen
Federal Republic of Germany

ABSTRACT

In this lecture we want to study how computer simulations are employed to investigate the solvation of ions in water on a microscopic level. First the basic concepts and assumptions of the Molecular Dynamics (MD) and Monte Carlo (MC) computer simulation methods are very briefly recalled. The intermolecular (and sometimes also intramolecular) microscopic interaction potentials, which are the essential input to such simulations, are discussed. Selected results, both structural and dynamical, are then succintly presented to demonstrate some of the quantities that can be studied. It is not the purpose of this paper to be another review article. All it attempts to do is to give a brief introduction into the field for readers who may not be familar with it, and to give a few hints to the recent literature.

1. Introduction

Ions are "hydrated" in water, everybody knows that. On the other hand, when it comes to describing this phenomenon on a microscopic level in any detail, a lack of knowledge immediately appears, in spite of the fact that aqueous ionic solutions have been the traditional focus of much research effort in chemistry and physics for more than half a century. It is just extremely difficult to build a consistent microscopic picture on the basis of macroscopic considerations. Computer simulations are in a certain sense the attempt to procede the other way around. Starting from the microscopic interactions between the various species in a chemical system, one tries to get access to macroscopic observables by generating statistical mechanical ensembles (or at least representations thereof) numerically and then computing the appropriate expectation values. One then compares the values obtained from the simulation with experimental ones for as many quantities as possible. Of course, agreement between the two is not a proof that the interaction potentials adopted to describe the microscopic interactions are the "correct" ones, but reproducing many observable quantities in a simulation is still an extremely stringent test, and a positive outcome certainly allows to call the underlying model a "reasonable" one. A negative outcome of such a comparison, on the other hand, is an indication of insufficient understanding and thus a challenge for improvement. We shall see some of both in the following.

The structure of an ionic solution, i.e. the average relative arrangement of ions and solvent molecules, will in general result from a delicate balance between ion-ion, ion-solvent and solvent-solvent interactions. It is clear that a concentration dependence is to be expected, especially when one remembers that some simple salts are soluble in water up to very high concentrations. Ionic concentrations may also be very high locally, e.g. in the vicinity of an interface such as an electrode. In water, the most important solvent, the presence of hydrogen bonds will be an important and non-negligible effect. Due to its strong orientational character and to a binding energy which may be of a magnitude comparable to that of intramolecular bonds, the hydrogen-bond often confers unusual properties to the chemical system where it occurs. All that has been said above will also be reflected in the dynamic properties.

2. Simulation Techniques

Monte Carlo (MC) and Molecular Dynamics (MD) computer simulations are very powerful tools for the study of liquids in general. Because of the extreme difficulties of an analytical treatment by integral equation methods, ionic solutions in hydrogen-bonded solvents are specially suited for simulation studies. We shall first very briefly review the principles of the classical computer simulation methods as applied to liquids. An excellent monograph on this subject is available [1], and the development of the field has been documented in various ways [2]. The application of simulation methods to aqueous ionic solutions and related systems has also been extensively described [3].

In the MD approach, the classical trajectories of a number of point masses or rigid bodies, representing atoms and molecules moving under the influence of their mutual interactions, are computed numerically. The total volume, the total number of particles, and total energy are usually kept constant during the simulation. The collection of trajectories resulting from the simulation can then be considered as the representation of a microcanonical ensemble. In the MC-case, a random walk process is used to construct a representation of the canonical ensemble. In both cases it is also possible to devise simulation schemes corresponding to other ensembles of statistical mechanics [2,4]. Averages are then computed from these ensembles, leading to the desired macroscopic quantities. In chemistry one is usually interested in the properties of 'large' systems (say something like 10^{20} particles), but since only a few thousand particles can at best be explicitly treated in the simulation, additional assumptions have to be made. The basis of these assumptions is that it suffices to treat explicitly particles separated by distances less than a correlation length. The method of 'periodic boundary conditions' together with the 'minimum distance convention', which is most often applied in simulations, and similar approaches, have been developed for this purpose.

The number of numerical integration steps (MD) or particle moves (MC) that can be carried out in a simulation is limited. The length of the individual timestep or move is again limited by many numerical and other practical requirements, and also depends on the model used. For liquids with rather intricate interactions like the ones that we want to study here, even with modern very large scale computers, it is difficult to exceed total simulated times of a few hundred picoseconds (or roughly equivalent amounts of MC moves). It is thus important to recognize that only phenomena occuring at this timescale are accessible to such simulations.

The microscopic interatomic or intermolecular interaction potentials are the most important 'input' to a computer simulation. They may be taken from ab-initio or other quantum chemical calculations, from experiments or from other ansätze according to chemical intuition. In some cases certain or all intramolecular degrees of freedom are also included in a simulation, and the corresponding potentials have to be provided as well. To be used in the numerical calculations, the potentials, and in the MD-case also the forces, have either to be tabulated with a sufficient accuracy, or have to be expressed analytically. It is usually assumed that the total intermolecular potential can be expressed as a sum of pair-potentials $V_{ij}(r_{ij})$:

$$V^{tot} = \sum_{i>j} V_{ij}(r_{ij}) \tag{1}$$

where i and j refer to so-called force-centers (or sites) on two molecules. A force-center is a location on a molecule, often coinciding with the position of an atom, where for instance a partial electric charge is assumed to represent an excess (positive or negative) electron density. Lennard-Jones potentials are often used to model an atom or even a group of atoms in a molecule. In intramolecular potentials, and in some instances also in intermolecular ones, three-body contributions V_{ijk}, e.g. depending on angles, have to be included in order to obtain satisfactory results from the simulation. The potentials together with the other parameters used to describe a molecule are often called the 'model' for this molecule.

3. The Models

It is little surprising that very many models have been developed for computer simulations of liquid water and other systems containing water [3]. In most cases the molecule is assumed to be indeformable (so-called rigid models) while in some instances the molecule is allowed internal deformations and motions (flexible models). In such a case an intramolecular potential must also be supplied. Such flexible models have been developed for instance by adding an intramolecular potential to an existing rigid model [5].

The most important feature to be reproduced by a model for water is of course the hydrogen bond. In virtually all models this feature is described by Coulomb interactions between positive partial charges (usually about one quarter to one half of an elementary charge) located at the positions of the hydrogens, and corresponding countercharges. These may be located in a way to form a distorted tetrahedron with the hydrogen positions (ST2 model [6]), representing in a way the lone pair orbitals of the molecule, or may be lumped together either at the position of the oxygen (CF, SPC and TIPS models [7]) or at a position not coinciding with an atom, on the H-O-H angle bisector (MCY, TIP4P models [8]). The dipole moment of the molecule results from the partial charges and the geometry and is often chosen to be somewhat higher than the observed gas phase dipole moment (1.86 D for H_2O). Of course, in flexible models, the deformation of the molecules due to interactions with their neighbors in the liquid will lead to fluctuations, and in general also to an average increase, of the molecular dipole.

The models use empirical functions of Lennard-Jones 12-6 type or other simple site-site pair potentials to describe the van-der-Waals part of the intermolecular interaction. The subtle interplay between the steric hindrances described essentially by these potentials and the Coulomb potentials allows to mimick the typical saturability of the hydrogen bond. Goldmann and Backx [9] have shown that such combinations of empirical potentials and charge distributions are a much more effective representation of real water than an equivalent empirical potential augmented by point multipoles at the center.

At this point it may be of interest to note that there are two quite different philosophies concerning the development of model potentials. In the first instance, quantum mechanical ab-initio calculations of very high quality are first performed for as many relative distances and orientations of two molecules as possible. The parameters of skillfully chosen functions are then fitted to these energies. If more than two molecules are included in the quantum mechanical treatment (which is very time consuming) this method allows in principle also to extract many-body contributions to the potential [10]. In the other instance, pair-potentials are refined by comparing results from simulations with experimental ones. The resulting potentials are then said to contain many-body effects in an "effective" way and are thus called effective potentials. The MCY potential and its successors are prominent examples of the first approach, while the ST2, TIP and CF models belong to the latter kind. We note also that the intramolecular potentials used in these classical simulations are in a certain sense also effective ones.

4. Results

The results for water and aqueous systems discussed here have been obtained using the rigid ST2 and TIP4P models, and a flexible one [11]. As examples, we shall now briefly look at results obtained from MD simulations of the following systems:

- pure water,
- aqueous solutions of simple salts like NaCl, $CaCl_2$ or $LaCl_3$ at various concentrations, and
- water and hydrated ions in the vicinity of an interface, here a model-biomembrane.

Details can be found in the literature.

4.1. Pure Water

These results are discussed with two goals in mind, firstly to briefy demonstrate the influence of the hydrogen bonding on the structural properties of the liquid, and secondly to use them later for comparison with the ionic solutions.

The structure of a liquid can be described by the set of atom-atom radial distribution functions (RDFs) for all kinds of atoms present. The RDF $g_{\alpha\beta}(r_{\alpha\beta})$ describes the variation of the local density of particles of type β around a particle

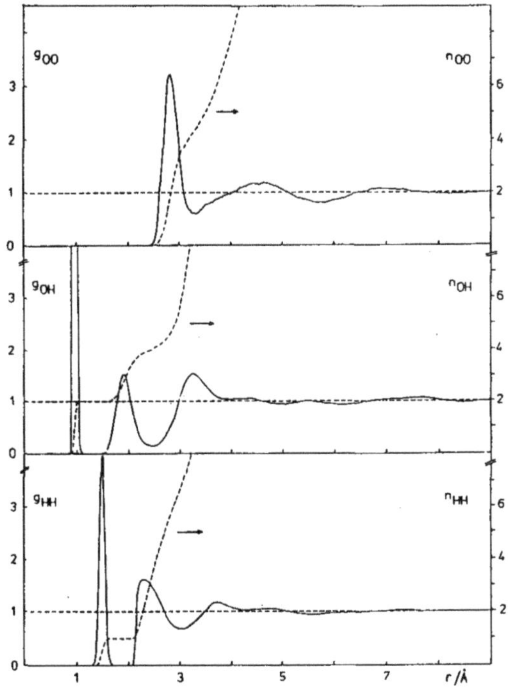

Figure 1. Radial pair distribution functions g_{OO}, g_{OH} and g_{HH} for pure water (T = 290 K, $\rho = 1g \cdot cm^{-3}$).

of type α relative to the average density expected for homogeneously distributed particles. The integral

$$n_{\alpha\beta}(r_{\alpha\beta}) = 4\pi\rho_\beta \int_0^{r_{\alpha\beta}} g_{\alpha\beta}(r'_{\alpha\beta}) r'^2_{\alpha\beta} dr'_{\alpha\beta} \tag{2}$$

gives then the total number of particles β around α as a function of their distance. It is sometimes called the running integral of the RDF. The RDFs often show several oscillations before they become equal to unity at large distances. The value of $n_{\alpha\beta}$ at the first minimum of $g_{\alpha\beta}$ is a possible rigorous definition of the number of next neighbors of a species, i.e. in our case, of the hydration number. RDFs are determined experimentally by X-ray, neutron and electron diffraction. The difficulty here resides in the fact that in a system containing N different kinds of atoms, there are $N \cdot (N+1)/2$ different atom-atom functions $g_{\alpha\beta}$. As many independent experiments are thus needed to determine them.

Figure 1 shows typical RDFs $g_{OO}(r_{OO})$, $g_{OH}(r_{OH})$ and $g_{HH}(r_{HH})$, obtained here from an MD simulation of pure liquid water slightly below room temperature. The agreement of the peak positions and heights with experimental results [12] is generally good. All models developed for water render these functions reasonably well, we shall thus not go into any detail, but rather highlight a few typical features. The results shown here are from a simulation at 290 K. Due to the low temperature, the structures in the RDFs are quite pronounced. A few remarks:

g_{OO}: The integral over the first peak (eq. 2) is slightly more than 4. This is much less than the number of 8 to 12 expected for densely packed spheres. The oxygen atoms are kept apart by the hydrogen bonds and the number of neighbors is thus reduced. Four neighbors is also what is expected for a tetrahedral coordination like the one found in ice. At low enough temperatures (like here) a weak second peak is visible in pure water, it disappears with increasing temperature. Generally speaking there is not much structure beyond 6 or 7 Å, the RDF is equal to 1. The pronounced short range order (the maximum value of g_{OO} is about 3) and the lack of long range order are typical of such liquids.

g_{OH}: The first peak (integral = 2) describes the distribution of the intramolecular O-H distances; this peak yields information about the deformation of the molecules due to the interactions with the neighbors. The next peak (integral also about 2) describes the positions of hydrogens of the two neighboring molecules which donate H-bonds to the central molecule. This peak is lower and much wider than the intramolecular peak; height and shape of this peak are quite sensitive to perturbations of the hydrogen bonds, for instance by solutes in the liquid. The third peak (at a little over 3 Å) shows that the positions of the second hydrogen atoms belonging to the bond-donating molecules, or of the hydrogen atoms belonging to molecules accepting H-bonds, are also correlated.

g_{HH}: Here again we have an intramolecular peak which yields information about the distorted molecular geometries in the liquid. The following, albeit small, oscillations are again in keeping with a mostly tetrahedral coordination of the water molecules.

At this point, a short remark is appropriate concerning the term 'hydrogen bond'. The existence of such a bond at a certain time during the simulation must be established from geometrical or energetical considerations. What is meant by the statement 'a hydrogen bond exists' is thus for instance that two oxygen atoms and one hydrogen atom are aligned, within certain tolerances, in the way that is expected for a hydrogen bond. Since one **always** has distributions of distances and angles, (and also of interaction energies) it is a matter of definition whether a certain geometry is considered a 'distorted bond' or a 'broken bond'. It is therefore interesting to compute different properties for 'bonded' and 'unbonded' molecules according to various definitions and to compare them with experimental information in an attempt to reach a consistent definition.

The tetrahedral coordination of a water molecule by its neighbors has now been mentioned several times, yet the evidence for such a structure presented up to now is only circumstantial. We have only seen that all the RDFs are compatible with such an arrangement. In the simulation the existence of a tetrahedral order in the system can be explicitly demonstrated, e.g. by a method proposed by Pálinkás et al. [13]. It is nevertheless difficult to quantify such results, and one often appeals to visualization techniques for the interpretation of such data. It is interesting that model potentials giving quite similar RDFs sometimes yield quite different three-dimensional patterns. The ST2 model is thus sometimes criticized for being "overtetrahedral'. By this one means that the distributions of say intermolecular O-H-O-angles may be too narrow, or roughly that the structure obtained may be in a certain sense too ice-like.

We shall now briefly consider the motions of the molecules in the liquid. The function which is used to describe them is the velocity autocorrelation function (vacf) $c_{vv}(t)$:

$$c_{vv}(t) = \frac{1}{NM}\sum_{i=1}^{N}\sum_{j=1}^{M}(\vec{v}_i(t_j)\vec{v}_i(t_j+t)) \tag{3}$$

or its normalized counterpart. Here $\vec{v}_i(t_j)$ is the velocity of particle i at the time t_j and N is the number of particles over which the average is taken. This can be all particles in the system or only selected particles fulfilling certain conditions. One can for instance compare the vacfs computed for the water molecules in the hydration shells of the ions with the ones obtained for water molecules farer away from the ions. M in equation 3 is the number of 'time origins' t_j over which the vacf is averaged. Instead of using $\vec{v}_i$ in equation 3, it is sometimes useful to correlate only certain components of the velocities, or combinations of velocities of several particles belonging to the same molecule. The Fourier-cosine transform of the normalized vacf

$$\hat{c}_{vv}(\omega) = \int_0^\infty \frac{c_{vv}(t)}{c_{vv}(0)}\cos(\omega t)dt \tag{4}$$

is easier to interpret and more instructive than the vacf itself. This function is directly related to the elastic neutron scattering. Under certain assumptions, it can also be compared with results from infrared and Raman experiments. We shall term $\hat{c}_{vv}(\omega)$ the spectral density of the motions.

Figure 2 shows such spectral densities for the center of mass motions of a water molecule in pure water. Two main bands are easily identified, one at about 50 cm^{-1} and one around 200 cm^{-1}. These two bands have been identified spectroscopically [14]. The band at lower frequencies is ascribed to motions of the oxygen atom perpendicular to the O-H...O direction of the bond, i.e. to a bending of the bond, while the band at high frequency is ascribed to the hindered translation of the molecule. The hindered rotations (librations) and the vibrations of flexible molecules in the liquid can also be studied. Approximations are usually introduced to decouple rotations and vibrations of flexible (and thus distorted) molecules.

Figure 3 shows the spectral densities obtained by such an approximate method for the librations around the three principal axes of inertia (see caption) of a water molecule in pure water. The librations around the x- and z-axes have their maximum at about 420 cm^{-1} while the libration around the y-axis is found at higher frequencies.

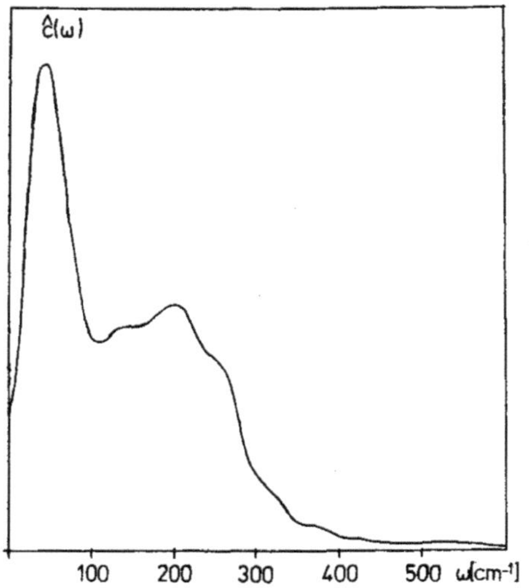

Figure 2. Spectral densities of the center of mass motions of a water molecule in pure water, in arbitrary units.

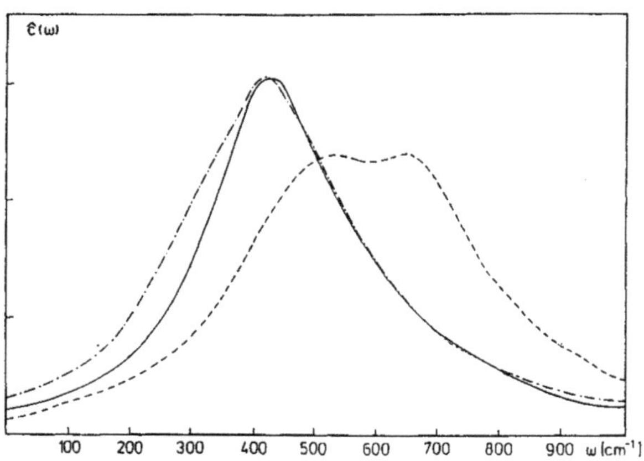

Figure 3. Spectral densities of the librations around the three principal axes of inertia for pure water, in arbitrary units. Solid line: around the axis parallel to the surface normal (x); dashed line: around the axis perpendicular to the dipole moment in the plane of the molecule (y), dash-dotted line: around the dipole axis (z).

The main reason for this is the low moment of inertia about this axis, but we note that the moments of inertia about the x- and z-axes are not identical. The fact that these two bands are very similar should not be interpreted as being the consequence of a homogeneous environment. Experimental values between 475 and 725 cm $^{-1}$ [14] are reported for these motions.

4.2. Ionic Solutions

The often peculiar properties of aqueous ionic solutions are the consequence of the competing influence of the ions and of the hydrogen bond network of the water. Only simple ions will be considered here, but more complex ones are also studied. Great care must be exercised in such cases since, as stated above, the microscopic interaction potentials must first be known and reliably parametrized. Cruder approaches than the ones sketched above must thus be used often.

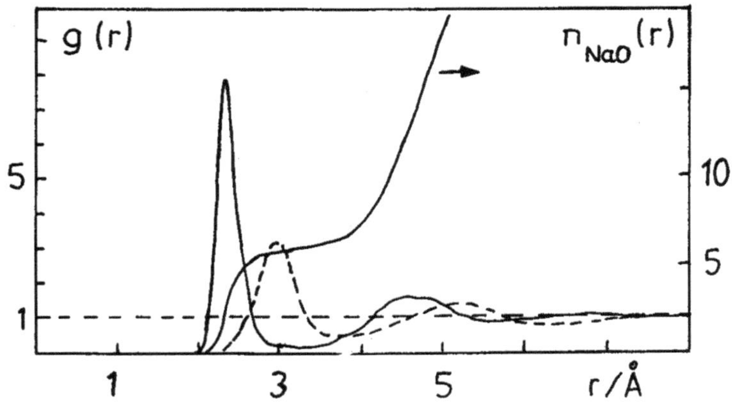

Figure 4. Ion-oxygen (solid line) and ion-hydrogen (dashed line) RDFs from a simulation of a 2.2m NaCl solution. The running integral n(r) refers to the ion-oxygen function.

There are ten atom-atom RDFs in a simple aqueous salt solution. Besides the three ones for water discussed above there are four functions describing the ion-water radial order and three ion-ion functions. As an example of ion-water RDFs (i.e. g_{Ion-O} and g_{Ion-H}) Figure 4 shows the functions obtained for the Na^+-ion in a 2.2 molal solution. The first peaks in both functions are higher than the ones found in pure water; this strongly coordinated layer of water around the ion is the hydration shell. Figure 4 also shows that the influence of this cation extends beyond the first hydration shell. Small and/or highly charged ions develop a stronger radial order than large and less charged ones. As an example a few characteristic values of the ion-oxygen RDFs for

Table 1. Characteristic values of selected cation-oxygen RDFs. R_n refers to the position where the RDFs becomes equal 1 for the n'th time, r_{Mn} and r_{mn} are the positions of the n'th maximum and minimum, respectively. From simulations of 2.2m (Na^+), 1.1m (Ca^{2+}) and 1m (La^{3+}) chloride solutions [15]; all distances are in Å.

ion	R_1	r_{M1}	$g(r_{M1})$	R_2	r_{m1}
Na^+	2.10	2.30	8.0	2.66	3.10
Ca^{2+}	2.23	2.39	14.0	2.72	3.2-3.7
La^{3+}	2.37	2.55	18.4	2.81	3.16-3.65

three cations of about equal 'size' are collected in Table 1. Ion-water potentials from ab-initio quantum mechanical calculations were used in all these simulations.

The height of the first peak increases markedly with increasing ion charge and the first minimum becomes wider. This widening gap is an indication that the hydration layer becomes more and more a separate entity with less and less exchange with the remainder of the water. The mean orientation of the water molecules with respect to the ion can indirectly be deduced from the ion-oxygen and ion-hydrogen RDFs (see e.g. Figure 4). This is the procedure one has to use experimentally. In the simulations, the distribution of orientations can be obtained directly and all necessary moments can be computed unambiguously. In the simulations discussed above it is found that the most probable orientation is the one where an ion is located on the prolongations of the water molecule dipole vectors. The distributions of the angles around this maximum have a non-negligable width, and are in some instances concentration dependent.

Generally speaking, anions are comparable to cations with respect to the way in which they structure their environment. In the case of Cl^- the peak height is, in keeping with the size of the ion, less than the one found for F^- or K^+ and comparable to the one found for Cs^+. The agreement between simulations with different models as to the hydration around this ion is at first glance less satisfactory than for cations. A comparison of simulations of several salt solutions with the same model, on the other hand, reveals that the height, and to a certain extent also the position of the first peak in g_{Cl-O}, are sensitive to the type of the counterion and to the concentration of the salt (see e.g. refs. 14).

The main structural difference between cationic and anionic hydration is the orientation of the water molecules in the hydration shells. Here we find that the water molecules in the first hydration shell tend to point one hydrogen toward the anion, i.e. they form an hydrogen bond type configuration with the ion. For Cl^- the

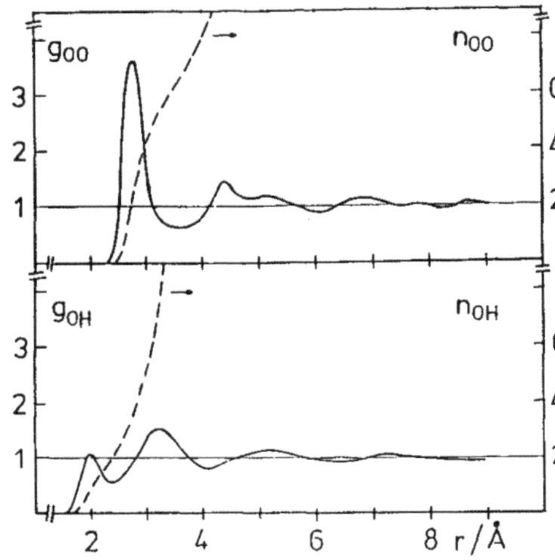

Figure 5. Intermolecular parts of the oxygen-oxygen and oxygen-hydrogen RDFs from a simulation of a 2m $LaCl_3$ solution.

distribution of the hydrogen bond angles is rather wide, and its shape and width are concentration dependent.

The presence of ions in finite concentrations of course also affects the oxygen-oxygen, oxygen-hydrogen and hydrogen-hydrogen RDFs. Figure 5 shows, as an example, functions obtained for a 2 molal $LaCl_3$ solution. The differences between these functions and the ones in Figure 1 are obvious (the same water model is used). At this concentration roughly 35 and 60 percent of the water molecules are in the first hydration shells of La^{3+} and Cl^-, respectively. Of course, the hydration shells do overlap. Due to the large number of water molecules in the first hydration shell of La^{3+}, the first peak in g_{OO} is shifted to smaller distances and a more pronounced second peak develops. On the other hand, the first peak in g_{OH} is reduced. This peak is to a certain extent a measure of the hydrogen bonding, as was discussed above. These changes in the RDF's are accompanied by changes in the average water-water interaction energy. In highly concentrated solutions these interactions can even be repulsive.

The influence of the ions on the motions of the water molecules is studied in Figure 6. The power spectra of the center of mass motions are plotted separately for the hydration water of Ca^{2+}, of Cl^- and the remainder of the water (called here bulk water) in a 1.1m $CaCl_2$ solution. The spectral densitiy of the bulk water resembles the one found for pure water (Figure 2). This is also largely true for the hydration water of the Cl^- ion. The only difference is an increase in intensity in the region around

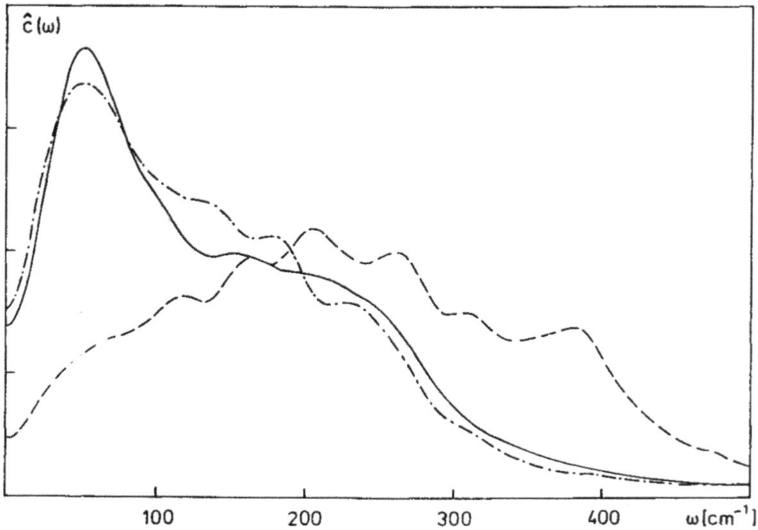

Figure 6. Spectral densities of the center of mass motions of the water molecules in a 2.2m $CaCl_2$ solution, in arbitrary units. Solid line: bulk water, dashed line: hydration water of Ca^{2+}, dash-dotted line: hydration water of Cl^-.

100 cm^{-1} for motions parallel to the plane of the molecule. From a dynamical point of view these water molecules thus are still part of the hydrogen bond network, in keeping with the orientation of the molecules discussed above. The water molecules in the hydration shell of Ca^{2+} display a different motional pattern. Broad bands appear with maxima at 275 cm^{-1} (parallel) and 200 cm^{-1}. Here the motions are dominated by the ion-water interactions and by the repulsion between water molecules in the hydration shell of the ions. The trends seen here are in agreement with the ones found in inelastic neutron scattering of such solutions and with results from Raman spectroscopy. We also note in Figure 6 that the spectral density at $\omega=0$, which is proportional to the self-diffusion coefficient, is much lower for the Ca^{2+} hydration water than for the two other subsystems. Trends for the spectral densities and for the self-diffusion similar to the ones discussed here have been found in simulations of salt solutions with other models (cf. e.g. ref. 16).

The librational motions of the water molecules are also sensitive to the molecular environment. The situation here is more complex than for the hindered translation, and we cannot go into any details. The trends with respect to the peak positions and the band shapes found for different water models are for instance sometimes different. Extensive work has also been undertaken with respect to the intramolecular vibrations [17]. Here the reader is also referred to the literature for details.

Studies similar to the ones discussed here have been undertaken for ionic solutions in other hydrogen bonded solvents like ammonia or alcohols. It should also be mentioned that besides the MD and MC simulations the method of Brownian Dynamics (BD) [18] has been employed to study ionic solutions. In this method the solute-solvent interactions are accounted for in a simplified way only as averages with random perturbations. The advantage is that the accessible timespan is longer than in MD simulations.

We conclude this section by stressing again that only a few examples could be given here and that many interesting aspects such as the dielectric behavior, transport and thermodynamic properties, and many others could not be covered. Some shortcomings of the models have been pointed out, and new models and refined approaches are continuously being put forward. A few recent examples are given in ref. 19.

4.3. Interfacial Systems

Many interfacial systems with hydrogen-bond forming molecules on one side have already been studied by computer simulations [20]. This approach has recently been extended to very simple model biosurfaces [21]. These studies have for instance been undertaken to study the typical phenomena due to the perturbation of the liquid in such an interfacial region. In the vicinity of an interface, several structuring effects will typically compete with each other in an aqueous solution: the effects of the (often charged) surface, the structure of the hydrogen bond network, and the structure of the hydration shell of the ions (or of whatever is left of the latter two). The mobility of atoms or groups on the surface, i.e. their ability to react to the influence of the liquid, are also important.

In the example given here, arrays of COO^- groups represented the hydrophilic headgroups of typical membrane molecules. They were free to move in the plane of the membrane and to rotate around their axis of symmetry. The density of the headgroups was varied between 0.042 $Å^{-2}$ and 0.05 $Å^{-2}$ (realistic values from experiment). The rigid TIP4P water model was used here; the headgroup-water and headgroup-ion potentials were parametrized as sums of Lennard-Jones and Coulomb potentials. Ion-water potentials were taken from fits to ab-initio calculations.

This stronly charged and corrugated surface induces density oscillations extending about 8 Å deep into the liquid. In cases where the heagroups are well localized in a plane, several layers of water can be distinguished. Figure 7 shows density profiles for water as a function of the distance from the surface for various concentrations of Na^+. The average density in the layer closest to the surface is about twice the normal water density. The water molecules in the interfacial region have also a long range spatial correlation parallel to the surface and are strongly oriented with respect to the surface. Bulk-like RDFs for water are recovered at distances of more than 10 Å from the surface. In some cases structureless transition regions are found between the bulk and the region of membrane induced structure.

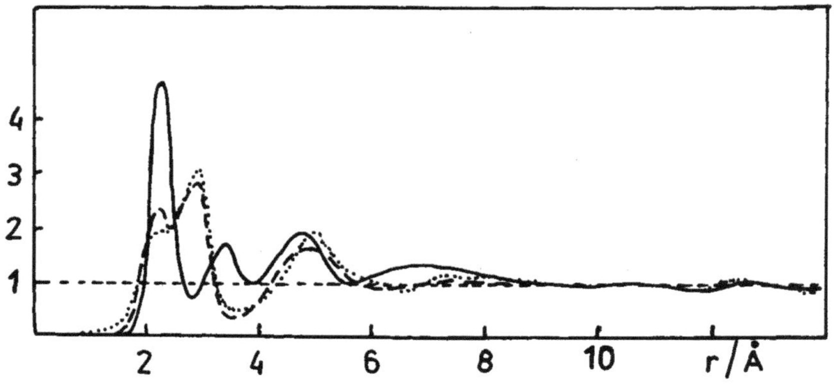

Figure 7. Density profiles of the oxygens, normalized to $\rho = 1\text{g} \cdot \text{cm}^{-3}$, as a function of the distance from the surface for three different concentrations of Na^+ ions, increasing from the solid to the dashed to the dotted line.

Interesting structures have also been found for the adsorption layer of the Na^+ ions. Due to the mobility of the groups forming the surface the ions are able to form favorable sites for adsorption after partial dehydration has taken place. The changes in the layering of the adsorbed water with concentration (see Figure 7) are due to the remainders of the hydration shells of ions adsobed in this way. Further ions are prevented from entering this region, which is thus saturated at about 50 % coverage (ions vs. COO^--groups).

5. Conclusions

People sometimes say that computer simulations are like the "Glasperlenspiel"[22], or at least more like it than any science ought to be. It was also the purpose of this contribution to try to convince the reader that there is more to this line of work than the fun that comes with doing it. Reality (as a physicist or chemist sees it) is represented by certain fundamental equations, which in the simulations are manipulated by numerical rather than by the usual analytical means. This procedure is often deemed less satisfactory than the traditional approach. Yet, for most systems of chemical, and more and more also of biochemical interest, this is the one way in which microscopic and macroscopic understanding can be connected with a minimum of assumptions. Furthermore, the numerical approach opens up a wealth of opportunities for systematic studies and thus provides a rigorous test of one's physical thinking or intuition. This is especially true in non-linear problems.

Finally, results from simulations, like e.g. correlation functions, are often necessary 'input' for detailed analytical theories.

In the present case, computer simulations have for instance allowed to move away from the limiting case of infinite dilution toward the study of solutions like they might occur in the lab (or sea-water, or body fluids). There have been particularly fruitful interactions with experiment in studies of the ionic hydration, and some aspects were pointed out here. In spite of the fact that the understanding is not yet perfect, these studies must now be extended, with all due care, to more complex systems like interfaces. Here there are challenges not only for the computational chemist, but also for the quantum chemist for the development of interaction potentials, and for the experimentalist for providing data allowing critical and unambiguous comparisons. I am sure that the reader will find examples of all this in the present volume.

6. Acknowledgement

It is a pleasure to thank Alexei Kornyshev for his invitation to join the Working Party on Electrochemistry which he organized at the ICTP in Trieste. Many of the simulations reported here were carried out on the computers of the Max-Planck Gesellschaft, the Hochschulrechenzentrum der Technischen Hochschule Darmstadt, the Institut für angewandte Mathematik (ZAM) and the Höchstleistungsrechenzentrum (HLRZ) at the Forschungsanlage Jülich (KFA), and at the Rheinisch-Westfälische Technische Hochschule Aachen. These grants of computer time and the help and support from the staffs at these sites are gratefully acknowledged. The author thanks the Deutsche Forschungsgemeinschaft for the award of a Heisenberg fellowship, and support by the Fonds der der Chemischen Industrie, Frankfurt, is also acknowledged.

7. References

1. M. P. Allen and D. J. Tildesley, *Computer Simulations of Liquids,* (Clarendon Press, Oxford 1987).
2. G. Ciccotti, D. Frenkel and I. R. McDonald, eds., *Simulations of Liquids and Solids, Molecular Dynamics and Monte Carlo Methods in Statistical Mechanics,* (North-Holland, Amsterdam 1987);
   D. K. Remler and P. A. Madden, *Mol.Phys.* **70** (1990) 921.
3. P. Bopp (1987) in: *Softwareentwicklung in der Chemie 1, Proceedings des Workshop "Computer in der Chemie"*, ed. J.Gasteiger, (Springer Verlag 1987);
   P. Bopp (1987) in *The Physics and Chemistry of Aqueous Ionic Solutions.* eds. M-C. Bellissent-Funel and G. W. Neilson, (Reidel Dordrecht 1987);
   K. Heinzinger, in *Computer Modelling of Fluids, Polymers and Solids* , ed. C. R. A. Catlow, (Kluwer Academic Publishers, 1990);
   P.Bopp, in *Molecular Dynamics (MD) Computer Simulations of Hydrogen-Bonded Liquids,* in *Intermolecular Forces –An Introduction on Modern Methods and Results*, eds. P. Huyskens, W.A.P. Luck and Th. Zeegers-Huyskens, (Springer Verlag, in press).

4. S. Nosé *J.Chem.Phys.* **81** (1984) 511.
5. K. Toukan and A.Rahman, *Phys.Rev.B* **31** (1985) 2643;
   G. Lie and E. Clementi, *Phys.Rev.A* **33** (1986) 2679;
   P. Demontis, G. B. Suffritti, E. S. Fois and A. Gamba, *Chem.Phys.Letters* **127** (1986) 456;
   L. X. Dang and B. M. Pettitt, *J.Phys.Chem.* **91** (1987) 3349.
6. A. Rahman and F. H. Stillinger *J.Chem.Phys.* **60** (1974) 545.
7. H. L. Lemberg and F. H. Stillinger *J.Chem.Phys.* **62** (1975) 1677;
   H. J. C. Berendsen, J. P. M. Postma, W. F. van Gunsteren and J. Hermans, in B. Pullman (ed), *Intermolecular Forces,* (Reidel 1981);
   J. Chanderasekhar and W. L. Jorgensen, *J.Chem.Phys.* **77** (1982) 579.
8. O. Matsuoka, E. Clementi and M. Yoshimine, *J.Chem.Phys..* **64** (1976) 1351;
   W. L. Jorgensen, J. Chanderasekhar, J. D. Madura, R. W. Impey and M. L. Klein, *J.Chem.Phys.* **79** (1983) 926.
9. S. Goldmann and P. Backx, *J.Chem.Phys.* **74** (1986) 2761.
10. G. Corongiu, M. Migliore and E. Clementi, *J.Chem.Phys.* **90** (1989) 4629;
    M. M. Probst, E. Spohr, K. Heinzinger, *Chem.Phys.Letters* **161** (1989) 405;
    M. M. Probst, E. Spohr, K. Heinzinger and P. Bopp *Molecular Simulation*, in press.
11. P. Bopp, G. Jancsó and K. Heinzinger, *Chem.Phys.Letters* **98** (1983) 129;
    G. Jancsó, P. Bopp and K. Heinzinger, *Chem.Phys.* **85** (1985) 377.
12. A. K. Soper and M. G. Philips, *Chem.Phys.* **107** (1986) 47.
13. G. Pálinkás, T. Radnai, W. Dietz, Gy. I. Szász and K. Heinzinger, *Z.Naturforsch.* **37a** (1982) 1049.
14. B. Curnutte and D. Williams, in *Structure of Water and Aqueous Solutions*, ed. W. A. P. Luck, (Verlag Chemie, Weinheim 1974);
    M. G. Sceats and S. A. Rice, *J.Chem.Phys.* **72** (1980) 3236, and references therein;
    G. E. Walrafen, M. R. Fischer, M. S. Hokmabadi and W. H. Yang, *J.Chem.Phys.* **85** (1986) 6970;
    G. E. Walrafen, M. S. Hokmabadi, W. H. Yang and G. Piermarini, *J.Phys.Chem.* **92** (1988) 4540.
15. G. Jancsó, K. Heinzinger and P.Bopp, *Z.Naturforsch.* **40a** (1985) 1235;
    M. M. Probst, T. Radnai, K. Heinzinger, P. Bopp and B. M. Rode, *J.Phys.Chem.* **89** (1985) 753;
    W. Meier, P. Bopp, M. M. Probst, E. Spohr and J-l. Lin, *J.Phys.Chem.* **94** (1990) 4672.
16. G. Pálinkás, E. Hawlicka, and K. Heinzinger, *J.Phys.Chem.* **91** (1987) 4334.
17. P.Bopp, *Chem.Phys.* **106** (1986) 205;
    E. Spohr, G. Pálinkás, K. Heinzinger, P. Bopp and M. M. Probst, *J.Phys.Chem.* **92** (1988) 6754;
    M. J. Wojcik, K. Hermansson, J. Lindgren and L. Ojamäe, *J. Chem.Phys.* submitted.
18. P. Turq, F. Lantelme and H. L. Friedmann, *J.Chem.Phys.* **66** (1977) 3039.
19. I. Ruff and D. J. Diestler, *J.Chem.Phys* **93** (1990) 2032;

U. Niesar, G. Corongiu, E. Clementi, G. R. Kneller and D. K. Bhattacharya, *J.Phys.Chem.* **94** (1990) 1138;
M. Prévost, D. van Belle, G. Lippens and S. Wodak, *Mol.Phys.* **71** (1990) 587;
K. A. Motakabbir and M. Berkowitz, *J.Phys.Chem.* **94** (1990) 8359.
Z. Slanina, *Chem.Phys. Letters* **172** (1990) 367;
J. Åquist, *J.Phys.Chem.* **94** (1990) 8021;
E. N. Brodskaya and A. I. Rusanov, *Mol.Phys.* **71** (1990) 567.

20. R. Sonnenschein and K. Heinzinger, *Chem.Phys.Letters* **102** (1983) 55;
E. Spohr and K. Heinzinger, *Electrochim. Acta* **33** (1988) 1211;
M. Matsumoto and Y. Kataoka, *J.Chem.Phys.* **88** (1988) 323;
O. A. Karim and A. D. J. Haymet, *J.Chem.Phys.* **89** (1988) 6889;
E. Spohr, *J.Phys.Chem.* **93** (1989) 6171;
E. Spohr, *Chem.Phys.* **141** (1990) 87.
21. E. Egberts and H. J. C. Berendsen, *J.Chem.Phys.* **89** (1988), 3718;
M. Schlenkrich, K. Nicklas, J. Brickmann and P. Bopp, *Ber.Bunsen. Phys.Chem.* **94** (1990) 133.
22. *Das Glasperlenspiel, Versuch einer Lebensbeschreibung des Magister Ludi Josef Knecht, samt Knechts hinterlassenen Schriften*, herausgegeben von Hermann Hesse, Zürich 1943, Suhrkamp Taschenbuch Verlag 1972.

NONLOCAL ELECTROSTATICS IN ELECTROCHEMISTRY

A.A.KORNYSHEV
The A.N.Frumkin Institute of Electrochemistry
of the Academy of Sciences of the USSR
Leninskii prosp.31, Moscow 117071, USSR

(EXTENDED ABSTRACT)

Some of the most wide-spread phenomena are determined by the the values of electric fields at the microscopic distances from the field sources comparable with the characteristic scales of the structure of the medium. This is particularly typical for electrochemistry which deals with dissolved ions or/and charged electrodes. There is a semiphenomenological approach to parameterization of microscopic effects in electrostatics. Yet limited to a linear response approximation, it is known as *nonlocal electrostatics*.[1-4] This is a version of electrostatics of continuous media which, however, operates not with a limited concept of a macroscopic dielectric constant, but with a more general nonlocal permittivity tensor $\varepsilon_{\alpha\beta}(\mathbf{r},\mathbf{r}')$. The latter figures as a kernel in the nonlocal constitutive relation between induction and electric field,

$$D_\alpha(\mathbf{r})=\Sigma_\beta \int d\mathbf{r}'\varepsilon_{\alpha\beta}(\mathbf{r},\mathbf{r}')E_\beta(\mathbf{r}').$$

The spatial dispersion of the dielectric permittivity is responsible for the structure of the medium. The particular dependence of $\mathbf{r}$ and $\mathbf{r}'$ is determined by system structural properties as manifested in the spatial correlation of the fluctuations of the medium bound charge.

In the bulk of the homogeneous medium $\varepsilon_{\alpha\beta}(\mathbf{r},\mathbf{r}') = \varepsilon_{\alpha\beta}(\mathbf{r}-\mathbf{r}')$. Here, it is convenient to deal with the Fourier transform of this function, $\varepsilon_{\alpha\beta}(\mathbf{k}) = \Sigma_{\alpha\beta}(k_\alpha k_\beta/k^2)\varepsilon_{\alpha\beta}(\mathbf{k})$, the quantity called the *nonlocal dielectric function*. This reduction is, generally impossible, at interfaces where the permittivity depends on $\mathbf{r}$ and $\mathbf{r}'$ separately.

The nonlocal electrostatics has been applied for a variety of phenomena in the bulk related to ionic solvation, interionic interactions and charge transfer kinetics (for a review see [1]). Most interesting effects of nonlocal electrostatics takes place at interfaces due to a competition, sometimes, striking, of the nonlocal screening in the neighboring components[3,4]. In a number of problems it was possible to obtain nontrivial effects (and approve it by comparison with experiments) for an arbitrary form of system $\varepsilon_{\alpha\beta}(\mathbf{r},\mathbf{r}')$ with definite limiting properties of this function in the bulk of the neighboring media. Information about $\varepsilon_{\alpha\beta}(\mathbf{r},\mathbf{r}')$ is accumulated here in certain moments of this function which figure as "parameters" of the results. Note, that the same "parameters" happen to appear in different problems. Therefore, having evaluated such a moment by a comparison with experiment of theoretical formula obtained for one phenomenon, it is possible to use this value to describe another phenomenon. Such *cross-correlations* enabled Vorotyntsev to build a practically modelless theory of specific ionic adsorption at the metal electrolyte interface with parameters taken from the double layer capacitance measurements (see his lecture in this issue). Results of this kind are the doubtless success of nonlocal electrostatics.

In a number of cases one has to deal with some approximate ansatz which expresses $\varepsilon_{\alpha\beta}(\mathbf{r},\mathbf{r}')$ through $\varepsilon(k)$ of the system components. After the expression of observables through $\varepsilon(k)$ further reduction, at least for bulk quantities, invokes some approximations for $\varepsilon(k)$; progress along these lines has also been reviewed [1-5], for some later applications see the original papers on the charge transfer reactions [6-9], time dependent fluorescence shifts[10], statistical mechanics of ionic ensembles at the interface of two dielectrics[12].

We know a good deal about $\varepsilon(k)$ of metals[13]. This is not so for polar solvents, water in particular. An approximation has been broadly used in the quoted papers (where necessary) for $\varepsilon(k)$ of an associated polar liquid, the simplest form of which is given by an expression

$$\varepsilon(k) = \varepsilon_* + (\varepsilon - \varepsilon_*)[1 + (\varepsilon/\varepsilon_*)k^2\lambda^2]^{-1}$$

which interpolates between the long range macroscopic (ε) and short range (ε_*) dielectric behavior with λ being the characteristic correlation length determining this crossover.

The focus of recent series of works [14-16] was to show that this approximation is inadequate for dipolar and water like liquid and to emphasize the importance of $\varepsilon(k)$ containing regions of negative values and points of divergence. Seminar paper by Fonseca in this volume is a contribution in this direction with an alternative view presented by Kornyshev, Kossakowski and Vorotyntsev. In addition to a general analysis the latter work tries to extract $\varepsilon(k)$ of water from experimental and computer-simulated partial structure factors (oxygen-oxygen, oxygen-hydrogen, hydrogen-hydrogen). It appears, however, that the values of $\varepsilon(k)$ at small k are very hard to obtain due to enormous sensitivity to the long distance behaviour of all the structure factors.

Applications of nonlocal electrostatics came to a turning point. Further progress is hardly possible until we come to certain conclusions about the behavior of $\varepsilon(k)$ for water and other polar solvents. Interest to this problem is so high now that in spite of big difficulties we could expect decisive results in the nearest future.

References

1. A.A.Kornyshev, in *The Chemical Physics of Solvation*, ed.R.R.Dogonadze, E.Kalman, A.A.Kornyshev and J.Ulstrup, Part A (Elsevier, Amsterdam, 1985),p.77.
2. A.A.Kornyshev, in *The Chemical Physics of Solvation*, ed. R.R.Dogonadze, E.Kalman, A.A.Kornyshev and J.Ulstrup, Part C (Elsevier, Amsterdam, 1988), p.355.
3. M.A.Vorotyntsev, *ibid.* p.401.
4. M.A.Vorotyntsev and A.A.Kornyshev, *Electrostatics of Media with spatial Dispersion*, (Nauka, Moscow,1991) in Russian;
5. A.A.Kornyshev and W.Schmickler, *J.Electroanal.Chem.* 202 (1986) 1.
6. A.A.Kornyshev and J.Ulstrup, *Chem.Phys.Lett.* 126 (1986) 74.
7. P.G.Dzhavakhidze,A.A.Kornyshev and L.I.Krishtalik, *J.Electroanal.Chem.* 228 (1987) 329.
8. D.K.Phelps,A.A.Kornyshev and M.J.Weaver, *J.Phys.Chem.* 94 (1990) 1454.
9. A.M.Kjaer and J.Ulstrup, *Inorg.Chem.* 25 (1986) 6448; 26 (1987) 2052.

10. A.A.Kornyshev, A.M.Kuznetsov, D.K.Phelps and M.J.Weaver, *J.Chem.Phys*.91 (1989) 7159.
11. S.Leikin and A.A.Kornyshev, *J.Chem.Phys*. 90 (1990) 6890.
12. M.A.Vorotyntsev and S.Ivanov, *Elektrokhimiya* 25 (1989) 554.
13. V.D.Gorobchenko and E.G.Maximov, *Usp.fiz.Nauk*. 130 (1980) 65.
14. Ph.Attard, D.Wei and G.N.Patey, *Chem.Phys.Lett*.172 (1990) 69.
15. F.O.Raineri, Y.Zhou, H.L.Friedman, G.Stell, *J.Chem.Phys*. (in Press).
16. T.Fonseca and B.Ladany, *ibid*.(in Press).

SCREENING IN DENSE IONIC FLUIDS

M. P. TOSI
Dipartimento di Fisica Teorica dell'Università di Trieste and International Centre for Theoretical Physics, I-34100 Trieste, Italy

ABSTRACT

There has been great progress in recent years in determining and understanding the structure of molten salts. I focus on molten alkali halides and discuss two main points concerning their liquid structure and its relationship with static electrical response in these dense ionic conductors. These are (i) the nature of screening and the related definitions and properties of the screening length and of the dielectric function, and (ii) developments in integral equations techniques for the evaluation of molten salt structure and static screening from given pair potentials.

1. Introduction

The liquid pair structure of a number of alkali halides and polyvalent metal halides has been determined in recent years by the multipattern neutron diffraction method based on isotopic enrichment, which was first developed for alloys by Enderby, North and Egelstaff[1]. The experimental effort has been accompanied, especially for alkali halides, by a great deal of computer simulation work based on the adoption of suitable models for the ionic interactions, using simple pair potentials as well as Hamiltonians which take account of the electronic polarizability of closed-shell ions. The availability of both experimental and simulation data has in turn stimulated theoretical studies of statistical mechanical models for dense ionic liquids, ranging from primordial models of charged hard spheres solved analytically in the mean spherical approximation to more realistic pair-potentials models solved numerically by refined integral equations methods. Most of this work is covered in recent reviews[2-4].

For my present purposes I reproduce in Fig. 1 the partial pair distribution functions $g_{+-}(r)$, $g_{++}(r)$ and $g_{--}(r)$ of molten NaCl near freezing, as obtained from neutron diffraction[5], computer simulation based on pair potentials[6] and liquid structure theory using the same pair potentials[7]. Some main features are immediately evident. Coulomb attractions and closed-shell overlap repulsions between Na^+ and Cl^- ions lead to a first-neighbour shell of unlike ions around any given ion, with a sharp region of excluded volume, whereas Coulomb repulsions between like (Na^+-Na^+ or Cl^--Cl^-) ions push them into second-neighbour shells having very similar shapes and less sharply defined regions of excluded volume. Alternation of the two ionic species in space is clearly being preserved to a good extent across melting, with the implication that the charge density around any given ion has an oscillatory profile in space. It will be the aim of section 2 below to discuss how

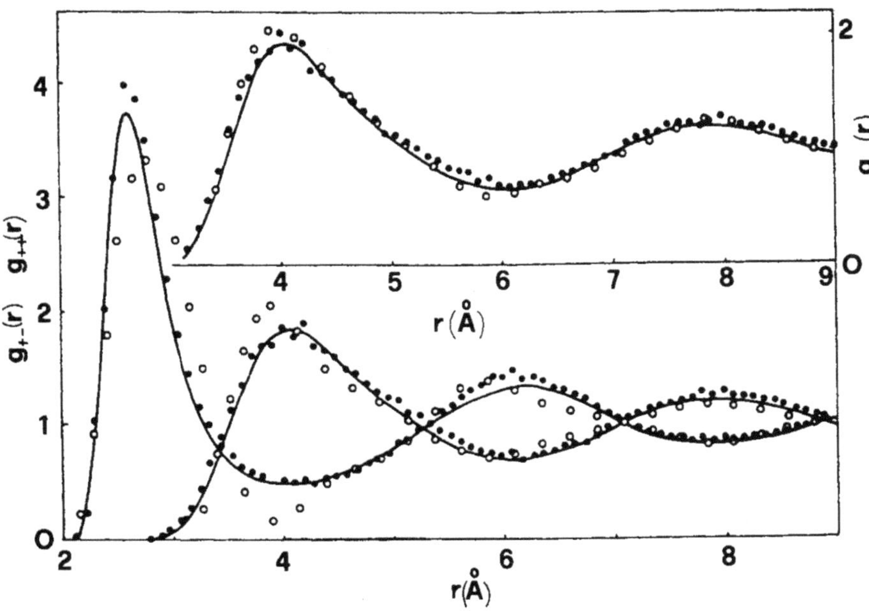

Fig. 1. Partial pair distribution functions of molten NaCl near freezing, from neutron diffraction data (circles), computer simulation (dots) and liquid structure theory (curves).

these structural features of a typical simple molten salt are reflected in the definitions and properties of its screening length and its static dielectric function. Illustrations of the usefulness of the screening length in this context will be given by reference to the capacitance of a model electrode/molten salt interface at the point of zero charge and to the stability of bound ionic complexes in liquid mixtures of polyvalent metal halides and alkali halides.

Returning to Fig. 1, we also note that the agreement between the results of liquid structure theory and the computer simulation data is almost fully quantitative. Section 3 will discuss the main features of such quantitative theory and stress their relevance in a classical problem in theoretical electrochemistry, i. e. the dipole layer formed by a dense ionic fluid facing a charged hard wall at high surface charge density. For completeness we may here remark on Fig. 1 that, whereas the agreement between a pair-potentials model and experiment on the real molten salt is quite satisfactory for both $g_{++}(r)$ and $g_{--}(r)$, there clearly are important quantitative discrepancies in the shape of the first-neighbour shell in $g_{+-}(r)$. This function from the diffraction data peaks at a larger interionic separation (2.78Å against 2.6Å) and is somewhat broader and at the same time more stable against ionic exchange with the surrounding fluid. Such residual discrepancies between model and real salt may be largely due to the electronic deformability of the ions. There still is room for improvement in simulation models to treat the contribution of closed electron shells polarization to structure and screening in molten salts[8].

2. Liquid structure and screening

Let us first recall some basic results of the early theory of static screening in a weakly coupled ionic conductor, considering for simplicity a classical plasma of identical point-like particles carrying charge e and embedded in a neutralizing background of density n. The electric potential $\phi(z)$ created in the plasma facing a charged planar hard wall at $z = 0$ is related to the charge density $n_Q(z)$ in the plasma by Poisson's equation

$$\frac{d^2\phi(z)}{dz^2} = -4\pi\, n_Q(z) \quad , \tag{1}$$

which is to be solved with the boundary condition relating the electric field at the wall to the charge ρ on the wall per unit area. The Gouy-Chapman theory[9] solves the problem analytically by relating the charge density to the potential by the Boltzmann formula,

$$n_Q(z) = ne\,\{\exp[-e\phi(z)/k_BT] - 1\} \ . \tag{2}$$

In particular, the capacitance C_o of the interface at the point of zero charge ($\rho \rightarrow 0$), per unit area, is

$$C_o = \frac{1}{4\pi L_D} \quad , \quad L_D = \left(\frac{4\pi n e^2}{k_BT}\right)^{-1/2} \ . \tag{3}$$

Of course, L_D measures the thickness of the dipole layer at the interface and is now known as the Debye screening length.

Debye and Hückel[10] recognized that the same approach can lead to the structure and thermodynamics of the bulk plasma, when an average plasma particle is viewed as a point-like "electrode" bearing a charge e. The charge density $n_Q(r)$ around the "electrode" is by definition equal to $ne[g_Q(r) - 1]$, $g_Q(r)$ being the radial distribution function of the plasma. This function may thus be evaluated together with the electric potential $\phi(r)$ from the linearized Poisson-Boltzmann approach, with the result

$$k_BT\,[1 - g_Q(r)] = e\phi(r) = \frac{e^2}{r}\exp(-r/L_D) \ . \tag{4}$$

The length L_D still is the thickness of the dipole layer screening the "electrode", since the contribution of the surrounding particles to the total potential drop is

$$\lim_{(r\rightarrow 0)} \left[\phi(r) - \frac{e}{r}\right] = -\frac{e}{L_D} \ . \tag{5}$$

Evidently, the quantity ($-e^2/L_D$) is the average Coulomb energy of interaction between a plasma particle and the rest of the plasma, and hence ($-e^2/2L_D$) is the "Madelung" internal energy of the plasma per particle.

As we shall see, a suitable extension of the above considerations yields a useful definition of the screening length for a dense ionic fluid. It will be convenient to start by discussing microscopic screening, i. e. the generalization of Eq. 4. This is conventionally done through the static dielectric function $\varepsilon(q)$, which is defined by writing the Fourier transform $\phi(q)$ of the electric potential as

$$\phi(q) = \frac{4\pi e}{q^2 \varepsilon(q)} . \tag{6}$$

Of course, for the weakly coupled plasma one has from Eq. 4

$$\varepsilon(q) = 1 + \frac{1}{L_D^2 q^2} . \tag{7}$$

In a dense ionic fluid, however, the low-q form of $\varepsilon(q)$, though again related to macroscopic electrical behaviour, does not provide a useful definition of the screening length.

2.1. Static dielectric function

On increasing the coupling strength in the plasma, the degree of short-range order increases and eventually $g_Q(r)$ acquires an oscillatory behaviour, instead of the monotonic approach to unity which is given in Eq. 4 by the Debye-Hückel theory. This process is illustrated after Fourier transform in the left-hand drawing in Fig. 2, which reports the structure factor $S(q)$ of the plasma as obtained from computer simulation[11] for three values of the coupling strength Γ. The structural evolution with increasing Γ may be visualized as associated with the emergence and growth of a well defined shell of first neighbours, leading to "overscreening" of the central particle by its first neighbours.

The static dielectric function $\varepsilon(q)$ is directly related to $S(q)$ by virtue of the classical fluctuation-dissipation theorem, which gives

$$\frac{1}{\varepsilon(q)} = 1 - \frac{4\pi n e^2}{q^2 k_B T} S(q) . \tag{8}$$

The drawing on the right-hand side of Fig. 2 shows that the transition from a monotonic $S(q)$ to a sharply peaked $S(q)$ is reflected in $\varepsilon(q)$ as a transition from the regime $\varepsilon(q) > 1$ to the regime $\varepsilon(q) < 0$.[12] Only the intermediate range $0 < \varepsilon(q) < 1$ is strictly excluded by the requirements of causality and stability.[13,14]

In particular, in the low-q region one finds

$$\lim_{(q\to 0)} \varepsilon(q) = 1 + \frac{1}{\alpha q^2} , \quad \alpha = (4\pi e^2)^{-1} \left(\frac{\partial \mu}{\partial n}\right)_{T,P} \tag{9}$$

and the thermodynamic derivative of the chemical potential μ entering the quantity α is found to become negative with increasing Γ.[11] Its equivalent in a liquid atomic mixture would be the thermodynamic derivative $(\partial\mu/\partial c)_{T,P}$, which must be positive for stability against phase separation of the components. However, a fluid of charged species is stabilized against their separation by their mutual Coulomb interactions, even if the appropriate thermodynamic derivative becomes negative.

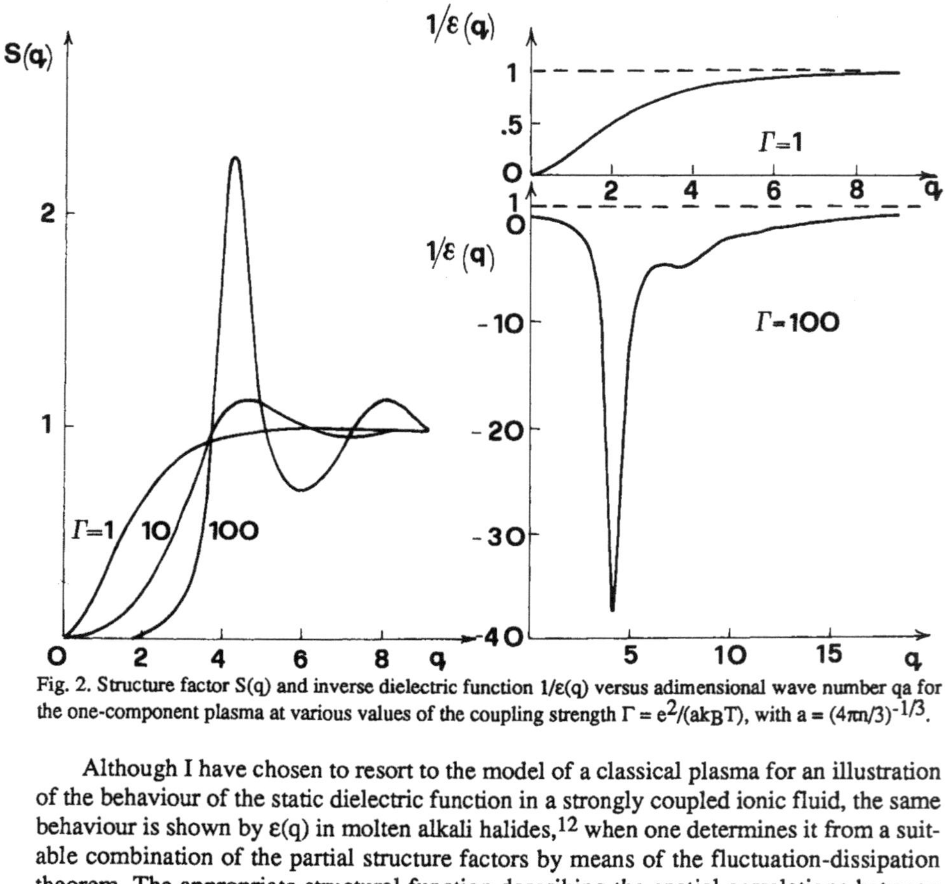

Fig. 2. Structure factor S(q) and inverse dielectric function 1/ε(q) versus adimensional wave number qa for the one-component plasma at various values of the coupling strength $\Gamma = e^2/(ak_BT)$, with $a = (4\pi n/3)^{-1/3}$.

Although I have chosen to resort to the model of a classical plasma for an illustration of the behaviour of the static dielectric function in a strongly coupled ionic fluid, the same behaviour is shown by ε(q) in molten alkali halides,[12] when one determines it from a suitable combination of the partial structure factors by means of the fluctuation-dissipation theorem. The appropriate structural function describing the spatial correlations between charge density fluctuations is

$$g_Q(r) = \frac{1}{2}[g_{++}(r) + g_{--}(r) - 2g_{+-}(r)] \tag{10}$$

and the structure factor to be used in Eq. 8 is

$$S_Q(q) = 1 + 4\pi n \int_0^\infty g_Q(r) \frac{\sin(qr)}{qr} r^2 dr \ , \tag{11}$$

n being the number of ion pairs per unit volume. It is clear from Fig. 1 that $g_Q(r)$ shows strong oscillations, starting from the first-neighbour shell and changing sign from shell to shell, so that $S_Q(q)$ is qualitatively similar to the structure factor of the plasma at $\Gamma = 100$ in Fig. 2. The behaviour of the quantity α in the low-q form for ε(q) in Eq. 9 has been evaluated and discussed in detail for the primitive model[15,16] (schematizing the molten alkali halide as a fluid of equi-sized charged hard spheres). The change in sign of α from positive to negative is indeed related to the appearance of oscillations in $g_Q(r)$ and occurs in the temperature - density plane in the neighbourhood of the critical density for the transition from an ionic gas to an ionic liquid.

2.2. Screening length

We must thus resort to a generalization of Eq. 5 for a useful definition of the screening length in a dense ionic fluid. The "Madelung" internal energy of a molten alkali halide per ion pair is

$$U_C = n\int dr \frac{e^2}{r} g_Q(r) = \int \frac{dq}{(2\pi)^3} \frac{4\pi e^2}{q^2} [S_Q(q) - 1] . \tag{12}$$

In order to arrive at simple and physically transparent analytic results, let us resort at this point to the primitive model as solved in the mean spherical approximation[17] (MSA). This yields

$$U_C^{MSA} = -\frac{e^2}{L + \frac{1}{2}\sigma} , \quad L = \sigma\,[(1 + 2\sigma/L_D)^{1/2} - 1]^{-1} \tag{13}$$

where σ is the hard-spheres diameter. By the same arguments that have accompanied Eq. 5, and noticing that the surface of the "electrode" lies at $r = \sigma/2$ rather than at $r = 0$, we can identify the length L in Eq. 13 as the screening length of the model molten salt.

In further elaboration, though still using the primitive model in the MSA, let us consider the addition of probe ions having charge Z'e, diameter σ' and density n' with $n'\sigma'^3 \to 0$. The potential drop from the surface of a probe ion to infinity is

$$\Delta\phi = \phi(\sigma'/2) - \phi(\infty) = \frac{Z'e}{\frac{1}{2}\sigma'} + 4\pi n \int_{\sigma'/2}^{\infty} r^2 dr\, [g_{p+}(r) - g_{p-}(r)] \frac{e}{r} \tag{14}$$

where the suffix p denotes the probe ions. The MSA yields

$$\Delta\phi = \frac{2Z'e}{\sigma'\,(1 + \frac{1}{2}\sigma'/L)} \tag{15}$$

and hence the capacitance C of the probe per unit area is

$$C = \frac{1}{\pi\sigma'^2} \left[\frac{\partial\Delta\phi}{\partial(Z'e)}\right]^{-1} = \frac{1}{4\pi}\left(\frac{1}{L} + \frac{2}{\sigma'}\right) . \tag{16}$$

The capacitance of a charged hard wall follows by taking the limit $\sigma' \to \infty$ and is

$$C_o = \frac{1}{4\pi L} \ . \tag{17}$$

This result, which as a consequence of the limitations of the MSA is independent of the charge Z' and is thus applicable only at the point of zero charge, is to be compared with Eq. 3. Of course, in the weak coupling limit ($L_D \gg \sigma$) we find from Eq. 13 that $L \to L_D$.

The above results are easily extended to the case of a model ionic liquid made of charged hard spheres (CHS) having arbitrary charges and diameters and floating in a uniform dielectric (to simulate the electronic polarizability of the ions). One obtains by the same line of argument a screening length which is in general a function of density, temperature, dielectric constant, charges and diameters, whence one may estimate the capacitance of electrode/molten salt interfaces.[18] This estimate yields the correct order of magnitude (e. g., one estimates $C_o = 25.6\ \mu F/cm^2$ for molten KCl at 800 C, while the measured value for the molten KCl/liquid Pb interface is[19] $28.8\ \mu F/cm^2$) and reproduces the observed trends for the dependence of C_o on the ionic species (sensitivity to the alkali and little sensitivity to the halogen). However, the model yields C_o^{-1} increasing slowly with T and thus fails to account for the observed[19] temperature dependence of the capacitance of metal/ molten salt interfaces. It is now established that this discrepancy between theory and observation is not due to the MSA treatment, since the same behaviour of C_o^{-1} with T has been reported from computer simulations of the hard wall/molten salt interface.[20]

2.3. Stability of complexes in molten salt mixtures

As an additional illustration of the usefulness of the above definition of screening length for a molten salt, I briefly recall at this point its use in developing a criterion for the stability of bound molecular ions ("complexes") in liquid mixtures of polyvalent metal and alkali halides.[21] The problem is, for instance, whether in the $CaCl_2.2KCl$ melt tetrahedral $(CaCl_4)^{2-}$ complexes will form and persist over times at least as long as typical diffusional times (of order 10^{-11} s). Some 140 such liquid mixtures have been studied over the years by appropriate experimental probes, involving diffraction measurements on a few systems but mostly Raman scattering, infrared emission and absorption, optical absorption due to electronic transitions, and thermodynamic measurements.[22]

The problem may be tackled theoretically by evaluating the reaction $(CaCl_4)^{2-} \Leftrightarrow Ca^{2+} + 4Cl^-$ in the dense liquid medium, the effect of the medium being primarily the screening of the bound $(CaCl_4)^{2-}$ complexes by the free K^+ ions. Appeal to Mott's criterion for the stability of a bound state in a conductor will immediately suggest that complexes are stable if the ratio L/d between the screening length L and the size d of the bound state (i. e. the cation-anion distance within a complex) is larger than some critical value. An MSA calculation estimates the critical value of L/d to be 1.6 for molten salt mixtures.

The adjoining Table 1 shows that the above critical ratio separates the alkaline earth - alkali chlorides into complex forming (Be and Mg) and non-forming (Sr and Ba), with rapid fluctuations between fourfold and higher coordinations being clearly expected in the systems close to it (Ca). The results of this stability sorting criterion are in agreement with

Table 1. Range of L/d for polyvalent metal-alkali chloride mixtures on varying the alkali A from Li to Cs

	$AlCl_3.ACl$	$BeCl_2.2ACl$	$MgCl_2.2ACl$	$CaCl_2.2ACl$	$SrCl_2.2ACl$	$BaCl_2.2ACl$
L/d	1.74÷1.80	1.79÷1.85	1.66÷1.72	1.56÷1.61	1.52÷1.57	1.48÷1.53

the available Raman scattering data.[23] The Table includes also the Al-alkali chloride mixtures, in which the $(AlCl_4)^-$ complex is a well-known stable species.

3. Liquid structure theory for molten alkali halides

The great advantage of using for molten salts the CHS model as solved in the MSA is that, once the screening length is determined, analytic expressions can be given for structural and thermodynamic functions.[17] Leaving for a moment aside the question of the appropriateness of the model, one may first question the quantitative reliability of the MSA predictions. Indeed, the MSA involves statistical mechanical assumptions which improve on the Debye-Hückel theory only insofar as finite sizes are allowed for the ions. As a consequence, the MSA is notoriously poor for dilute ionic fluids, where ionic contact is a rare event and hence size effects are secondary to Coulombic interactions. However, size effects are crucial at molten salt densities and the MSA yields for these systems, with a judicious choice of ionic diameters, results of very reasonable semiquantitative value.[24]

At a more refined level one would wish to start with reliable pair potentials and use a fully reliable statistical mechanical approach. The quality of the results that can be obtained for molten alkali halides has already been illustrated in Fig. 1. Let us briefly discuss here the second aspect of the problem, aiming at comparison between theory and simulation for given pair potentials rather than at the much more delicate problem of microscopic ionic interactions and comparison with experiment.

A good starting point for a theory of structure in fluids with long-range forces is commonly regarded to be the hypernetted chain approximation (HNC). The HNC combines the exact relationship between the partial correlation functions $h_{\alpha\beta}(r) = g_{\alpha\beta}(r) - 1$ and the partial Ornstein-Zernike direct correlation functions $c_{\alpha\beta}(r)$,

$$h_{\alpha\beta}(r) = c_{\alpha\beta}(r) + \sum_{\gamma} n_\gamma \int d\mathbf{r}' \, c_{\alpha\gamma}(|\mathbf{r} - \mathbf{r}'|) \, h_{\gamma\beta}(r') \quad , \tag{18}$$

with an approximate closure of the hierarchy,

$$g_{\alpha\beta}(r) = \exp\left[-\phi_{\alpha\beta}(r)/k_B T + h_{\alpha\beta}(r) - c_{\alpha\beta}(r)\right] . \tag{19}$$

In the above equations Greek suffixes denote the ionic species, at partial densities n_α, and $\phi_{\alpha\beta}(r)$ are the pair potentials. In fact, a diagrammatic analysis of the pair correlation functions leads to the exact relation

$$g_{\alpha\beta}(r) = \exp\left[-\phi_{\alpha\beta}(r)/k_B T + h_{\alpha\beta}(r) - c_{\alpha\beta}(r) - b_{\alpha\beta}(r)\right] , \tag{20}$$

where the so-called bridge functions $b_{\alpha\beta}(r)$, which are set to zero in the HNC, are defined in terms of a series of increasingly complex diagrams. Much of the recent effort to improve liquid structure theory has gone into the assessment and evaluation of bridge functions.

A very important aspect is the so-called thermodynamic consistency of the theory, i. e. the requirement that alternative routes to the evaluation of the isothermal compressibility (in particular, the route using the virial theorem for the pressure and the route using the Ornstein-Zernike relation for the structure factor in the limit $q \to 0$) should lead to the same result. This point was first stressed for one-component fluids in the work of Rosenfeld and Ashcroft,[25] who showed that dramatic improvements in the calculated shape of the first-neighbour shell follow from enforcing the requirement of thermodynamic consistency in the construction of bridge functions. This is also true for molten salts, as far as the first-neighbour (unlike ions) shell is concerned. In addition, however, an account of the intermediate-r behaviour of the bridge functions is needed in relation to the shape of the second-neighbour (like ions) shells.[7] Such an account was taken in the calculations leading to the results in Fig. 1 through the leading diagram for the bridge functions, which is schematically shown in the left-hand drawing in Fig. 3. The same type of diagram has also been found to be crucial in dealing with the dipole layer formed by a model electrolyte solution facing a charged hard wall, at large values of the wall charge density.[26] The appropriate diagram for the bridge functions in the latter problem is shown in the right-hand drawing in Fig. 3. It is generally the case that the evaluation of interfacial structures is much more sensitive than that of bulk liquid structure to one's liquid structure theory.

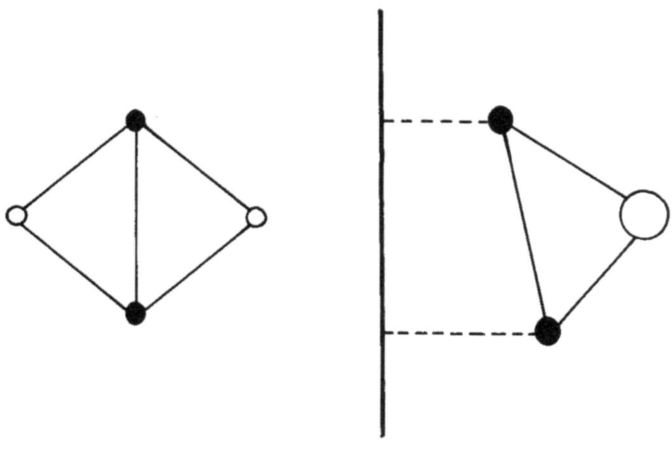

Fig. 3. Leading diagrams for the bridge functions for the correlations between a pair of particles (left-hand drawing) and between a particle and a hard wall (right-hand drawing). The diagrams account for the effect of two further intervening particles, shown by filled circles. The full lines represent $h_{\alpha\beta}(r)$ functions and the broken lines represent single-particle density profiles facing the wall.

Table 2. Internal energy U, pressure P and isothermal compressibility K of molten NaCl at 1148 K and at the measured standard-pressure volume

	U(kJ/mol)	P(kbar)	K(10^{-12} cm^2/dyn)
HNC	- 706.5	7.2	22÷66
HNC + bridges	- 713.8	1.2	31.7
Simulation	- 716.1	1.2	----
Experiment	- 714.3	0	32.4

As a further illustration of the quality reached by the theory for bulk properties of molten alkali halides, I report in Table 2 the values of thermodynamic properties for molten NaCl in comparison with simulation results and with experimental data.

4. Concluding remarks

The main emphasis in this short review has been on the static electrical response of molten alkali halides and its relationship to their liquid structure. I have stressed that the unusual behaviour of the static dielectric function in a dense ionic fluid is associated with the qualitative nature of microscopic screening and structure and has its own relevance in connection with their change across the transition from an ionic gas state to the liquid phase, but is irrelevant to a proper definition of the screening length. The concept of screening length can be introduced through the evaluation of the Coulombic energy of the liquid and remains useful in a number of physical contexts.

Refined theories for liquid structure and screening from given pair potentials are now available. The main quantitative difficulty, especially for salts other than the alkali halides, arises from the lack of appropriate models for the ionic interactions. One may, of course, wonder whether fully quantitative calculations would really be more useful than a broad systematization of the evidence and qualitative understanding, such as was illustrated in Table 1 above.

Acknowledgements

I acknowledge continued sponsorship by the Ministero dell'Università e della Ricerca Scientifica e Tecnologica.

REFERENCES

1. J. E. Enderby, D. M. North and P. A. Egelstaff, Phil. Mag. **14** (1966) 961.
2. M. Rovere and M. P. Tosi, Rep. Progr. Phys. **49** (1986) 1001.
3. R. L. McGreevy, Solid State Phys. **40** (1987) 247.
4. J. E. Enderby and A. C. Barnes, Rep. Progr. Phys. **53** (1990) 85.
5. S. Biggin and J. E. Enderby, J. Phys. **C15** (1982) L305.
6. J. W. Lewis, K. Singer and L. V. Woodcock, J. Chem. Soc. Faraday Trans. II **71** (1975) 41.
7. P. Ballone, G. Pastore and M. P. Tosi, J. Chem. Phys. **81** (1984) 3174.
8. M. Dixon and M. J. L. Sangster, Phil. Mag. **35** (1977) 1049; M. Dixon and M. J. Gillan, Phil. Mag. **B43** (1981) 1099.
9. G. Gouy, J. Chim. Phys. **29** (1903) 143; G. Gouy, J. Physique **9** (1910) 457; D. L. Chapman, Phil. Mag. **25** (1913) 475.
10. P. Debye and E. Hückel, Z. Phys. **24** (1923) 185, 305.
11. M. Baus and J. P. Hansen, Phys. Rep. **59** (1980) 1.
12. A. Fasolino, M. Parrinello and M. P. Tosi, Phys. Lett. **66A** (1978) 119.
13. P. C. Martin, Phys. Rev. **161** (1967) 143.
14. O. V. Dolgov, D. A. Kirzhnits and E. G. Maksimov, Rev. Mod. Phys. **53** (1981) 81.
15. G. Stell, K. C. Wu and B. Larsen, Phys. Rev. Lett. **37** (1976) 1369.
16. M. Rovere, R. Miniero, M. Parrinello and M. P. Tosi, Phys. Chem. Liquids **9** (1979) 11.
17. A. Waisman and J. L. Lebowitz, J. Chem. Phys. **56** (1972) 3086, 3093; L. Blum and J. S. Høye, J. Phys. Chem. **81** (1977) 1311.
18. K. R. Painter, P. Ballone, M. P. Tosi, P. J. Grout and N. H. March, Surf. Sci. **133** (1983) 89; P. Ballone, G. Pastore, M. P. Tosi, K. R. Painter, P. J. Grout and N. H. March, Phys. Chem. Liquids **13** (1984) 269.
19. A. D. Graves, J. Electroanal. Chem. **25** (1970) 349, 357.
20. R. M. Esnouf, A. C. D. Smith and P. J. Grout, Phil. Mag. **A58** (1988) 27.
21. Z. Akdeniz and M. P. Tosi, Phys. Chem. Liquids **17** (1987) 91; Z. Akdeniz, Wang Li and M. P. Tosi, Europhys. Lett. **5** (1988) 613.
22. For a review see Z. Akdeniz and M. P. Tosi, J. Phys.: Cond. Matter **1** (1989) 2381.
23. S. Sakai, T. Nakamura, N. Umesaki and N. Iwamoto, Phys. Chem. Liquids **14** (1984) 67.
24. M. C. Abramo, C. Caccamo, G. Pizzimenti, M. Parrinello and M. P. Tosi, J. Chem. Phys. **68** (1978) 2889.
25. Y. Rosenfeld and N. W. Ashcroft, Phys. Rev. **A20** (1979) 1208.
26. P. Ballone, G. Pastore and M. P. Tosi, J. Chem. Phys. **85** (1986) 2943.

A MOLECULAR PICTURE FOR $\epsilon(k)$ AND ITS CONSEQUENCES FOR THE ENERGETICS OF IONIC SOLVATION

TERESA FONSECA and BRANKA M. LADANYI
Department of Chemistry, Colorado State University
Fort Collins, CO 80523, USA

ABSTRACT

We present the results for the wavevector-dependent dielectric permittivity tensor, $\epsilon(k)$, of a model fluid resembling methanol and we investigate its consequences for the energetics of ionic solvation. Molecular dynamics computer simulation and reference hypernetted chain (RHNC) integral equation methods were used to evaluate $\epsilon(k)$; we find that the longitudinal component of $\epsilon(k)$, $\epsilon_L(k)$, has a pole and becomes negative for fairly small values of k. If we use the resulting $\epsilon_L(k)$ to evaluate free energies of ionic solvation by the nonlocal electrostatics approach, we find no improvement upon the Born approximation. A reason for the failure of nonlocal electrostatics is the fact that it completely ignores the short range solute-solvent repulsion. When this is taken into account, much improved values are obtained for the free energies. However, this approach has other serious limitations since it neglects the effects of the ion on the structure of the surrounding solvent. Our RHNC calculations show that due to these effects the ionic solvation free energy is not simply proportional to the square of the ionic charge, but exhibits additional dependence on the sign and magnitude of the charge.

Introduction

Interactions of polar and ionic solutes with polar solvents play an important role in chemistry. Solubility and chemical equilibria are influenced by the structure of the solvent shells surrounding the solute, while the dynamics of structural rearrangements within these shells influence the rates of chemical reactions that involve charge redistribution in the solute.

An important quantity in the equilibrium theory of electrolyte solutions is the Helmoltz free energy ΔF_{solv}, defined as the free energy necessary to charge up a solute molecule in a polar medium. For an ion of charge q,[1]

$$\Delta F_{solv} = \int_0^q (U_{is}^{es}/q)\, dq\,, \tag{1}$$

where U_{is}^{es} is the contribution to the internal energy per ion from ion-solvent electrostatic interactions. The simplest approach to ΔF_{solv} and U_{is}^{es} treats the solvent as a dielectric continuum with static dielectric constant ϵ . ΔF_{solv} can then be evaluated by using the familiar Born equation[2]

$$\Delta F_{solv} = -\frac{1}{2\epsilon_0}\int d\mathbf{r}\,\left(1-\frac{1}{\epsilon}\right)|\mathbf{D}(\mathbf{r})|^2\,, \tag{2}$$

where $\mathbf{D}(\mathbf{r})$ is the dielectric displacement vector produced by the ion. For a spherical ion of radius a and charge q Eq. (2) reduces to:

$$\Delta F_B = -\frac{q^2}{8\pi\epsilon_0 a}\left(1-\frac{1}{\epsilon}\right). \tag{3}$$

Experimental evidence indicates that the Born approximation, ΔF_B, overestimates free energies of solvation. This is a consequence of the breakdown of the continuum description for the solvent molecules in the vicinity of the ion. Nonlocal electrostatics[3] has been one of the approaches used to attempt to correct this problem.

In this approach one starts again with a Born-type equation for the free energy of solvation,

$$\Delta F_{solv} = -\frac{1}{2\epsilon_0}\int d\mathbf{r}\,\mathbf{D}(\mathbf{r})\cdot\mathbf{P}(\mathbf{r})\,, \tag{4}$$

where now $\mathbf{P}(\mathbf{r})$ is the polarization of the medium, i.e., the response of the solvent to the electric field generated by the ion. Note that, for a spherical ion, the displacement $\mathbf{D}(\mathbf{r}) = D(\mathbf{r})\hat{\mathbf{r}}$ is in the radial direction, so only $P_L(\mathbf{r}) = \mathbf{P}(\mathbf{r})\cdot\hat{\mathbf{r}}$, the longitudinal component of the polarization, contributes to ΔF_{solv}. To proceed further, it is assumed that linear response is valid and $P_L(\mathbf{r})$ is then given by:[3,4]

$$P_L(\mathbf{r}) = \int d\mathbf{k}\, e^{-i\mathbf{k}\cdot\mathbf{r}}\left(1-1/\epsilon_L(\mathbf{k})\right) D(\mathbf{k})\,, \tag{5}$$

where $\epsilon_L(\mathbf{k})$ is the wavevector-dependent longitudinal dielectric permittivity of the *pure solvent*. Its presence in Eq. (5) is a consequence of using the linear response approximation, which assumes that the solvent structure is not affected by the presence of the ion. Nonlocal electrostatics introduced in this way refers to the inclusion *solely* of the bulk solvent structure. While it might be true that at large distances from the ion the solvent structure is the bulk one, we expect that strong ion-solvent interactions will significantly alter the solvent structure in the vicinity of the ion. This should have important consequences for ΔF_{solv}. Another problem with Eq. (5) is that it involves $\epsilon_L(\mathbf{k})$: this quantity cannot be experimentally determined and has only recently become accessible through the use of molecular theory and computer simulation. Prior to these developments, several functional forms for $\epsilon_L(\mathbf{k})$ had been proposed[3] and are still in use. Their validity will be discussed further on in this paper.

The outline of the remainder of this paper is as follows. In the next section we briefly introduce the theoretical framework used to evaluate $\epsilon_L(\mathbf{k})$ via computer simulation and from the solution of integral equations for the intermolecular pair correlation function. Then we present our results for the longitudinal and transverse components of $\boldsymbol{\epsilon}(\mathbf{k})$ of a model of liquid methanol. We attempt to provide a molecular picture for $\epsilon_L(\mathbf{k})$ and discuss the consequences of our results to the energetics of ionic solvation. We compare the results for ΔF_{solv} obtained using $\epsilon_L(\mathbf{k})$ to the value of this quantity calculated from the structure of a dilute ionic solution and discuss the limitations of the nonlocal electrostatics approach.

Theoretical Framework

The dielectric permittivity of a polar solvent can be related to the susceptibility, $\boldsymbol{\chi}^0(\mathbf{k})$, which determines the polarization of the medium in the presence of an external field, $\mathbf{E}_0(\mathbf{k})$,

$$\mathbf{P}(\mathbf{k}) = \epsilon_0 \boldsymbol{\chi}^0(\mathbf{k}) \cdot \mathbf{E}_0(\mathbf{k})\,. \tag{6}$$

For a system of cubic or spherical symmetry $\boldsymbol{\chi}^0$ is a symmetric second-rank tensor with the symmetry axis along **k**. We can thus define two components of the susceptibility,[4] longitudinal,

$$\chi_L{}^0(\mathbf{k}) = 1 - 1/\epsilon_L(\mathbf{k}) \tag{7a}$$

and transverse,

$$\chi_T{}^0(\mathbf{k}) = \epsilon_T(\mathbf{k}) - 1, \tag{7b}$$

where ϵ_L and ϵ_T are, respectively, the longitudinal and transverse components of the dielectric permittivity. Like $\boldsymbol{\chi}^0(\mathbf{k})$, $\boldsymbol{\epsilon}(\mathbf{k})$ is a second-rank tensor with the symmetry axis along **k** at finite **k**, but becomes a scalar ϵ in the limit $\mathbf{k} \to \mathbf{0}$. For a solvent of rigid nonpolarizable molecules, linear response allows us to evaluate $\boldsymbol{\epsilon}(\mathbf{k})$ in terms of Fourier transforms of spatial correlations of molecular dipoles.[4] Specifically,

$$\epsilon_L(\mathbf{k}) = 1/(1 - 3yS_L(k)) \tag{8a}$$

and

$$\epsilon_T(\mathbf{k}) = 1 + 3yS_T(k) \tag{8b}$$

where

$$y = \frac{\rho\mu^2}{9k_B T\epsilon_0} \tag{9}$$

measures the average strength of dipolar interactions in a fluid at temperature T with number density $\rho = N/V$ and molecular dipole moments of magnitude μ. $S_L(k)$ and $S_T(k)$ are often called dipolar "structure factors" and can be evaluated according to

$$S_L(k) = \frac{3\langle |\mathbf{M}_L(\mathbf{k})|^2\rangle}{N\mu^2} \tag{10a}$$

and

$$S_T(k) = \frac{3\langle |\mathbf{M}_T(\mathbf{k})|^2\rangle}{2N\mu^2} \tag{10b}$$

where $\mathbf{M}_L = (\mathbf{M}\cdot\hat{\mathbf{k}})\hat{\mathbf{k}}$ and $\mathbf{M}_T = \mathbf{M} - \mathbf{M}_L$ are the longitudinal and transverse components of

$$\mathbf{M}(\mathbf{k}) = \sum_{j=1}^{N} \boldsymbol{\mu}_j \exp(i\mathbf{k}\cdot\mathbf{r}_j), \tag{11}$$

the Fourier transform of the dipole moment density. In Eq. (11) $\boldsymbol{\mu}_j$ and $\mathbf{r}_j$ denote the dipole moment and the position of center of mass of molecule j. With the help of this expression, the dipolar structure factors can be expressed in terms of intermolecular pair correlations:

$$S_L(k) = 1 + 3(N-1)\langle(\hat{\mathbf{u}}_1\cdot\hat{\mathbf{k}})(\hat{\mathbf{u}}_2\cdot\hat{\mathbf{k}}) \exp(i\mathbf{k}\cdot\mathbf{r}_{12})\rangle \tag{12a}$$

and

$$S_T(k) = 1 + \tfrac{3}{2}(N-1)\langle[(\hat{\mathbf{u}}_1\cdot\hat{\mathbf{u}}_2) - (\hat{\mathbf{u}}_1\cdot\hat{\mathbf{k}})(\hat{\mathbf{u}}_2\cdot\hat{\mathbf{k}})] \exp(i\mathbf{k}\cdot\mathbf{r}_{12})\rangle, \tag{12b}$$

where $\hat{\mathbf{u}}_j$ is the unit vector along the dipole of molecule j and $\mathbf{r}_{12} = \mathbf{r}_1 - \mathbf{r}_2$.

Molecular dynamics (MD) computer simulations and integral equation approaches can be used to evaluate the intermolecular pair correlations. Eqs. (12) above can be used to compute the dipolar structure factors and Eqs. (8) provide a way to determine the longitudinal and transverse components of the dielectric permittivity tensor. In the next section we present our results for $\epsilon(\mathbf{k})$ for a model polar fluid resembling methanol at room temperature. MD is used to calculate $\epsilon(\mathbf{k})$ at small $\mathbf{k}$ values and the reference hypernetted chain (RHNC) integral equation method[5] is used to extend this calculation over the whole range of $\mathbf{k}$ values.

Numerical Results for $\epsilon(\mathbf{k})$

In the MD simulation we have used a three-site model of liquid methanol. In this model the oxygen atom, O, and the hydroxyl proton, H_O, constitute two of the sites while the methyl group, CH_3, treated as a unit symbolized by C, accounts for the third site. The molecules are rigid and the site-site intermolecular potential, $u_{\alpha\beta}(r)$, is a sum of Coulombic and short-range Lennard-Jones interactions,

$$u_{\alpha\beta}(r) = q_\alpha q_\beta/(4\pi\epsilon_0 r) + 4\epsilon_{\alpha\beta}\left[(\sigma_{\alpha\beta}/r)^{12} - (\sigma_{\alpha\beta}/r)^{6}\right] \tag{13}$$

where r is the site-site separation, q_α is a point charge located at site α and $\epsilon_{\alpha\beta}$ and $\sigma_{\alpha\beta}$ are, respectively, the Lennard-Jones well depth and collision diameter. The model is characterized by the following values of the parameters:[6,7]

$q_C = 0.297$ e , $\quad q_O = -0.728$ e , $\quad q_H = -(q_C + q_O)$;

$\epsilon_{CC}/k_B = 91.2$ K, $\quad \epsilon_{OO}/k_B = 87.9$ K, $\quad \epsilon_{HH} = 0.0$;

$\sigma_{CC} = 3.861$, $\quad \sigma_{OO} = 3.083$ Å , $\quad \sigma_{HH} = 0.0$.

The Lennard-Jones parameters for unlike interactions have been computed using the Lorentz-Berthelot combining rules. The molecular geometry of this model is based on the gas-phase values of the intramolecular bond lengths and angles for methanol:[8]

$$d(C\text{-}O) = 1.43\ \text{Å}\ ,\quad d(O\text{-}H_O) = 0.9451\ \text{Å}\ ,\quad <COH_O = 108.53^\circ .$$

In the MD simulation we have used periodic boundary conditions, with 256 molecules in the original box whose side equals 25.98 Å. The simulation was performed in the NVE ensemble, with an average temperature of 298 K, and the equations of motion were integrated by using SHAKE[9a] and Verlet leap-frog[9b] algorithms with a time step of 8 fs. The Ewald sum method[10] has been used to treat the long-ranged electrostatic interactions. Other relevant details of this simulation are given elsewhere.[7]

To use the RHNC integral equation approach we adopted a simplified interaction model for methanol: the molecules were represented as hard spheres of diameter σ with dipoles of magnitude μ and axial quadrupoles of magnitude Q. Specifically, the following parameters were used:[7]

$$\sigma = 3.861\ \text{Å}\ , \quad \mu = 2.337\ \text{D} \quad \text{and} \quad Q = 2.97\ \text{DÅ}\,.$$

The value of μ is the same as for the model used in the MD simulation and the value of σ is consistent with an effective hard-sphere diameter for methanol estimated from compressibility data.[11] Q was employed as an adjustable parameter to obtain the dielectric constant close to that found in the MD simulation. For a fluid of axially symmetric molecules, the RHNC theory gives the pair correlation function $h(12) = h(\hat{\mathbf{u}}_1,\hat{\mathbf{u}}_2,\mathbf{r})$ as a series of products of rotational invariants $\Phi^{mn\ell}(\hat{\mathbf{u}}_1,\hat{\mathbf{u}}_2,\hat{\mathbf{r}})$ and radial coefficients $h^{mn\ell}(r)$. The series is in principle infinite, but is in practice truncated at the values of m, n, and ℓ consistent with the strength of multipolar interactions and with the physical properties under consideration. The series used here includes terms for $m,n \le 2$ and $\ell \le 4$. Eqs. (12) are expressed in this representation as

$$S_L(k) = 1 + \frac{\rho}{3}\left(2\tilde{h}^{112}(k) + \tilde{h}^{110}(k)\right) \tag{14a}$$

and

$$S_T(k) = 1 + \frac{\rho}{3}\left(\tilde{h}^{110}(k) - \tilde{h}^{112}(k)\right), \tag{14b}$$

where $\tilde{h}^{110}(k)$ and $\tilde{h}^{112}(k)$ are Fourier-Hankel transforms[5] of radial coefficients of rotational invariants $\Phi^{110}(12) = \hat{\mathbf{u}}_1\cdot\hat{\mathbf{u}}_2$ and $\Phi^{112}(12) = 3(\hat{\mathbf{u}}_1\cdot\hat{\mathbf{r}})(\hat{\mathbf{u}}_2\cdot\hat{\mathbf{r}}) - \hat{\mathbf{u}}_1\cdot\hat{\mathbf{u}}_2$.

Figure 1 depicts our results for the structure factors $S_L(k)$ and $S_T(k)$. The dots represent the MD results while the lines represent the RHNC results. We note that due to the periodic boundary conditions used in the MD simulation we are restricted to values of $\mathbf{k} = (\ell,m,n)2\pi/L$ where ℓ, m and n are integers and L is the length of the box. In contrast to the MD simulation, the RHNC theory allows us to analyze the entire range of $\mathbf{k}$ values. Given that the intermolecular potential models used in the MD simulation and in the RHNC calculations differ at short range, the two sets of structure factors are expected to differ at large k. However, as can be seen from Figure 1, the agreement at small to moderate k is excellent.

Having evaluated the dipolar structure factors we can compute the dielectric permittivity of our system. The results for $\epsilon_L(\mathbf{k})$ and $\epsilon_T(\mathbf{k})$ are presented in Figure 2. The interesting fact emerging from these results is that $\epsilon_L(\mathbf{k})$ has a pole and becomes negative for fairly small values of $\mathbf{k}$. This behavior has also been found for several other polar fluids[13-15] and, as we shall argue in the next section, is expected for moderately to highly polar fluids at room temperature and liquid densities.

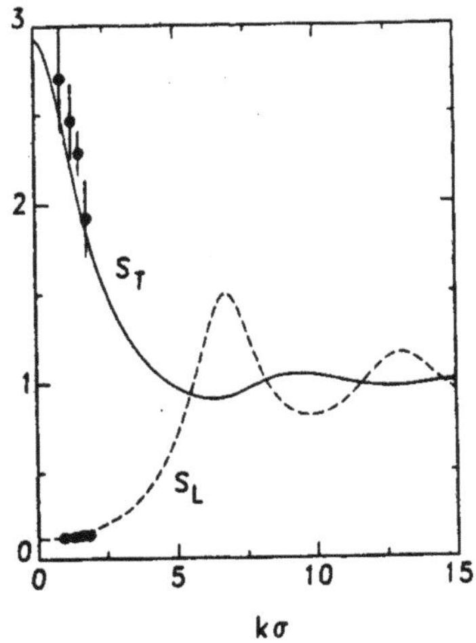

Figure 1. Longitudinal and transverse dipolar structure factors $S_L(k)$ and $S_T(k)$ of methanol plotted vs wavevector k in units of σ^{-1} = 0.259 Å^{-1}. Filled circles are the MD simulation results and lines the RHNC theory results for $S_T(k)$ (———) and $S_L(k)$ (– – –). Error bars are shown for the MD results for $S_T(k)$. For $S_L(k)$, the error bars are smaller than the circles depicting the data.

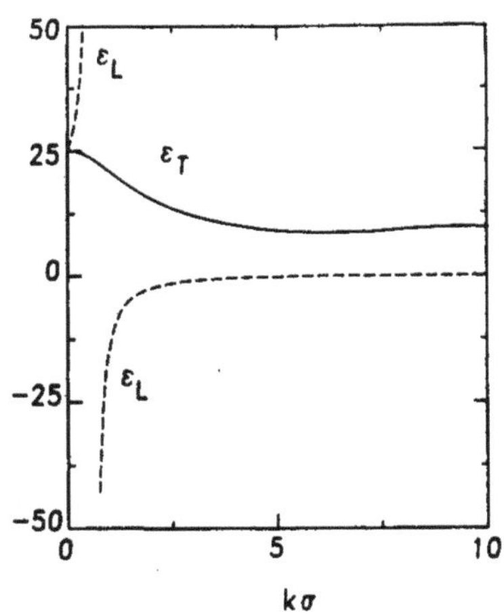

Figure 2. RHNC theory results for the longitudinal, $\epsilon_L(k)$ (– – –), and transverse, $\epsilon_T(k)$ (———), components of the dielectric permittivity tensor of methanol plotted vs wavevector k in units of σ^{-1} = 0.259 Å^{-1}.

A Molecular Picture for the Longitudinal Component of $\epsilon(\mathbf{k})$

Since we are interested in exploring the implications of our findings to the energetics of ionic solvation, we restrict ourselves to the discussion of the behavior of $\epsilon_L(\mathbf{k})$ as this is the only quantity needed to evaluate the free energy of solvation using nonlocal electrostatics (see Eqs. (4) and (5) above). Analogous considerations apply to $\epsilon_T(\mathbf{k})$.

We start by analyzing the limiting values of $S_L(k)$ for small and large values of k. In the limit $k \to 0$, $\epsilon_L(\mathbf{k})$ tends to ϵ, the static dielectric constant, which leads to

$$3yS_L(0) = (1 - 1/\epsilon) . \qquad (15)$$

The large k limit is obtained by noting that contributions to Eq. (12a) from intermolecular pair correlations vanish as $k \to \infty$, leading to

$$S_L(\infty) = 1 . \qquad (16)$$

One should, however, interpret this limit with caution. The practical limit of large k values for the present formulation of dielectric properties corresponds to $k = 2\pi/\sigma$, where σ is the effective molecular diameter. At larger k values the details of the charge distribution within the molecule should be taken into account; when this is done one obtains $S_L(\infty) = 0$ or, equivalently, $\epsilon_L(\infty) = 1$, and another pole is found in $\epsilon_L(\mathbf{k})$ for k values around $2\pi/d$, where d is the charge separation distance.[12]

In considering intermediate k values it is useful to note that, as k becomes smaller, we can visualize the intermolecular interactions as point dipole interactions. Then, let us focus on a single dipole μ_1 and examine its interactions with the other dipoles μ_j at a distance r_{1j}. In the direction along μ_1, dipoles μ_j tend to align parallel to μ_1; dipoles in the two directions orthogonal to μ_1 tend to be oriented antiparallel to μ_1. This is depicted in Figure 3. To analyze $S_L(k)$ we now assume that μ_1 is along $\mathbf{k}$. The phase factor $\exp(i\mathbf{k}\cdot\mathbf{r}_{1j})$ has no effect on the dipoles orthogonal to $\mathbf{k}$, but its oscillatory nature diminishes the net contribution from dipoles along $\mathbf{k}$; the resulting $S_L(k)$ has been presented in Figure 1. According to Eq. (8a), if $S_L(k) > 1/3y$ $\epsilon_L(\mathbf{k})$ becomes negative. Our results show that this already happens at very small values of k: also, this behavior is not specific to our model and should in fact be found for most polar liquids, i.e. for all fluids for which $y > 1/3$. The location of the pole is largely determined by the dipolar strength parameter y. In the present case, $y = 2.74$. For water at room temperature, $y \approx 8$, and the pole occurs at an even smaller value of k.[15] For larger k values, $S_L(k)$ oscillates around 1: this means that for certain k values the phase factor can decouple all intermolecular correlations or can turn them into positive

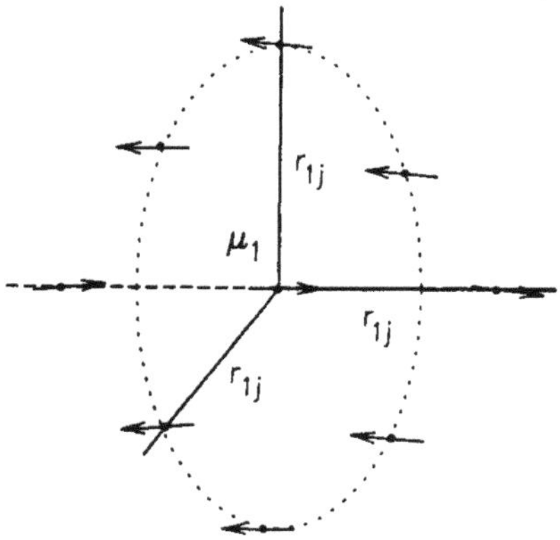

Figure 3. A schematic representation of μ_1, the dipole moment of molecule 1, and of the surrounding molecular dipoles at a distance r_{1j}.

correlations. Let us now examine the consequences of this behavior of $\epsilon_L(\mathbf{k})$ for the energetics of ionic solvation.

Nonlocal Electrostatics and Free Energies of Solvation

We dedicate the remainder of this paper to the analysis of the use of nonlocal electrostatics to evaluate free energies of solvation. To assess the validity of this approach we compare the resulting ΔF_{solv} values to the ones obtained by calculating the solvation free energy from the intermolecular structure of a dilute ionic solution, obtained by solving the RHNC equations for this system.

In the case of a spherical ion of charge q and radius a, the nonlocal electrostatics expression for ΔF_{solv} becomes (see Eqs. (4) and (5)):[3]

$$\Delta F_{solv} \simeq \Delta F_{NL} = \frac{-q^2}{4\pi^2\epsilon_0}\int_0^\infty dk\,\frac{\sin^2 ka}{(ka)^2}\,\left[1 - 1/\epsilon_L(\mathbf{k})\right]. \tag{17}$$

When this equation was first proposed, several functional forms were assumed for $\epsilon_L(\mathbf{k})$ with the common feature that $\epsilon_L(\mathbf{k}) \geq 1$ for all k values and that $\epsilon_L(\mathbf{k})$ would monotonically decrease with increasing k. Such functional forms lead to $|\Delta F_{NL}| < |\Delta F_B|$, a trend that is in agreement with experimental evidence. However, as our results (and the results of other groups[12-15]) indicate, these forms are incorrect for polar fluids where $\epsilon_L(\mathbf{k})$ actually becomes negative at intermediate k values.

A more accurate approach to ΔF_{solv} is obtained by returning to Eq. (1) and evaluating U_{is}^{es} by averaging the electrostatic ion-solvent potential over the structure of the dilute ionic solution. We adopt this approach here and use the RHNC theory to evaluate the ion-solvent pair correlation function needed to carry out the required average. Specifically, for a spherical ion of charge q in our model solvent, the electrostatic ion-solvent potential is a sum of ion-dipole (id) and ion-quadrupole (iQ) interactions. This leads to

$$\begin{aligned} U_{is}^{es} &= N\left[\langle u_{id}(12)\rangle + \langle u_{iQ}(12)\rangle\right] \\ &= \frac{Nq}{4\pi\epsilon_0}\left[-\mu\langle\Phi^{011}(12)/r^2\rangle + \frac{Q}{2}\langle\Phi^{022}(12)/r^3\rangle\right], \end{aligned} \tag{18}$$

where $\Phi^{011}(12) = \hat{\mathbf{r}}\cdot\hat{\mathbf{u}}_2$ and $\Phi^{022}(12) = 3(\hat{\mathbf{r}}\cdot\hat{\mathbf{u}}_2)^2 - 1$. After the averages over molecular orientations have been carried out

$$U_{is}^{es} = \frac{q\rho}{\epsilon_0}\int_0^{\infty} dr\left[-\frac{\mu}{3}h_{is}^{011}(r) + \frac{2Q}{5r}h_{is}^{022}(r)\right], \tag{19}$$

where $h_{is}^{0\ell\ell}(r)$ is the radial coefficient of the rotational invariant $\Phi^{0\ell\ell}(12)$ in the expansion of the ion-solvent pair correlation function.

Using Eq. (19), we have evaluated U_{is}^{es}/q for $-e \leq q \leq e$ and for two values of the ionic diameter $\sigma_i = 2a$, and have then obtained the resulting ΔF_{solv} by numerical integration of Eq. (1). In Table I some of these results are compared with the corresponding nonlocal electrostatics solvation free energy, ΔF_{NL}, obtained using our RHNC theory results for $\epsilon_L(\mathbf{k})$. All the results are in units of the Born free energy, ΔF_B, given by Eq. (3). The RHNC ionic solution results, denoted by ΔF_{solv}, are expected to be accurate within the range of q and σ_i values given in Table I and can be regarded as the 'exact' molecular theory results that ΔF_{NL} is supposed to approximate.

Table I - Solvation free energies for different ions in methanol

σ_i/σ	q/e	$\Delta F_{solv}/\Delta F_B$	$\Delta F_{NL}/\Delta F_B$	$\Delta F_{NL}^{is}/\Delta F_B$
1.0	0.5	0.776	2.271	0.852
1.0	-0.5	0.768	2.271	0.852
1.0	1.0	0.685	2.271	0.852
1.0	-1.0	0.832	2.271	0.852
1.4	0.5	0.914	2.072	0.938
1.4	-0.5	0.755	2.072	0.938
1.4	1.0	0.811	2.072	0.938
1.4	-1.0	0.841	2.072	0.938

As can be seen from the table, the molecular theory correctly predicts that $\Delta F_{solv}/\Delta F_B < 1$. In addition, it exhibits interesting features absent from ΔF_B and ΔF_{NL}, both of which are linear in q^2. It is apparent from the data of Table I that ΔF_{solv} obtained from the RHNC theory is not linear in q^2, given that $\Delta F_{solv}/\Delta F_B$ varies with the sign and magnitude of q. This is due to two molecular aspects of the ion-solvent interactions. First, when $|q|$ is small, the ion is repelled by the solvent, since solvent-solvent electrostatic interactions are more attractive than ion-solvent interactions. Second, the solvent has an axial quadrupole in addition to the dipole moment and its charge distribution therefore lacks a center of symmetry. This leads to different values of U_{is}^{es} for anions and cations. In the present case, since μ and Q are both positive U_{is}^{es} is more attractive for fully charged anions. This is further illustrated in Figure 4, in which the integrand of Eq. (19), specifically,

$$h_{is}^{es}(r) = \frac{q}{e}\left(h^{011}(r) - \frac{6Q}{5\mu}h^{022}(r)/r\right) \tag{20}$$

is plotted against r for ions of charges $\pm e$ and $\sigma_i = 1.4\sigma$. This quantity determines how the solvent structure around the ion influences the ion-solvent contribution to the internal energy. The figure shows that the first solvation shell is very sharply

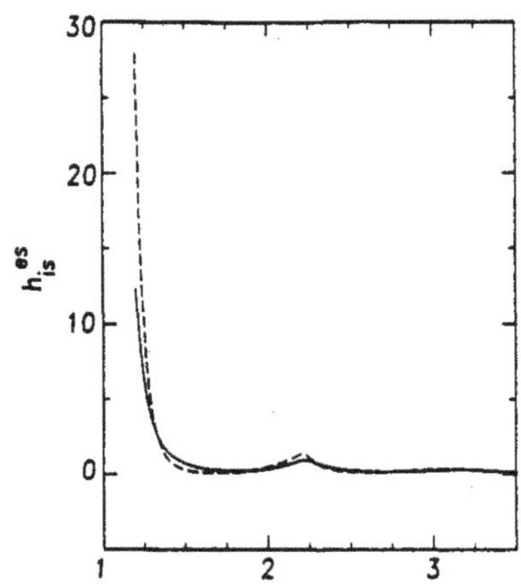

Figure 4. RHNC theory results for the ion-solvent correlation $h_{is}^{es}(r)$, Eq. (20), plotted vs r in units of methanol diameter σ = 3.861 Å. Shown are the results for ions of diameter σ_i = 1.4σ and charges +e (———) and -e (– – –).

defined for both anions and cations, indicating that solvent molecules in this shell are responsible for a large portion of U_{is}^{es}. The figure also shows pronounced differences in the structures of the first two solvation shells for anions and cations, with anions exhibiting the effects of more attractive electrostatic interactions.

Unlike the molecular theory, the nonlinear electrostatics approach does not improve upon the Born approximation, but instead results in *larger* solvation energies. In fact, we find that for both ion diameters $\Delta F_{NL}/\Delta F_B > 2$. The reasons for this are apparent from Figure 2 and Eq. (17). From Eq. (17) we see that because of the modulation by $\sin^2(ka)/(ka)^2$, only small k values contribute significantly to ΔF_{NL} unless the ion radius a is small relative to the size of the solvent molecules. In this range of k values our model for $\epsilon_L(\mathbf{k})$ is accurate, and, as Figure 2 shows, leads to $1 - 1/\epsilon_L(\mathbf{k}) > 1 - 1/\epsilon$. Thus the use of $\epsilon_L(\mathbf{k})$ in Eq. (17) cannot lead to the desired result, i.e., a decrease below ΔF_B in the magnitude of the solvation free energy.

While it is clear that nonlinear aspects of ΔF_{solv} such as its dependence on the sign of q cannot be predicted by nonlocal electrostatics, this approach can be improved to yield a more reasonable estimate of the solvation free energy. To do this we note that two important aspects of the molecular nature of the solvent surrounding the ion lead to differences from the continuum model. First, electrostatic interactions of the ion with the nearby solvent molecules are screened much less than one would predict on the basis of the value of ϵ and are therefore more attractive. Taking this effect into account would increase the magnitude of ΔF_{solv} above ΔF_B. Second, short range solute-solvent repulsion makes the minimum distance of approach between the ion and a solvent molecule larger than a. In our molecular model of the ionic solution this distance is $\sigma_{is} = a + \frac{1}{2}\sigma$, clearly larger than a. Taking into account this larger effective size of the ions would decrease the magnitude of ΔF_{solv}. It is this second aspect of ion-solvent interactions that is clearly ignored in the nonlocal electrostatics approach. Its absence explains why a net increase in the magnitude of ΔF_{solv} is predicted. This problem can, however, be easily remedied by replacing a in Eq. (17) by σ_{is}. The last column in Table I contains these results, labelled as ΔF_{NL}^{is}. Including the effects of ion-solvent repulsion substantially improves the nonlocal electrostatics results: One finds that $\Delta F_{NL}^{is}/\Delta F_B < 1$ and one also finds smaller deviations from the Born model for the larger ions, in agreement with molecular theory.

Conclusion

Using MD simulation and molecular theory we have calculated the wavevector-dependent dielectric tensor of a model of liquid methanol. We have also extended our RHNC integral equation calculations to the structure and thermodynamics of dilute ionic solutions in order to compare the results of the molecular theory to the nonlocal electrostatics approach to the solvation free energy of ions. The solvent structural input into ΔF_{NL} is $\epsilon_L(\mathbf{k})$, which has a pole at a relatively small k value and subsequently becomes negative for our model of methanol as well as for other model polar solvents[12-15] which have recently been studied. We have shown that, as a result of this behavior of $\epsilon_L(\mathbf{k})$, ΔF_{NL} seriously overestimates ΔF_{solv} and is a poorer approximation to it than the simple continuum Born theory. We have argued that this is a consequence of the fact that nonlocal electrostatics ignores short range solute-solvent repulsion and have shown that a much improved value of ΔF_{NL} is obtained by using the ion-solvent distance of closest approach as the effective ion radius. However, this approach still has serious limitations, since it neglects the effects of the ion on the structure of the surrounding solvent. For our model of ionic solution these effects lead to significant charge-dependence of $\Delta F_{solv}/\Delta F_B$.

Acknowledgments

This work was supported in part by the NIH Grant No. GM27945 and the NSF Grant No. CHE-9019707. Acknowledgment is also made to the donors of the Petroleum Research Fund, administered by the American Chemical Society, for partial support of this research.

References

1. D.Y.C. Chan, D.J. Mitchell and B.W. Ninham, *J. Chem.Phys.* **70** (1979) 2946.
2. M. Born, *Z. Phys.* **1** (1920) 45.
3. (a) A.A. Kornyshev, *Electrochim. Acta* **26** (1981) 1; (b) R.R. Dogonadze and A.A. Kornyshev, *J. Chem. Soc. Faraday* **II70** (1974) 1121.
4. (a) G.S. Stell, G.N. Patey and J.S. Høye, *Adv. Chem. Phys.* **48** (1981) 183; (b) P.A. Madden and D. Kivelson, ibid. **56** (1984) 467.
5. (a) P.H. Fries and G.N. Patey, *J. Chem. Phys.* **82** (1985) 429; (b) J.S. Perkyns, P.H. Fries and G.N. Patey, *Mol. Phys.* **57**(1986) 529.
6. M. Haughney, M. Ferrario and I.R. McDonald, *J. Phys. Chem.* **91** (1987) 4934.
7. T. Fonseca and B.M. Ladanyi, *J. Chem. Phys.* **93** (1990) 8148.
8. R.M. Lees and J.G. Baker, *J. Chem. Phys.* **48** (1968) 5299.
9. (a) G. Ciccotti and J.-P. Ryckaert, *Comput. Phys. Rep.* **4** (1986) 345; (b) L. Verlet, *Phys. Rev.* **159** (1967) 98; R.W. Hockney, *Methods comput. Phys.* **9** (1970) 136.
10. M.P. Allen and D.J. Tildesley, *Computer Simulation of Liquids*, (Clarendon, Oxford, 1987).
11. D. Ben-Amotz and D.R. Herschbach, *J. Phys. Chem.* **94** (1990) 1038.
12. F.O. Ranieri, Y. Zhou, H.L. Friedman, and G. Stell, preprint.
13. M. Neumann, *Mol. Phys.* **57** (1986) 97.
14. (a) A. Chandra and B. Bagchi, *J. Chem. Phys.* **90**(1989) 1832; (b) *ibid.* **91**(1989) 3056.
15. P. Attard, D. Wei and G.N. Patey, *Chem. Phys. Lett.* **172** (1990), 69.

ON THE ANALYTICAL PROPERTIES OF THE STATIC DIELECTRIC FUNCTION OF WATER

A.A.KORNYSHEV[1,2], D.A.KOSSAKOWSKI[1,3], M.A.VOROTYNTSEV[1]

1) The A.N.Frumkin Institute of Electrochemistry of the Academy of Sciences of the USSR, Leninskii prosp.31, 117071 Moscow, USSR

2) Medical Biophysics Research Laboratory, University Clinics r.d.I., Technical University of Munich, Ismaningerstr.22, D-8000 Munich 80, FRG

3) Department of physical and chemical biology, Moscow Physico-Technical Institute, Institutskii per.9, 141700 Dolgoprudny, USSR

ABSTRACT

We discuss the behaviour of static dielectric function, $\varepsilon(k)$, of water and other polar solvents on the basis of its relation to partial atomic structure factors with an emphasis on the problem of singularities on the real axis (divergence and negative values) discussed recently by various authors. We show that the behaviour of $\varepsilon(k)$, particularly in the range of small k, is extremely sensitive to the form of a small difference between the structural factors at large distances, and to the values of parameters characterizing the intramolecular charge distribution. We show the origin of singular behaviour of $\varepsilon(k)$, claimed in the mentioned works. We assert that the conclusions about the universality of the singular behaviour are premature in view of the instability of results for the structure factors in the range where they are essentially inaccurate, both in the experiment and computer simulation.

1. INTRODUCTION

1.1. *Historical remarks.*

Nonlocal electrostatics is an efficient tool to invoke the short range structure of the system components into electrostatic calculations by introduction of nonlocal dielectric functions[1,2]. Among the new results so far obtained the most valued are those which do not use the specific form of the dielectric response of the medium but only its general properties. While there is, indeed, a number of such results, in many cases one has to address the particular form of the nonlocal dielectric response.

When considering the static properties of a polar liquid in the bulk one deals with the longitudinal component of the static dielectric tensor, $\varepsilon(k)$[1]. In application of nonlocal electrostatics to various solvation problems[1] the Lorentzian form

$$\varepsilon(k)= \varepsilon_*+(\varepsilon-\varepsilon_*)/(1+k^2\lambda^2\varepsilon/\varepsilon_*) \qquad (1)$$

was broadly used as a simplest approximation interpolating between the short range (ε_*) and long range (ε) dielectric behaviour. For instance for water the macroscopic dielectric constant $\varepsilon \approx 80$ while ε_*, the dielectric constant of "infrared modes", ≈ 5, λ being of the order of correlation range of the intermolecular orientations. Application of this simple form helped to illustrate the ideas of nonlocal electrostatics and to obtain in analytical form a number of results. Many of them were experimentally verified or are under experimental test at present (see e.g.[1-21]).

Recently, a number of publications has appeared in which $\varepsilon(k)$ of model polar liquids was calculated by means of different approximate integral equations schemes [24-28] or computer simulation[23,27]. Simple equation (1) was not approved and all the results which are the consequences of this formula were naturally questioned.

According to these works the behaviour of $\varepsilon(k)$ is like the one shown in Fig.1b [23-28] or 1c[28] but not in Fig.1a. It thus exhibits a pole on the real axis. Attard et.al.[22] emphasize that "the physical significance of this behaviour is not obvious although its theoretical basis is clearly sound."

In this note we shall try to give an explanation to this singularity both from the physics standing behind it and the features of the mathematical model. We then discuss whether such an "exotic" [24] $\varepsilon(k)$ behaviour does appear in nature or not and study the properties of $\varepsilon(k)$ at small k.

1.2. *FDT, $\varepsilon(k)$ and polarization-polarization correlation function*

For a classical subsystem (the modes of which have the frequencies smaller than $kT/\hbar$) of a polar medium the fluctuation-dissipation theorem takes the form[30]

$$C(k) = 1/\varepsilon_*(k) - 1/\varepsilon(k) = (4\pi/\theta)S(k) \tag{2}$$

where $\varepsilon_*(k)$ is the high frequency wave number dependent dielectric constant (responsible for the modes of frequencies higher than $\theta/\hbar$, θ=kT) and

$$S(k) = \Sigma_{\alpha\beta}(k_\alpha k_\beta/k^2)\int dR\; e^{ikR}\; S_{\alpha\beta}(R) \tag{3}$$

If one defines the longitudinal ($||$) and transversal ($\perp$) components (in R-space) of the static correlation function of polarization fluctuations

$$S_{\alpha\beta}(R) = \langle P_\alpha(R)P_\beta(0)\rangle = S_{||}(R)\; R_\alpha R_\beta/R^2 + S_\perp(R)\; (\delta_{\alpha\beta} - R_\alpha R_\beta/R^2) \tag{4}$$

then

$$S(k) = \int_0^\infty dR\; 4\pi R \left[S_{||}(R)\frac{\sin kR}{kR} - \frac{2}{(kR)^2}\,[\,S_{||}(R) - S_\perp(R)\,]\,\{\frac{\sin kR}{kR} - \cos kR\,\}\right] \tag{5}$$

Thus, the knowledge of $S_{||}(R)$ and $S_\perp(R)$ - either from experiments or simulation, or from model theoretical treatment - would provide us with the result for $C(k)$. If the medium is purely classical as in the models of rigid dipolar-multipolar spheres,

$$\varepsilon_*(k) = 1 \tag{6}$$

and this scheme thus gives directly $\varepsilon(k)$. When one is interested in the wave numbers smaller than those, high ones, for which $\varepsilon_*(k)$ exhibits dispersion, then $\varepsilon_*(k)\simeq\varepsilon_*\simeq$const, ε_* can be evaluated through dielectric loss spectra as the dielectric constant of the upper bound of Debye region (for water 5).

Possible difference between the longitudinal and transversal components of the $S_{\alpha\beta}(R)$ tensor was discussed long ago[30]. Obviously, they must not be the same, in particular, because two dipoles interact differently depending on whether they are lying along the axis passing through their centers or oriented perpendicular to the axis. Of course, it is hard to presume whether this difference be important in a system which is more complicated than a liquid of dipolar-multipolar spheres. However, not this difference is responsible for the "anomalies" in $\varepsilon(k)$.

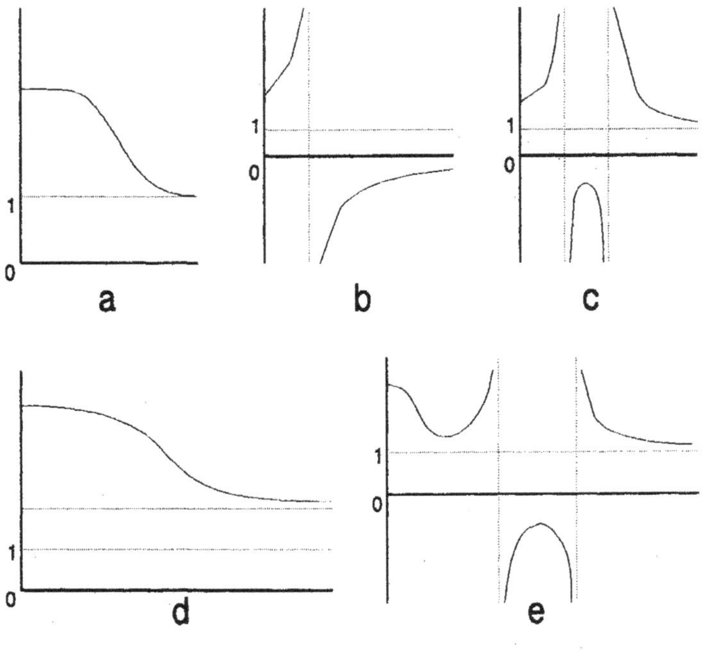

Fig.1 Different variants of the behaviour of dielectric function $\varepsilon(k)$ discussed in the text

1.3. *The short-wavelength limit of S(k) and the behaviour of ε(k)*

Let the system be purely classical (i.e. Eq.6 holds). Then

$$\varepsilon(k) = [1-(4\pi/\theta)S(k)]^{-1} \tag{7}$$

According to Kirzhnitz et.al.[29] the only constraint on $\varepsilon(k)$ is the forbidden range $0<\varepsilon(k)<1$ so that $S(k)>0$.

In particular when $4\pi\, S(k)/\theta = 1$, $\varepsilon(k)$ diverges. This may never happen if $S(k)$ were a monotonically decreasing function: $S(k)= S_0 f(k)$ where $f(k)$ decreasing from $f(k=0)=1$ to $f(k\to\infty)=0$, because

$$(4\pi/\theta)S_0 = 1-\varepsilon^{-1} < 1 \tag{8}$$

In this case the plot of $\varepsilon(k)$ looks like that of Fig.1(a). However, when $f(k)$ is nonmonotonic, increasing at small k, it may easily happen that the denominator in Eq.7 would turn to zero and $\varepsilon(k)$ diverge. The value of $S(k\to\infty)$ is also important. In particular, $S(k\to\infty)=\mathrm{const}\neq 0$ when $S_{\alpha\beta}(R)$ has a δ-function contribution. The contribution $\delta(R)$ is always present in a system of particles with point dipoles:

$$\begin{aligned}\langle P_\alpha(R)P_\beta(0)\rangle &= V^{-1}\int dr\; P_\alpha(R+r)P_\beta(r) = \\ V^{-1}\sum_{ij} \mu_i^\alpha\, \mu_j^\beta &\int dr\; \delta(r-r_i)\delta(r+R-r_j) = \\ N\, V^{-1}\langle\mu^2\rangle\; \delta(R) &+V^{-1}\sum_{i\neq j}\mu_i^{\;\alpha}\, \mu_j^{\;\beta}\; \delta(R-r_{ij})\end{aligned} \tag{9}$$

where V is the volume of the system and N, the total number of particles.

If the dipoles are not point-like, all the δ-functions above "broaden", the broadening determined by the characteristic separation of the charges, d, within the particle. In this case $S(k\to\infty) \to 0$.

Let the dipoles be point ones, so that $S(k\to\infty)=\mathrm{Const}\neq 0$. If the $\mathrm{Const} > S_0$ and $(4\pi/\theta)\mathrm{Const} > 1$, there is a region of negative values of $\varepsilon(k)$ and the point of divergence in between (just as in the case shown in

Fig.1 of Ref.22). However, when the dipoles are not point-like the plot will be different, namely as the one sketched in Fig.1c. On the other hand, if the Const < S_0, $\varepsilon(k)$ will have a form shown in Fig.1d (the proof was given in Ref.31; but note obvious misprints in the two resulting formulae).

It is, finally, possible that S(k) initially decreases and then increases to provide a pole in $\varepsilon(k)$. Then instead of the plot in Fig.1c we shall deal with that of Fig.1e.

1.4. *Physical meaning of divergence and negative values of $\varepsilon(k)$. Do they occure for dense polar liquids?*

For a system of hard spheres with point dipoles an appearance of nonzero $S(k\to\infty)$ is inevitable. In the case of large enough dipole moments μ and particle densities the value of $S(k\to\infty)$ may happen to be sufficiently high to lead to a plot of Fig.1b. However, for non-pointlike dipoles the situation is much more complicated. It will be the subject of the present paper.

The physical meaning of the singular points on the real axis and the range of negative values is better understood in the value of $1/\varepsilon(k)$ which determines a response to induction

$$E_k=D_k/\varepsilon(k), \quad E^{(m)}=4\pi P_k= [1-1/\varepsilon(k)]\ D(k) \tag{10}$$

When $\varepsilon(k)\leq 0$, $|E_m|>|D(k)|$, i.e. the reaction field will be greater than the induction field. How this can happen? Imagine there is a certain preorganized "oscillating" structure in a medium polarization having a period $2\pi/k$. The reaction to an induction $D_k e^{ikr}$ may be very strong depending on the amplitude of the preorganized "oscillation". Following Kirzhnitz et.al.[29] this phenomenon is called *overscreening*. For the range of negative values of $\varepsilon(k)$ to be broad enough, a region of k is needed for which such oscillation structures are favourable.

Range of negative $\varepsilon(k)$ was found initially both theoretically[32] and in simulation[33] for molten salts, treated as a mixture of hard balls with positively or negatively charged centers. It might be interesting to check whether this range would retain in a model with charge distribution on the spheres.

While the oscillating structures are probable for simple "model" polar fluids (nonassociated by hydrogen bonds), will this be true for real water?

2. Dielectric function for a molecular liquid with intramolecular charge distribution

Thus far we discussed the behaviour of $\varepsilon(k)$ in terms of the polarization-polarization correlation function. Now we want to apply an approach based on the relationship between the dielectric function and the binary correlation functions of a system of neutral molecules composed of charged fragments[34,35].

2.1. *Basic expressions*

Consider a system with classical degrees of freedom. In terms of fluctuations of the bound charge of the medium, $\rho = -\mathrm{div}\, P$, the FDT for such a system takes the form

$$1 - \varepsilon^{-1} = (4\pi/\theta)k^{-2}S_{\rho}(k) \tag{11}$$

$S_{\rho}(k)$ being the bound charge correlation function :

$$S_{\rho}(k) = \int d^3R\, e^{ikR} \langle\rho(R)\rho(0)\rangle \tag{12}$$

Let an electroneutral molecule consist of charged groups with fixed centers; the distribution of a charge q_i around its centre is

$$q_i(R)=q_i\bar{f}_i(R) \tag{13}$$

For a liquid consisting of such molecules,

$$S\;(k) = \sum q_i q_j f_i(k) f_j(k)\; n\; [\delta_{ij} + (1-\delta_{ij})\sin(kl_{ij})/(kl_{ij}) + n\int_0^{\infty} dR\; 4\pi\, R^2 \frac{\sin(kR)}{kR}\, h_{ij}(R)\;] \tag{14}$$

where i is the index of the charge group within the molecule, $f_i(k)$ is the charge distribution form-factor

$$f_i(k) = \int d^3R\, e^{-ikR}\, \tilde{f}_i(R) \qquad (15)$$

l_{ij} and $h_{ij}(R)$ are, respectively, the intramolecular distances between the charge centers and binary correlation functions of the centers of the charges belonging to different molecules, n is the density of the liquid.

For point charges $f_i(k)=1$ and $S_\rho(k)$ coincides with the expression given in Ref.34, 35.

Thus having data for $h_{ij}(R)$ it is generally possible to calculate the dielectric function. These data may be obtained either from experiment or from computer simulation. The latter provides a completely "self-consistent" approach as the pair correlation functions are simulated for the specified charge distribution of the molecule.

As for the treatment of the data obtained in the real (diffraction) experiment, it requires a model for the charge distribution. Indeed, in order to extract partial structure factors for the molecular components one has to combine different diffraction techniques (neutron, electron, x-ray scattering) which probe density distribution (neutrons) and electron density distribution[36].

2.2. *3 point charges model of the water molecule.*

Now, let us focus on a particular model to exemplify the suggested method of calculation of $\varepsilon(k)$ and problems met on this way. We consider the model used in the computer simulation by Malenkov[37] to deal with his results for binary correlation functions.

Geometry: the molecule is rigid, l_{OH} distance is 0.98 Å and the HOH angle is 109.5°; point charges location: negative charge is located at the "oxygen", $q_-=-2\,q_+$; two positive charges $q_+= q = 0.34e$ are located at the each proton (e is the charge of the electron).

For this model Eq.14 takes the form

$$S_\rho(k)= 2nq^2\left[3 + \frac{\sin(kl_{HH})}{kl_{HH}} - 4\,\frac{\sin(kl_{OH})}{kl_{OH}} + 2n\int_0^\infty dR\; 4\pi R^2\,\frac{\sin(kR)}{kR}\,H(R)\right] \qquad (16)$$

Here H(R) is the combination of the binary correlation functions

$$H(R) = h_{OO}(R) + h_{HH}(R) - 2h_{OH}(R) \qquad (17)$$

2.2.1 *Asymptotic behaviour of $\varepsilon(k)$.*

Continuum description of the dielectric properties of the liquid is meaningful up to the values of k not much greater than the inverse intraatomic distances. Thus, the most physically interesting (and valuable for various applications) is the small k range and the macroscopic k=0 limit. Eqs.11,16,17 give the small k expansion of the $1-\varepsilon^{-1}(k)$ function

$$1-\varepsilon^{-1}(k) \simeq A\,k^{-2} + B + C\,k^{2} + O(k^{4}) \qquad (18a)$$

$$A = \frac{(8\pi n)^2 q^2}{\theta} R_2 \qquad (18b)$$

$$B = \frac{4\pi n q^2}{3\theta}(4\,l_{OH}^2 - l_{HH}^2 - 8\pi n R_4) \qquad (18c)$$

$$C = \frac{\pi n q^2}{15\theta}(l_{HH}^4 - 4\,l_{OH}^4 + 8\pi n R_6) \qquad (18d)$$

Here R_n stands stands for the n-th order moment of the H(R) function

$$R_n = \int_0^\infty dR\, R^n\, H(R) \qquad (19)$$

Specifying the parameters of the model we obtain (all R_n in Å^{n+1}):

$$A=44.5\,R_2\;;\quad B=11.468-7.418\,R_4;\quad C=1.281+0.371\,R_6 \qquad (20)$$

Due to the definition of the binary correlation function $R_2=0$ and there is no singularity at k = 0.

The value of the dielectric constant $\varepsilon(0)$ is determined by the sum of a constant arising from the intramolecular parameters and the fourth moment of the H(R) function. These contributions (relatively large, ≈ 10) must very accurately compensate each other to give a small value (viz.

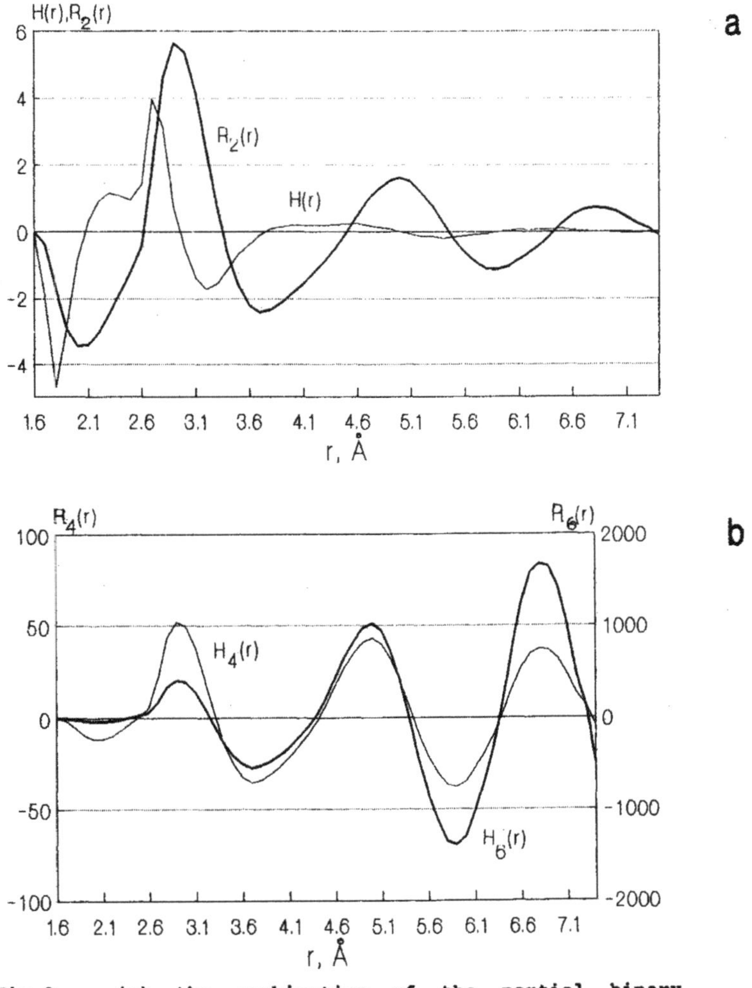

Fig.2. (a) the combination of the partial binary correlation functions, H(R), and its second moment as a function of the upper limit of the integration; (b) the fourth and the sixth moments.

0.987) to provide $\varepsilon(0) = 80$. The difference in the third digit of R_4 leads to the change of the dielectric constant from 80 (R_4=1.413) to infinity (R_4=1.411) ! So to obtain some meaningful value of $\varepsilon(0)$ one must know R_4 with a great precision.

The term of the second order in k in Eq.18a determines the behaviour of the dielectric function in the small k limit. It depends on the sixth moment of H(R). For R_6<-3.453 $\varepsilon(k)$ decreases ,otherwise it increases.

There is one more factor which influences the behaviour of $\varepsilon(k)$. If the charges where not point-like ($f_i(k)\neq$const) there is an additional term of the second order d^2k^2, d being the characteristic radius of the charge distribution around the centre, the quantity of atomic scale. Though it cannot influence essentially the small k behaviour of the curve, it will be important at large k, particularly when $f_1(k)\neq f_2(k)$.

Attard et.al.[22] show that for the system of hard spheres with point dipoles the analytical solution within the mean spherical approximation gives positive second order correction $\varepsilon(k) = \varepsilon(0)+a^2k^2+O(k^4)$. But it is still a question would this result be true for more realistic models.

2.2.2. *Numerical results*

In order to demonstrate how the general requirements for the H(R)-moments affects the calculation of the dielectric function let us try to use Malenkov data. The H(R) function and its second, fourth and sixth moments are shown in Fig.2. The last point is at R=7.5 Å. This is due to a limited number of water molecules involved in the simulation and the traditional interest not to the details of the binary correlation functions long-range approach to unity , but rather to characteristic points in the region of the first and second coordination shells. On the contrary, for our purposes the long-range behaviour is of crucial importance: we see that though H(R) tends to zero,the given range is insufficient to obtain the values of the higher moments. For the given data near the upper cutoff they oscillate changing sign !

So, how can we treat these data ? Of course, it is possible to cut the data off at large distances in order to satisfy the restrictions on the moments: R_2=0 and 1.411<R_4<1.546 to provide positive values of $\varepsilon(0)$.

Suppose that we cut the data off at a point where R_2=0 (see Fig.2). As a result we obtain $\varepsilon(0)$ < 1 which is nonsense ! It is also useless to fit R_4 to get $\varepsilon(0)$ (assuming $R_2=0$), because this moment oscillates with the amplitude 10 upon 0.05 Å variation of the cutoff, while the value of $\varepsilon(0)$ depends on the third digit in the value of R_4.

Note that another source of the inaccuracy in determination of the dielectric constant is the inaccuracy of the intramolecular parameters. It is of no importance in the computer simulation which is performed self-consistently for given values of the intramolecular constants, but may have far leading consequences in the treatment of the experimental data, where we are to specify the charge distribution. Indeed, an 0.03 Å variation of some intramolecular distance may change the sign of $\varepsilon(0)$!

Thus, it is impossible to get the dielectric constant from the given data. In solving this task on the basis of the binary correlation functions one needs the data at distances up to 100 Å.

The cutoff procedure influences also the behaviour of the dielectric function . Assuming R_2 is put to be zero and R_4 is forced to give sufficiently large value of $\varepsilon(0)$, e.g. $\varepsilon(0)=80$, and varying the cutoff distance in the calculation of R_6 we obtain absolutely different curves at small k (Fig.3).

All the above problems arise considering only the asymptotic behaviour when $k \ll 1Å^{-1}$. As one can see from Fig.4 various cutoff distances affect but weakly the behaviour of $S_\rho(k)$ at $k \gg 1Å^{-1}$, but the curves differ strongly in the range of $k < 1Å^{-1}$ (unphysical values $S_\rho(k)<0$ are due to the lack of data). This tendency is even stronger pronounced in the plots of $S_\rho(k)/k^2$; we do not display them here because an arbitrary cutoff does not provide $S_\rho(k)_{k\to 0}k^2$, and therefore the curve contains singularity at k=0.

Various authors[22-28] usually demonstrate plots for the structure factors in the k-space for the range $0 < k\sigma < 20$, σ being the size of the particle (~3Å). They were obviously to meet the difficulties, discussed above, in small k limit, which, unfortunately have never been discussed. As we just saw, the fitting is a dangerous procedure in this limit. It can end up with any type of the small k behaviour of the dielectric function one would like to get.

To summarize our findings we stress that in order to obtain reliable results using the atom-atom binary correlation functions we need to have data at very large distances (up to 100Å)! We furthermore need to know them with very high precision because these data should give us the values of a quantity which is a small difference between these functions.

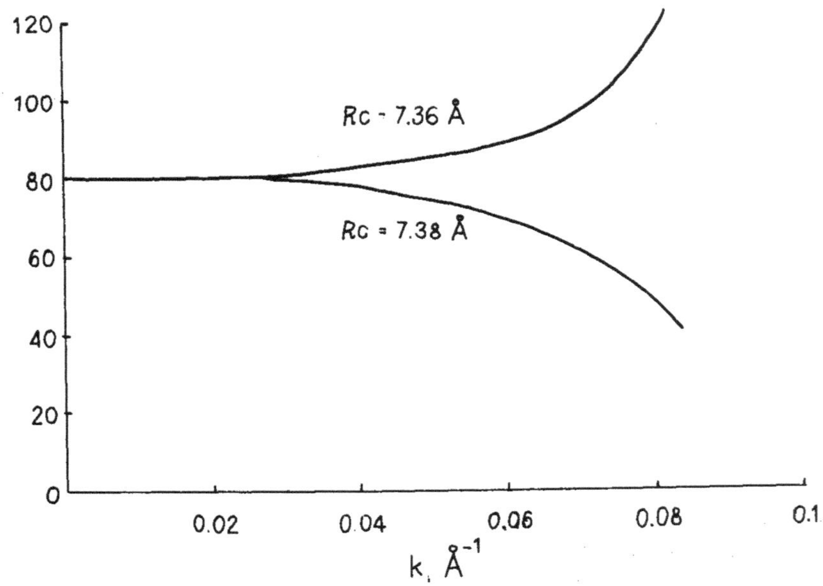

Fig.3. The behavior of ε(k) at small k for various cutoff distances, R_c; R_2 is put to be zero and R_4 is assumed to be 1.413 to satisfy the requirement ε(0)=80.

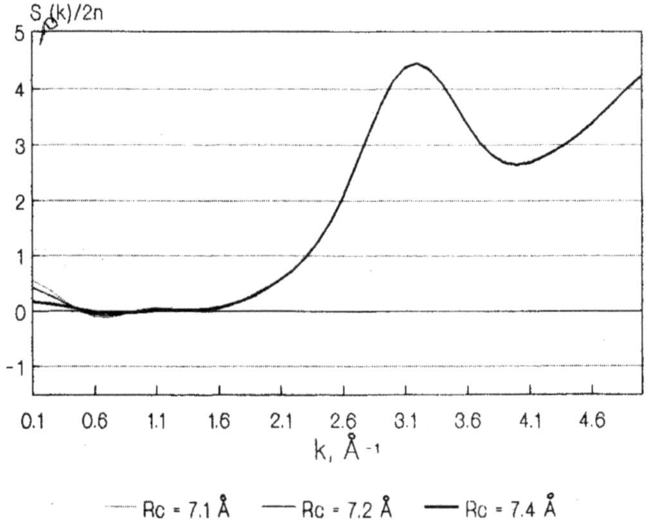

Fig.4. The structural factor $S_\rho(k)$ (without any fitting) for various cutoff distances.

Calculating the Fourier transforms when cutting the binary correlation functions off at some large distance one has to be sure that the value of the moments of H(R) (at least R_4 and R_6) become stable with respect to variation of the cutoff distance. This is true no matter what is the source of the data, be it integral equations calculations, computer simulation data or diffraction measurements.

3. Electrostatic contribution to solvation energy

The electrostatic part of the free energy of solvation of an ion with a radius a is given by Dogonadze-Kornyshev equation[1]

$$W(a) = \frac{z^2e^2}{\pi}\int_0^\infty dk\frac{\sin^2 ka}{k^2a^2}\,(1-\varepsilon^{-1}(k)) \tag{21}$$

which gives the classical Born formula

$$W(a) = \frac{z^2e^2}{2a}\,(1-\varepsilon_s^{-1}) \tag{22}$$

in the limit of a greater than any characteristic $\bar{k}^{-1}$ of $\varepsilon(k)$ variation.

The results of application of this equation to the data reported by Attard et.al.[22] and Raineri et.al.[28] are shown in Fig.5.

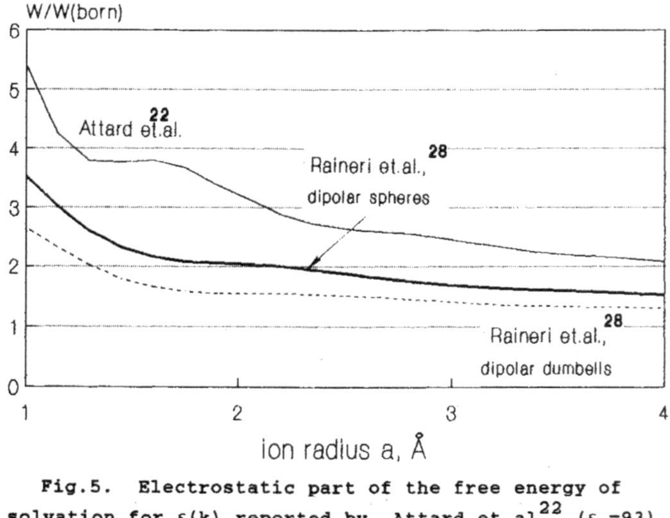

Fig.5. Electrostatic part of the free energy of solvation for $\varepsilon(k)$ reported by Attard et.al.[22] (ε_s=93) and Raineri et.al.[28] (ε_s=29.5) compared with that calculated according to Born formula.

The obtained energies essentially exceed those which are prescribed by the classical Born formula. Note that experimental data never exceed the Born energies; in water the absolute values are typically from 50% to 30% smaller than Born energies, ranging from $a \cong 1$Å to $a \cong 2$Å, respectively.

Of course, one may suppose that the reported $\varepsilon(k)$[22,28] are "good" but the Dogonadze-Kornyshev equation is "bad". The latter is the result of the linear response approximation, i.e. it does not take into account the dielectric saturation which must reduce the excess over Born energies. It furthermore ignores the absence of the solvent inside the space occupied by the ion, the feature, however of minor importance for radii of 1÷2 Å [1]. The compensation of the large energies can be due to Ulig terms [38] (the work on creating a cavity in a liquid to be filled by the ion) which, however, start to be considerable only for large ions ($a \geq 3$ Å)[1,38].

Anyway, the demonstrated excess seems to be too large to be attributed to possible drawbacks of Eq.21.

5. Conclusion

We regard the conclusions of Ref.22 - about the inadequacy of the simple Lorentzian formula for $\varepsilon(k)$ at small k - as premature, when applied to water. Negative $\varepsilon(k)$ values may appear to be an artifact due to an employed multipolar expansion. They, furthermore, lead to free energy of ionic solvation considerably larger than the Born values. This is in contradiction with the experiments even for nonassociated polar liquids, not speaking of water.

The lack of direct measurements of $\varepsilon(k)$ is now severely felt. Once we yet do not have them, some progress may be achieved by treating directly the k-space data of the diffraction experiments. This would allow us to avoid errors arising from the transition from the k-space to the R-space and backwards.

While this paper contains rather negative than positive results, we still believe that the presented analysis would be helpful for further understanding of $\varepsilon(k)$ of polar liquids and would stimulate new research. We plan ourselves to use more precise data for the partial structure factors together with the models of the charge structure of water molecules in order to get some new information on the behavior of $\varepsilon(k)$.

6. Acknowledgements

The authors are thankful to Professors L.I.Kristalik, G.G.Malenkov, and A.M Kuznetsov for an interest to this work and numerous valuable conversations. A.A.K. and M.A.V. appreciate useful discussions with Prof.M.F.Golovko and and Dr.F.Raineri. This work has been initiated during the stay of A.A.K. as a guest at Purdue University (USA) and finalized during his stay at the Technical University of Munich (FRG) as a Humboldt Prize *Geistwissenchaftler*; he is thankful to Prof.M.J.Weaver at Purdue who has drawn his attention to the problem of singular $\varepsilon(k)$ behavior, to *Alexander von Humboldt Stiftung* for the financial support, and to Prof.G.Cevc for generous hospitality which A.A.K. encountered at the Laboratory of Medical Biophysics in Munich.

R E F E R E N C E S

1. A.A.Kornyshev, In: *The Chemical Physics of Solvation*, Eds. R.R.Dogonadze, E.Kalman, A.A.Kornyshev,and J.Ulstrup, Part A (Elsevier, Amsterdam, 1985)p.77; Part C(Elsevier, Amsterdam, 1988)p.355.
2. M.A.Vorotyntsev ,*ibid.*,Part C,p.401.
3. M.A.Kuznetsov, J.Ulstrup, and M.A.Vorotyntsev, *ibid.*,p.163.
4. E.M.Itskovich, A.M.Kuznetsov,and J.Ulstrup,*ibid.*,p.275.
5. A.A.Kornyshev and J.Ulstrup *Chem.Phys.Lett.*126 (1986) 74.
6. J.Ulstrup, *J.Phys.Chem.*91 (1987) 5153.
7. A.A.Kornyshev and W.Schmickler, *J.Electroanal.Chem.*202 (1986) 1.
8. P.G.Dzhavakhidze, A.A.Kornyshev and L.I.Krishtalik *J.Electroanal.Chem.* **228** (1987) 329.
9. D.L.Phelps, A.A.Kornyshev and M.J.Weaver *J.Phys.Chem.* **94** (1990) 1454.
10. A.A.Kornyshev, *J.Electroanal.Chem.* **204** (1986) 79.
11. M.A.Vorotyntsev, *Theory of specific ionic adsorption at a non-transitional metal surface* (this issue).
12. A.A.Kornyshev, G.I.Tsitsuashvili and A.E.Yaroschuk, *Elektrokhimiya*, **25** (1989) 1027, 1037.
13. A.M.Berezhkovskii, S.A.Kabakchi, A.A.Kornyshev and G.E.Chudinov, *Dokl.AN SSSR.* 307 (1989) 625.
14. A.A.Kornyshev, A.M.Kuznetsov, D.K.Phelps, M.J.Weaver, *J.Chem.Phys.*

91 (1989) 7159.
15. G.Cevc, *Chem.Scripta.* **25** (1985) 96.
16. M.L.Belaya, M.V.Feigelman and V.G.Levadny, *Chem.Phys.Lett.* **126** (1986) 361.
17. S.Leikin and A.A.Kornyshev, *J.Chem.Phys.* 90 (1990) 6890.
18. A.A.Kornyshev and J.Ulstrup, *Chem.Scripta* 25 (1985) 58; *J.Electroanal Chem.* 183 (1985) 387.
19. A.A.Kornyshev, *J.Electroanal.Chem.* 225 (1988) 297.
20. M.A.Vorotyntsev and S.N.Ivanov, *Elektrokhimia* 23 (1987) 215, 774.
21. M.A.Vorotyntsev and S.N.Ivanov, *ibid.* **24** (1988) 805, **25** (1989) 554.
22. P.Attard, D.Wei and G.N.Patey, *Chem.Phys.Lett* **172** (1990) 69.
23. M.Neumann, *Mol.Phys.* 57 (1986) 97.
24. A.Chandra and B.Bagchi, *J.Chem.Phys.* 90 (1989) 1832, 7338; **91** (1989) 3056.
25. R.F.Loring and S.Mukamel, *J.Chem.Phys.* 87 (1987) 1271.
26. D.Wei and G.N.Patey, *J.Chem.Phys.* **91** (1989) 7113; **93** (1990) 1399.
27. T.F.Fonseca and B.M.Ladanyi, *J.Chem.Phys.* 93 (1990) 8148.
28. F.O.Raineri, Y.Zhou, H.L.Friedman and G.Stell, *J.Chem.Phys.* (1990), in press.
29. O.V.Dolgov, D.A.Kirzhnitz and E.G.Maximov, *Rev.Mod.Phys.* **53** (1981) 81.
30. R.R.Dogonadze, A.A.Kornyshev and A.M.Kuznetsov, *Teor.Mat.Fiz.* **15** (1973) 127 (English transl.: *Theoretical & Mathematical Physics* (USSR) **15**, No.3).
31. K.Holub and A.A.Kornyshev, *JCS Faraday II.* **76** (1980) 67.
32. M.F.Golovko and I.R.Yukhnovskij, In: *The Chemical Physics of Solvation*, Eds. R.R.Dogonadze, E.Kalman, A.A.Kornyshev and J.Ulstrup, Part A (Elsevier, Amsterdam, 1985) p.207.
33. M.Revere, R.Miniero, M.Parinello and M.P.Tosi, *Phys.Chem.Liq.* 9 (1979) 11.
34. J.Høye and G.Stell, *J.Chem.Phys.* **65** (1976) 18; **67** (1977) 1176.
35. G.Stell, G.N.Patey and J.Hoye, *Adv.Chem.Phys* **38** (1981) 183.
36. G.Pǎlinkǎs, E.Kǎlmǎn and P.Kovǎcs, *Mol.Phys.* **34** (1977) 505, 525.
37. G.G.Malenkov, *The Structure of Aqueous Systems* (Nauka, Moscow, 1991) in Russian.
38. A.A.Kornyshev and A.G.Volkov, *J.Electroanal.Chem.* **180** (1984) 363.

INFLUENCE OF DIELECTRIC FRICTION AND DIELECTRIC SATURATION ON THE SPECIFIC IONIC CONDUCTANCE

P. BRITO
Instituto de Fisica, Facultad de Ciencias Exactas y Tecnologia, Universidad Nacional de Tucuman, Av. Independencia 1800, 4000 Tucuman, Argentina.
SAPIU, CONICET, Argentina.

ABSTRACT

The Zwanzig and Hubbard-Onsager's expressions for the dielectric friction coefficient are deduced and directly compared using a continuum model. The influence of dielectric saturation on the permittivity of the solvent is explicitly taken into account in the calculation of the electric drag force on the ion. This permits to give an explanation of the curious hydrodynamic saturation phenomenon founded by Hubbard and Onsager and also to obtain a good description of experimental behaviour of the specific ionic conductance in various solvents as a function of the inverse of the ion's radious.

1. Introduction

The specific ionic conductance defined as

$$\lambda_o = \frac{(ze)^2 N_A}{\xi} \tag{1}$$

for a ion of charge $q = |ze|$ shows, for a given solvent, a nonlinear dependence with the inverse of the ion's radious[1] R_i. In equation(1), N_A is the Avogadro number and ξ is the total friction coefficient. Inmediatly after the Walden rule was setled[2], the total friction coefficient was identified with the hydrodinamic or viscous friction coefficient deduced by Stokes[3]

$$\xi_o = 4\pi(6\pi)\eta\, R_i \tag{2}$$

for slip (stick) boundary condition, being η the viscosity of the solvent. This approximation led to a good description of the experimental results for big ions but large deviations were detected for small or multivalent iones. In 1920 Born[4] suggested that such deviations may be reduced if the electrical interaction between the ion and the surrounding solvent is explicitly included in the calculus of the total drag force. Afterwards several attemps have been made being the more important those developped by Zwanzig[5,6] and Hubbard and Onsager[7,8] in the continuous approach and by Wolynes[9] in the statistical mechanical one The main difference between Zwanzig and Hubbard-Onsager's theories is usually atributed to the different flow velocity expressions used. In the first the calculations were made by taking the velocity solution of the ordinary Navier-Stokes equation whereas the solution obtained from the electrohydronamic one, in which the force due to the electrical ion-solvent interaction is included, was taken by the last authors. The Hubbard-Onsager's results give a better description of the experimental data of λ_o in the whole range of ionic radious and lead to the prediction of a curious effect of hydrodynamic saturation in the limit of a punctual ion($R_i \to 0$).

Extensions of the original Hubbard-Onsager's theory have been made[10-11] in order to include effects such as rotational viscosity and dielectric saturation.

When both hydrodynamic and dielectric drag forces are considered, the total friction coefficient is usually put in the following form

$$\xi = \xi_o + \xi_D \tag{3}$$

where ξ_D is the dielectric friction coefficient . A direct comparison of the values of this coefficient from the expressions deduced in the above mentioned theoretical works shows that $\xi_D(Z) \gg \xi_D(HO)$.

2. Proposed Model and Theoretical Considerations

A continuum model is considered in which the ion of electric charge $q = + e$ is represented by a dielectric sphere of static permittivity ε_a and radious R_i inmersed in an infinite homogeneous solvent whose low and high frequency permittivities are ε_s and ε_i. In order to evaluate the changing electric field in the surrounding medium, when the ion moves throughout the solvent with a stationary velocity $\vec{u}$,a multipole expansion[12] is used. It is considered that the electric drag force on the ion at time t is due to the reaction of the solvent molecules polarized by the ionic field at previous ion's positions. It is assumed that the field at any ionic positions at time $t' < t$ can be described, from the instantaneous one at time t, by a superposition of the fields of a monopole and a punctual dipole both situated at this last position. The dipole vector points out in the opposed sense to the ionic displacement, $\vec{\delta r}$, and it is defined as

$$\vec{m} = - e\, \vec{\delta r} \qquad (4)$$

The reaction of the external medium to the monopole gives a zero field at the ion's position as can be deduced from symmetry considerations but due to the reaction of the solvent polarized by the dipole exists a nonzero field at the centre of the dielectric sphere. This reaction field is parallel to the dipole vector $\vec{m}$ and in the present paper it will be considered the cause of the electric drag force on the ion. This force can then be written as

$$\vec{F_R} = e\, \vec{E_R} = - \xi_D\, \vec{u} \qquad (5)$$

The dielectric friction coefficient is thus defined by an equation similar to that used for the hydrodynamic one. The expression for E_R can be obtained by solving Laplace equation or from the Coulomb law if the surface, σ_p, and volume, ρ_p, densities of polarization charge are given and it is[13]

$$\vec{E_R} = \frac{2(\varepsilon_s - \varepsilon_a)}{3\, \varepsilon_a\, \varepsilon_s\, R_i^3}\, \vec{m_{ex}} \qquad (6)$$

being

$$\vec{m_{ex}} = \frac{3\, \varepsilon_s}{(2\, \varepsilon_s + \varepsilon_a)}\, \vec{m} \qquad (7)$$

the external dipole moment associated to the ionic sphere.

The results deduced by Zwanzig and Hubbard and Onsager for the dielectric friction coefficient can be easily obtained after the dipole moment $\vec{m_{ex}}$ is defined in each case. Thus ,

i) if the value of the external dipole is given by

$$m_{ex} = e\, u\, \tau_L \qquad (8)$$

where τ_L is the longitudinal relaxation time of the solvent, it can be seen that the volume density ρ_p becomes zero and then the reaction field is determined only by the surface density σ_p. Under the condition $\varepsilon_a = \varepsilon_i = 1$, the result obtained by Zwanzig in 1963 is recovered.

ii) if the value of the external dipole is given by

$$m_{ex} = \frac{3\, \varepsilon_s}{(2\, \varepsilon_s + \varepsilon_a)}\, e\, V(r)\, \tau_L \qquad (9)$$

where V(r) is the solution for the flow velocity derived from the hydrodynamic Navier-Stokes equation, it can be seen that now the surface density σ_p becomes zero for both stick

and slip boundary condition. Then the reaction field is due to the volume distribution of polarization charge. Again under the condition $\varepsilon_a = \varepsilon_i = 1$, the results obtained are identical to those deduced by Zwanzig in 1970 for the stick and slip dielectric friction coefficients.
iii) if the external dipole moment is defined as

$$m_{ex} = (1/2)\, e\, V_z(r)\, \tau_L \qquad (10)$$

where $V_z(r)$ is the z-component of the flow velocity solution of the electrohydrodynamic Navier-Stokes equation for $R_i^4 >> R_{HO}$, the reaction field can be calculated after the surface and volume densities of polarization charge are determined. Finally under the condition $\varepsilon_a = \varepsilon_i$ and taking into account only ρ_p the expressions of the dielectric friction coefficient for both stick and slip boundary condition deduced by Hubbard and Onsager are also recovered[14].

A comparison of the results obtained in i) to iii) shows that not only different expressions for the flow velocity have been used in the definition of m_{ex} but also different screening factor. If in ii) the calculation of the dielectric friction coefficient is made taken the screening factor in eq.(9) equal to one, then it results

$$\zeta_D^{st}(Z) = \frac{17}{272}\,\frac{(\varepsilon_s-1)\, e^2}{\varepsilon_s^2\, R_i^3}\,\tau_D \qquad \text{(stick)} \qquad (11a)$$

$$\zeta_D^{sl}(Z) = \frac{2}{16}\,\frac{(\varepsilon_s-1)\, e^2}{\varepsilon_s^2\, R_i^3}\,\tau_D \qquad \text{(slip)} \qquad (11b)$$

As can be seen the stick coefficient is now almost equal to the corresponding Hubbard-Onsager one, but still the dielectric friction coefficient is nearly twice $\zeta_D(HO)$ in the slip case. This difference can be reduced if the part of the reaction field due to the surface density, σ_p, which is nonzero in the slip case, is included in the calculation developped in iii). The resulting expression for the slip dielectric friction coefficient is then

$$\zeta_D^{sl}(HO) = \frac{2}{15}\,\frac{(\varepsilon_s - \varepsilon_i)\, e^2}{\varepsilon_s^2\, R_i^3}\,\tau_L \qquad (12)$$

which is now only slighter different from $\zeta_D^{sl}(Z)$. The correction introduced increases the slip dielectric friction coefficient with respect to that deduced by Hubbard and Onsager which is somewhat small when compared with the experimental results in pure[15] and mixture[16] polar solvents.

As a first conclusion it can be said that the main difference between Zwanzig and Hubbard-Onsager's results seems to be more closely related with the screening factor than with the velocity solution used in the calculation. This means that all the mathematical troubles involved in the resolution of the the modified Navier-Stokes equation can be avoided, in a first order velocity approximation, when dielectric friction effect is studied.

3. Dielectric Saturation Effect

Using the same continuum model, the dielectric saturation effect is included in the calculation of the dielectric friction coefficient by assuming that the medium outside the ionic sphere is nonlinear with a permittivity given by

$$\varepsilon = \varepsilon_s(1 - \lambda\, E_o^2) \qquad (13)$$

where λ is the nonlinear Piekara factor[17] and E_o the ionic field. In order to find σ_p and ρ_p, the solutions of the Laplace[12] equation for the linear medium and of that one derived by Buckingham[18] for a nonlinear dielectric have been obtained after a large and tedious calculation[19] using the perturbation approximation proposed by Bordewijk[20]. By assuming that the permittivity expression, eq.(13), holds in the whole range of ionic radious, it can be seen that there is a certain radious, R_m, for which the permittivity of the near solvent becomes equal to ε_a. This means that in this case there is not difference between the electrical responses of the dielectric sphere and the layer of solvent just outside the ion. Afterwards the reaction field is determined, the following expressions for the dielectric friction coefficient are obtained[21] assuming that $\varepsilon_a = \varepsilon_i$,

For $R_i > R_m$

$$\zeta_D^{st} = \frac{3}{40}\,\frac{(\varepsilon_s-\varepsilon_i)\,e^2}{\varepsilon_s^2\,R_i^3}\,\tau_D\left(1 + g\,\frac{Y_1(\varepsilon_s,\varepsilon_i)\,e^2}{Y_2(\varepsilon_s,\varepsilon_i)\,R_i^4}\right) \qquad (14a)$$

$$\zeta_D^{sl} = \frac{3}{20}\,\frac{(\varepsilon_s-\varepsilon_i)\,e^2}{\varepsilon_s^2\,R_i^3}\,\tau_D\left(1 + g\,\frac{Y_3(\varepsilon_s,\varepsilon_i)\,e^2}{Y_2(\varepsilon_s,\varepsilon_i)\,R_i^4}\right) \qquad (14b)$$

For $R_i < R_m$

$$\zeta_D^{st} = \frac{1}{120}\,\frac{(\varepsilon_s-\varepsilon_i)\,e^2}{\varepsilon_s^2\,R_m^3}\,\tau_D\left(1 + g\,\frac{S_1(\varepsilon_s,\varepsilon_i)\,e^2}{S_2(\varepsilon_s,\varepsilon_i)\,R_m^4}\right) \qquad (15a)$$

$$\zeta_D^{sl} = \frac{1}{60} \frac{(\varepsilon_s-\varepsilon_i)\ e^2}{\varepsilon_s^2\ R_m^3} \tau_D \left(1 + g \frac{S_3(\varepsilon_s,\varepsilon_i)\ e^2}{S_4(\varepsilon_s,\varepsilon_i)\ R_m^4} \right) \qquad (15b)$$

with

$$Y_1 = 78\varepsilon_s^2 - 567\varepsilon_s\varepsilon_i + 102\varepsilon_i^2$$
$$Y_2 = 630\varepsilon_s^2(\varepsilon_s-\varepsilon_i)(2\varepsilon_s+\varepsilon_i)$$
$$Y_3 = 10\varepsilon_s^2 - 705\varepsilon_s\varepsilon_i + 50\varepsilon_i^2$$

$$S_1 = (\varepsilon_s-\varepsilon_i)^2(2\varepsilon_s+\varepsilon_i)^2\ R_s^3\ e^2\ (-\ 3600\varepsilon_s\varepsilon_i + (30\varepsilon_s^2+3285\varepsilon_s\varepsilon_i+150\varepsilon_i^2)X-(48\varepsilon_s^2 - 252\varepsilon_s\varepsilon_i + 50\varepsilon_i^2)X^3)$$
$$S_2 = 42\ kT\ \varepsilon_s(\varepsilon_s-\varepsilon_i)(2\varepsilon_s+\varepsilon_i)R_m^4\ M_2\ (40-33X+2X^3)$$
$$S_3 = (\varepsilon_s-\varepsilon_i)^2(2\varepsilon_s+\varepsilon_i)^2\ R_s^3\ e^2\ (-\ 1800\varepsilon_s\varepsilon_i + (10\varepsilon_s^2+1095\varepsilon_s\varepsilon_i+50\varepsilon_i^2)X\ )$$
$$S_4 = 42\ kT\ \varepsilon_s(\varepsilon_s-\varepsilon_i)(2\varepsilon_s+\varepsilon_i)R_m^4\ M_2\ (20-11X\ )$$
$$M_2 = 1240\varepsilon_s^4+512\varepsilon_s^3\varepsilon_i-408\varepsilon_s^2\varepsilon_i^2+535\varepsilon_s\varepsilon_i^3+146\varepsilon_i^4$$
$$X = R_i/\ R_m \qquad (16)$$

Equations (14) and (15) become identical at $R_i = R_m$. This equations have been obtained after the expression for the Piekara factor deduced by Grosse and Brito[22], in which an ajustable parameter g was added, has been used. It can be seen that now the dielectric friction coefficient show an explicit dependence with the size of the solvent molecules (R_s is the solvent molecular radious) as well as with the temperature.

It must be noted that in the limit of a punctual ion both stick and slip expressions reduce to the same one as it was founded by Hubbard and Onsager. This leads to think that their strange hydrodynamic saturation effect is due to the influence of the dielectric saturation phenomenon.

4. Discussion of Results

Combining eqs.(1), (14) and (15) the corresponding expressions for the specific ionic conductance can be obtained. In Figures 1 and 2 the variation of those theoretical results with R_i^{-1} for the solvents acetonitrile and pyridine are showed. Also experimental values are indicated and it can be detected a very good agreement between them. The curves determined by the values obtained by Stiles et al[11] when dielectric saturation was included in the Hubbard-Onsager model, are also given.

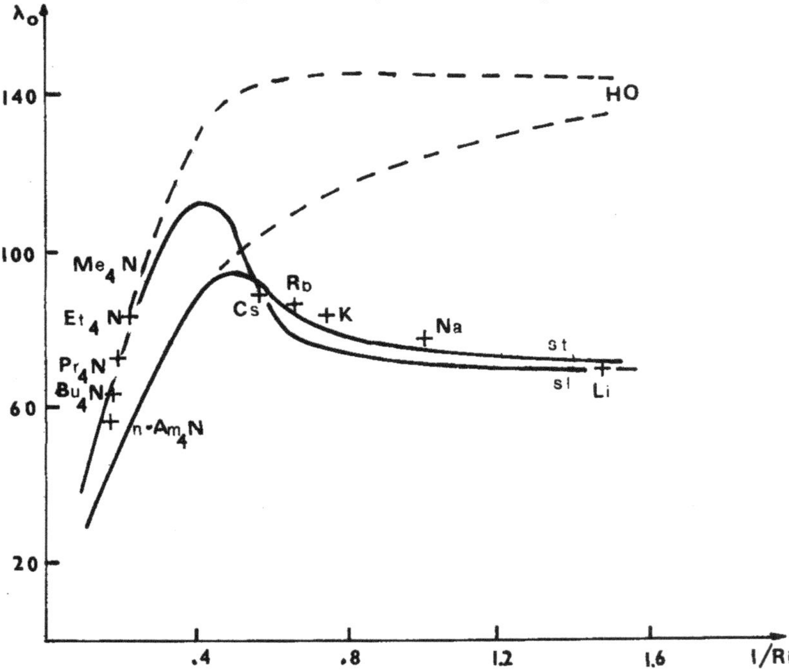

Figure 1 : Comparison of the theoretical λ_o ($ohm^{-1}cm^2mol^{-1}$) vs. Ri^{-1}(A^{-1}) and experimental values. Solvent : acetonitrile (g = 0.20).

Ibuki et al[16] have studied the variation of the friction increment defined by

$$\Delta\xi = \xi - \xi_o \qquad (17)$$

as a function of concentration for solutions whose solvent is a mixture of polar liquids. The value of the total friction coefficiente ξ are determined from experimental data of specific conductance. Figures 3 and 4 shows the curves $\Delta\xi$ vs. volume concentration Φ obtained using the slip dielectric friction coefficients given in eqs.(14) and (15) for the monovalent iones Li and Na in the solvent water-ethanol under the assumption that $\Delta\xi = \xi_D$. Again a good agreement with the experimental results can be seen[23]. In all figures the worse descriptions are given by the curves corresponding to Hubbard-Onsager theory.

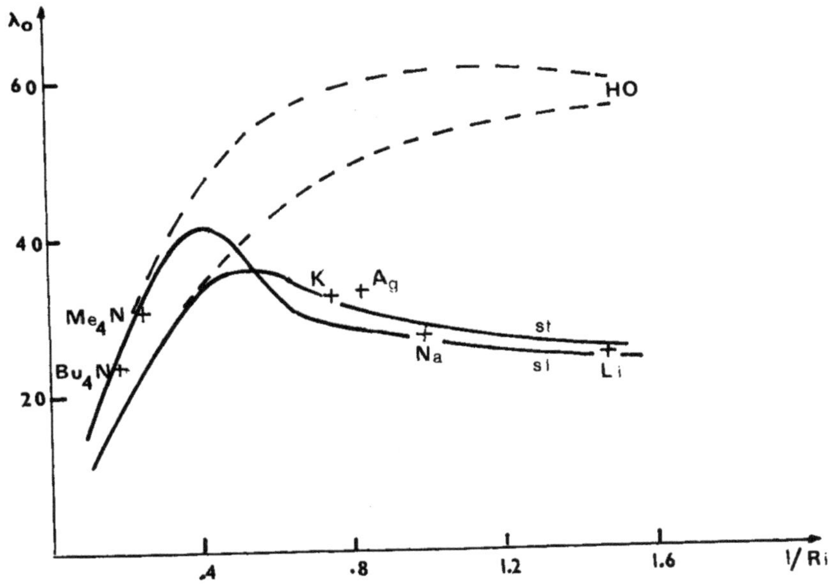

Figure 2 : Comparison of the theoretical λo ($ohm^{-1}cm^2mol^{-1}$) vs. $Ri^{-1}(A^{-1})$ and experimental values. Solvent : pyridine (g = 0.09).

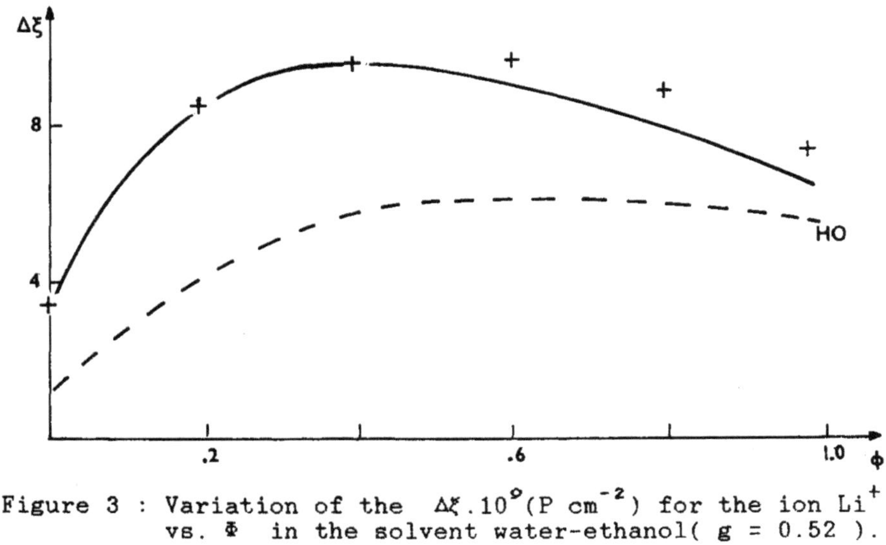

Figure 3 : Variation of the $\Delta\xi.10^9$($P\ cm^{-2}$) for the ion Li^+ vs. Φ in the solvent water-ethanol(g = 0.52).

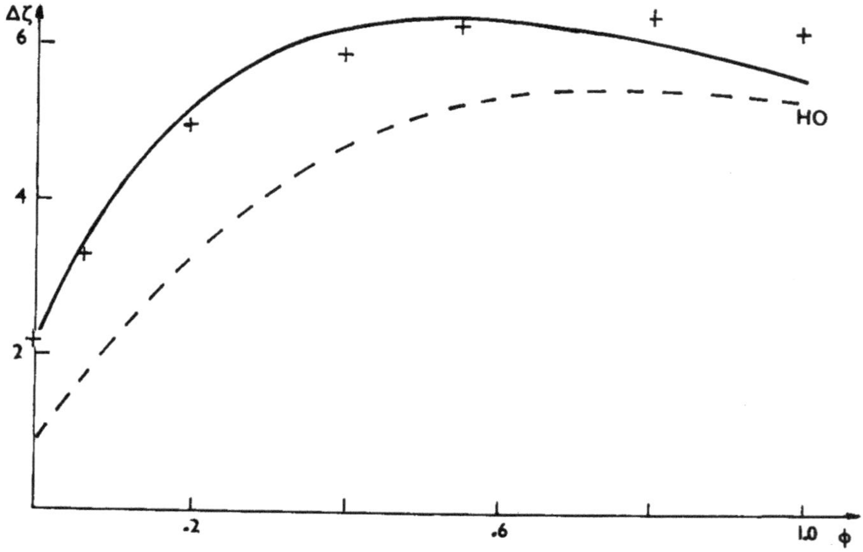

Figure 4 : Variation of the $\Delta\zeta.10^{9}$(P cm^{-2}) for the ion Na^{+} vs. Φ in the solvent water-ethanol(g = 0.52).

5. References

1. H.S.Harned and B.B.Owen, *The Physical Chemistry of Electrolytical Solutions*, (Reinhold, N.York,1958).
2. P.Walden, *Z.Phys.Chem.*, **55** (1906) 207.
3. L.Landau and E.M.Lifshitz, *Fluid Mechanics*, (Addison-Wesley Pub., London,1959).
4. N.Born, *Z.Physik*, **1** (1920) 221.
5. R.Zwanzig, *J.Chem.Phys.*, **38** (1963) 1603.
6 R.Zwanzig, *J.Chem.Phys.*, **52** (1970) 3625.
7. J.B.Hubbard and L.Onsager, *J.Chem.Phys.*, **67** (1977) 4580.
8. J.B.hubbard, *J.Chem.Phys.*, **68** (1978) 1649.
9. P.G.Wolynes, *J.Chem.Phys.*, **68** (1978) 473.
10. J.B.Hubbard and R.F.Kayser, *J.Chem.Phys.*, **74** (1981) 3535.
11. P.J.Stiles, J.B.Hubbard and R.F.Kayser, *J.Chem.Phys.* **77** (1982) 6189.
12. J.A.Stratton, *Electromagnetic Theory*, (McGraw-Hill, N.York,1941).
13. C.J.F.Böttcher, *Theory of Electric Polarization*, I, (Elsevier, Amsterdam, 1973).
14. P.Brito, (to be sent to publish)

15. D.F.Evans, T.Tominaga, J.B.Hubbard and P.G.Wolynes, *J.Phys.Chem.*, **83** (1979) 2669.
16. K.I.Ibuki and M.Nakahara, *J.Chem.Phys.*, **84** (1986) 6979.
17. A.Piekara, *C.R.Acad.Sci.*, **204** (1937) 1106.
18. A.D.Buckingham, *J.Chem.Phys.*, **25** (1956) 428.
19. P.Brito, *Thesis*, (Universidad Nacional de Tucuman, 1988).
20. P.Bordewijk, *J.Chem.Phys.*, **58** (1973) 1220.
21. P.Brito, (to be sent to publish).
22. C.Grosse and P.Brito, *Can.J.Phys.*, **56** (1978) 216.
23. P.Brito and M.Mechetti, (to be sent to publish).

THE INTERFACE: STRUCTURE AND ELECTRIC PROPERTIES

STATISTICAL MECHANICS OF INHOMOGENEOUS POLARS LIQUIDS AND IONIC SOLUTIONS

J.P. BADIALI

Structure et réactivité aux interfaces, université P. et M. Curie F(74), 4 place Jussieu 75230, Paris Cedex 05, France.

ABSTRACT

Some structural aspects concerning the distribution of polar molecules and ions near a neutral wall are investigated. We try to relate some results of statistical mechanics and well known concepts currently used in Electrochemistry (image potential, electrostriction, effective dielectric constant, ...). By such a physical analysis we hope to suggest new approaches in the case of the real electrochemical interface.

1. Introduction

During the last twenty years very important progress have been made in the theory of simple liquids [1], molecular liquids [2] and in the case of fluids in interfacial regions [3]. In parallel the theory of the electrical double layer has been revisited and many excellent reviews exist in this field [4-6]. One part of these papers is generally devoted to a discussion about the primitive model of electrolyte solutions in which the solvent is replaced by a dielectric continuum. Other fundamental aspects which has been discussed concern the validity of the Gouy-Chapman theory and the splitting inner layer/diffuse layer which is generally used in the analysis of experimental data [7]. The aim of this seminar is totally different. First of all, we should like to introduce, as far as it is possible, the basic tools of statistical mechanics in focusing on their physical content. We intend to discuss the assumptions in connection with very well-known concepts in Electrochemistry as for example: the image potential, the solvation shell, the electrostriction or the dielectric constant near a wall. Thus, this paper is very far from being an exhaustive review on the statistical mechanics of the electrochemical double layer. It is rather an introduction to this field. Many technical details which have been discussed in other papers will be not reproduced here.

In Section 2 we present the basic tools of statistical mechanics. Section 3 is devoted to an analysis of the profile for a polar fluid near a wall. In Section 4 we discuss some aspects of the pair correlation function between two molecules adsorbed onto a wall. Finally, in Section 5 we analyze the potential of mean force for one ion in a polar solvent near a wall.

2. The basic tools

In order to characterize the structure of inhomogeneous polar fluids we consider the density-orientational profile $\rho(1)$ which gives the probability of finding the molecule 1 at a given position with a given orientation and the pair correlation function $\rho(1,2)$ which determines the probability of finding the molecules 1 and 2 in specified states. It is usual to introduce a total correlation function h(1,2) by the relation $\rho(1,2) = \rho(1)\,\rho(2)\,(1 + h(1,2))$. The functions $\rho(1)$ and h(1,2) can be obtained from the partition function Z by functionnal derivations [1]. Z is essentially a product of Boltzmann factors $\exp - \beta V(i,j)$ where V(i,j) is the interaction pair potential between the particles i and j while β is $1/k_B T$. Instead of the Boltzmann factor it is more efficient to consider the Mayer function $f(i,j) = 1 - \exp - \beta V(i,j)$. Now Z, $\rho(1)$ and h(1,2) can be represented graphically as a combination of bonds f(i,j) according to precise rules [1, 8]. For instance in Fig. 1a) we give the first graphs of h(1,2). The white points represent the molecules 1 and 2 which are in a given state while each black point "i" is a vertex of weight $\rho(i)$. In the evaluation of graphs we have to perform an integration over the coordinates which define the state of i.

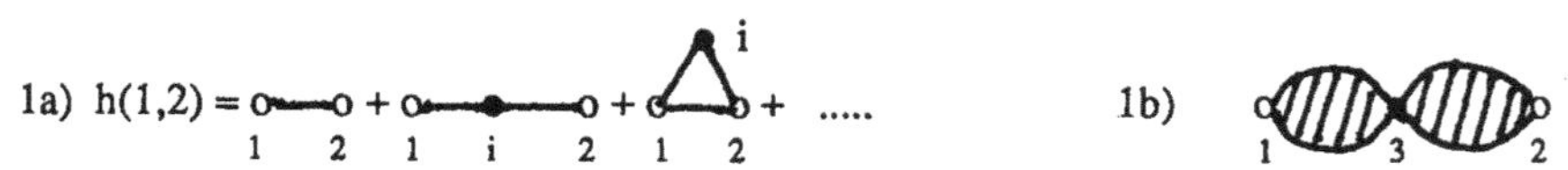

Figure 1. 1a) First graphs in the expansion of h(1,2)

1b) Diagrammatic structure including a nodal point (3)

We may consider h(1,2) as an expansion relative to the number of vertices. Then, in the bulk phase, h(1,2) appears as a power series of the density number ρ. In this virial expansion the first graphs may reveal the physical ideas which determine h(1,2) as we shall see in section 4. At low densities an exact expression of h(1,2) can be obtained.Unfortunatly this expansion can not predict quantitative results for densities corresponding to the liquid state where we must consider an infinite sum of graphs. A study of h(1,2) can be made by using topological arguments.

In Fig. 1b) we consider a graph in which the particles 1 and 2 are connected via 3. Each hatched part represents a given combination of bonds. In such a structure the vertex 3 has a paticular status, if this point is removed 1 and 2 are then disconnected. A vertex such as 3 is called a nodal point. In order to understand the physical meaning of a nodal point we may consider the case where 1 is an ion while all others vertices represent molecules. Then Fig. 1b) represents a part of the correlation function between one ion 1 immersed in a solvent and the molecule 2. The bonds connecting 2 and 3 ignore the presence of the ion, they represent a property of the pure solvent while those between 1 and 2 represent a part of the solvation shell. Then the graph drawn in Fig.1b) represents a convolution, over the spacial coordinates, between a part of the solvation shell and a property of the pure solvent. This kind of convolution is typical from the existence of nodal points.

The graphs pertaining to h(1,2) can be separated in two parts with respect to the existence of nodal points. The part of h(1,2) which is free from nodal point is called the direct correlation function c(1,2). Its definition is also [1]

c(1,2) = f(1,2) + [the sum of all graphs in which the points 1 and 2 are connected by two paths, at least] (1)

For very large distances r_{12} between 1 and 2, c(1,2) behaves as $-\beta V(1,2)$ (except near a second order phase transition). Thus, in general, the long range behavior of c(1,2) is exactly known. This point is extremely important in the case of polar fluids where the range of the dipole-dipole interaction leads to some surface contributions. In opposite the long-range behavior of h(1,2) requires a sophisticated calculation as we shall see in Section 4.

By an analysis in terms of nodal points we can write h(1,2) as

$$h(1,2) = c(1,2) + \int c(1,3)\, \rho(3)\, h(3,2)\, d3 \qquad (2)$$

This is the Ornstein-Zernike (OZ) equation. It represents a first relation between h(1,2) and c(1,2). At least formally, this integral equation can be solved by iteration giving

$$h(1,2) = c(1,2) + \int c(1,3)\, \rho(3) c(3,2)\, d3 \;+ \int c(1,3)\, \rho(3) c(3,4)\, \rho(4) c(4,2)\, d3d4 \;+ .... \qquad (3)$$

From topological arguments [1] we can show that h(1,2) and c(1,2) are also related by

$$1 + h(1,2) = \exp\,[-\beta V(1,2) + h(1,2) - c(1,2) + B(1,2)] \qquad (4)$$

where a new quantity B(1,2) which is called the " Bridge function " has been introduced. It represents a combination of graphs that can never be reduced to a convolution over the spatial coordinates. The main approximations in the liquid state consists in neglecting B(1,2). The assumption B(1,2) = 0 leads to the HNC closure

$$1 + h(1,2) = \exp\,[-\beta V(1,2) + h(1,2) - c(1,2)] \qquad \text{(HNC)} \qquad (5)$$

In addition, if we expand the exponential linearly in terms of h(1,2) - c(1,2) we get the Percus-Yevick closure

$$1 + h(1,2) = \exp\,[-\beta V(1,2)][\,1 + h(1,2) - c(1,2)] \qquad \text{(PY)} \qquad (6)$$

In another closure, the Mean Spherical Approximation (MSA), we assume that c(1,2) is given by its long range behavior $c(1,2) = -\beta V(1,2)$. In the case of point-ions in a dielectric continuum this closure leads to the Debye-Huckel theory. For fluids having a hard core the MSA is given by $c(1,2) = -\beta V(1,2)$ outside the core and $h(1,2) = -1$ inside the core. These two conditions are sufficient for obtaining both h(1,2) outside the core and c(1,2) inside.

In various situations the MSA gives some analytical results which are realistic although no very accurate. Since the MSA is well defined it is possible to build up a new theory in which the MSA becomes the basic ingredient. The pair correlation function given by the MSA appears as a renormalized potential which replaces V(1,2) [9]. In this new expansion the convergence is very efficient because the renormalized potential is much smaller than V(1,2). Many numerical results presented below are obtained with this approximation. They represent a real improvement compared to those obtained with the standard MSA.

3. The density-orientational profile of polar fluids near a neutral wall

Polar fluids are complicated systems for several reasons. The dipolar potential is long-ranged, this means that some quantities may depend on the geometry of the sample. Accordingly, in order to find the intrinsic properties of the material we have to extract properly the surface contributions. As an example, it is not so easy to prove that the dielectric permittivity or the Kerr constant are intrinsic properties of the material under consideration [10-12]. A second difficulty with polar fluids is the existence of a non-radial pair potential. This leads to expand systematically the components of each vector $\hat{v}$ in terms of spherical harmonics $Y_l^m(\hat{v})$. For instance, we write the profile as

$$\rho(1) = (\rho/4\pi)(\, 1 + \sum \rho_l(z)\, Y_{2l}^{m}(\hat{\mu}_1)) \tag{7}$$

and h(1,2) as

$$h(1,2) = \sum h_{l1,l2,l3}^{m1,m2,m3}(r_{12}, z_1, z_2)\, Y_{l1}^{m1}(\mu_1)\, Y_{l2}^{m2}(\mu_2)\, Y_{l3}^{m3}(\mathbf{r}_{12}) \tag{8}$$

where $\hat{\mu}_1$ is the unit vector in the direction of the dipole moment of the molecule 1 and $\hat{\mathbf{r}}_{12}$ the unit vector in the direction joining 1 and 2. In Eqs. 7and 8 the sums run on all values of the integers m and l. For homogeneous systems Eq. 8 can be made more tractable by introducing the rotational invariants [13,14,15]. Now, the introduction of Eqs. 7, 8 and a similar expansion for c(1,2) in the OZ equation (2) leads to a set of coupled equations relating the expansion coefficients. The isotropic part of $\rho(1)$ *i.e.* $(\rho/4\pi)(\, 1 + \rho_0(z)\, Y_0^0(\mu_1))$ will be noted $\rho(z)$ where z is the distance to the wall. The symmetry of the dipolar interaction leads to selection rules which produce large cancellations in the calculation [16].

As an illustration we can study the solvation of one hard-sphere (noted o) immersed in a polar solvent (labeled by s) by using the MSA. The solvent molecules will be represented by a hard-sphere with a permanent dipole at the center. The total correlation function $h_{os}(1,2)$ between the hard-sphere o and the solvent molecule 1 is given by the following OZ equation

$$h_{os}(0,1) = c_{os}(0,1) + \rho \int c_{os}(0,2)\, h_{ss}(2,1)\, d2 \tag{9}$$

Here, the direct correlation function $c_{os}(0,1)$ between o and s is of hard-sphere type *i.e.* isotropic. Then, in the total pair correlation function between solvent molecules $h_{ss}(2,1)$ we select the isotropic part. In the MSA , this part of $h_{ss}(2,1)$ is the pair correlation function of a pure hard-sphere fluid. Accordingly, in the MSA the distribution of polar molecules around a test particle which is a hard-sphere ignores the polarity of the solvent.

If the diameter of the test particle goes to infinite $h_{os}(0,1)$ gives the profile $\rho(1)$ for polar molecules against a pure hard wall. Such a procedure for generating an interface from a mixture has been proposed in [17,18]. In this limit, the MSA gives for $\rho(1)$ the profile of a pure hard-sphere fluid. However $\rho(1)$ must depend on the polarity of the solvent, at least for two reasons.

a) At a given density, the introduction of a permanent dipole in a hard sphere fluid is equivalent to switch on an attractive pair potential between the molecules. This potential decreases the internal pressure p[1]. From the so-called contact theorem[19] we know that the value of the density profile against the wall $\rho(z{=}0)$ is related to p by $p = k_BT\rho(z{=}0)$. Thus $\rho(z)$ must be different from the hard-sphere profile.

b) More generally we may expect that $\rho(1)$ exhibits an orientational structure. In order to understand this point we first use a model " à la Onsager ". We assume that the interaction of 1 with the remaining part of the liquid can be replaced by a dielectric continuum of permittivity ε_1 which covers the half space filled by the liquid. The other side, the wall, has a permittivity ε_2 . From electrostatics we know that the molecule 1 is then subject to a potential w(1) which is nothing else than the classical image potential

$$w(1) = (\mu^2/16)\,[(\varepsilon_1 - \varepsilon_2)/\,\varepsilon_1(\varepsilon_1 + \varepsilon_2)\,]\,(1 + \cos^2\theta)/\,z^3 \qquad (10)$$

where θ is the angle of the dipole moment relative to the normal at the interface and μ is the magnitude of the dipole moment. It is clear that w(1) which has the meaning of a potential of mean force[20] can not determine the profile near the wall where the packing effects due to the hard-core are important. However, w(1) could be a realistic approximation of the potential of mean force far from the wall.

For moderate polar fluids ($\varepsilon \lesssim 30$) $\rho(1)$ has been calculated by Monte-Carlo simulations[21] and by a perturbation theory based on an expansion in terms of renormalized potential[22]. The agreement between the two approaches is rather satisfying. In Eq.7 the function $\rho_2(z)$ determines the first orientational part of the profile, it describes the alignment of the molecule relative to the normal at the interface. $\rho_2(z)$ is given in Fig. 2. This function shows the existence of a layering in the orientational structure. Near the wall the molecules tend to orient their dipole in a direction parallel to the wall. When the sign of $\rho_2(z)$ is positive we have a region where the favoured orientation is normal to the wall and so on. Similar results have been also obtained in [23] by using the same kind of perturbation theory.

Far from the wall an exact expression of $\rho(1)$ can be obtained in very general conditions[24,25]. For example the density profile can be written

$$\rho(z) = (\rho/4\pi)\ \left[1 - (1/16\pi)\ [(d\,\varepsilon_1/d\rho)\ \rho\chi/\beta]\ [(\varepsilon_1 - \varepsilon_2)/\,\varepsilon_1(\,\varepsilon_1 + \varepsilon_2)]\,/\,z^3\ \right] \qquad (11)$$

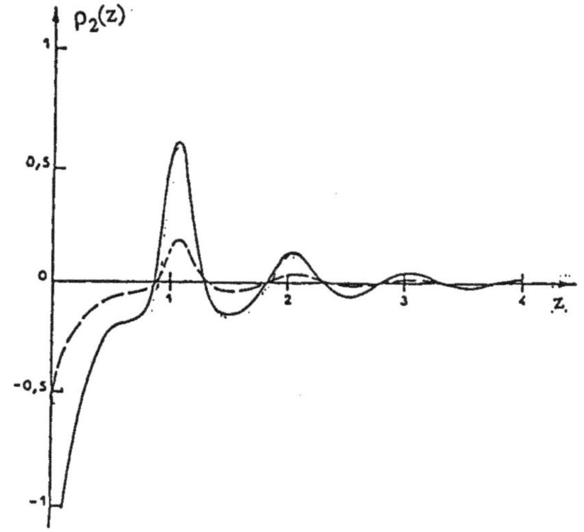

Figure 2. Representation of the function $\rho_2(z)$ which determines the alignement of the molecule relatively to the direction normal to the wall. The liquid is a dipolar hard-sphere fluid (dipole moment μ, hard sphere diameter σ). The curve (– — — –) corresponds to a liquid of reduced dipole moment $\mu^*= \beta\mu^2/\sigma^3$ =1, the dielectric constant is about16. The curve (---------) corresponds to μ^*= 2, the dielectric constant is about 30. z is in σ unit.

In this expression all quantities are known. First of all, we recognize the combination of dielectric constant and the spatial dependence which exist in Eq. 10 and we can say that it is the image potential which creates the particular behavior of the profile. In the derivation of Eq. 11 the dielectric constant is introduced at a microscopic level, it appears as a given sum of graphs. In Eq. 11 also appear the isothermal compressibility χ and the derivative $(d\varepsilon_1/d\rho)$. The product $[(d\varepsilon_1/d\rho)\rho\chi/\beta]$ is nothing else than the electrostriction constant K in bulk phase [26]. When a field E is switched on, at constant chemical potential, we observe a variation of the density number $\delta\rho$ given by $\delta\rho = K\,E^2$. Thus the symmetry breaking at the interface relates $\rho(z)$ to a property of the bulk phase which is non linear in terms of the applied field. The special behavior of $\rho(z)$ can be understood like this [25]. The molecule 1 may "learn " the existence of the surface by three different ways. First of all the molecule 1 is in a uniform solvent, then the contribution to Eq. 11 is simply the classical image potential given by Eq. 10. Secondly a molecule 2 in the neighborhhood of 1 creates an image potential which acts on 1. This part can be interpreted in term of local balance of force, it introduces χ . The last contribution which gives rise to $d\varepsilon_1/d\rho$ represents a very intricate coupling between the molecule 1 and two molecules of its the neighborhhood. This effect is beyond a local description of the interface. In order to get $\rho(z)$ we must take into account an infinite number of bridge diagramms.

A similar interpretation concerning the orientational part of the profile part can be done. In order to illustrate this point we consider axially symmetric molecules "i" for which the polarizability tensor $\alpha(i)$ is given by

$$\alpha(i) = \alpha_0 U + \gamma \hat{\mu}_i \cdot \hat{\mu}_i \qquad (12)$$

where α_0 is the mean or scalar polarizability and γ measures the anisotropy of polarizability. Now the interface becomes optically birefringent. At each poin z, the refractive index $n_z(z)$ in the direction normal to the interface becomes different from $n_x(z)$ which corresponds to a direction parallel to the interface. In very general conditions we can get [27]

$$n_z(z) - n_x(z) = (4\pi / 3)\,((n_0^2+2)^2/6n_0)\,\gamma\,\rho_2(z) \qquad (13)$$

where n_0 is the bulk refractive index and $\rho_2(z)$ is defined in Eq. 7. If we assume that the orientational structure of the liquid is primarily determined by the existence of a permanent dipole , as a first approximation we can calculate $\rho_2(z)$ in the case of a polar/non-polarizable fluid. Then Eq. (13) becomes

$$n_z(z) - n_x(z) = (1/2\beta)\,[(\varepsilon_1 - \varepsilon_2)/\varepsilon_1(\varepsilon_1 + \varepsilon_2)]\,K_a / z^3 \qquad (14)$$

where we may rigorously identify K_a with the Kerr constant in bulk phase [28]. Hence the image potential creates at the interface a " natural " Kerr effect. By natural we mean that the birefringence appears without any external field. It is the image potential which drives this effect. As the electrostriction, the birefringence in homogeneous systems is a quadratic function of the applied field. Note that the quantity $n_z(z) - n_x(z)$ appears in the ellipsometry measurements [29].

The exact asymptotic expression of the profile can be also used for checking the approximate closures [30]. It is interesting to note that the HNC approximation which is very efficient in bulk phase, particularly in the case of polar fluids [31], gives very poor results at the interface when it is used as a closure of the so-called Henderson-Abraham-Barker equation [18].

Beyond the case of the neutral wall, the structure of polar fluids in presence of an electric field has been investigated by the MSA and some simple versions of the HNC [32-35]. In that case we have no Monte-Carlo simulation for testing the accuracy of these closures. A comparison Monte-Carlo / integral equations has been done in the case of a polar fluid in presence of an adsorbing wall [36].

4. The potential of mean force between two molecules adsorbed against a wall

For describing the structure of the solvent near an electrode, the classical models of the electrochemical interface introduce the concept of dielectric constant in a monolayer or effective dielectric constant [7]. This constant appears as a fitting parameter which has a value smaller than the bulk permittivity (typically a value of about 6 is well accepted). Here, we will discuss the introduction of an effective dielectric constant in the light of exact results in statistical mechanics. The discussion will be restricted to the case of polar/non-polarizable molecules.

The total pair correlation function h(1,2) can be written as [20] $h(1,2) = \exp -\beta w(1,2) - 1$ where w(1,2) is the potential of mean force between the molecules 1 and 2. w(1,2) contains the direct pair potential V(1,2) between 1 and 2 plus an indirect potential which represents the mean effect of all the other molecules. In general w(1,2) is a very complicated function. However when the distance r_{12} is very large we may expect that w(1,2) takes a simple form. We first consider the case of homogeneous liquids.

When r_{12} is very large V(1,2) is dominated by the dipolar interaction $V_{dd}(1,2)$. The first intuitive idea consists in assuming that the effect of the surrounding produces a screening characterized by the dielectric permittivity ε of the medium. A first approximation of w(1,2) is certainly given by $V_{dd}(1,2)/\varepsilon$. In a second step, around each molecule, we introduce an excluded volume related to the molecular hard-core. Assuming that the dielectric continuum exists only outside this volume we can still calculate w(1,2) from electrostatics. This new approximation of w(1,2) is $[3\varepsilon/(\varepsilon+2)]^2 V_{dd}(1,2)/\varepsilon$. This result is not totally exact since a uniform dielectric constant outside the core has been assumed. A new correction taking into account the particular structure in the neighborhood of 1 and 2 must be considered. It introduces the so-called g-Kirkwood factor. Finally, from statistical mechanics it can be shown that the exact result is given by [10,37]

$$w(1,2) = [3\varepsilon g/(\varepsilon+2)]^2 V_{dd}(1,2)/\varepsilon = [(\varepsilon-1)/3y]^2 V_{dd}(1,2)/\varepsilon \tag{15}$$

where $y = 4\pi\rho\beta\mu^2/9$. In addition, in this case an exact expression for the dielectric constant is $[(\varepsilon-1)(\varepsilon+2)/3\varepsilon] = 3yg$ [38]. Eq. 15 which is exact if r_{12} is very large can be reinterpreted in two different ways. We can say that we have two point-dipoles immersed in a solvent of effective dielectric constant $\varepsilon/[(\varepsilon-1)/3y]^2$. However, since $V_{dd}(1,2)$ is proportional to μ^2 we can also interpret Eq. 15 as the pair potential between two point-dipole of effective dipole moment $\mu[(\varepsilon-1)/3y]$ immersed in a solvent of permittivity ε. The long-range behavior $h_L(1,2)$ of h(1,2) is simply $h_L(1,2) = -\beta\, w(1,2)$.

An analogous result has been found in the case of inhomogeneous liquids [39,40]. Before discussing the exact expression of $h_L(1,2)$, we may try to understand the result from a physical point of view. We first consider a molecular liquid in front of a pure hard wall (its dielectric permittivity ε_2 is one) as shown in Fig. 3a). Since we are interested in $h_L(1,2)$, it is natural to separate f(1,2) a long-range part $f_L(1,2)$ determined by $-\beta V_{dd}(1,2)$. The remaining part $f_s(1,2) = f(1,2) - f_L(1,2)$ is short ranged.

In order to calculate $h_L(1,2)$ we consider the first graph given in Fig. 1. It gives the zero order density expression of $h_L(1,2) = -\beta V_{dd}(1,2)$. With the second graph given in Fig. 1 we consider a first effect of the liquid via the molecule 3. Now, we must consider f(1,3), f(3,2) and the profile $\rho(3)$ which is the weight of the vertex 3 (see Fig. 3a)). A graph in which all the molecules are connected by short-range bonds can not contribute to $h_L(1,2)$.

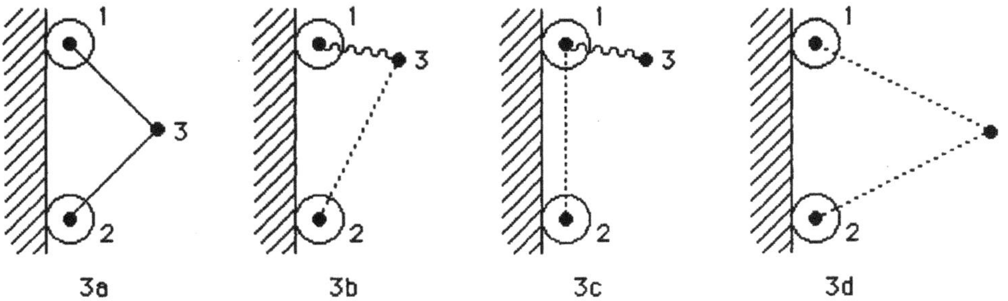

Figure 3. Various steps in the calculation of $h_L(1,2)$. (-----)represents a f(i,j) bond, (- - - - -) is the long range part $f_L(i,j)$ of f(i,j), the wavy line is $f_S(i,j)$. The hatched part represents the hard-wall.

In Fig. 3b, between 1 and 3 we have $f_s(1,3)$ while $f_L(3,2)$ relates 3 and 2. In this case 3 must be closed from 1 while the distance r_{32} is not very different from r_{12}. Accordingly, we can calculate this graph by moving 3 independently of 2. The molecule 3 explores the surrounding of 1 as indicated in Fig. 3c. The net result is still a dipolar interaction between 1 and 2 but the dipole moment of 1 is changed due to the presence of 3. In Fig. 3d, we only consider the long range parts $f_L(1,3)$ and $f_L(3,2)$. When 3 is far from the interfacial region each $f_L(i,j)$ is small but 3 may cross the overall half space. As a consequence of the slow decreasing of the dipolar interaction the weakness of $f_L(i,j)$ is compensated by the possibility of exploring a very large region. The net result of this graph is first proportional to $-\beta V_{dd}(1,2)$ but the restriction imposed to 3 of staying in the half space also introduces a new effect which is proportional to the classical image potential. When 3 remains in the interfacial region the weakness of the dipolar interaction is not compensated by a large integration and the net result is short ranged compared to the dipolar interaction.

Thus, two effects appear in the calculation of $h_L(1,2)$. One comes from the region far from the interface, it introduces some bulk properties and can be related to classical electrostatics. The second one takes into account the specific structure around the molecules 1 and 2. It changes μ_i into an effective dipole moment **M**(i). This intuitive picture can be justified by using a rigorous diagrammatic expansion taking into account an infinite sum of graphs. The exact result is

$$h_L(1,2) = -\beta\,[\mathbf{M}(1).\nabla_1]\,[\mathbf{M}(2).\nabla_2]\left[\,1/\varepsilon_1 r_{12} + (\varepsilon_1-\varepsilon_2)/\varepsilon_1(\varepsilon_1+\varepsilon_2)\,r_{12}{}^*\right] \qquad (16)$$

where the gradient operator ∇_i acts on the spatial coordinates of i, $r_{12}{}^*$ is obtained from r_{12} by mirroring the position of 1 relative to the plane z=0 and ε_2 is the dielectric constant of the wall. Hence, $h_L(1,2)$ only contains the dielectric constants ε_1 and ε_2 relative to homogeneous regions. In the derivation of (16) the introduction of ε_1 has been done rigorously at a microscopic level, ε_1 appears as a sum of graphs relative to the bulk phase. **M**(i) is the effective dipole moment associated to the molecule i. If **M**(i) is replaced by the true dipole moment μ_I of the molecule i then $h_L(1,2)$ is simply the pair potential between two point dipoles immersed in a dielectric medium of uniform dielectric permittivity ε_1 in front of another uniform dielectric of permittivity ε_2. Thus, in the calculation of the potential of mean force between two molecules adsorbed onto a dielectric wall we must essentially calculate the effective dipole moment **M**(i) all the other quantities are explicitly known.

At the present time, very few things are known about **M**(i) [39,40]. The exact expression of **M**(i) at a microscopic level has been established. **M**(i) is related to μ_I by a tensor **g**(i) which generalizes the g-Kirkwood factor in inhomogeneous situations. **g**(i) is determined by the profile and the short-range part of the inhomogeneous correlation function h(1,2). Finally, the relation between the components of **g**(i) and those of **M**(i) is non-local. This means that these components are related via an integral equation. The introduction of an effective dielectric constant for characterizing the monolayer in contact with the wall is thus a model of **M**(i) which gives approximate answers to these relations. To illustrate the situation, an oversimplified model has been discussed in [40]. To our opinion it is extremely dangerous to use such a local dielectric constant. For example, in general conditions we can not replace systematically ε_1 by this effective dielectric constant. In order to go beyond a formal discussion, more works in this important field will be very useful.

5. The potential of mean force for one ion in a polar solvent near a wall

In this section we calculate the free energy w(z) that we must spend for bringing an ion from the bulk solvent to the interface. The solvent is formed of dipolar hard-sphere of diameter σ having permanent dipole at their center and the ion is a hard sphere of diameter σ_i bearing a charge Z.

The potential of mean force w(z) can be separated in two parts $w_o(z)$ and $w_{el}(z)$. The first contribution $w_o(z)$ comes solely from steric effects. As a first approximation it can be calculated by using the PY approximation. Then $w_o(z) = -(1/\beta)\ln g_o(z)$ where $g_o(z)$ is the profile for hard sphere of diameter σ_i immersed in a hard sphere solvent of diameter σ when the number density of the solute species "i" tends to zero. $w_{el}(z)$ appears when the dipole-dipole and ion-dipole interactions are switched on. As previously, we can first calculate $w_{el}(z)$ by considering the ion immersed in a uniform dielectric constant which fills the half space. From classical electrostatics we get a result similar to the one given in Eq.10

$$w_{el}(z) = [(\varepsilon_1 - \varepsilon_2)/\varepsilon_1(\varepsilon_1 + \varepsilon_2)]\,(Z^2/4z) \qquad (17)$$

In opposite to the case of polar fluids this is the exact result far from the wall [30]. Here there is no correction related to the compressibility or the electrostriction phenomena. The particular structure which exists around an ion, its solvation shell, has no effect on the long range behavior of $w_{el}(z)$. This results because a solvent molecule located in the neighborhood of the ion acts on this ion via an image potential which decreases as $1/z^3$, this effect is negligible compared to one predicted by Eq.17 which varies as the inverse of the distance to the wall.

By using a perturbation theory in terms of renormalized potential $w_{el}(z)$ has been calculated for different values of the ratio (σ_i / σ) [41,42]. Particular results are shown in Fig.4.

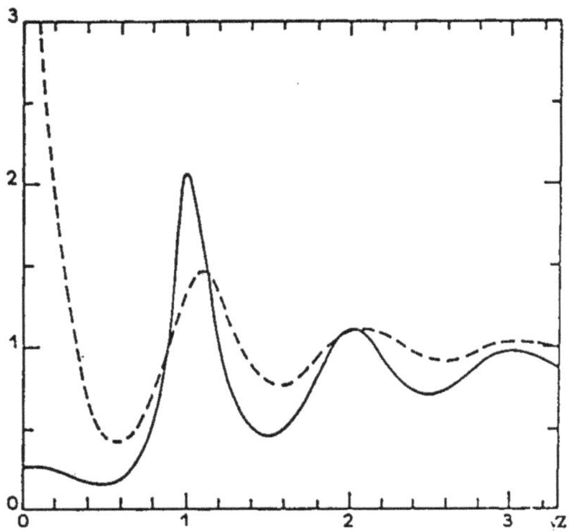

Figure 4. The "profile" $g_i(z)$ (——) compared to the pure hard-sphere profile $g_0(z)$ (– – –). Molecules and ion have the same diameter ($\sigma_i = \sigma$). z is in σ unit.

We have represented $g_i(z) = \exp - \beta w(z)$ in the case $\sigma_i / \sigma = 1$ and $g_o(z)$ which is associated with the steric effects. Far from the wall, $g_i(z)$ decreases as 1/z but the exact behavior given by Eq. 17 is not reproduced. Near the wall we observe large oscillations due to the packing effect. The first peak of $g_i(z)$ is located near σ_i. As a consequence of the electrostatic interactions we observe that the contact value $g_i(z{=}o)$ is very small compare to $g_o(z{=}0)$, roughly speaking the ion is excluded from the layer in contact with the wall. Two effects can contribute to this depletion, one comes from the image potential while the other one is related to the existence of a solvation shell. This point has been discussed carefully in where it appears that, in this model the image effect is dominating [43].

In table 1, we give $w_o(z{=}o)$, $w_{el}(z{=}o)$ and $w(z{=}o)$ as a function of (σ_i / σ) for one ion of given charge. $w(z{=}o)$ is the free energy needed for bringing an ion from the bulk phase to the contact with the wall. For large values of (σ_i / σ), $w(z{=}o)$ is negative as a consequence of steric effects and the ion is adsorbed. Then $g_i(z{=}o) = \exp- \beta w(z{=}o)$ is very large compared to unity. This is in agreement with the classical observation that the ionic adsorbability generally increases with the ionic radius [44]. In opposite for small values of (σ_i / σ) the electrostatic effects are dominating, w(z=o) is positive. The ion prefers to stay in the bulk solvent.

(σ_i/σ)	$w_0(z=o)$	$w_{el}(z=o)$	$w(z=o)$
1/2	-1.1	5.0	3.9
1	-1.5	2.8	1.3
2	-1.9	1.6	- 0.3
3	-2.3	1.0	-1.3

Table 1: Effect of the ratio (σ_i / σ) on $w_0(z=o)$, $w_{el}(z=o)$ and $w(z=o)$. These energies are in k_BT units. The charge of the ion is fixed.

6. Conclusion

In this paper we have discussed some structural aspects concerning the distribution of polar molecules and ions near a neutral wall. The discussion has been presented in order to relate some basic ideas currently used in Electrochemistry with recent results derived by statistical mechanics. We hope that such a physical description may suggest some simple approaches in the case of the real electrochemical interfaces.

References

1. J.P. Hansen and I.R. McDonald, *Theory of simple liquids*, (Academic Press, New York, 1986)

2. C.G. Gray and K.E. Gubbins, *Theory of Molecular Fluids*, (Clarendon Press, Oxford, 1984)

3. J.S. Rowlinson and B. Widom, *Molecular Theory of Capillarity*, (Clarendon Press, Oxford, 1982)

4. S.L. Carnie and G. M. Torrie, *Adv. Chem. Phys.* **56** (1984) 141. S. L. Carnie, *Ber Bunsenges. Phys. Chem.* **91** (1987) 262

5. L. Blum, in *Fluid Interfacial Phenomena*, ed. C.A. Croxton(A. Wiley , New York, 1986)

6. D. Henderson, in *Trends in interfacial Electrochemistry*, ed. A. Fernando Silva, C **179**, 473, (Reidel, Dordrecht. 1986).

7.See for instance : R.M. Reeves *in Modern Aspects of Electrochemistry*, ed. B.E. Conway and J.O'M. Bockris **Vol. 9** (Plenum Press New York 1974) and M.A. Habib *in Modern Aspects of Electrochemistry*, ed. B.E. Conway and J.O'M. Bockris **Vol. 12** (Plenum Press New York, 1977).

8. G. Stell, in *The Equilibrium Theory of Classical Fluids*, ed. H.L Frisch and J.L. Lebowitz, (Bemjamin, New York, 1964).

9.H.C. Andersen and D. Chandler, *J.Chem. Phys.*, **57** (1972) 1918.

10. G. Nienhuis and J.M. Deutch, *J. Chem. Phys.*, **55** (1971) 4213.

11. G. Nienhuis and J.M. Deutch, *J. Chem. Phys.*, **56** (1972) 235.

12. G. Nienhuis and J.M. Deutch, *J. Chem. Phys.*, **56** (1972) 1819.

13. L. Blum and A.J. Toruella, *J. Chem. Phys.*, **56** (1972) 303.

14. L. Blum, *J. Chem. Phys.*, **57** (1972) 1862.

15. L. Blum, *J. Chem. Phys.*, **58** (1973) 3295.

16. G.S. Rushbrooke, *Mol. Phys.*, **37** (1979)761.

17. H. Reiss, H.L. Frisch and J.L.Lebowitz, *J. Chem. Phys.*, **31** (1959) 369.

18. D. Henderson, F.F. Abraham and J.A. Barker, *Mol. Phys.*, **31** (1976) 1291.

19. L. Blum and D. Henderson, *J. Chem. Phys.*, **74** (1981) 1902.

20. T.L. Hill, *Statistical Mechanics*, (McGraw-Hill, New York, 1956)

21. D. Levesque and J.J. Weis, *J. Stat. Phys.*, **40** (1985) 29.

22. J.P.Badiali, *Mol. Phys.*, **55** (1985) 939.

23. M.F. Golovko, S.N. Blotsky and O.A. Pizio, Electrochim. Acta, **34** (1989) 63 and also S.N. Blotsky, M.F. Golovko, O. A. Pizio and E. Spohr, Preprints of the Institute of Theoretical Physics , ITP-89- 20E , Kiev, 1989

24. J.P. Badiali, *J. Chem. Phys.*, **90** (1989) 4401.

25. Q. Zhang, J.P. Badiali and W.H. Su, *J. Chem. Phys.*, **92** (1989) 4609.

26. L.D. Landau and E.M. Lifshitz, *Electrodynamics of continuum media*, (Pergamon, London ,1960)

27. Q. Zhang and J.P. Badiali, (to be published)

28. see for instance S. Kielich, in *Dielectric and Related Molecular Processes*, Specialist Periodical Reports Vol. 1 (The Chemical Society, Washington D.C. 1972)

29. D. Beaglehole, in *Fluid Interfacial Phenomena*, ed. C.A. Croxton, (A. Wiley , New York, 1986)

30. Q. Zhang, J.P. Badiali and M. L. Rosinberg, *J. Mol. Liquids*, (in press)

31. P.H. Fries and G.N. Patey, *J. Chem. Phys.*, **92** (1985) 429.

32. D.J. Isbister and B.C Freasier, *J. Stat. Phys.*, **20** (1979) 331.

33. J.M. Eggebrecht, D.J. Isbister, and J.C. Rasaiah, *J. Chem. Phys.*,**73** (1980) 3980.

34. J.C. Rasaiah, D.J. Isbister and J.M.Eggebrecht, *J. Chem. Phys.*, **75** (1981) 5497.

35. J.C. Rasaiah, D.J. Isbister and G. Stell, *Chem. Phys. Lett.*, **79** (1981) 189.

36. V. Rusier, M.L. Rosinberg, J.P. Badiali, D. Levesque and J.J. Weis., *J. Chem. Phys.*, **87** (1987) 5012.

37. J.S. Hoye and G. Stell, *J. Chem. Phys.*, **61** (1974) 4213.

38. J.M. Deutch, in *Ann. Rev. Phys. Chem.*, **24** (1973) 301.

39. J.P. Badiali, J. Chem. Phys., **89** (1988)2397.

40. J.P. Badiali and F.Forstmann, *Chem. Phys.*, **141**(1990) 63.

41. J.P. Badiali, M.L. Rosinberg and V. Russier, *Mol. Phys.*, **56** (1985)105 .

42. V. Russier, J.P. Badiali and M.L. Rosinberg, *J. Electroanal. Chem.*, **220** (1987) 213 .

43. D. Wei, M.L. Rosinberg, A. Perera and G. Patey, *J. Chem. Phys.*, **89** (1988) 4994.

44. J.O'M. Bockris, M.A. V. Devanathanand K. Muller, *Proc. R. Soc. London Ser A*, **274** (1963) 55.

METAL ELECTRONS IN THE DOUBLE LAYER THEORY

A.A.KORNYSHEV
The A.N.Frumkin Institute of Electrochemistry
of the Academy of Sciences of the USSR
Leninskii prosp.31, Moscow 117071, USSR

(Extended Abstract)

Present status of the theory of the response of the metal/electrolyte interface to charging and its manifestation in the electric and optical properties of the interface are discussed *along the lines of the previous review article*[1]. As in Ref.1, the emphasis is given to the density functional theory and simulation of the electronic subsystem of the metal in interaction with the bounding electrolyte.

Now, the response of the free electrons of smooth homogeneous metal surfaces seems more or less clear (up to the "subtleties" due to single crystal orientation). It is thus timely to return to the response of the solution side[2]. Advances in this field[3] are limited to severe idealization of the molecular pattern of the solution. Therefore, the efforts to extract an information about the response of the solution side by subtracting the metal contribution from the compact layer capacitance data as done by Badiali and Amokrane (see their paper in this issue), are most valuable. They show, in particular, that the developed picture of interfacial relaxation[1], i.e. the variation of the distance of closest approach of solvent molecules to the electrode self-consistently with the deformation of the surface electronic profile) does relax many of the requirements on the expected reorientational response of the molecules neighboring the electrode. Indeed, what is left for reorientations can be rationalized by means of rather elementary models.

Of new topics (not contained in Ref.1) we dwell on a jellium theory of optical second harmonic generation (SHG) from the metal electrolyte interface [4]. This theory is limited to smooth, laterally continuous interface. Its results enables us to explore what is the nonlinear optical response of such a surface and how does it vary with the charge of the electrode. Presently the main interest to experimental signals of SHG at the metal electrolyte interface is due to "asimuthal flowers" indicative of ordered surface structures of certain symmetry; the ornament changes drastically with the polarization of the electrode which is an indication of surface reconstructions. These "hits" of SHG in electrochemistry are beyond the jellium consideration. Thus, our results may be instructive for understanding the *background signal* due to nonlinear polarizability of the metallic electrons at the "homogeneous" interface.

References

1. A.A.Kornyshev, *Electrochim.Acta* **34** (1989) 1829.
2. A.A.Kornyshev, in *The Chemical Physics of Solvation*, ed. R.R.Dogonadze, E.Kalman, A.A.Kornyshev, and J.Ulstrup, Part C (Elsevier, Amsterdam, 1988), p.335.
3. W.Schmickler and D.Henderson, *Prog.Surf.Sci.* **22** (1986) 323.
4. P.G.Dzhavakhidze, A.A.Kornyshev, A.Liebsch and M.I.Urbakh, in the *Proceedings of the Interdisciplinary Conference on Electrified Interfaces*, ed. D.H.Henderson and O.Melroy, *Electrochim.Acta* (1991) in Press; *Phys.Rev.* **B** (1991) in Press.
5. L.J.Simpson, Yeke Nang and T.E.Furtak, *ibid.*

Modeling of the metal/water interface

Rolando Guidelli and Giovanni Aloisi

Dipartimento di Chimica,Universita' di Firenze,Via G.Capponi,9 - 50121Firenze - Italy

Abstract

The metal/water electrified interface, both in the absence and in the presence of a neutral solute, is modeled on the basis of a three dimensional lattice of TIP4P water molecules in the pair correlation approximation of Kirkwood's variation method. Comparison with experiment is examined.

Electrified interfaces between several *sp* metals and dilute solutions of nonspecifically adsorbed 1-1 valent electrolytes exhibit a number of common features which can be summarized as follows:

1) The differential capacity C_i corrected for the diffuse layer on the basis of the Gouy-Chapman (GC) theory is independent of the electrolyte concentration; it either shows a maximum at a positive charge (Hg[1], Ag[2]) or increases towards positive charges in the proximity of zero charge (Ga[3], Cd[4], Bi[4,5], Pb[6], In[7,8], Tl[8]), when the metal is electro-oxidized at positive charges.

2) The entropy of formation of the interface (namely the difference in the entropy of the components when they are present at the interface and when they are present in the bulk of the respective phases), as corrected for the diffuse layer on the basis of the GC theory, is independent of the electrolyte concentration and shows a maximum at a negative charge (Hg[9], Ag[10], Au[11]).

We may further add that several pieces of experimental evidence combined with a few moderate extrathermodynamic assumptions strongly suggest that:

3) The potential difference $\phi^{M\text{-}S}$ across the metal/water interface at zero charge is negative[12].

The first feature, pointed out first for the mercury/water interface in 1954 by Grahame[1], had far reaching consequences in the understanding of electrified interfaces. Thus, on the one hand it proved that the GC theory, in spite of the rough assumptions underlying it (modeling of an electrolytic solution as a system of point charges embedded in a dielectric continuum), was quantitatively accurate. On the other hand, the successful applications of the GC theory opened the discussion on the physical meaning of the resulting GC-corrected quantities. With a heuristic approach it can be stated that, whenever the GC correction yields a pseudo-thermodynamic quantity which is independent of the electrolyte concentration, then it practically eliminates from this quantity the contribution due to ions, so that the resulting GC-corrected quantity can be ascribed to the reorganization which the pure solvent and the metal undergo when they are brought into contact to form the interface[13]. In the early literature this reorganization was confined to the monomolecular layer of solvent molecules adjacent to the electrode surface, called the *inner layer*[14].

Speaking in more general terms, the above three experimental features may be explained by water reorganization at the interface, by metal reorganization at the interface, essentially in the form of electron spillover, or by both these causes. In the early literature electron spillover was completely disregarded. However, during the last decade a new school of thought has taken up an opposite position[15-19], by trying to explain the features of C_i almost exclusively on the basis of electron spillover. We will examine the role of the metal later on. For the moment we will focus our attention on the behaviour of a fluid of water molecules in contact with a charged wall and on how this behaviour may affect the above interfacial properties.

Modeling of Water Reorganization at the Interface

The problem of water reorganization at an electrified interface can be tackled by a Monte Carlo (MC) simulation of a system of soft water spheres enclosed between two soft parallel charged walls, at a distance of 30 Å from each other[20]. Water-water interactions are expressed by the TIP4P intermolecular potential[21]. This consists of a Lennard-Jones 12-6 spherical potential plus three partial charges which are incorporated in the sphere to reproduce H-bonding. Two equal positive charges +q are located in a peripheral position whereas a negative charge -2q is close to the center of the sphere, albeit slightly shifted towards the two positive charges. This distribution of partial charges was chosen in such a way as to account for as many thermodynamic and structural properties of bulk water as possible[21].

The potential-distance profile between the two walls is shown in Fig. 1 for different values of the charge density σ on the left wall. It should be noted that at zero charge the potential difference between any of the two walls and the bulk of the liquid is negative. This negative potential difference, which correctly accounts for experimental point (3), stems in a natural way from the asymmetrical distribution of the partial charges inside the water molecule.

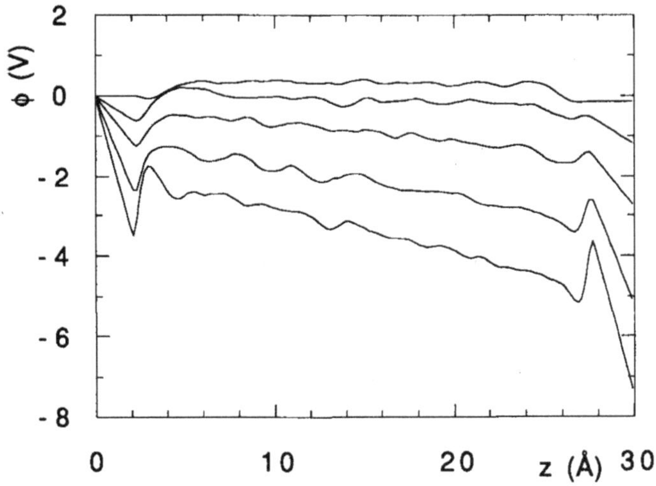

Fig. 1. Potential vs. the distance (z) from the left-hand charged wall, for different surface charges. The surface charge is equal to 0, 2.5, 5, 10 and 15 μC cm^{-2}, starting from the upper curve.

At zero charge the adsorbed water molecules tend to assume compromise orientations which favour H-bond formation parallel to the electrode surface, due to the impossibility of H-bonding in the direction towards the metal, while still retaining a sufficiently high H-bonding capability in the direction towards the bulk liquid. On the average, these compromise orientations have the more peripheral partial charges (namely the positive ones) slightly turned towards the bulk liquid. At charges different from zero the electric potential shows a rapid change within the *vacuum* region enclosed between the charged wall and the equilibrium distance of closest approach of the centers of the water molecules. Here, the potential gradient is equal to the opposite of the electric field *in vacuo* between the two charged walls, i.e. $4\pi\sigma$. At larger distances from the wall we first have an abrupt decrease in the absolute value of the electric potential, due to the orientation polarization of the water molecules in the immediate vicinity of the electrode, and then small oscillations which are mainly due to the noise in the numerical calculations involved in the MC simulation. Apart from this noise, we may observe that in the intermediate region between the two walls the electric potential varies almost linearly with distance, exhibiting a slope which is proportional to the charge density σ. Such a slope is clearly equal to the electric field $4\pi\sigma/\varepsilon$, and hence provides a value for the bulk dielectric constant ε which is consistent with the choice of the TIP4P water-water potential. The potential-distance profile also shows that the wall-water correlation tends to vanish at distances greater than one, or at most two, molecular diameters; beyond this distance the wall practically "sees" the water molecules as a dielectric continuum. If the water behaved as a dielectric continuum throughout the whole region between the walls, as predicted by the GC theory at infinite dilution, then the potential profile would show the slope $4\pi\sigma/\varepsilon$ starting from the wall surface, as shown by curve **a** in Fig. 2. It is therefore evident that the vertical separation between curve **a** and curve **b** in Fig. 2 measures the contribution ϕ_w to the wall-liquid potential difference from water reorganization.

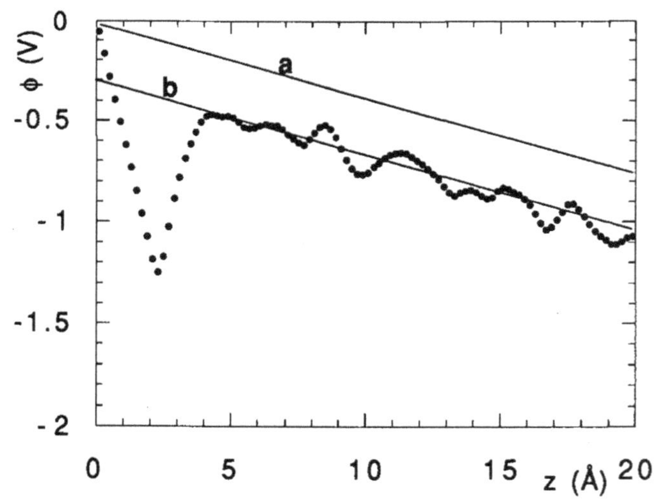

Fig. 2. Potential vs. distance for 5 μC cm^{-2}: MC results (dots), GC assumption (a) and extrapolation of the bulk potential profile to the surface (b).

If we calculate ϕ_W as a function of the charge density σ on the wall and we differentiate σ with respect to ϕ_W, we obtain the differential capacity C_i in Fig. 3, which shows a maximum at a small positive charge, in agreement with experimental point (1). The curve also shows an increase in capacity at extreme positive and negative charges, once again in agreement with the experimental behaviour of the mercury/water interface[1]. According to the simulation, the latter increase in capacity is due to a concomitant increase in the number density of the water molecules in the immediate vicinity of the charged wall.

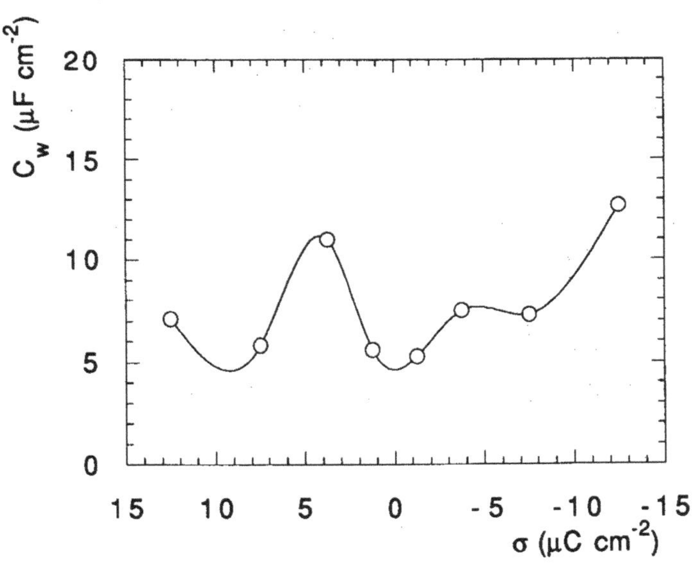

Fig. 3. Differential capacity vs. charge from MC simulation.

To overcome the problems related to the noise in the MC simulation as well as computational costs, without losing the main advantageous features of the model, the treatment can be simplified by adopting a three-dimensional lattice model of water molecules interacting according to the TIP4P intermolecular potential[22,23]. The statistical-mechanical treatment of this regular lattice is carried out by accounting for local order on the basis of a first-order approximation, namely the pair correlation approximation of Kirkwood's variation method[24,25]. The notable decrease in computational time attained by the use of a three-dimensional lattice model of water molecules is due to the freezing of the translational degrees of freedom as well as to the ready singling out of the *most probable* orientations, namely the orientations which make by far the major contribution to the partition function of the whole system. These are indeed the orientations which may form H-bonds towards any two of the first nearest neighbouring sites. After having ascertained that the lattice geometry does not affect the qualitative conclusions drawn from the model, provided the density of the lattice sites reflects the experimental density of bulk water, we will report here the results obtained with a body-centered cubic lattice. The water orientations considered are shown in Fig. 4. Twelve orientations have the hydrogens directed towards the ends of any of the diagonals of the 6 faces of the elementary cubic cell (**a**). The other 12 orientations have the hydrogens approximately directed towards the ends of any of the 12 edges of the cubic cell (**b**). The statistical mechanical calculation was carried out by accounting explicitly for the interactions of any water orientation with those occupying the first 14 neighbouring sites, and by accounting for longer range interactions self consistently on the basis of the mean field approximation.

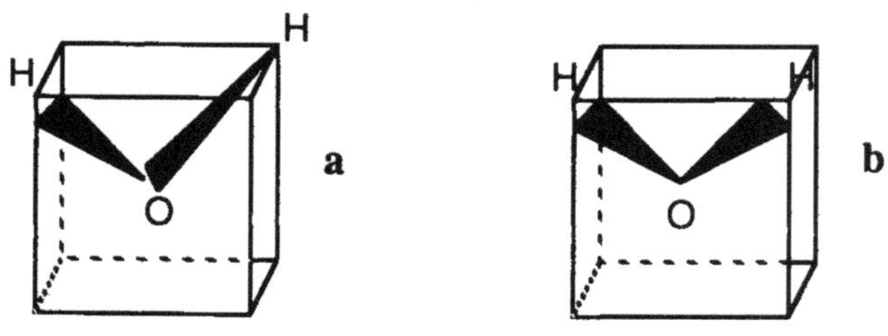

Fig. 4. Water orientations considered in the modeling of the interface.

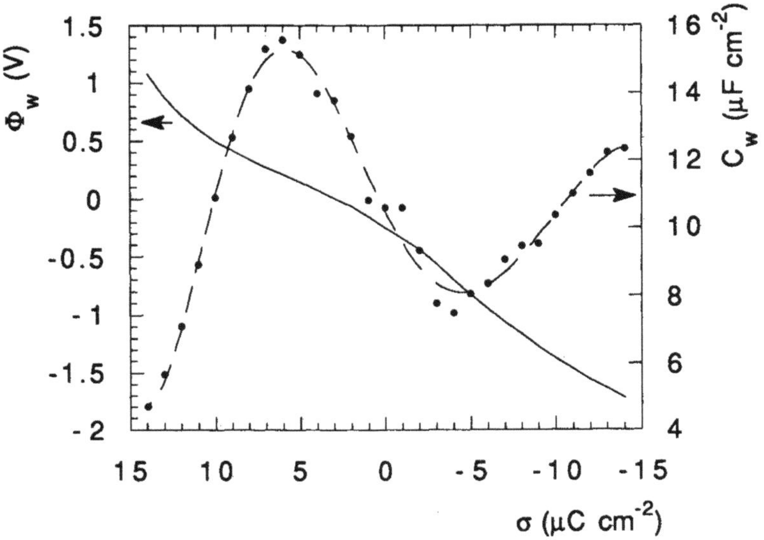

Fig. 5. Potential surface excess ϕ_W and differential capacity C_W vs. surface charge σ

Figure 5 shows plots of ϕ_W and of $C_W = \partial\sigma/\partial\phi_W$ vs. σ. It can be seen that ϕ_W is negative at zero charge, in agreement with experimental point (3). The curve of the differential capacity $C_W = \partial\sigma/\partial\phi_W$ against σ shows a maximum value at a positive charge $\sigma \approx +5\ \mu C\ cm^{-2}$, in agreement with experimental point (1). Finally, Fig. 6 shows the calculated curve of the entropy surface excess ΔS_W against σ, which exhibits a maximum at a negative charge $\sigma \approx -14\ \mu C\ cm^{-2}$, in accordance with experimental point (2). It is therefore evident that the present model accounts satisfactorily for all salient features of experimental behaviour without the introduction of adjustable parameters.

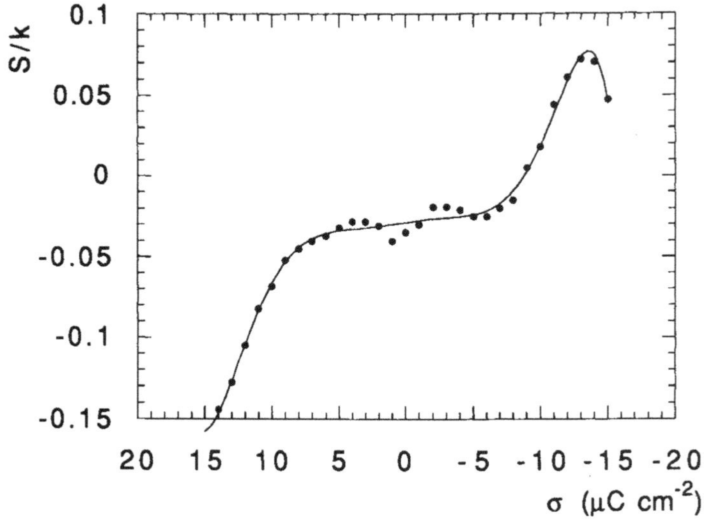

Fig. 6. Entropy surface excess vs. surface charge σ

Consideration of Electron Spillover

Let us now examine how the consideration of electron spillover can modify the above picture. If electron spillover is accounted for on the basis of the *jellium* model, then it is possible to show that the integral capacity K_j of the inner layer, with the exclusion of the contribution from water polarization, takes the simple form of the capacity across the plates of a parallel-plate capacitor[16]:

$$K_j = 1/[4\pi(d-x'_\sigma)] \qquad (1)$$

Here x=d is the equilibrium distance of closest approach of water molecules to the jellium edge, whereas $x=x'_\sigma$ is the *image plane*, namely the plane where the image of any charge placed at a sufficient distance from the jellium is located. Now, the image plane stands in front of the jellium edge and is predicted to move outwards as σ becomes progressively more negative. Hence, if the position x=d of closest approach of the water molecules to the jellium edge is kept constant, then the gradual shift outwards of the image plane as σ becomes progressively more negative, causes the two plates of our ideal capacitor to approach each other with a consequent increase in the integral capacity K_j. Since the experimental value of the integral capacity in the proximity of the zero charge increases in the opposite direction, namely towards positive charges (see experimental point 1), this behaviour can only be explained on the basis of water polarization[16].

However, the predicted behaviour can be notably modified if the x=d plane is allowed to move with varying the charge density σ on the metal[16,19]. Indeed, the equilibrium position of closest approach of water molecules to the jellium edge can be considered to result from a balance of opposing pressures, namely:

(I) a Van der Waals pressure attracting the water molecules towards the metal;

(II) a strongly repulsive pressure resulting from the overlapping of the metal electronic tail with the electronic closed shell of the water molecules;

(III) an electrostatic pressure resulting from the attraction of the diffuse-layer ions towards the metal charge σ, which pressure can be shown on general grounds to be equal to $2\pi\sigma^2/\varepsilon$, where ε is a dielectric constant[26].

Pressure (II) tends to produce a progressive repulsion of the water molecules, and hence of the x=d plane, as the image plane $x=x'_\sigma$ moves outwards, thus causing the x=d plane to move roughly parallel to the image plane. On the other hand, the electrostatic pressure (III), provided its magnitude is comparable with that of the other pressures, tends to push the x=d plane progressively towards the jellium edge as the absolute value of the charge density σ increases. As a result of these two pressures, the x=d plane is expected to attain a maximum distance from the jellium edge at a negative charge and to approach gradually the jellium as we depart from this charge (see Fig. 7). Hence, in the proximity of zero charge, the separation between the image plane $x=x'_\sigma$ and the x=d plane will decrease when proceeding towards positive charges, causing the integral capacity K_j to increase in the same direction, in agreement with experimental point (1).

It should be noted, however, that the Van der Waals attractive pressure (I) between metals and water molecules, as estimated on the basis of electrodynamical considerations[19,27], turns out to be relatively high. Hence, in order to attain a physically significant equilibrium distance of closest approach x=d, this attractive pressure must be opposed by a high repulsive pressure (II), which can be generated, for instance, by a potential barrier fixed with respect to the water molecules and repelling the electron tail. Under these conditions the electrostatic pressure (III) due to the diffuse-layer ions will be comparable with the two preceding pressures (I) and (II), giving rise to effect represented in Fig. 7, only if the dielectric constant ε is ascribed a value close to unity, as done by Amokrane et al.[19].

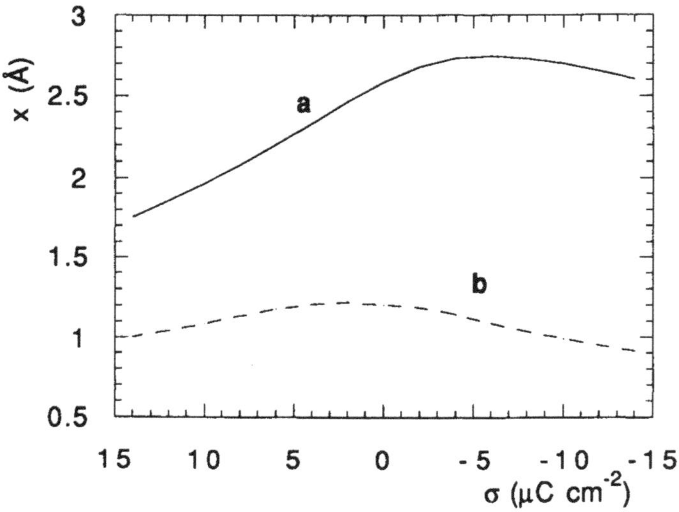

Fig. 7. Positions of the plane of closest approach of the solvent molecules (**a**) and of the image plane (**b**), as a function of surface charge, for $\varepsilon = 1$.

However, the expression $2\pi\sigma^2/\varepsilon$ can also be directly derived from the Gouy-Chapman theory[26]; hence, if we accept the validity of this theory by comparing theoretical predictions with the experimental GC-corrected quantities as obtained in the limit of infinite dilution of the electrolyte, then the ε value is not a matter of choice, since the GC theory yields satisfactory results only provided ε is given the value, $\approx$ 80, of bulk water. With this ε value, the electrostatic pressure (III) is more than one order of magnitude lower than the two opposing pressures (I) and (II), and therefore exerts no appreciable effect upon the location of the x=d plane.

Figure 8 shows plots of $x=x'_\sigma$ and x=d vs. σ as obtained by combining the three-dimensional lattice model of TIP4P water molecules with a simple one-parameter jellium model[28,29], and by determining the equilibrium distance x=d *via* an iterative self-consistent procedure based on a balancing of the three previous pressures, from (I) to (III). In practice, once we set $\varepsilon = 80$, only the two opposite pressures I and II bear a significant contribution, and both planes $x = x'_\sigma$ and $x = d$ move progressively away from the jellium edge with increasing negative charge.

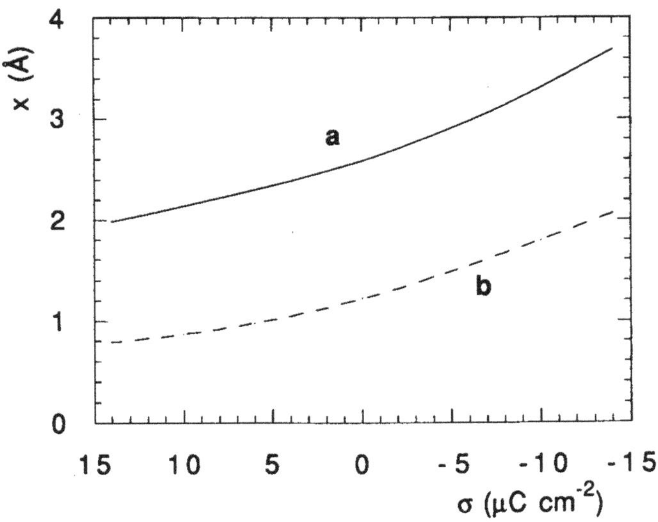

Fig. 8. Positions of the plane of closest approach of the solvent molecules (a) and of the image plane (b), as a function of surface charge, for $\varepsilon = 80$.

In view of Eq.1 the contribution from the (jellium + electron gas) to the σ-dependence of the integral capacity is small, because the separation between the plates of our ideal capacitor varies only slightly with varying σ. The contributions to the overall potential difference $\phi^{M\text{-}S}$ across the interface as obtained from this model are shown in Fig. 9. It can be seen that the potential difference $4\pi\sigma d$ created by σ in the absence of polarization effects does not show a perfect linear dependence upon σ because now d depends upon charge; as opposed to this potential difference, the surface dipole potential differences ϕ_{el} and ϕ_w, due to the polarization of water molecules and of the inhomogeneous electron gas, respectively, become progressively more positive as σ shifts towards negative values. The algebraic sum of these three contributions yields the whole potential difference ϕ^{M-S} across the interface, whose plot against σ shows a minimum slope at positive charge densities; this corresponds to a maximum in the differential capacity $C_i = \partial\sigma/\partial\phi^{M-S}$. It is readily seen that such a maximum is determined primarily by water polarization, which opposes the external electric field $4\pi\sigma$ more effectively just at these positive charge densities.

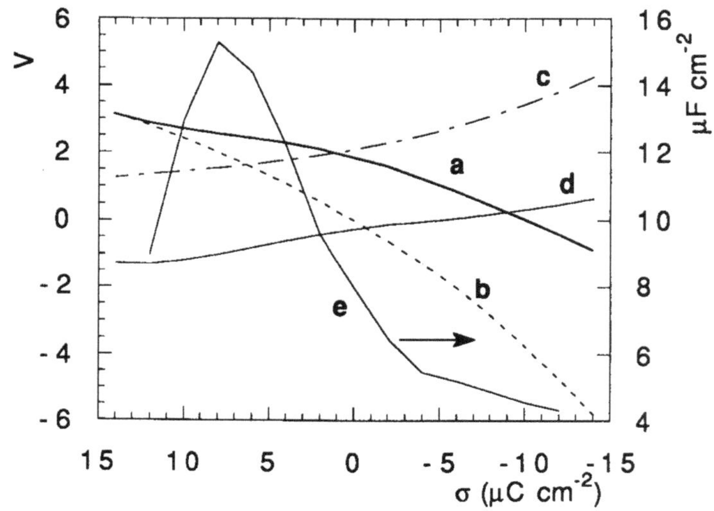

Fig. 9. Contributions to the overall potential difference ϕ^{M-S} (a) across the interface: 'condenser' contribution $4\pi d\sigma$ (b), jellium contribution ϕ_{el} (c), and solvent contribution ϕ_w (d). Curve (e) is the differential capacity $C_i = \partial\sigma/\partial\phi^{M-S}$.

Modeling of the Metal-Water Interface in the Presence of Noninteracting Solute Molecules

Let us now generalize the three-dimensional lattice model of TIP4P water molecules so as to include the presence of solute molecules. The predictions of this generalized model may then be compared with the experimental adsorption behaviour from water on mercury of a broad class of weakly interacting neutral solutes, namely monofunctional aliphatic compounds[30,31]. For simplicity, we will assume that the solute molecules have the same size as the water molecules and that they interact neither between themselves nor with the water molecules. In this way direct solute-solute interactions will be excluded, and only the solute-solute interactions induced by water reorganization at the interface, namely the so called "hydrophobic interactions", will be left. The calculated adsorption behaviour of this hypothetical solute will thus serve to single out the role played by hydrophobic interactions.

Adsorption isotherms were obtained by equilibrating the water and solute molecules in the first lattice plane adjacent to the wall with those in the remaining lattice planes[23]. The solute was assumed to be present in the bulk water phase at such a low concentration as to permit us to disregard its presence in all lattice planes except for that immediately adjacent to the electrode surface plane, where adsorption takes place. With this assumption the solute behaves ideally in the bulk, and hence its chemical potential increases linearly with kT ln c, where c is the bulk solute concentration. Adsorption isotherms at constant charge were obtained by increasing gradually the chemical potential of the solute in the bulk, μ=(constant+ kT ln c), and by estimating the corresponding increase in the solute concentration in the first lattice plane, namely the solute surface concentration.

Figure 10 shows plots of the surface coverage θ vs. the solute chemical potential μ as calculated for this monomeric solute at different charge values. It can be seen that the surface coverage tends to an apparent limiting value less than unity which is lower the more the charge departs from its value of maximum adsorption. Such a behaviour is actually shown by the adsorption isotherms at constant charge of the majority of monofunctional aliphatic compounds[32-37]. According to the model, this behaviour is explained by considering that the removal of a water molecule from the adsorbed state by one solute molecule involves both the breaking of H-bonds and the overcoming of the electric field strength which anchors the water molecules to the adsorbed state. In the proximity of zero charge the external electric field $4\pi\sigma$ is small and the breaking of H-bonds plays a major role. Under these circumstances, as the surface coverage θ increases, the progressive interposition of noninteracting solute molecules between adsorbed water molecules decreases the probability of H-bond formation between the latter molecules. Hence, the residual water molecules will be more easily displaced from the adsorbed state by the adsorbing solute molecules the higher is θ, and θ will increase with log c faster than predicted by the Langmuir isotherm, in which lateral interactions between adsorbed solvent and solute molecules are ignored. As the absolute value of the charge density σ increases, however, water molecules tend to reorient under the electric field in order to minimize their electrostatic potential energy compatibly with the randomizing forces of thermal agitation. Hence, they will be more strongly anchored to the electrode surface, and the slope of the plot of θ vs. ln c at constant charge will decrease accordingly.

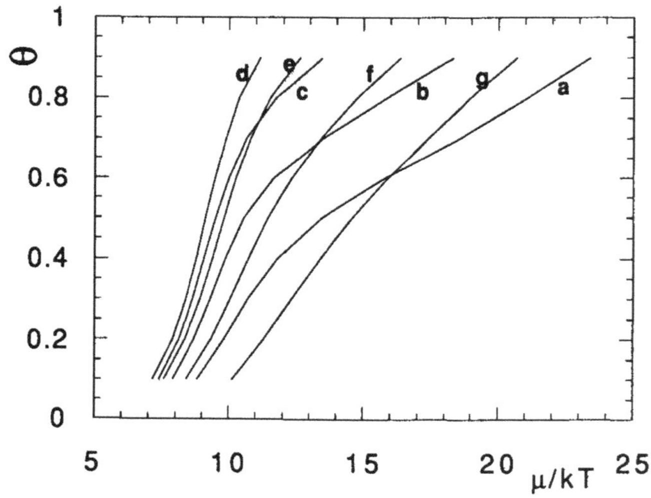

Fig. 10. Adsorption isotherms at constant surface charge. Charge values are: -12 (a), -8 (b), -4 (c), 0 (d), 4 (e), 8(f), 12(g) $\mu C\ cm^{-2}$.

The standard Gibbs energy of adsorption at zero coverage as calculated from this model shows a parabolic dependence upon the charge density σ (see Fig.11). Such a parabolic dependence agrees with the experimental behaviour of practically all aliphatic compounds[30-37].

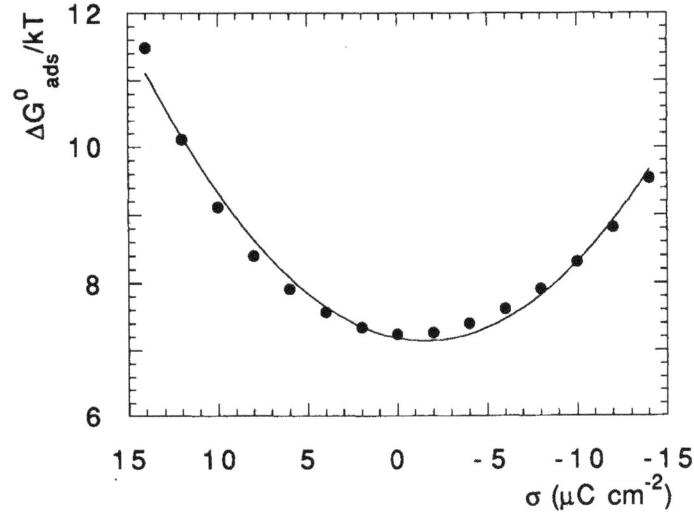

Fig. 11. Standard Gibbs Energy of adsorption of the solute as a function of surface charge (dots). The solid curve is a least-squares fit to a second degree polinomial.

It should also be noted that the minimum in the calculated curve of ΔG°_{ads} vs. σ, which corresponds to a maximum of adsorptivity, lies at a small negative charge of about -1 – -2 μC cm^{-2}. This prediction is particularly significant, because the charge of maximum adsorption for the majority of aliphatic compounds actually takes values ranging from -1 to -3 μC cm^{-2} , independent of the particular nature of the functional group[30,38]. It is therefore useful to point out the rationale behind such a modelistic prediction. Since a nonpolar solute molecule interacts neither with its neighbours nor with the external electric field, its behaviour is induced by water molecules; in particular, a maximum in adsorptivity corresponds to a minimum in the resistance of adsorbed water molecules to displacement. Now, we have seen that at zero charge water molecules have an average orientation with the negative end of the dipole slightly turned towards the metal. Hence an average horizontal orientation of the water molecules will be attained at a somewhat negative charge. It is between this negative charge, where the average electrostatic energy of the adsorbed water molecules equals zero, and the zero charge, where the external electric field $4\pi\sigma$ equals zero, that the electrostatic energy of adsorbed water molecules will attain a minimum value, which will favour their displacement by adsorbing solute molecules.

So far we have disregarded the effect of the nature of the metal on adsorption, although it is well known that the adsorptivity of a given surfactant from water is somewhat different on different *sp* metals even at far negative charges, where anionic specific adsorption and metal surface oxidation can be ruled out. Such a difference in behaviour can be ascribed, at least partially, to a different extent of electron spillover from the metal into the water phase. Electron spillover creates an electric field which tends to oppose the spillover itself and which is therefore constantly directed from the metal towards the solution. This electric field is superimposed on the uniform electric field $4\pi\sigma$ created by the charge density σ on the metal surface but, as distinct from the latter, it becomes vanishingly small at distances greater than 1-2 Å from the metal surface plane. Hence, it will only act upon the first lattice plane of water and solute molecules, or, to be more precise, upon the dipoles of those bonds of adsorbed solvent and solute molecules which are closer to the electrode surface. In particular, it will repel dipoles oriented with the positive end towards the metal, whereas it will attract those with an opposite orientation. This may explain why the adsorptivity of aliphatic alcohols is greater on Hg than on Ga under otherwise identical conditions even at far negative charges, where water chemisorption on Ga can be ruled out[39]. In fact, the molecules of aliphatic alcohols, independent of whether they are adsorbed in a flat or a vertical orientation, come into contact with the metal with their C-H dipoles, whose positive ends (which lie on the hydrogens) are closer to the electrode surface than the corresponding negative ends (which lie on the carbons). Moreover, the extent of electron spillover at a given charge density is more pronounced on Ga than on Hg, because the bulk electron density in Ga is about twice as high as in Hg. Hence, at any given charge density, the electric field due to electron spillover on Ga will be greater than on Hg and will exert a stronger repulsive effect on the adsorbed alcohol molecules, thus decreasing their adsorptivity with respect to mercury. On the contrary thiourea, which is adsorbed in a fixed orientation with the negative end of the C=S group towards the metal, is more strongly adsorbed on Ga than on Hg at far negative charges[40,41], in accordance with the above qualitative predictions.

References

1. D.C. Grahame, J. Am. Chem. Soc. **79**,2093(1957).
2. G. Valette and A. Hamelin, J. Electroanal. Chem. **45**,301(1973).
3. I.A. Bagotskaya, A.M. Morozov and N.B. Grigoryev, Electrochim. Acta **13**,873(1968).
4. B.B. Damaskin, U.V. Pal'm and M.A. Sal've, Elektrokhimiya **12**,232(1976).
5. U.V. Pal'm, M.P. Pyarnoya and M.A. Sal've, Elektrokhimiya **13**,873(1977).
6. K.V. Rybalka and D.I. Leikis, Elektrokhimiya **3**,383(1967).
7. N.B. Grigor'ev, I.A. Gedvillo and N.G. Bardina, Elektrokhimiya **8**,409(1972).
8. N.B. Grigor'ev, V.A. Bulavka and Yu. M. Loshkarev, Elektrokhimiya **11**,1404(1975).
9. J.A. Harrison, J.E.B. Randles and D.J. Schiffrin, J. Electroanal. Chem. **48**,359(1973).
10. A. Popov, O. Velev, T. Vitanov and D. Tonchev, J. Electroanal. Chem. **257**,95(1988).
11. A. Hamelin, L. Stoicoviciu, and F. Silva, J. Electroanal. Chem. **229**,107(1987).
12. For a review see S. Trasatti, *Modern Aspects of Electrochemistry*, Vol.13, edited by B.E. Conway and J.O'M. Bockris (Plenum Press, New York, 1979).
13. G. Aloisi, M.L. Foresti and R. Guidelli, J.Chem. Phys. **89**,1141(1988).
14. For a review, see R. Guidelli, *Trends in Interfacial Electrochemistry*, edited by A.F. Silva (Reidel, Dordrecht, 1986).
15. A.A. Kornyshev and M.A. Vorotyntsev, J. Electroanal. Chem. **167**,1(1984).
16. A.A. Kornyshev, Electrochim. Acta **34**,1829(1989).
17. J.W. Halley, B. Johnson, D. Price and M. Schwalm, Phys. Rev.B **31**,7695(1985).
18. J.W. Halley and D. Price, Phys. Rev.B **35**,9095(1987).
19. S. Amokrane, V. Russier and J.P. Badiali, Surf. Sci. **210**,251(1989).
20. G. Aloisi, M.L. Foresti, R. Guidelli and P. Barnes, J. Chem. Phys. **91**,5592(1989).
21. W.L.Jorgensen, J. Chandrasekhar, J.D. Madura, R.W. Impey and M.L. Klein, J. Chem. Phys. **79**,926(1983).

22. G. Aloisi and R. Guidelli, J. Electroanal. Chem. **260**,259(1989).
23. G. Aloisi and R. Guidelli, J. Chem. Phys., submitted.
24. A.R. Allnatt, Mol. Phys. **14**,289(1968).
25. J.A. Barker, J. Chem. Phys. **44**,4212(1966).
26. D. Henderson, L. Blum and J.L. Lebowitz, J. Electroanal. Chem. 102,**315**(1979).
27. J.F. Annett and P.M. Echenique, Phys. Rev.B **32**,6853(1986).
28. J.R. Smith, Phys.Rev. **181**,522(1969)
29. M. Partenskii and Ya. Smorodinskii, Sov.Phys. Solid State **16**,423(1974)
30. B.B. Damaskin, O.A. Petrii and V.V. Batrakov, *Adsorption of Organic Compounds on Electrodes*, Plenum Press, New York, 1971.
31. B.B. Damaskin and V.E. Kazarinov, *Comprehensive Treatise of Electrochemistry*, Vol.1, edited by J.O'M. Bockris, B.E. Conway and E. Yeager (Plenum, new York, 1980).
32. A. De Battisti, B.A. Abd-El-Nabey and S. Trasatti, J. Chem. Soc. Faraday Trans.1 **72**,2076(1976).
33. H. Nakadomari, D.M. Mohilner and P.R. Mohilner, J. Phys. Chem. **80**,1761(1976).
34. B.B. Damaskin and S.L. Dyatkina, Elektrokhimiya **14**,152(1978).
35. M. Carla', G. Aloisi, M.L. Foresti and R. Guidelli, J. Electroanal. Chem. **197**,123(1986).
36. M.L. Foresti, M.R. Moncelli, G. Aloisi and M. Carla', J. Electroanal. Chem. **255**,267(1988).
37. M. Carla', G. Aloisi, M.R. Moncelli and M.L. Foresti, J. Colloid Interface Sci. **132**,72(1989).
38. H. Raous, D. Schuhmann, E. Tronel-Peyroz and P. Vanel, J. Electronal. Chem. **137**,393(1982).
39. G. Pezzatini, M.R. Moncelli, M. Innocenti and R. Guidelli, J. Electroanal. Chem. **295**,275(1990).
40. I.A. Bagotskaya, S.A. Fateev, N.B. Grigor'ev and N.G. Bardina, Elektrokhimiya **6**,369(1970).
41. G. Pezzatini, M.R. Moncelli and R. Guidelli, J. Electroanal. Chem. in press.

THE CAPACITANCE OF AN IDEALLY POLARIZED ELECTRODE: ROLE OF THE METAL, SOLVENT AND TEMPERATURE EFFECTS

J.P. BADIALI

and

S. AMOKRANE

Structure et réactivité aux interfaces, université P. et M. Curie F(74), 4 place Jussieu 75230, Paris Cedex 05, France.

ABSTRACT

A new analysis of capacitance data for an ideally polarized electrode without ionic adsorption is proposed. The contribution arising from a non-traditional description of the electrode is obtained via a theoretical calculation, leading to a semi-empirical determination of the solvent contribution. The role of the metal and of the solvent and the effect of temperature are discussed.

1. Introduction

In the last decade, important concepts which formed for long years the basis of the traditional description of the capacitance[1] of the metal-electrolyte solution interface have been revisited. On the one hand, the extension to the interface of the statistical mechanical methods (integral equations and computer simulations) used previously for the study of homogeneous ionic solutions cast some doubt on the validity of the classical separation of the solution into inner and diffuse layer[2]. On the second hand, a description of the electrode surface beyond the classical picture - structureless plane with ideal polarizability - has been (re)introduced. Then it followed from all recent models[3-6] sharing a minimum set of *a priori* reasonable ingredients that a non negligeable contribution to the capacitance must be associated to this new picture of the electrode surface. This conclusion is in sharp contrast to the basic assumption made in the traditional approach.

This paper will be concerned only with this second aspect. In the case of an ideally polarized electrode without ionic adsorption, we shall illustrate the new analysis of the capacitance which follows from this improved description of the metal surface. In order to discuss some of the ingredients entering these non-traditional models, we shall first recall some well established elements in the physics of metal surfaces which are relevant in the modelling of the metal at the interface. Then we shall examine some aspects of the interaction of a solvent molecule with a metal surface. From this, we propose a method for describing the coupling between the metal and the liquid phase.

From a formal description of the charge distribution and the polarization accross the interface, a formal splitting of the capacitance is obtained. In this splitting, one part, C_s, reduces to the only contribution considered in the traditional approach. The remaing part, C_m, arises from the aspects which improve upon the classical description of the electrode and its coupling with the solvent. We shall first indicate how C_m can be obtained via a theoretical calculation[7]. Once it has been calculated, one obtains a semi-empirical estimation of C_s as $C_s^{-1} = C_i^{-1} - C_m^{-1}$, where, as usual, C_i is deduced from experiment by correcting the measured capacitance from the Gouy-Chapman contribution. This new method has the advantage of avoiding the difficult task of a complete calculation of the capacitance from a fully microscopic model of the whole interface.

This method and the subsequent analysis of experimental data will be illustrated in the case of various interfaces. For aqueous solutions, we shall discuss the effect of changing the nature of the metal or the temperature[8]. Then we shall discuss some results on the role of the solvent we have obtained in a recent work[9]. Finally, we shall make some concluding remarks on the use of capacitance curves as a probe of the interactions and the structure at the interface.

2. The metal-vacuum interface

From the experimental point of view, a number of technics of the physics of surfaces are available for probing the properties of the metal surface in vacuum. Among these, the surface energy and the work function have been the most widely used for testing the predictions of the theoretical calculations. These calculations followed from the pioneering work[10] of Lang and Kohn (LK) in 1970 who combined the jellium model of the metal and the density functional formalism [11]. It is not the purpose of this work to detail the vast body of theoretical work devoted to metal surfaces (see for e.g. 12 and refs. therein). We shall only consider relatively "light technics" which have been utilized for the study of s-p and noble metal surfaces in models of the interface.

The widely popular jellium model and its various extensions is the prototype of the models of the metal used at the electrochemical interface. In the pure jellium, the semi-infinite lattice is ignored. A uniform positive background with step profile is introduced in order to neutralize the negative charge due to the conduction electrons. This introduces considerable simplifications such as the fact that the problem becomes a one-dimensional one. This is also the case when ionic pseudo-potentials averaged in planes parallel to the interface are used in order to mimic the effect of discrete ions. Another type of extended jellium is obtained when an effective valence is introduced in order to take approximately into account the effect of d electrons in the case of noble metals.

In these models, the main quantity is the electronic density profile n(z) which gives the probability of finding an electron at a distance z from the jellium edge. n(z) can be first obtained by solving a one electron Schrödinger-like equation as in the work of LK. Within the local density approximation for the exchange and correlation energy, the LK scheme gives an essentially exact solution for the density profile in the jellium model. A more practical (and less accurate) method consists in directly minimizing the ground state energy of the electron gas, $E[n(z)]$, written as a functional of $n(z)$, where $n(z)$ is chosen in a given family of trial functions[13].

From the knowledge of $n(z)$ and the associated jellium profile $n\theta(z)$ (= n if $z\leq 0$, 0 otherwize) the potential drop $\Delta\phi = \phi(\infty) - \phi(-\infty)$ can be obtained by solving the Poisson equation. The first step towards the metal-solution interface is to consider that the metal bears a charge σ per unit area, induced by a charged plane located at L far outside the metal, in vacuum. Then $\Delta\phi$ can be written as :

$$\Delta\phi = 4\pi\sigma(L - z_0) \tag{1}$$

where by defintion

$$z_0 = (1/\sigma)\int_{-\infty}^{+\infty} z[n\theta(z) - n(z)]dz \tag{2}$$

To this system we can associate a differential capacitance C by the ususal defintion : $1/C = d\Delta\phi/d\sigma$. From eqs. 1-2, we have:

$$1/C = 4\pi[L - X_0(\sigma)] \tag{3}$$

where

$$X_0(\sigma) = \partial\{\int_{-\infty}^{+\infty} z[n\theta(z) - n(z)]dz\} / \partial\sigma \tag{4}$$

For an ideal metal, the induced charge σ is supposed to be uniformly distributed on a geometrical plane at z=0, so that $X_0(\sigma) = 0$ and the inverse capacitance is simply $1/C= 4\pi L$. For the real metal and near zero charge, X_0 is the position of the image plane as shown by Lang and Kohn[14]. Then the capacitance in eq.3 can also be associated to a plane condenser, provided that the "electrode plane" is located not at z=0 but instead at a positive value $X_0(\sigma)$ as shown by LK.

The fact that $X_0(\sigma)$ is non zero is the immediate consequence of the departure of the actual distribution of the induced charge on the metal surface from its idealization in classical electrostatics. Then to this term can be associated a finite capacitance $C_m = -(4\pi X_0)^{-1}$. Remains its magnitude. All calculations, among them those of Gies and Gerhardt[15], who used at a charged surface the "exact" self-consistent LK scheme, lead to values for C_m of about 10 $\mu F/cm^2$ near the pzc, that is the same order of magnitude than that observed for C_i in the case of Hg. This observation and the expectation that a weak interfacial coupling should not modify this order of magnitude, motivated the introduction in the early eighties of the jellium model for studying the metal-solution interface. In order to go further, we discuss now some aspects of the theoretical description of the coupling at the interface.

3. A model for the metal-solvent molecule interaction

It is generally accepted from conventional electrochemical methods[16] that most solvents used in capacitance studies interact rather weakly with the electrode, in the sense that the molecules do not form a chemical bond with the metal. In the case of water, this observation agrees with adsorption studies on solids in UHV. Especially in the case of monocrystalline silver, the binding energy is estimated as equal at most to the water-water interaction[17]. In the absence of more detailed knowledge, one may conjecture that such a binding should be mostly due to the gas of conduction electrons and a possible interaction due to orbital mixing with the ions in the metal should be unimportant for an extended surface (see the work on small metal clusters[18] for comparison). At present, it does not seem unreasonable to model the interaction of a single molecule with the metal surface in a way similar to the interaction of a rare gas atom with the surface, although it is clear that the binding energy is much smaller in the latter case. Then we have considered a model-potential of the physisorption type. Following the work of Zaremba and Kohn[19], the interaction potential V(Z) of a rare-gas -like molecule located at a distance Z from the metal surface is written as:

$$V(Z) = V_{HF}(Z) + V_{cor}(Z) \tag{5}$$

The first term, $V_{HF}(Z)$, is a short range repulsion due to the exchange between the conduction electrons of the metal and the closed shell orbitals of the molecule. It corresponds to a treatment in the Hartree Fock approximation. Harris and Liebsh[20] have extended the treatment of Zaremba and Kohn in order to include correlations between the metal electrons. They introduced a pseudo-potential V^{ps} which they treated by a perturbation theory. They have shown that a first approximation is to consider V^{ps} as a local operator. Then instead of $V_{HF}(Z)$ they obtain a repulsive term in the form:

$$V_R(Z) = \int n(\mathbf{r}) V^{ps}(\mathbf{r},\mathbf{R})\, d\mathbf{r} \tag{6}$$

where n(**r**) is the electron density profile of the metal in vacuum and **R** the coordinate of the adatom. Note that they mention that this local approximation may not be sufficient. In our case, however, we may still use it for a qualitative purpose. In eq. 5, the second term $V_{cor}(Z)$ is an attractive contribution related to long range electronic correlations between the metal and the atom. It corresponds to the presence of dispersion forces like in the Van der Waals potential. Zaremba and Kohn have shown that the calculation of $V_{cor}(Z)$ requires the kowledge of the dynamic response function of the metal and the atomic polarizability. An expansion of the general expression of $V_{cor}(Z)$ in inverse powers of Z gives as the leading term the Van der Waals energy. In our work, we have followed the alternative formulation due to Annett and Echenique[21]. In this formulation $U_{VdW}(Z)$ can be evaluated in a closed form. The interested reader is refered to their paper for details. Note that the total potential V(Z) presents the typical shape of a Lennard-Jones potential, whose depth depends on the precise value of the parameters entering V_R and U_{VdW}. If a given molecule is subject to this potential, we expect that at a non zero temperature, it will explore the potential up to a distance Z=d where the potential V(d) is approximately equal to its mean kinetic energy.

To summarize the important points in the preceeding sections, we may say first that in the presence of an external field, the metal surface does not behave like the surface of an ideal conductor. This non-ideality is associated to a partial contribution to the capacitance which cannot be neglected compared to those typical of the metal-solution interface. This contribution eventually depends on the surface charge. The second important aspect is that the interaction of a weakly bound molecule is associated to a Lennard-Jones type potential. Then one may expect that the solvent molecules will approach the surface at a distance d which will depend on the thermodynamic state of the liquid phase. As we can see it in eq.6, this distance may also depend on the charge since the electron density profile will eventually depend on the charge.

4. The metal-solvent coupling

The preceeding section concerned the interaction of an isolated molecule with the surface. In order to deal with the coupling of the metal with the solution, we have to consider both the effect of the solvent and that of the electrolyte. Since we deal with interfaces with no ionic adsorption, the interaction with the electrolyte is purely electrostatic. In addition, the distance of closest approach of non-asorbed ions to the surface is of the order of one molecular diameter while the width of n(z) is much smaller. Then to a good approximation, one can ignore the details of the charge distribution in the solution when dealing with the density profile n(z). In this step, the charge $-\sigma$ in the solution can be assumed to be uniformly distributed on a charged plane located at D from the metal surface.

The coupling with the solvent is much more difficult to treat in detail. In principle one should use the potential V(Z) defined in eq.5 and make a self consistent treatment of both the statistical mechanics of the solution side and the quantum mechanical description of the metal as detailed in the second section. In this case, for a given configuration $\{\mathbf{R}_N\}$ of the N molecules, one must add to the energy functional E[n(z)] a coupling term given by:

$$U_{sol}[\{\mathbf{R}_N\}, n(z)] = \Sigma_I \int n(z) V^{ps}(\mathbf{r},\mathbf{R}_I)\, d\mathbf{r} \tag{7}$$

As discussed in our previous work[22], one may expect that if the coupling is weak, the discrete sum can be replaced by an integral over the average distribution ρ(Z) of the molecules, making thus a mean-field treatment of the functional E[n(z)]. Strictly speaking, this amounts to consider only the first term in a perturbation expansion about a reference system characterized by the average distribution ρ(Z). The coupling term is then replaced by :

$$U_{sol} = \int n(z)\, \rho(Z) V^{ps}(\mathbf{r},\mathbf{R}) d\mathbf{r} dZ \tag{8}$$

The practical scheme is as follows. One starts with a reference distributions ρ(Z), which for convenience we take as that of hard spheres in the presence of a hard wall at a distance d from the surface, ρ(Z- d). Then one minimizes E[n(z)] including U_{sol}. This allows the evaluation of a new V(Z) as given in eq. 5. An intermediary step consists in replacing V(Z) by an equivalent hard wall located at a new value of d which is such that the force equilibrium at the interface is verified[23]. The whole process must of course be iterated to self consistency. At self consistency, we obtain a unique value of n(z), V(Z) and d. This final value of d is by definition the metal-solvent distance of closest approach and has the physical meaning detailed previously in the case of one molecule. We can now evaluate the relevant properties of the interface.

5. Electrical properties of the interface

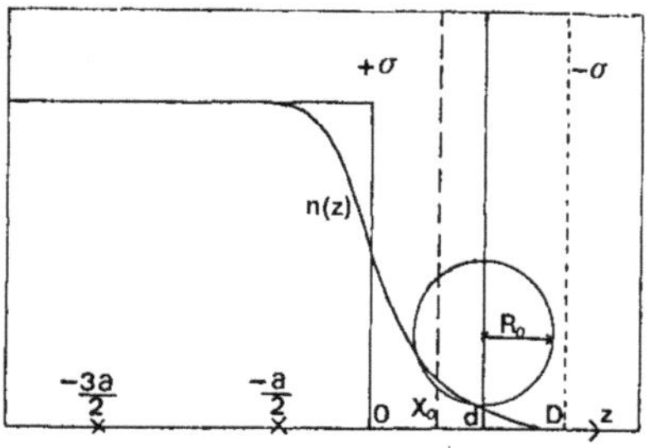

Figure 1. Schematic representation of the interface. Crosses on the z axis indicate the position of the ionic planes in the metal. n(z) is the electronic density profile and σ the charge per unit area on the electrode. a, X_0, d, R_0 and D are respectively the interplanar spacing in the direction normal to the interface, the position of the image plane, the distance of closest approach of the solvent molecules, the molecular radius and the position of the outer Helmholtz plane.

A schematic representation of the interface is given in fig.1 where we detail the characteristic distances. In particular, the excess charge in the solution is located at $z \geq D$, which we choose as:

$$D = d + R_0 \tag{9}$$

where R_0 is the molecular radius. Formally, we may write the total potential drop as:

$$\Delta\phi_s^m = \phi(-\infty) - \phi(+\infty) = -4\pi\left[\int_{-\infty}^{+\infty} z\rho_m(z)dz + \int_{-\infty}^{+\infty} z\rho_s(z)dz + \int_{-\infty}^{+\infty} P(z)dz\right] \tag{10}$$

where $\rho_m(z)$ and $\rho_s(z)$ are the charge distributions averaged in the (x,y) plane in the metal and in the solution respectively and P(z) the polarisation in the z direction. From eq.10, a formally exact expression of the capacitance can be derived :

$$1/C = 4\pi[X_s(\sigma) - X_0(\sigma)] - 4\pi\partial/\partial\sigma\int_{-\infty}^{+\infty} P(z)dz \tag{11}$$

where similarly to eq.4 we defined

$$X_s(\sigma) = -\partial/\partial\sigma\int_{-\infty}^{+\infty} z\rho_s(z)dz \tag{12}$$

It is convenient to write $\rho_s(z)$ as:

$$\rho_s(z) = -\sigma\delta[z-(d + R_0)] + \tilde{\rho_s}(z), \tag{13}$$

In this expression, we have isolated the charge $-\sigma$ peaked on the plane $z = (d+R_0)$. $\tilde{\rho_s}(z)$ is the remaining part of $\rho_s(z)$, so that the net charge corresponding to $\tilde{\rho_s}(z)$ vanishes. Then, by using eq.12 and the formal splitting of $\rho_s(z)$ in eq.13, we get:

$$X_s(\sigma) = d + R_0 + \sigma\partial d/\partial\sigma + \tilde{X_s}(\sigma) \tag{14}$$

where $\tilde{X_s}(\sigma)$ corresponds to $\tilde{\rho_s}(z)$. We can now rewrite eq. 11 as follows:

$$1/C = 4\pi[\, d - R_0 + \sigma\partial d/\partial\sigma - X_0(\sigma)\,] + 4\pi(2R_0 + \tilde{X_s}(\sigma) - \partial/\partial\sigma\int_{-\infty}^{+\infty} P(z)dz \tag{15}$$

From eq.15, we defined the metal contribution as

$$C_m^{-1} = 4\pi[\, d - R_0 + \sigma\partial d/\partial\sigma - X_0(\sigma)\,] \tag{16}$$

and the remaining part of C^{-1} as C_s^{-1}:

$$C_s^{-1} = 4\pi(2R_0 + \tilde{X_s}(\sigma) - \partial/\partial\sigma\int_{-\infty}^{+\infty} P(z)dz) \tag{17}$$

At the present stage these expressions are simple definitions, since only the total capacitance is a measurable quantity. However, the quantities C_m and C_s correspond both to a well defined physical content. C_s contains the same ingredients than those invoked in the classical models of the interface. In these models, one deals with the inner layer capacitance C_i deduced from the measured capacitance C after correction from the contribution of the diffuse layer as given by the Gouy-Chapman theory. Then C_i is associated to a monolayer of solvent between two charged planes. The first one being the metal surface, the distance d is just R_0. The second one located at $2R_0$ from the first one, *ie* at $(d + R_0)$, is the outer Helmholtz plane, representing the distribution of the excess charge in the solution at high electrolyte concentration. From the definition of $\tilde{\rho_s}(z)$ in eq.13, this assumption means that $\tilde{X_s}(\sigma)=0$. If we further assume that P(z) is limited to a monolayer between $d-R_0$ and $d+R_0$ and split it as usual into orientational and a distorsional contribution characterized by a uniform dielectric constant ε_d, a simple calculation shows that eq. 17 reduces to :

$$C_s^{-1} = 4\pi\,[2R_0 - \partial/\partial\sigma\int P(z)\,dz\,] = 4\pi e/\varepsilon_d - 1/C_{dip}, \tag{18}$$

where $e = 2R_0$ is the thickness of the monolayer, a value of about 6 is taken for the "dielectric constant of the inner layer" ε_d and where C_{dip} is associated to the orientation of the permanent dipoles.

Within the assumptions of the classical model, the contribution C_m as defined in eq.16, is such that $C_m^{-1} = 0$. In this equation, we see that C_m is determined by $X_0(\sigma)$, i.e. by the relaxation with charge of the surface dipole of the metal. This is an intrinsic property of the metal, though it must be calculated in the presence of the solvent. The last two quantities d and its variation with charge $\sigma\partial d/\partial\sigma$ reflect the coupling at the interface. This coupling involves the interaction of the metal with the electrolyte and with the solvent.

Two different routes for investigating the capacitance follow from eqs. 15 to 17. The first one would consist in calculating directly C_m given by eq. 16 and C_s by eq. 17 from a fully microscopic model of the interface and then compare the resulting C in eq. 15 to experimental data. This is indeed a formidable task and none of the models proposed so far has yet achieved this goal, with good agreement between theory and experiment. The second one consists in calculating either C_m or C_s. From experimental data and the calculated term, one obtains a semi-empirical estimate of the other term, whith the expectation that this term which contains more specific information will be easier to analyze.

At present, it is much easier to obtain reliable results of C_m from presently available models of the metal and its coupling with the solvent. Then, from the knowledge of C_m, we defined[7] a semi-empirical estimation of the solvent contribution as:

$$C_s^{-1} = C_i^{-1} - C_m^{-1}, \tag{19}$$

where C_i is the inner layer capacitance obtained by correcting the measured capacitance from the Gouy -Chapman contribution. Hence, following the classical terminology, C_s should not contain a contribution of the diffuse layer.

Eq. 19 is the basis of the new analysis of the capacitance we have recently proposed[8-9]. We stress that the splitting of C_i into two contributions is in no way unique. Besides the physical meaning of each term in such an arbitrary splitting, the ultimate test of its adequacy is its ability to provide a coherent explanation of the greatest possible number of experimental facts. This we examine in the following section.

6. Results and discussion

In this section we present some selected results which will illustrate the method presented above. Details may be found in our previous work.

6-1. *Calculation of C_m for a silver-water interface*

In refs 7 and 22, we have calculated C_m for low index faces of a monocristalline silver electrode by using an extended jellium model of the metal. In this extension, the electron-ion interaction was modelled by the Ashcroft empty core pseudo-potential with parameter R_c and the s-d hybridization was considered by the introduction of an effective valence Z_s=1.5 as in ref 24. The calculation was made by direct minimization of the metal energy functional E[n(z)]. The different terms entering E[n(z)] have been detailed elsewhere. In $V_R(Z)$, the short range electron-molecule repulsion was taken as a constant barrier of height V_0 within a sphere of radius R_0 taken as that of the water molecule. The Van der Waals attraction was calculated with parameters appropriate for water.

For a set of values of the surface charge σ, we calculate $X_0(\sigma)$ by using an appropriate version of the change in self-consistent field method[25] and evaluate $d(\sigma)$ as indicated above. The C_m^{-1} curve calculated in this way is given in fig. 2, together with the curves estimated for

Au, Hg and Ga. The overall shape of C_m^{-1} as a function of charge results from the combination of a monotonous variation of $X_0(\sigma)$ and a non-monotonous variation of $d(\sigma)$. This second observation is qualitatively important. As a consequence of a weak coupling with the metal surface, the solvent molecules can approach the electrode up to a mean distance which depends on charge. This contrasts with the case of chemisorbed molecules for which one would expect that due to the strong interaction, the adsorbed molecules will stay at a fixed position, irrespective of the value of the interfacial field.

It is not useless to recall the reasons of this important feature. It can be understood by considering the main forces which determine d: electrostatic pressure due to the electrical field *via* the solvated ions and short range repulsion due to the electrons (the Van der Walls term itself depends little on charge). While the electrostatic attraction results always in a decrease of d, the overall magnitude of the repulsion depends on the sign of the charge. This is a consequence of the variation of the electron density profile with charge: at positive charging, the electrons are pulled back to the metal interior. A solvent molecule at a given position Z will then feel less repulsion from the electron cloud. The converse effect occurs at negative charging: a larger spill over of the electrons from the surface means a stronger repulsion felt by the molecules. This behaviour seems very general and we have checked that it does not depend on the precise value of the parameters of the model. This conclusion has been made by several authors[5-6,8] and should be a characteristical property of physisorption at a charged surface. Note for e.g. that the variation of the metal-solvent distance seems to be important for the interpretation of ellipsometric data [26].

The direct output of the method[7] is to obtain C_s for water by using eq. 19 and the calculated C_m for silver in conjunction with innner layer capacitance curves C_i for a silver-aqueous electrolyte interface (Ag/ NaF and Ag/ KPF_6 [27]). The $C_s(\sigma)$ curves are given in fig.3.

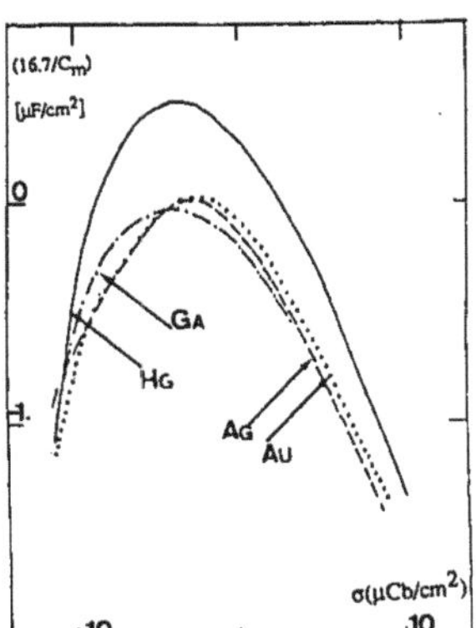

Figure 2. C_m^{-1} curves for Ag, Au, Hg and Ga. The curve for Ag is the result of the calculation. The curves for Au, Hg and Ga are estimated from eq. 20.

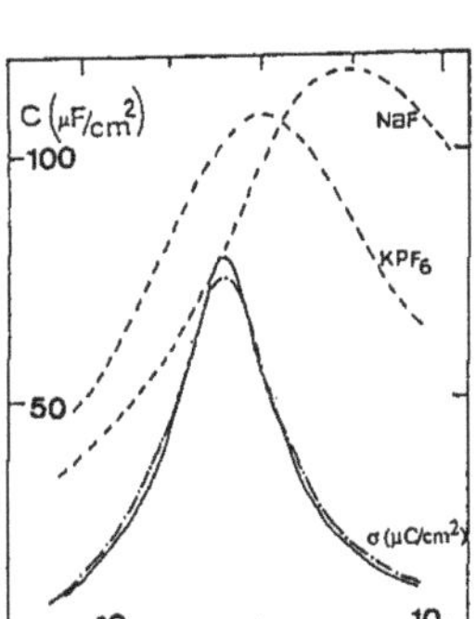

Figure 3. C_s curves for aqueous solutions. Ag/NaF (dash-dotted) and Ag/ KPF_6 (full curve). Dashed curves are the corresponding C_i curves.

The important points in fig.3 is that the C_s curves have a very simple bell shape and present *a maximum at a negative charge* σ_m, in agreement with the well accepted fact that at the pzc, the water molecule is preferentially oriented with the oxygene towards the metal[16]. If we recall the long debated subject of water orientation at the interface, this observation is the first piece of evidence that this analysis does indeed introduce more coherence in the interpretation of experiments.

6-2. *Role of the metal*

If we want to discuss other metals than silver, we should in principle calculate C_m and use again eq.19 with the respective experimental C_i curves. For the case of gold, mercury and gallium we want to discuss now, we adopted a different approach for two reasons. On the one hand, we do not have a satisfactory theoretical description of the electronic properties of these three metals. This problem is further complicated for liquid Hg and Ga, whose interfacial structure is unknown. The second and main reason is that we ultimately found out that this is even unnecessary, at least at the present stage of the theory, as we detail it now. If we compare C_i curves for the same solvent but with different metals, it is natural as a first approximation to consider that C_s is the same in all cases. This assumed "universal" behaviour is the natural consequence of the way we *defined* C_s, since the effect of the coupling via the distance d and its variation with charge is included in C_m. In this case, the difference between interfaces with the same solvent being firstly due to C_m, one may estimate it for a metal M' and a solution S' if it is known for another interface with metal M and solution S by using:

$$1/C_m(M'/S') = 1/C_m(M/S) + [1/C_i(M'/S') - 1/C_i(M/S)] \qquad (20)$$

which is obtained from eq.19 by writting that C_s is the same for the two interfaces. The C_m^{-1} curves given in fig 2 were obtained in this way. While the corresponding C_i are very dissimilar, all C_m^{-1} curves in fig.2 have the shape of an inverted parabola, asymmetric wr to zero charge, just like the curve calculated for silver.

A detailed discussion of the variation of C_m from metal to metal has been presented in ref 28 to which we refer the interested reader. We briefly mention here that in our model, the metal specificity apprears in the density of conduction electrons n, the electron-ion pseudopotential and the surface structure (interplanar spacing in the direction normal to the surface for a solid, ionic density profile for a liquid). n determines directly X_0 and indirectly d, through the metal solvent-interaction. The surface structure plays a role both for X_0 and d. From the arguments presented in ref. 28, the sequence in fig.3 is in complete qualitative agreement with known surface properties of these metals. Note that the maximum variation of C_m^{-1} at the pzc corresponds roughly to a difference of about 0.2Å in the quantity $(d - X_0)$. This quantitative difference is also compatible with the known variation with the nature of the metal of the related physical properties[28].

These observations provide an alternative interpretation of the role of the metal on the capacitance.They show that the difference between metals may be explained largely by well established physical properties of these metals. However, this variation being relatively small, it is difficult to describe it quantitatively at present.

This interpretation is to be compared with the conventional one. As well known, the metal affects the capacitance in the classical analysis only *via* a "chemical" interaction which favours a given orientation of the molecular dipole at the pzc. The strength of this interaction depends on the nature of the metal in the same sequence as in the so called "hydrophilicity scale"[16]. Of course, our interpretation does not exclude the idea that such an interaction should be considered in parallel with the previously mentioned effects. Let us consider now how it would enter our calculation. Since it involves energies negligeable compared with the typical energy experienced by the electrons, such a"residual interaction" will add negligeable terms in the electronic energy functional. Hence, for a given metal, the calculation of C_m can be made without considering the effect of this additional coupling. Then one may expect that an accurate calculation of C_m for different metals, will give for C_s estimated from eq.19, a shift of the position σ_m of its maximum from the pzc, just as we found it in the case of the silver-water interface (see fig. 3). This shift will possibly depend on the nature of the metal. We may say *a posteriori* that the observed variation of C_i curves from metal to metal may reflect both the difference in the physical properties of these metals *via* C_m as discussed above and the effect of the "residual interaction" on the solvent polarization, *via* C_s. At present, we cannot determine the relative importance of the additional specifity embodied in the prefered orientation. However, we may already say that a metal dependent value of σ_m should be in a very narrow range. Indeed, the splitting of C_i into C_m and C_s as in eq. 19 results always in very stable C_s curves, which do not depend much on details of C_m as, for e.g., those which would result from a calculation for different metals. On the contrary, as we discussed it in refs. 8 and 28, the C_i curves obtained by combining different C_m and C_s curves are very sensitive to minor variations in C_m and/or C_s. The physical effects we have discussed above were therefore sufficient to produce the observed variation from metal to metal, still by using the same C_s in all cases. From basic considerations in the physics of metals surfaces, we know that these effects must exist. This strongly suggests that a possible additional effect due to the "residual"interaction should be at most of the same importance.Therefore, attributing the metal specificity only to its influence on the polarization of the solvent is a simplification which, at least, ignores important properties of the electrode surface.

Finally, it seems *a posteriori* quite natural that the difference among various interfaces with the same solvent and different metals should be attributed firstly to specific properties of the metals. Even though a more detailed discussion should include all possible aspects of this specificity, these observations on the role of the metal confirm the efficiency of the analysis which follows from the splitting of the capacitance embodied in eqs. 15-17 and 20.

6-3. *The effect of temperature on* C_s

If we consider the classical data of Grahame for Hg/NaF[29], we may estimate the effect of temperature on C_s by using eq.19 and C_m^{-1} for Hg given in fig. 2, assumed to be independent of the temperature. The curves obtained in this way are given in fig. 4.

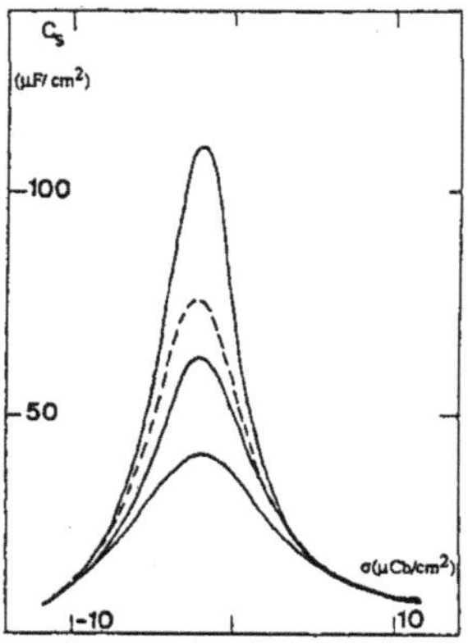

Figure 4. C_s curves at different temperatures. From top to bottom: t =5, 25, 45 and 85 °C. The dashed curve is the C_s curve for Ag/NaF given in fig. 3

Since C_m is assumed as temperature independent, C_s curves in fig.4 are nothing else than a different representation of the experimental C_i curves. However, in contrast to the complicated aspect of C_i curves, the behaviour of C_s is very simple. At each temperature, $C_s(\sigma)$ is a simple bell shaped curve. At most charges now, $C_s(T)$ decreases with temperature. At all T, the rapid fall off of C_s from the maximum, may reflect saturation of the polarization.

This behaviour of C_s seems again a natural consequence of the competition of the orienting effect of the electrical field on the solvent dipoles and their thermal motion. In ref. 8, we have checked this feature on a model for the polarization of the Langevin type. While no direct experimental proof of the behaviour of C_s exists, the effect of temperature in fig.4 confirms again that experimental facts are better rationalized when one analyzes C_s rather than C_i.

6-4. *Role of the solvent*

As a final piece of evidence of the adequacy of the general analysis of the capacitance we propose, we consider now the effect of changing the nature of the solvent. Details can be found in our recent work[9]. In this paper, we developed the arguments showing that for a qualitative purpose, one may first use the same C_m for analyzing C_i data for non-aqueous solvents at the mercury electrode. From inner layer capacitance curves $C_i(\sigma)_{Hg/sol}$ for different

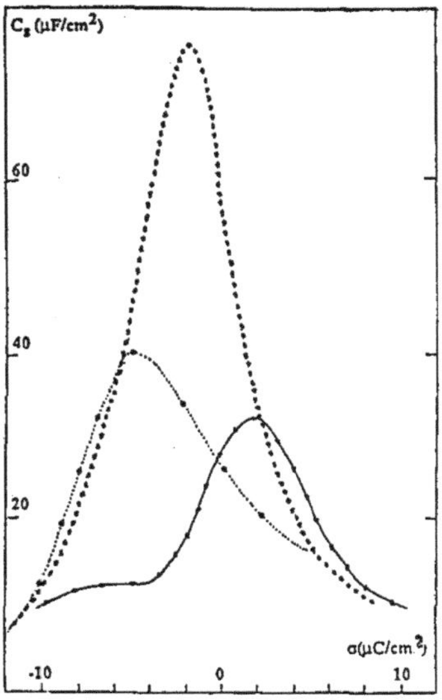

Figure 5a. C_s curves for water (- - - - - - - - - -) methanol (—•—•—•—) and formamide (··•······•······•··)

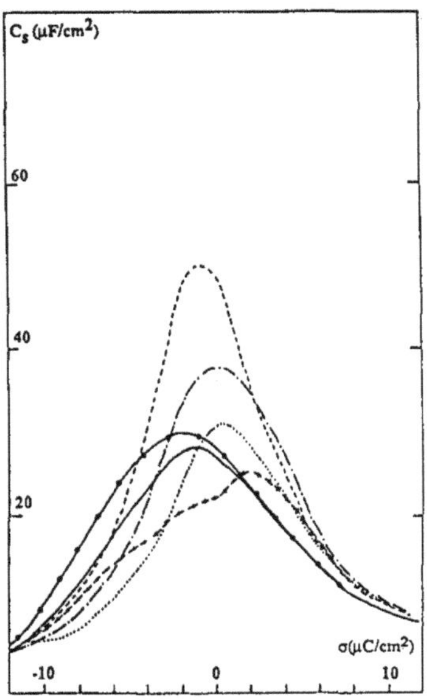

Figure 5b. C_s curves for dimethylformamide (—·—·—·—) dimethylsulfoxide (- - - - - - - - -) acetone (··············) ethhylene carbonate (—•—•—•—) propylene carbonate (————) and acetonitrile (—+ —+ —+ —+)

solvents and with the $C_m^{-1}(\sigma)$curve in fig. 3., we defined similarly to eq.19 the solvent contribution as:

$$C_s^{-1}(\sigma)_{sol} = C_i^{-1}(\sigma)_{Hg/sol} - C_m^{-1}(\sigma)_{Hg} \qquad (21)$$

$C_s(\sigma)$ obtained by using eq.21 is obviously a very indirect estimate of the solvent contribution so that details of the $C_s(\sigma)$curves might not be reliable. However, we found that the curves present some general features which can be very hardly thought of as artifacts of the method.

The $C_s(\sigma)$ curves in fig.5 were obtained by using eq.21 and capacitance data for nine solvents. While their overall shape is the same, the $C_s(\sigma)$ curves depend, as expected, on the nature of the solvent by their relative magnitude and by the charge σ_m at which they have their maximum. For formamide, H_2O, PC, EC, DMSO, σ_m is negative. It is almost zero for DMF and acetone and positive for MeOH and ACN.

We can discuss now the kind of information we may derive from σ_m. It is known that models based on a simple mechanism for the solvent orientation as a function of charge, predict that the maximum of $C_i(\sigma)$ corresponds to the maximum randomness in the orientation of the dipoles. In this class of models, the sign of the charge at which the theoretical $C_i(\sigma)$ is maximum is directly related to the assumed preferential orientation of the dipoles at the pzc. In these models, the maximum of the capacitance is also at the same charge than the maximum of the entropy. The latter is a more direct probe of solvent orientation, although the analysis of entropy data requires also a model. However, if one considers a more complex mechanism as for e.g. the presence of more than one orienting species, the capacitance hump in the theoretical C_i does not correlate with the maximum of the entropy. For water for e.g., this led to the reconciliation of the fact that the maximum of the entropy is at a negative charge and the hump in the experimental C_i at a positive charge.

More generally, it was concluded that maxima or humps in capacitance curves give no information on solvent orientation at the pzc[16]. Since no criterion alone allows a definitive conclusion, the prefered orientation has to be assessed by confronting different experiments. From a coherent set of observations most of the solvents in fig.5 are generally considered as being oriented at the pzc with the negative end of the molecular dipole towards the electrode (se for e.g. refs. 29-30). On figs. 5, we observe that the position $\sigma_m$of *the maximum of* $C_s(\sigma)$ is negative or very close to zero. Then with a possible exception for ACN and MeOH, these curves show a correlation between the position of the maximum of $C_s(\sigma)$ and the generally accepted prefered orientation at the pzc.

In our method for estimating $C_s(\sigma)$, the precise value of $\sigma_m$depends on the details of the slopes of $C_i(\sigma)$ and $C_m(\sigma)$ at moderate charges. Then a definitive conclusion about the value of σ_m, especially when it is close to zero, should require accurate values of $C_m(\sigma)$. In any case, the preceeding discussion illustrates very well the fact that genuine information present on the level of the "real" contribution of the solvent, $C_s(\sigma)$ are simply masked on the level of $C_i(\sigma)$ by the presence of the contribution $C_m(\sigma)$.

A further discussion of $C_s(\sigma)$ curves can be made from the following considerations. Since they were found to have a common shape, almost symmetrical wr to the position of the single maximum, this suggested that the relevant physical phenomenon is the variation of C_s vs. the effective charge $\sigma^* = \sigma - \sigma_m$. Then while the width of the curves is roughly the same, the important feature is the sequence of amplitude at $\sigma^*=0$. In order to investigate the factors which govern this behaviour, we used eq. 18. In terms of the charge σ^*, we obtained the dipolar contribution C_{dip} to C_s as :

$$C_{dip}^{-1}(\sigma^*) = 4\pi e/\varepsilon_d - C_s^{-1}(\sigma^*), \qquad (22)$$

We recall here that C_{dip} is the central quantity in most models of the traditional approach. It characterizes the non-linear response of the orientational polarization of the solvent to the applied field. In order to go further, we need the ratio $4\pi e/\varepsilon_d$. A detailed discussion of this term and its analogy with the classical K_{ion} can be found in refs 9, 28. We mention here that we took the value $\varepsilon_d=2$ for all solvents, and e as equal to the molecular diameter as a first step in a systematical analysis. We choose to determine this diameter by taking a fixed packing fraction η in the bulk for all solvents: $\eta = \pi\rho e^3/6 = 0.45$, from the theory of liquids ,with ρ equal to the bulk density. $C_{dip}(\sigma^*)$ curves obtained in this way are plotted in fig. 6, from which me make the following observations:

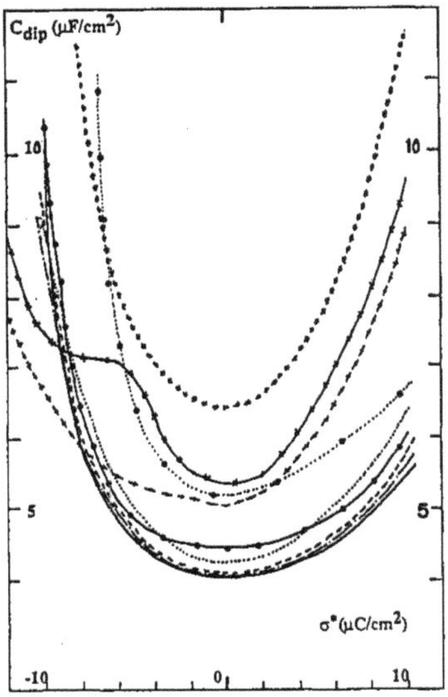

Figure 6. The contribution C_{dip} of the permanent dipoles as a function of the effective charge σ^*. Same caption as in figs. 5.

(1) around the pzc, the maximum variation of $C_{dip}(\sigma^*)$ for different solvents is strikingly reduced, when compared to that of $C_s(\sigma^*)$ in figs 5. This suggests that the dominant factor which explains the magnitude of C_s at the maximum is the molecular size. For e.g. the high value of C_s for water is due to its smaller size. The importance of the size of the solvent molecule has been advocated many times (see for e.g. the review by Payne[31]). However no correlation can be drawn[9] if we use $C_i(\sigma)$ in eq.15.

(2) at moderate σ^* the curves for DMSO, DMF, PC, EC and AC almost merge into a single curve, while those for H_2O, MeOH and Formamide are clearly separated. Note that the curve for ACN is also separated.

(3) at higher magnitude of σ^*, the saturation of the polarization depends on the solvent.

Point (2) and point (3) emphasize the role of intermolecular interactions[9]. Here, we recall that C_s curves and hence C_{dip} have been obtained with the same C_m. Then while the magnitudes at the pzc in fig. 6 cannot be considered as quantitatively accurate, the sequence will not be changed if we use the value of C_m appropriate for each solvent. The sequence near the pzc can be understood qualitatively, from a competition between the orienting effect imposed by the external field and the effect of intermolecular interactions which may oppose it. In what concerns the field, simple considerations show that $C_{dip}(\sigma^*)$ is proportional to $(e/\mu)^2/\partial\langle\cos\theta\rangle/\partial\xi$, where ξ is a reduced quantity defined by $\xi=\mu E^*/kT$. $E^*=E-E^r$ is the effective field, which reflects the competition between the applied field E and the residual field E^r. μ is the dipole moment and θ the angle of the dipole with the normal to the electrode surface.

For aprotic solvents, intermolecular interactions are essentially dipolar. An idea of the strength of dipolar interactions is the ratio $\upsilon_{dd}=(\mu^2/kTe^3)$ which is roughly constant (see table 1). Note that in the fluid, one should take into account some screening of the bare interaction. In addition, the coefficient $(e/\mu)^2$ does not differ much from unity. For these solvents with similar values of the dipole moment and of the radius, we expect similar values of C_{dip}. We indeed observe in fig. 6 that with the exception of ACN, the values of C_{dip} differ little for these solvents at moderate charges. The relatively weak differences in C_{dip} observed in fig. 6 should then indicate that the screening of the dipole-dipole interaction is also similar near the $\sigma^*=0$. Note that the position ACN which has a relatively smaller radius can be understood if ones considers the effect of the radius R_0 on C_m[9]. A close inspection of table 1 shows that the position of acetone results from the combined effect of its larger value of $(e/\mu)^2$ and smaller value of υ_{dd}.

For associated solvents, the H-bond, corresponding to energies in the range 5-20 kT, ie larger than $\upsilon_{dd}kT$, will dominate intermolecular interaction. Then these solvents are expected to be more difficult to orient than unassociated ones and hence should have much smaller values of $\partial\langle\cos\theta\rangle/\partial\xi$. Note that except formamide, they also have larger values of the ratio

$(e/\mu)^2$. Then as observed in fig. 7, $C_{dip}(\sigma^*)$ should be larger for water, formamide and methanol than for aprotic solvents. At higher charges, the competition between the external field and intermolecular interactions makes that the rise of $C_{dip}(\sigma^*)$ depends on the solvent. A detailed discussion is however difficult since the precise value of $C_{dip}(\sigma^*)$ is very sensitive at these charges to uncertainties in e/ε_d and $C_s(\sigma^*)$.

	H_2O	MeOH	Form.	AC	EC	PC	DMSO	DMF	ACN
e(A)	3.	3.86	3.83	4.71	4.55	4.94	4.66	4.78	4.21
μ(D)	1.84	1.66	3.66	2.88	4.87	4.94	3.96	3.82	3.92
$(e/\mu)^2(A/D)^2$	2.65	5.40	1.09	2.67	.87	1.	1.38	1.56	1.15
υ_{dd}	3.02	1.15	5.75	1.91	6.07	4.88	3.73	3.22	4.96
$\varepsilon_d/4\pi e$ ($\mu F/cm^2$)	5.89	4.58	4.61	3.75	3.88	3.57	3.79	3.70	4.20

Table 1. Solvent physical parameters. The dipole moments are taken from the literature. $4\pi e/\varepsilon_d$ is calculated with the tabulated values of the solvent diameter, e, and with $\varepsilon_d = 2$. υ_{dd} is half the maximum dipole-dipole interaction energy in vacuum.

In conclusion to this section, we may say now that when $C_m(\sigma)$ is extracted from $C_i(\sigma)$ a clear picture of the behaviour of different solvents emerges. It can be rationalized largely in terms of their physical parameters (dipole moment, molecular size) and the nature of intermolecular interactions. This is to be compared with the often adopted classification in different groups[32] according to the shape of $C_i(\sigma)$.

7. Conclusion

We first summarize the important results obtained from this new analysis of experimental data. The role of the solvent may be characterized by : (1) a contribution to the capacitance, C_s, which has a very simple variation with the electrode charge, (2) the natural variation of C_s with temperature as infered from the data on water, (3) the existence of a correlation between the position of the maximum of C_s and the prefered orientation at the pzc. In addition, the solvent specificity can be analyzed in a very simple way at the level of the contribution C_{dip} due to the orientation of the permanent dipoles in an electrical field.

In its turn, the metal is characterized by its electrostatic non-ideality and its interaction with a weakly adsorbed solvent. From well accepted concepts in the modern description of the physiscs of metal surfaces, its predicted influence on the capacitance is significant.

More generally, an important conclusion of the previous analysis is that inner layer capacitance curves $C_i(\sigma)$ corresponding to a wide variety of ideally polarizable interfaces can be discussed in a common physical framework. Other phenomena than those discussed in this work should determine the "fine structure" of the $C_i(\sigma)$ curves. Moreover, C_i being the result of C_s and C_m connected in series, its magnitude at a given charge results often for different interfaces from the magnification of relatively minor variations in C_s and C_m. A practical consequence of this situation is that very few informations at a microscopic level can be derived directly from the inner layer capacitance curve. In order to obtain specific information, one should work on a less global quantity than $C_i(\sigma)$.

Our analysis relies of course on a calculation of C_m based on a model which is in no way definitive and should be improved in several ways. However, it is based on presently well accepted ingredients. Then unless the importance of the contribution $C_m(\sigma)$ is at present considerably overestimated, because of a still undiscovered failure of the theory, both in our work and in the work of several others, all the features discussed above form a very coherent set of observations in favour of the correctness of the general analysis we have proposed. It remains indeed an "analysis" in the sense that we propose an alternative interpretation of known experimental facts. It is already rewarding that it brings some clarity in a long debated subject. Its predictive power is at present limited, since none of the main quantities it considers is amenable to direct experimental test. However, we already expect that the variation of the metal solvent distance with charge may play a role in other areas than capacitance studies, as it is suggested in the interpretation of optical data[26]. In the field of capacitance studies, a work currently in progress will be devoted to the consequences of this variation in the case of simple adsorbed ions.

References

1. R.M. Reeves, in *Modern Aspects of Electrochemistry*, ed. B.E. Conway, J.O'M. Bockris, Plenum Press N.Y., **V9**, 239 (1974). M.A. Habib, *ibid*, **V12**, 131 (1977)

2. D. Henderson, *in Trends in interfacial Electrochemistry*, **C 179**, 473, Reidel, Dordrecht. ed. A. Fernando Silva (1986). S.L. Carnie and G. M. Torrie, *Adv. Chem. Phys.* **56**, 141 (1984). S. L. Carnie, *Ber Bunsenges. Phys. Chem.* **91**, 262 (1987)

3. J.P. Badiali, M.L. Rosinberg and J. Goodisman, *J. Electroanal. Chem.* **150**, 25 (1983)

4. W. Schmickler and D. Henderson, *J. Chem. Phys.* **80**,3381 (1984).

5. A.A. Kornyshev and M.A. Vorotyntsev, *J. Electroanal. Chem.* **167**, 1 (1984)

6. J.W. Halley, B. Johnson, D. Price and M. Schwalm, *Phys. Rev.* **B 31**,7695 (1985).

7. S. Amokrane, V. Russier and J.P. Badiali, *Surf. Sci.* **217**, 425 (1989)

8. S. Amokrane and J.P. Badiali, *J. Electroanal. Chem*, **266**, 21 (1989)

9. S. Amokrane and J.P. Badiali, *J. Electroanal. Chem*, in press.

10. N.D. Lang and W. Kohn, *Phys. Rev.* **B1**,4555 (1970)

11. P. Hohenberg and W. Kohn, *Phys. Rev.* **136 B**, 864 (1964). W. Kohn and L.J. Sham, *Phys. Rev.* **140 A**, 1133 (1965)

12. N.D. Lang, *Solid State Phys.* **28**, 225 (1973). J.E. Inglesfield, Rep. Prog. Phys., **45**, 2223 (1982). R.O. Jones and O. Gunnarsson, *Rev. Modern Phys.* **61**, 689 (1989)

13. J.R. Smith, *Phys. Rev.* **181**, 522 (1969)

14. N.D. Lang and W. Kohn, *Phys. Rev.* **B8**, 6010 (1973)

15. P. Gies and R.R. Gerhardt, *Phys. Rev.* **B31**, 6843 (1985) *ibid* **B33**, 982 (1986).

16. S. Trasatti, in *Modern Aspects of Electrochemistry*, **V13**, 81, ed. B.E. Conway, J. O'M. Bockris, Plenum Press New York (1979).

17. M. Klaua and T.E. Madey, *Surf. Sci.* **136**, L42 (1984). P. A. Thiel and T.E. Madey, *Surf. Sci. Rep.* 7, 211 (1987)

18. J.E. Müller, *Surf. Sci.* **178**, 589 (1986)

19. E. Zaremba and W. Kohn, *Phys. Rev.* **B13**, 2270 (1976). *ibid*, **B15**, 1769 (1977)

20. J. Harris and A. Liebsch, *J. Phys.* **C15**, 2275 (1982)

21. J.F. Annett and P.M. Echenique, *Phys. Rev.* **B34**, 6853 (1986)

22. S. Amokrane and J.P. Badiali, *Electrochimica Acta* **34**, 39 (1989)

23. J.P. Badiali and M.E. Boudh-hir, *Mol. Phys.* **53**, 1399 (1984)

24. V. Russier and J.P. Badiali, *Phys. Rev.* **B 39**, 13193 (1989)

25. R. Monnier, J.P. Perdew, D.C. Langreth and J.W. Wilkins, *Phys. Rev.* **B18**, 656 (1978)

26. F. Chao, M. Costa, J. Lecoeur and J.P. Bellier, *Electrochim. Acta* **34,** 12, 1627 (1989).

27. G. Valette and A. Hamelin, *J. Electroanal. Chem.* **45**,301 (1973). G. Valette, *J. Electroanal. Chem.* **122,** 285 (1981). A. Hamelin, T. Vitanov, E. Sevastyanov and A. Popov, *J. Electroanal. Chem.* **145**, 225 (1983).

28. S. Amokrane and J.P. Badiali, to appear in *Modern Aspects of Electrochemistry.*

29. S.Trasatti, *Electrochim. Acta* **32**, 843 (1987)

30. S. Borkowska, *J. Electroanal. Chem.* **244**, 1 (1988)

31. R. Payne, in *advances in electrochem.. and electrochem. Engen.*, **V7,** 1 (1970)

32. R. Parsons, *J. Electroanal. Chem.* **59**, 229 (1975)

THE SEMICONDUCTOR - ELECTROLYTE INTERFACE

HEINZ GERISCHER
Fritz-Haber-Institut der Max-Planck-Gesellschaft,
Faradayweg 4-6, D-W-1000 Berlin 33, F. R. G.

Abstract

Three areas are reviewed where semiconductor electrodes differ the most from metal electrodes. One of these is the charge distribution in the electric double layer at the interface to an electrolyte which has drastic consequences for the other two selected areas, namely charge transfer reactions of electrons or ions and photoprocesses like photocurrents and photoluminescence. These topics are analyzed and represented by selected examples.

1. Introduction

There are pronounced differences between metal and semiconductor electrodes in contact with electrolytes. Some topics will be discussed here which should be of particular interest for physicists. One topic is the structure of the electrical double layer, where a space charge is found in the semiconductor like at solid state contacts. The presence of this space charge layer changes the kinetics of charge transfer reactions drastically in comparison with analogous reactions at metals. Another qualitative difference is the generation of large photocurrents and photovoltages at semiconductor electrodes and other phenomena connected with the long lifetime of excited electronic states. This can result in luminescence.

2. Structure of the Electrical Double Layer

The response upon a periodic or stepwise accumulation of small amounts of excess charge of opposite sign at the interface between an n-type semiconductor and an electrolyte can be represented by three capacitors in series as shown in figure 1. The three capacitors represent: C_{SC} the semiconductor space charge capacity; C_H the Helmholtz double layer capacity and C_G the space charge capacity of the electrolyte, the so-called Gouy layer [1,2]. All these capacities vary with the voltage drop in these layers. This variation is depicted in figure 2. The voltage drop is therein represented in dimensionless units with the thermal voltage equivalent kT/e_0 as the unit.

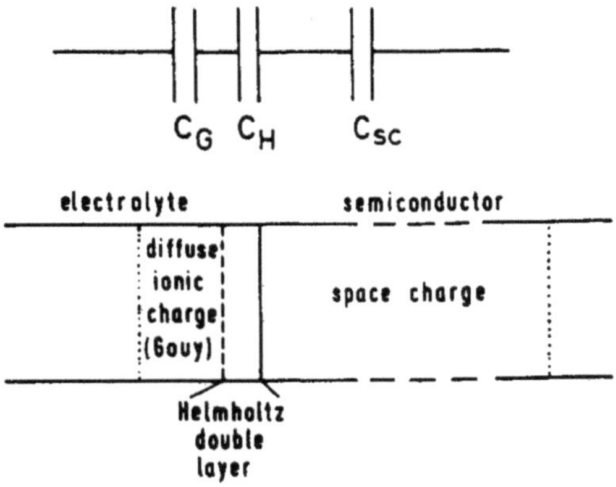

Fig. 1: Structure of the electrical double layer at the semiconductor-electrolyte interface and its representation by capacitors.

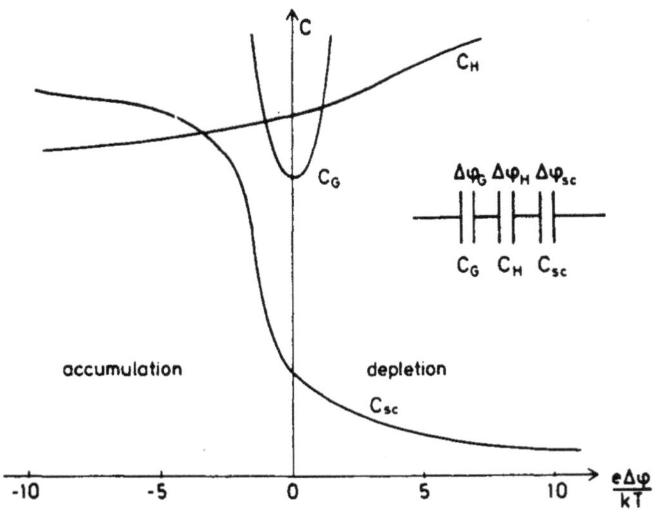

Fig. 2: The differential capacitance of the three capacitors of fig. 1 in dependence of the voltage drop in these parts of the double layer (specific adsorption of ions excluded).

In a series of capacitors, the response is controlled by the smallest one. Figure 2 indicates that, for positive (anodic) bias at the n-type semiconductor where a depletion layer is formed, the space charge capacity of the semiconductor controls the overall capacitance. C_G depends on the electrolyte concentration. It can be neglected in concentrated electrolytes where the value at the minimum ($\Delta\phi_G = 0$) is much larger than assumed in figure 1. In figure 1 and figure 2 it has been assumed that there is no specific adsorption of ions at the semiconductor interface. Otherwise, there would be a voltage difference in the Helmholtz double layer even if $\Delta\phi_{SC} = 0$ [3,4]. However, since C_H is and remains large with specific ion adsorption, C_{SC} still controls the capacitance behavior in the depletion region and C_H changes also in this case very little or not if the bias is varied in the anodic range. The bias where no excess charge is on the semiconductor is the so-called "flat band potential" because there is no electric field between the surface and the bulk of the semiconductor and the energy of the band edges is constant (flat) in the semiconductor (no band bending). The flat band potential can be derived by an extrapolation of the differential capacities in the depletion region [5,6].

The nearly ideal situation of figure 2 could be experimentally verified with semiconductors having very inert surfaces without specific interaction with the electrolyte. Such materials are layered semiconductors being exposed to the electrolyte with an intact van der Waals surface. Figure 3 represents such a measurement in comparison with the theory [7].

Specific ion adsorption affects the potential drop in the Helmholtz double layer. On semiconducting oxides, H^+ ions and OH^- ions are strongly interacting with the surface. The result is that the flat band potential varies nearly linearly with the pH, following the thermodynamic equilibrium of ion adsorption at interfaces. Details can be found elsewhere [8,9].

An interesting question from the physical point of view is the role of electronic surface states for the charge distribution at the interface. In vacuum, electronic surface states are usually dangling bonds or states split from the bulk states by surface reconstruction. They affect the charge distribution drastically if their energy is located in the band gap. The result is Fermi level pinning if their concentration is large enough [10]. In electrochemical systems, Fermi level pinning has often been assumed [11,12]. However, the requirements for Fermi level pinning in electrochemical systems are much more stringent than in vacuum. The reason is that the counter charge compensating the excess charge in surface states can not only be located in the space charge layer of the semiconductor as in the vacuum case but also in the Helmholtz layer of the electrolyte. This is demonstrated by the comparison shown in figure 4 and figure 5. The necessary density of surface states, acting here as acceptor states, which are negatively charged as long as $E_F \geqq E_{SS}$, is in the vacuum case compensated by the electric charge in the space charge layer. This controls the potential drop $\Delta\phi_{SC}$. An external bias can hardly change $\Delta\phi_{SC}$ due to the small capacity between a counter electrode in vacuum and the semiconductor. The band bending can only be affected by the doping concentration.

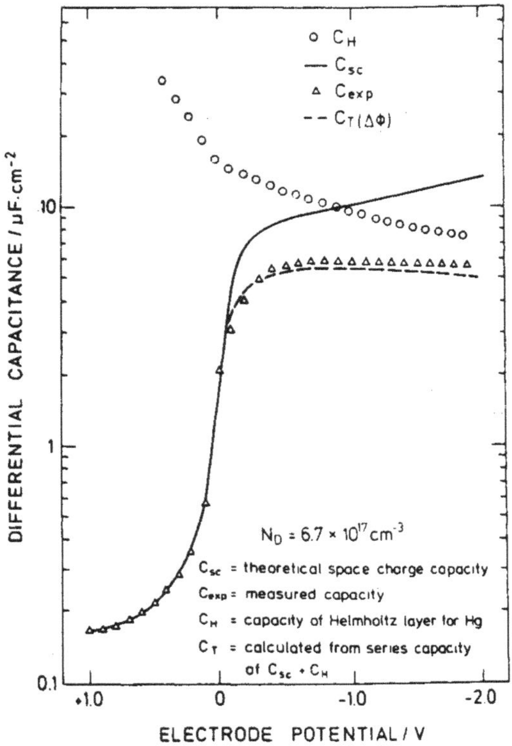

Fig. 3: Differential capacitance of an n-WSe_2 electrode in a solution of 0.1 M $(C_3H_7)_4NBF_4$ in acetonitrile in dependence of the bias versus the flat band potential. Experimental data compared with calculated data from the model of fig. 2 [7].

In the electrolyte case, the charge in surface states can be compensated by counter ions of the electrolyte if a bias is applied. In order to keep $\Delta\phi_{SC}$ constant, the charge in surface states has to be so large that the bias change occurs predominantly in the Helmholtz double layer which has a much larger value than the space charge capacity. Therefore, Fermi level pinning between bulk and surface of the semiconductor over a wide range of external bias requires a rather large concentration of surface states in the order of 10^{14} cm^{-2} eV^{-1} (or 10^{13} cm^{-2} in a narrow energy interval of 0.1 eV or less).

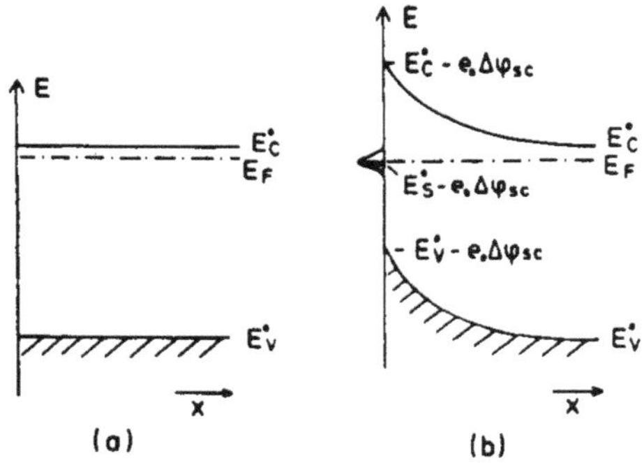

Fig. 4: Fermi level pinning at a semiconductor-vacuum (or gas) contact. (a) Ideal situation without surface states; (b) negative charge in surface states is compensated by positive space charge and the energy level of the surface state relative to the band edge controls the band bending, $e_o \cdot \Delta\phi_{SC}$.

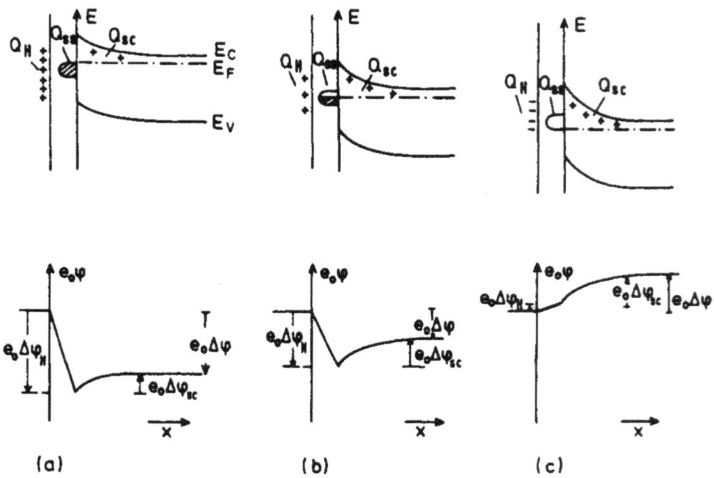

Fig. 5: Fermi level pinning in an electrolyte keeps the band bending $e_o \cdot \Delta\phi_{SC}$ nearly constant if the bias is varied from situation a → b → c in anodic direction. The charge Q_{SS} is negative. It has its maximum in situation (a), half of the maximum in (b) and reaches zero in (c). The variation of the bias changes predominantly $\Delta\phi_H$.

Surface states at electrochemical contacts are usually not dangling bonds since such bonds will normally react with constituents of the electrolyte. They are connected with structural imperfections, chemical impurities or the result of chemical reactions. They may even extend somewhat into the bulk in the case of a chemical transformation of the surface. Therefore it is extremely difficult to make a definite assignment to the origin of surface states at electrochemical contacts of semiconductors.

3. Charge Transfer Reaction

Like in the case of metals, one has to distiguish between electron transfer and ion transfer reactions. The kinetics of both types of reactions are at semiconductors very different from metals.

Electron transfer reactions

Electron transfer can occur in two different energy ranges due to the separation between the conduction and the valence band [8,13,14]. Since the electron transfer is a fast process and occurs between an initial and final state of the same energy, correlations between the energy levels of electron donors or electron acceptors in the electrolyte and the energy levels of the semiconductor are decisive for the kinetics. A distinction between electron transfer via the conduction band and via the valence band, the latter corresponding to a transfer of holes, can be made experimentally in accordance with the theory. Electrons and holes can be treated in such reactions like individual participants in the charge transfer reactions as indicated in the following reaction equations:

$$Ox^{+} \cdot solv + e_{c.b.} \rightleftharpoons Red \tag{1}$$

$$Ox^{+} \cdot solv \rightleftharpoons Red + h^{+}_{v.b.} \tag{2}$$

This means that electron transfer between electrons of the conduction band needs the presence of electrons at the interface and an encounter of this electron there with an acceptor molecule Ox^{+}. This is a cathodic process which has to be compensated at equilibrium by the reverse anodic process. The latter requires the presence of a donor molecule, Red, the interface and the availability of vacant electronic states in the conduction band into which an electron can be injected. As long as the electron concentration at the surface is not as high that degeneracy occurs, vacant states are available at the surface to a large and practically constant extent independent of the bias applied. In contrast, electrons are scarce and their concentration varies with the biaas exponentially. Consequently, the cathodic electron transfer in the conduction band is very much dependent on the bias, the anodic electron injection, however, is not. The result is a strongly rectifying current voltage curve.

The theoretical description of the electron transfer reaction via the conduction band can be given by the following equations [8,15].

$$j_c = j_c^+ - j_c^- = k_c^+ \cdot N_c \cdot N_{red} - k_c^- \cdot n_s \cdot N_{ox} \tag{3}$$

where k_c^+ and k_c^- are rate constants, N_{red} and N_{ox} the concentrations of the redox species, N_c the effective density of states in the conducton band and n_s the electron concentration at the surface. This concentration varies with the potential drop $\Delta\phi_{SC}$ in the space charge layer by

$$n_s = n_o \exp\left(- \frac{e_o \Delta\phi_{SC}}{kT} \right) \tag{4}$$

n_o is the concentration in the bulk.

At equilibrium, $\Delta\phi_{SC}$ has a definite value $\Delta\phi_{SC}^{\circ}$, where $j_c^+ = j_c^- = j_c^{\circ}$, which is the exchange current density for the particular concentrations, N_{ox} and N_{red}. If the external bias, ΔV, exclusively affects the potential drop in the space charge

$$\Delta V = \Delta\phi_{SC} - \Delta\phi_{SC}^{o} \tag{5}$$

eq. (3) can be written:

$$j_c = j_c^o \left| 1 - \exp\left(- \frac{e_o \Delta V}{kT} \right) \right| \tag{6}$$

This is the equation for an ideal diode.

In reality, there are deviations from this ideal behavior [16]. If the exchange current is small, because the concentration of the electronic charge carriers which react with the redox species is very small at equilibrium, one can measure the current in forward direction (i.e. with the majority carriers) over several orders of magnitude without limitations by the transport of the redox species to the interface. This can be demonstrated for the cathodic current at n-type semiconductors. The increase of the cathodic current remains exponential, however, the exponent is smaller than $e_o \Delta V/kT$ by a factor $1/f$ with $f > 1$. This is the same situation as in the forward current at solid state Schottky barriers [17,18]. Figure 6 gives an example from our own measurements with TiO_2 electrodes (unpublished). It shows the reduction of $[Fe(CN)_6]^{3-}$ ions in dependence of the potential applied in KCL solutions at different pH. The pH was controlled by addition of small amounts of HCl or NaOH to the 1 M KCl which served

$$j_v = j_v^+ - j_v^- = k_v^+ \cdot p_s \cdot N_{red} - k_v^- \cdot N_v \cdot N_{ox} \qquad (7)$$

In this case which can be studied at p-type semiconductors k_v^+, k_v^- are rate constants, N_v is the effective density of states in the valence band, and p_s is the hole concentration at the surface. In the bias range where the surface is non-degenerate and electronic equilibrium is maintaining, p_s is related to the band bending by

$$p_s = p_o \cdot \exp\left(\frac{e_o \, \Delta\phi_{SC}}{kT}\right) \qquad (8)$$

p_o is the hole concentration in the bulk. At equilibrium for the redox reaction, $j_v^+ = j_v^- = j_v^\circ$ and $\Delta\phi_{SC} = \Delta\phi_{SC}^\circ$.

If in the ideal case the bias only affects $\Delta\phi_{SC}$ (cf. eq. (4)), the current voltage curve of eq. (7) can be transferred to

$$j_v = j_v^o \left| \exp\left(\frac{e_o \, \Delta V}{kT}\right) - 1 \right| \qquad (9)$$

Deviation from this ideal diode equation at p-type semiconductors have the same reasons as pointed out for n-type semiconductors. Measurements at p-type materials are, however, difficult because most of the semiconductors decompose chemically if holes are accumulating at the contact to an electrolyte.

The magnitude of the exchange current densities in the energy range of the conduction band or the valence band is controlled by the correlations between the electronic energy levels of the semiconductor and the redox species Ox^+ and Red. This determines the activation energy. As a general tendency, it can be stated that an electron exchange with the conduction band is the more likely the more negative the redox potential and, vice versa, an electron exchange with the valence band is only to be expected for redox couples with a very positive redox potential. Another question is whether the electron transfer is an adiabatic process or not. This is decided by the matrix element for the electronic interaction between the electronic states of the semiconductor and the redox species. The detailed theory is found in the literature [8,19-21].

Ion transfer reactions

Semiconductors have strongly oriented bonds between neighboring atoms with a more or less covalent character. The number of next neighbors is much smaller than in metals. Ion transfer means that atoms in kink sites on the surface of the solid pass the interface and form solvated ions in solution. This is a process in which strong bonds have to be broken within a few steps. This process has to overcome high activation barriers. The same is true for the reverse process, the incorporation of ions from the solution into the crystal lattice with simultaneous association of electrons and the loss of strongly bound solvent molecules. Such reactions are very irreversible and equilibrium for ion exchange between the lattice and the electrolyte cannot be established. Only the anodic oxidation of semiconductors has been studied intensively because this process can be easily enforced by external means in contrast to the cathodic deposition. It is a model for bond breaking at the surface of a solid.

Elemental semiconductors like germanium and silicon have four covalent bonds between the neighbors of each atom in the bulk. At an atom in a kink site two bonds are left over. The electrons in these bonds are in electronic states of the valence band, their energy somewhat modified by the exposed location of the atom and the loss of neighbors. If holes are present at the surface, they will statistically reside in such bonds or even with preference there if these states are energetically more favorable. This, however, means that the bond to the neighbors is weakened. In the chemical picture of localized bonds, this state contains only one unpaired electron. At the same time, the atom has an excess positive charge and can interact much more strongly with nucleophilic reagents of the electrolyte than before. Such reagents are anions and the solvent itself, particularly if the solvent is water. The polar H_2O molecule can directly form a ligand or may dissociate and leave OH^- as a ligand. In this way, the intermediate state with a trapped hole at an atom in a kink site will be stabilized and the electronic state of the unpaired electron will be split off from the delocalized valence band states. It becomes a localized state with a higher energy, a level above the valence band edge. Such a state can efficiently trap holes. If this has happened, this bond is fully dissolved and the kink site atom is left with only one bond to the lattice. This bond can be broken much more easily by trapping of holes. Figure 7 shows the two steps in the oxidation of the first bond of a kink site atom in a simplified bond structure representation which can be used for materials with a diamond structure. Figure 8 describes the reaction path in terms of an energy profile for the consecutive steps [22].

It has been found that the unpaired electron can also be injected into the conduction band during the interaction with a reactant from the electrolyte, before a hole has been captured. This happens with a high probability at germanium [23] and at silicon in contact with fluoride solutions [24-26], to a very little extent at gallium arsenide [27]. Obviously, this thermal activation of the electron in the intermediate trap state is the more unlikely the larger the band gap of the semiconductor, as figure 8 suggests.

The kinetics of ion transfer in the anodic oxidation of semiconductors are controlled by the concentration of holes at the surface, and the first step of bond breaking is rate determining. P-type samples can therefore be easily oxidized by an anodic bias, while n-type samples can only be oxidized if holes are generated. The anodic current corresponds at n-type specimens to the rate of thermal generation of holes. The generation of holes can be accelerated by a high electric field at the surface at high bias [28] or this can be achieved by light absorption. Such photocurrents and their application for kinetic studies will be discussed in the next section.

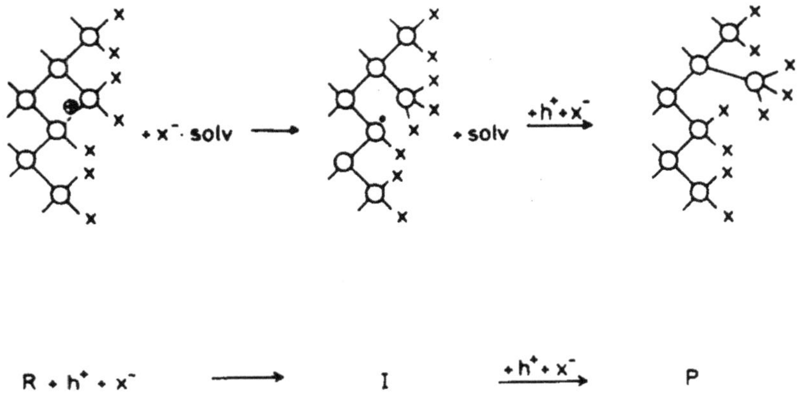

Fig. 7: Model for the breaking of a covalent bond at a kink site in two steps by holes.

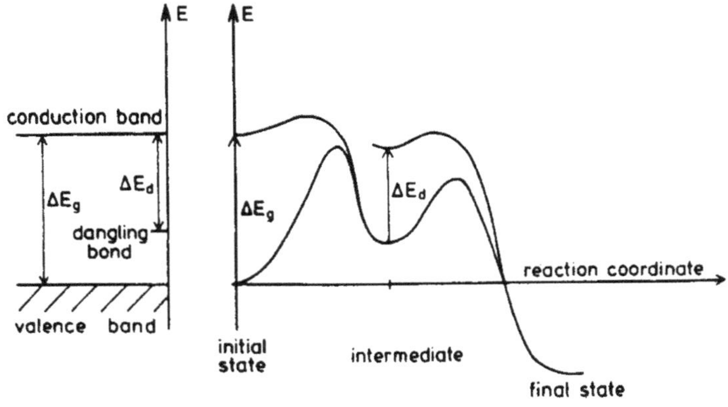

Fig. 8: Potential energy profile for the two steps of bond breaking by the interaction with a hole (upper curve) or by thermal electron injection into the conduction band (lower curve). On the left hand side, the energy levels in the semiconductor.

As the accumulation of holes at the surface weakens the bonds between atoms on the surface and the lattice and results in an increase of the reaction with nucleophilic reagents, the accumulation of electrons at the interface has a similar consequence for the interaction with electrophilic reagents. This can result in a reductive decomposition of semiconductors, particularly of compound semiconductors. It is in this case the more electronegative component of the semiconductor which passes the interface as an anion. An example is the decomposition of CdSe which forms Se^{2-} ions and a porous cadmium layer at cathodic bias in parallel to hydrogen evolution.

The principles of electrolytic decomposition of semiconductors have been analyzed on the basis of thermodynamics and kinetics in much detail [29,30]. A general representation of the bond breaking steps by holes or electrons is seen in figure 9.

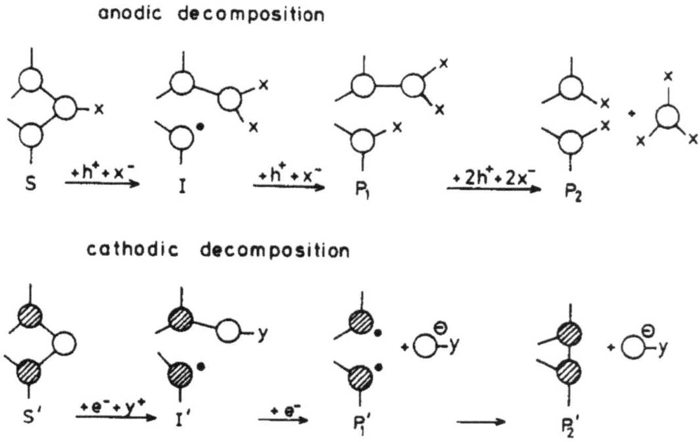

Fig. 9: Schematic description of the bond breaking steps on a semiconductor surface for the oxidation by the accumulation of holes or the reduction by accumulation of electrons.

4. Photoeffects and Luminescence

If light is absorbed in a semiconductor, electron hole pairs are generated. Their concentration increases until, in the steady state, as many electron hole pairs as are generated disappear by recombination. This is the situation for a semiconductor electrode at open circuit. If the semiconductor is part of a closed circuit, electrons and holes can pass the two contacts at different amounts and a photocurrent is generated. This happens particularly if the light is absorbed close to a contact and an electric field exists there due to the presence of a space charge layer. The situation at the semiconductor-electrolyte contact is in so far the same as at a semiconductor-metal contact. Figure 10 shows this situation for an n-type and a p-type semiconductor in contact with an electrolyte where the bias is positive for the n-type and negative for the p-type sample.

The quantum yield of the photocurrent depends on the competition with all types of recombination between excess electrons and holes. Recombination can occur in the bulk, in the space charge layer and at the surface. All these processes can be radiationless or radiative. The latter one is seen as luminescence. The experience has shown that surface recombination is normally radiationless or so weak in radiation that it can hardly be detected. An exact theoretical description of the kinetics of all these processes and the influence of the electric field in the space charge layer upon them is extremely complicated and has not been achieved yet. However, a simplified model represents the feature of this competition quite satisfactorily for semiconductor electrodes in a wide range of external bias. This model will be explained for n-type semiconductor electrodes, because only these can be stable in the dark over a wide range of bias. P-type semiconductors decompose at anodic bias and can only be studied at large enough cathodic bias. However, the same model can in principle be used for p-type electrodes in the range of their stability.

The simplest version of this model which can be applied for n-type electrodes to the range of anodic bias where a depletion layer exists, is the dead layer model [31,33]. It is assumed here that recombination occurs only in the bulk outside the space charge layer. All minority carriers which reach the space charge layer are driven to the surface and disappear there either by a photocurrent or by surface recombination. Another assumption is that all electron hole pairs generated within the depletion layer

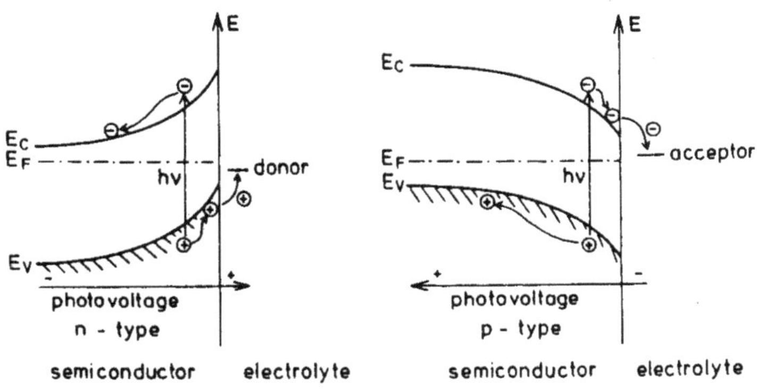

Fig. 10: Generation of photocurrents in a depletion layer of n-type (anodic) or p-type (cathodic) semiconductors.

are separated therein by the electric field and the holes being driven to the surface also react there with electron acceptors or disappear by recombination. This means that the concentration of the minority carriers drops to zero at the end of the space charge layer. It is further assumed that the charge carriers entering the depletion layer or being generated therein have a negligible effect on the charge distribution in the space charge layer. This means that their concentration remains small in comparison to the concentration of ionized donors and the extension of the space charge layer varies with the bias as in the dark. The latter assumption obviously requires low illumination intensities.

One can calculate the distribution of the minority carriers outside the space charge layer for this model [32,34]. The result is:

$$p(x, w) = \frac{\alpha I_0 \tau}{1 - \alpha^2 L^2} \left\{ \exp\left(-\alpha (x - w) \right) - \exp\left(-\frac{x - w}{L} \right) \right\} \quad (10)$$

x is the distance from the surface and w the exension of the space charge layer. w depends on the potential drop in this layer, $\Delta\phi_{SC}$, as

$$w = \left(\frac{2 \varepsilon \varepsilon_0}{e_0 N_D} \Delta\phi_{SC} \right)^{1/2} = w_0 \, \Delta\phi_{SC}^{1/2} \quad (11)$$

where ε is the relative dielectric constant, ε_0 the permittivity of vacuum and N_D the donor concentration.

α in eq. (10) is the absorption coefficient, τ the lifetime of the minorities, L the mean diffusion length, $L = (D\tau)^{1/2}$. The concentration profile varies with w as shown in figure 11.

If the bias is so high that the concentration of electrons at the surface approaches zero, surface recombination vanishes and the holes reaching the surface contribute only to the photocurrent. In this case a simple relation between photocurrent and photoluminescence can be obtained, provided recombination in the bulk occurs to some extent radiatively. This model is shown in figure 12.

The photocurrent, i_{ph}, results from the diffusion of holes into the space charge layer and the light absorption in this layer:

$$i_{ph} = A I_0 \left\{ 1 - \exp(-\alpha w) + \frac{\alpha L}{1 + \alpha L} \exp(-\alpha w) \right\}$$

$$= A I_0 \left\{ 1 - \frac{1}{1 + \alpha L} \exp(-\alpha w) \right\} \quad (12)$$

I_0 is the illumination intensity and A a conversion factor into electrical units.

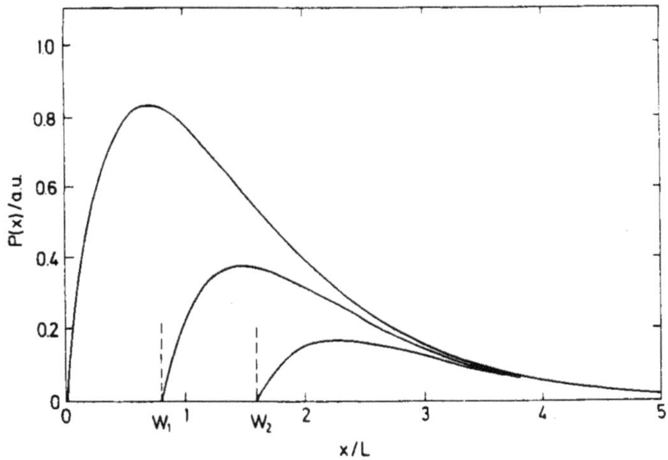

Fig. 11: Concentration profile of the minority carriers in a semiconductor at the steady state under illumination for different extensions of the depletion layer, w, calculated from the dead layer model with infinite rate of the consumption of the charge carriers at the interface.

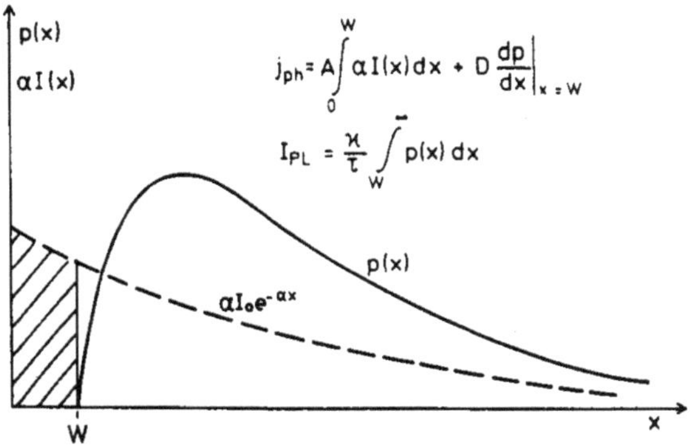

Fig. 12: Minority carrier concentration profile and rate of light adsorption in a depletion layer of a semiconductor (n-type) in the steady state of illumination according to the dead layer model as the basis for the calculation of the photocurrent and the photoluminescence intensity.

The photoluminescence intensity I_{PL} comes only from the field free bulk [34]:

$$I_{PL} = \kappa I_0 \frac{1}{1 + \alpha L} \exp(-\alpha w) \qquad (13)$$

κ is the fraction of radiative recombination, multiplied with a geometrical factor for the light recording instrument. One sees that the luminescence contains the same exponential factor for its decrease with w, or the bias respectively (cf. eq. (11)), as the photocurrent for its increase. If αL is large, the variation of the photocurrent with the bias becomes very small, while the relative change of the photoluminescence remains considerable. Photoluminescence is therefore much more sensitive to a variation of w than the photocurrent. This is demonstrated in figure 13 for a CdS electrode [36].

This model can be modified for the condition that the minority charge carriers do not react extremely fast at the interface and their concentration at the end of

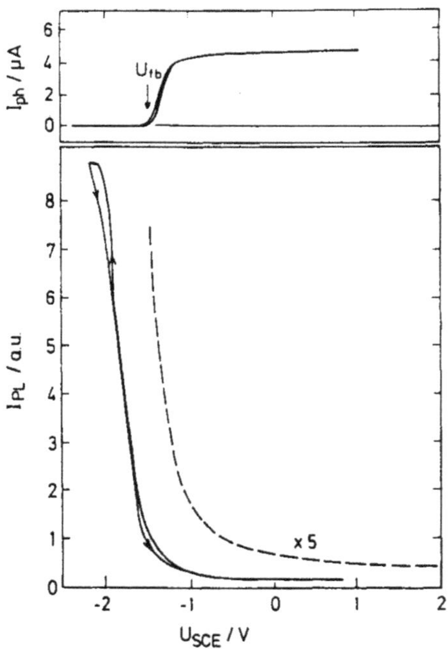

Fig. 13: Photocurrent and photoluminescence at a CdS electrode in 1 M Na_2S_x (x = 1.3) + 0.1 M NaOH solution in dependence of the bias. Flat band potential at -1.5 V vs. SCE. Excitation with light of λ = 474 nm [36].

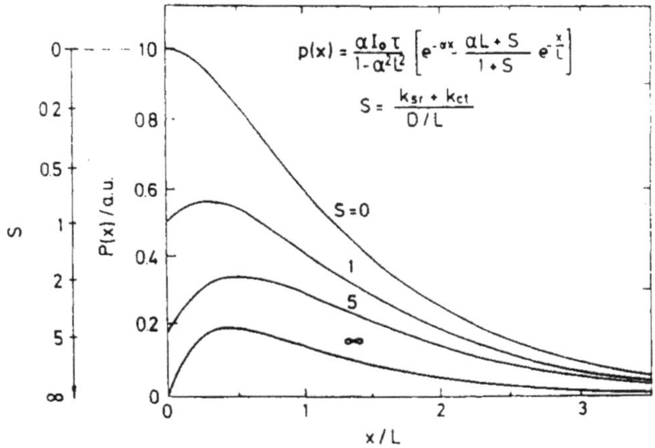

Fig. 14: Minority carrier concentration profiles at the flat band potential for different rates of their consumption at the interface by either recombination or electron transfer.

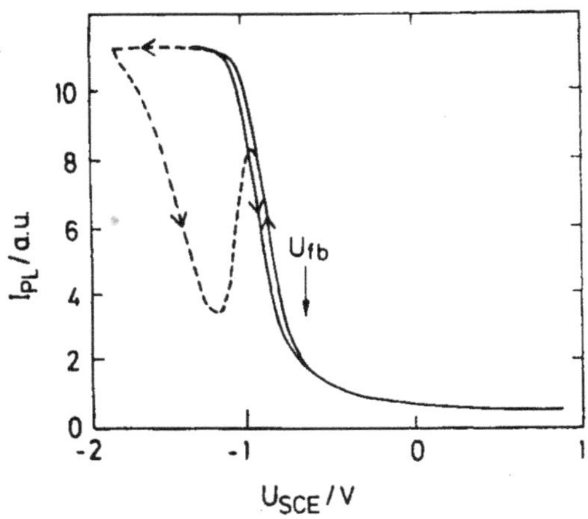

Fig. 15: Photoluminescence on a CdSe electrode in 0.1 M KOH + 0.1 M K_3/K_4 $Fe(CN)_6$ in dependence of the bias. Broken line shows a hysteresis indicating a change in surface composition of the semiconductor if the electron accumulation at the interface exceeds a critical concentration [36].

the space charge layer does not drop to zero. The concentration profile outside the space charge layer follows in this case the relation [32,34,35]:

$$p(x, w) = \frac{\alpha I_0 \tau}{1 - \alpha^2 L^2} \left\{ \exp\left(-\alpha (x - w) \right) - \frac{\alpha L + s}{1 + s} \exp\left(-\frac{x - w}{L} \right) \right\} \quad (14)$$

where s is a dimensionless reaction rate, the quotient between the sum of the virtual rate constants for charge transfer, k_{ct}, and surface recombination, k_{sr}, and the rate constant equivalent for the transport to the surface, D/L.

$$s = \frac{k_{ct} + k_{sr}}{D/L} \quad (15)$$

With $s \to \infty$, the ideal dead layer model is achieved (cf. eq. (10)). Such concentration profiles are shown in figure 14 for the flat band potential, w = 0.

One interesting possibility has been found with regard to photoluminescence at semiconductor electrodes. It is possible to reach a maximum of photoluminescence by totally suppressing surface recombination [36]. This occurs in the bias range where an accumulation layer is formed. Even if the majority carriers accumulate at the surface up to degeneracy, there still remains a space charge layer with 10 - 30 Å extension or even more, depending on the doping level. This is connected with an energy barrier for the minority carriers on their path to the surface. If this barrier is high enough, the minority carriers are so strongly repelled that they do not reach the surface any longer. In the steady state they can only disappear by recombination in the bulk. Figure 15 gives an example for such a situation where the photoluminescence reaches a plateau at cathodic bias which corresponds to this maximal luminescence. This can hardly be achieved with solid state Schottky barriers or in contact with gases nor in vacuum. It is an example how solid state properties can be analyzed by electrochemical experiments. If the potential turns too cathodic, the semiconductor surface starts being reduced and Cd clusters are formed thereon. This changes the surface dipole and with it the accumulation layer height. It also lets the rate of surface recombination increase. Such changes can clearly be detected in the photoluminescence intensity as this example demonstrates.

5. References

1. Yu. V. Pleskov, in *Comprehensive Treatise of Electrochemistry,* Vol. 1, eds. J. O'M. Bockris, B. E. Conway and E. Yeager, (Plenum Publ. Co., New York, 1980), p. 291.
2. H. Gerischer, *Electrochim. Acta* **35** (1990) 1677.
3. W. H. Laflere, F. Cardon and W. P. Gomes, *Surf. Sci.* **44** (1974) 541.
4. H. Gerischer, *Electrochim. Acta* **34** (1989) 1005.

5. J. F. Dewald, *Bell Syst. Techn. J.* **39** (1960) 615.
6. W. P. Gomes and F. Cardon, *Progr. in Surf. Sci.* **12** (1982) 155.
7. H. Gerischer and R. McIntyre, *J. Chem. Phys.* **83** (1985) 1363.
8. H. Gerischer, in *Physical Chemistry, An Advanced Treatment,* Vol. 9, ed. H. Eyring, (Academic Press, New York, 1970), p. 463.
9. A. Hamnett, in *Comprehensive Chemical Kinetics,* Vol. 27, ed. R. G. Compton, (Elsevier, Amsterdam, 1987), p. 242.
10. W. E. Spicer, I. Lindan, P. Skeath, C. Y. Su and P. Chye, *Phys. Rev. Lett.* **44** (1980) 420.
11. A. J. Bard, A. B. Bocarsly, Fu-Ren F. Fan, E. G. Walton and M. S. Wrighton, *J. Am. Chem. Soc.* **102** (1980) 3671.
12. R. L. Van Meierhaeghe, F. Cardon and W. P. Gomes, *J. Electroanal. Chem.* **188** (1985) 287.
13. H. Gerischer, *Z. Phys. Chem. (Frankfurt)* **27** (1961) 48.
14. J. F. Dewald, in *Semiconductors,* ACS Monography 140, ed. N. B. Hanney, (Reinhold Publ. Co., New York, 1959), p. 727.
15. H. Gerischer, in *Advances of Electrochemistry and Electrochemical Engineering,* Vol. 1, ed. P. Delahay, (Interscience, New York, 1961), p. 139.
16. R. Memming, *Ber. Bunsenges. Phys. Chem.* **91** (1987) 353.
17. V. R. Rideout, *Thin Solid Films* **48** (1978) 261.
18. E. H. Rhoderick and R. H. Williams, *Metal-Semiconductor Contacts,* 2nd ed., (Clarendon Press, Oxford, 1988).
19. R. R. Dogonadze and Yu. A. Chizmadzev, *Dokl. Akad. Nauk SSSR* **145** (1962) 848.
20. R. A. Marcus, *J. Phys. Chem.* **94** (1990) 1050.
21. H. Gerischer, *J. Phys. Chem.* **95** (1991) in press.
22. H. Gerischer and W. Mindt, *Electrochim. Acta* **13** (1968) 1329.
23. H. Gerischer and F. Beck, *Z. Phys. Chem. (Frankfurt)* **24** (1960) 378.
24. M. Matsamura and S. R. Morrison, *J. Electroanal. Chem.* **144** (1983) 113; **147** (1983) 157.
25. H. Gerischer and M. Lübke, *Ber. Bunsenges. Phys. Chem.* **91** (1987) 394.
26. L. M. Peter, D. J. Blackwood and S. Pons, *Phys. Rev. Lett.* **62** (1989) 308.
27. D. van Maekelbergh and J. J. Kelly, *J. Electrochem. Soc.* **136** (1989) 108.
28. H. Gerischer, D. Ross and M. Lübke, *Z. Phys. Chem. (Frankfurt)* **139** (1984) 1.
29. H. Gerischer, *J. Electroanal. Chem.* **82** (1977) 133; *J. Vac. Sci. Technol.* **15** (1978) 1422.
30. A. J. Bard and M. S. Wrighton, *J. Electrochem. Soc.* **124** (1977) 1706.
31. U. Langmann, *Appl. Phys.* **1** (1973) 219.
32. K. Mettler, *Appl. Phys.* **12** (1977) 75.
33. A. B. Ellis, in *Chemistry and Structure of Interfaces,* eds. B. Hall and A. B. Ellis, (Verlag Chemie, Weinheim, 1986), p. 254.
34. B. Smandek, G. Chmiel and H. Gerischer, *Ber. Bunsenges. Phys. Chem.* **93** (1989) 1094.
35. P. B. Johnson, C. S. McMillan, A. B. Ellis and W. S. Hobson, *J. Appl. Phys.* **62** (1987) 4903.
36. G. Chmiel and H. Gerischer, *J. Phys. Chem.* **94** (1990) 1612.

ADSORPTION

ADSORPTION AT ELECTRODES

Arthur T. Hubbard*

and

Ping Gao

Department of Chemistry
University of Cincinnati
Cincinnati, OH 45221-0172

ABSTRACT

Surface electrochemical studies are described and summarized in which atomic, ionic or molecular layers were allowed to form from aqueous solutions at well-defined Pt(111) surfaces. The resulting adsorbed layers were chemisorbed in most cases and stable in vacuum, permitting identification and quantitation by Auger spectroscopy, EELS, LEED and electrochemistry. Adsorbed atomic, ionic, or molecular layers formed at metal-solution interfaces frequently display long-range order. Molecular properties of the adsorbed layers correlat with their electrochemical properties. The molecular orientation of organic adsorbates was deduced from packing density measurements, supplemented with vibrational spectra. Interfacial variables such as electrode potential have a strong influence on interfacial structure along with the nature and mode of surface attachment of adsorbates. The angular distribution of Auger electron emission from metal single crystals and atomic adsorbed layers has proved to be useful for direct imaging of surface crystal and interfacial structure.

Introduction

Surface electrochemistry, along with other interfacial sciences, is enjoying a renaissance brought about by new surface characterization techniques which have emerged and proliferated during the past two decades. By employing modern surface spectroscopic techniques in conjunction with electrochemical methods, surface electrochemical studies are revealing aspects of the nature of adsorbed species and reactivity previously unseen [1-2]. By using a multitechnique approach, molecular level information about chemisorbed species at the metal-solution interface of well-defined electrode can be obtained, opening a new avenue for investigation of electrochemistry.

Adsorbed layers formed at well-defined electrodes in solution generally involve chemical reactions with the electrode surface, and are stable in vacuum [1-2]. Thus, surface electron spectroscopic methods in ultrahigh vacuum are applicable for the identification of chemisorbed species at such surfaces. Well-defined surfaces have been studied in order to eliminate many of the structural imperfections which would otherwise complicate the results due to the presence of a mixture of adsorption states. As a result of studies of such well-defined molecular systems, clear correlations between the adsorbed state and surface chemical or electrochemical reactivity are beginning to emerge.

The investigation of interfacial structure formed at a single-crystal electrode concerns the molecular orientation, mode of surface attachment, spectroscopic characteristics and chemical/electrochemical reactivity of adsorbed ions and molecules[1-2]. Because of the inherent difficulty of such studies, a combination of surface analytical techniques are necessary rather than a single technique. Electrochemical methods are widely utilized to form adsorbate layers on electrodes and to monitor their electrochemical activities before and after surface characterization in the UHV system. Several electron spectroscopic techniques have been employed for surface characterization: low-energy electron diffraction (LEED), for both substrate and adsorbate crystallographic structure determination; Auger electron spectroscopy for surface elemental analysis and quantitative measurement of surface coverage; and high resolution electron energy loss spectroscopy (EELS) for surface vibrational information and mode of bonding [1-2]. As a result of recent experimental and theoretical progress, angular distribution Auger microscopy (ADAM) has emerged as a very powerful technique for imaging atomic and molecular structure at the interface, providing direct information regarding surface structure, bonding, and molecular orientation [3].

The present paper provides a discussion of some representative findings from our recent studies of adsorbed layers on well-defined electrode surfaces. Adsorption of halide electrolytes (including solvents, coadsorbed ions etc.) is discussed first, followed by adsorbed organic layers of pyridine and its derivatives, and finally some illustrative examples of the new information provided by the recent application of ADAM. We limit the presentation here to a brief review; experimental details are given in the references cited.

Surface Ionic Layers

A primary thrust of surface spectroscopy at metal-solution interfaces is to gain insight into the structure and the nature of surface bonding. Of particular electrochemical interest are surface-adsorbate interactions for halides and pseudohalides in view of their structural simplicity, strong adsorption at metal surfaces, and common use as electrolytes. These studies have illustrated that electrode potential has a determining influence on the structure formed by adsorption of various anions at single–crystal surfaces [4].

Figure 1A shows a typical cyclic voltammogram of a Pt(111) surface containing a layer of Br atoms adsorbed from a slightly acidic bromide solution [4]. The adsorption of Br^- is a redox process:

$$Br^- = Br(ads) + e^-$$

LEED studies reveal that the sharp peaks (2 and 2' in Figure 1A) near -0.1V correspond to reductive desorption, accompanying a transition between a Pt(111)(3x3)-Br adlattice containing $\theta=4/9=0.444$ Br atoms per surface Pt atom(Figure 1B) and a Pt(111)(4x4)-Br adlattice containing $\theta=7/16=0.4375$ (Figure 1C). Br^- is not strongly adsorbed from alkaline solutions primarily due to strong competition from oxygenous adsorbates [4].

Relative affinities of various cations towards retention of water at such surfaces are explored [4]. Retention of water by the surface under vacuum correlates with Ca^{2+} retention and varies from 5 to 15 water molecules per Ca^{2+} cation, approaching a maximum as the solution pH increases. In contrast, K^+ ions, being less strongly hydrated, do not retain detectable amounts of water under vacuum.

Iodide adsorption at Pt(111) displays analogous structural transitions with electrode potential [5]. Emersion of Pt(111) from aqueous iodide solutions at controlled potential results in an ordered layer of iodine atoms:

$$I^- = I\ (ads) + e^-$$

A Pt(111)(3x3)-I [$\theta=4/9$] adlattice is present at moderately positive potentials, a Pt(111)($\sqrt{7}$x$\sqrt{7}$)R19°-I [$\theta=3/7$] adlattice in mid-range, and a Pt(111)($\sqrt{3}$x$\sqrt{3}$)R30°-I [$\theta=1/3$] adlattice at moderately negative potentials, followed by reductive desorption of I^-anions at negative extremes of potential. Information on iodine packing densities obtained from Auger data for these systems is summarized in Figure 2, along with the packing densities of K^+ and O (due to H_2O, OH etc.). Iodine adsorption shows a slight pH dependence, in contrast to the behavior of bromide [4] and chloride [7], presumably because iodine is covalently bonded to the electrode surface and suppresses adsorption of OH and related oxygen species. A remarkable property of iodide-treated surfaces is that the Pt(111)(3x3)-I and Pt(111) ($\sqrt{7}$x$\sqrt{7}$)R19°-I adlattices are hydrophobic whereas the Pt(111)($\sqrt{3}$x$\sqrt{3}$)R30°-I adlattice is hydrophilic [5]. Transition from ($\sqrt{7}$x$\sqrt{7}$) to ($\sqrt{3}$x$\sqrt{3}$) adlattice is signalled by a reversible, pH-independent spike in the cyclic voltammogram (Figure 2), as well as a change from hydrophobic to hydrophilic behavior [5]. Evidently, an abrupt change in electronic charge density at the interface accompanies the ($\sqrt{7}$x$\sqrt{7}$) to ($\sqrt{3}$x$\sqrt{3}$) structural transition. Retention of K^+ and oxygen species (such as H_2O, OH, etc.) by the surface is linked, but varies inversely with the iodine coverage (Figure 2). The iodine adlattices, (3x3), ($\sqrt{7}$x$\sqrt{7}$)R19° and ($\sqrt{3}$x$\sqrt{3}$)R30° are also observed when Pt(111)

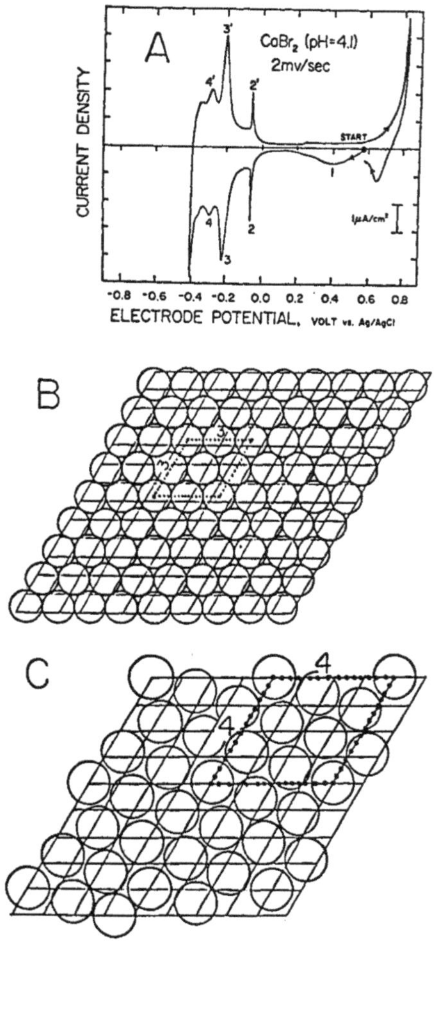

Figure 1. (A) Cyclic voltammogram of Pt(111) in 5 mM $CaBr_2$ (pH 4.1); scan rate, 2 mV/s; (B) Model structure of Pt(111)(3x3)-Br; Γ_{Br} =4/9 Br atoms per surface Pt atom; (C) Model structure of Pt(111)(4x4)-Br; Γ_{Br} = 7/16.

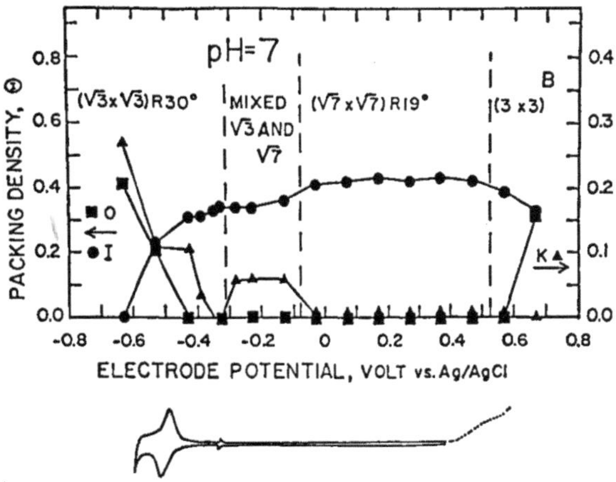

Figure 2. Adsorption profiles of I, K, and O from 0.1 mM KI (pH 7) at Pt(111) as a function of electrode potential.

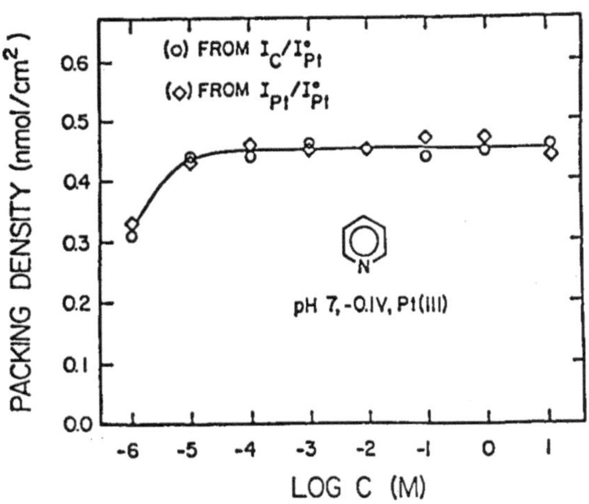

Figure 3. Packing density of PYR at Pt(111) versus concentration. Experimental conditions: electrode potential, -0.1 V vs. Ag/AgCl; electrolyte, 10 mM KF adjusted to pH 7 with KOH; rinsed with 0.1 mM KF (pH 7).

is treated with HI or I_2 vapor in vacuum [5-6]. Similar structural transitions are observed for chloride [7], cyanide, and thiocyanate [8] at Pt(111), as well as for chloride, bromide, and iodide at Ag(111) [9]. Adsorbed cyanide and thiocyanate layers also retain cations at Pt(111) surfaces [8] to neutralize surface charge due to the adsorbed ionic layer. The amounts of cations retained and the relative affinities of surface layers for cations are controlled by the size and strength of solvation of the particular cation, as well as by the pH of solution.

Surface Organic Molecular Layers

A major objective of the study of the electrode-solution interface is to understand the surface structure of organic molecular layers including surface bonding, and molecular orientation and the correlation between the surface structures and electrochemical activities. Pyridine (PYR) and its derivatives are of interest in view of their simple structures and interesting surface properties associated with the possible involvement of σ or/and π-bonding to the surface.

The pyridine ring is bonded to the Pt(111) surface through the N atom of the ring as indicated by the EELS spectra of adsorbed PYR, in which a peak near 416 cm^{-1}, attributed to the Pt-N stretching mode, is always observed [10].

Packing densities of PYR adsorbed on Pt(111) at different concentrations at a constant electrode potential, -0.1 V, obtained from the Auger data, are shown in Figure 3 [10]. A striking characteristic of the isotherm is the virtual constancy of packing density over a wide range of concentrations from 10^{-5} M up to neat PYR (12.4 M). The molecular packing density at the saturation limit is 0.45 nmol/cm^2, based upon ($I_C/I^{\bullet}_{Pt}$), or 0.46 nmol/cm^2 from ($I_{Pt}/I^{\bullet}_{Pt}$) [10]. These packing density data suggest a tilted orientation with the N atom bonded to the surface and a 71° angle between the plane of the ring and the Pt surface. The theoretical packing density in this configuration is 0.450 nmol/cm^2, based upon covalent and van der Waals radii [11]. Vertically oriented PYR would result in a much higher packing density, 0.73 nmol/cm^2, than the observed value of 0.45 nmol/cm^2, while the horizontal orientation would lead to only 0.38 nmol/cm^2 [10]. A similar angle of 74°±10° was estimated for adsorbed PYR under vacuum at room temperature on Pt(111) by use of NEXAFS data [12]. Also a tilted structure has been proposed for PYR adsorbed on Ag(111) from vacuum based on EELS studies [13].

Similar LEED patterns are observed for adsorbed PYR at concentrations between 10^{-5} M and neat PYR, although the sharpest pattern resulted from 1 mM PYR solution (shown in Figure 4A). A simulation of the LEED pattern with unit mesh vector lengths of 3.324 and 4.378, respectively, an included angle of 77.1°, and a rotational angle of 34.0°, is shown in Figure 4B. The simulated LEED pattern is superimposed upon the observed pattern (Figure 4C), and shows close agreement. Based upon three PYR molecules per (3.324x4.738, 77°)R34°-PYR unit cell, the calculated packing density for this structure is 0.421 nmol/cm^2, consistent with the measured packing density, 0.45 nmol/cm^2. Combining the EELS, Auger, and LEED observations leads to the surface

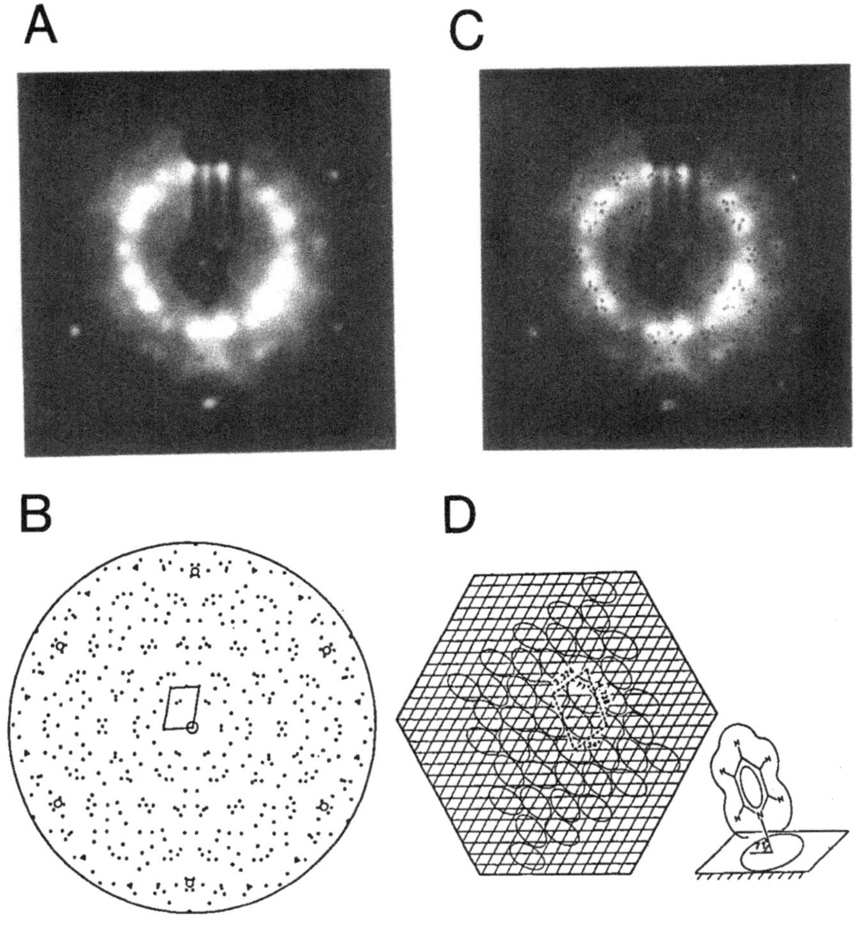

Figure 4. (A) Observed LEED pattern of adsorbed PYR on Pt(111); (B) Computer simulated LEED pattern of Pt(111)(3.324 x 4.738, 77.1°)R34°-PYR; (C) Comparison of simulated and observed LEED patterns; (D) Model structure of Pt(111)(3.324 x 4.738, 77.1°)R34°-PYR.

configuration for adsorbed PYR: Pt(111)(3.324x4.738, 77.1°)R34.0°-PYR, having an oblique, incommensurate mesh as illustrated in Figure 4D. The constant packing density seen from the isotherm curve and the incommensuracy of the adsorbed PYR layer suggest that the angle of 71° between the pyridine plane and the Pt surface results from a balancing of forces between σ-donor bonding of the nitrogen atom to Pt, and π back-donation from Pt into the aromatic ring [10].

Monosubstituted pyridine derivatives are also examined with the substituent groups chosen to alter the adsorption sterochemistry and/or provide possible alternative binding groups to the metal surface [10].

The influence of electrode potential upon the nature of adsorbed molecular species and the mode of surface attachment is illustrated by the EELS spectra of nicotinic acid(NA) (3-pyridinecarboxylic acid), Figure 5 [10]. The EELS spectrum of an adsorbed layer of NA formed from acidic solutions at a negative potential (-0.2 V) (Figure 5A) shows an intense O-H stretching mode (3566 cm^{-1}) attributed to the carboxylic acid group of NA. However, this band vanishes when adsorption is carried out at sufficiently positive potential (+0.6 V) (Figure 5B). A similar trend is observed for the 1733 cm^{-1} band (Figure 5A,B), the carboxyl C=O stretching mode, while other features remain essentially the same. This observation is a clear indication that the carboxylate moiety is bonded to the surface when the electrode potential is shifted from negative to positive values. Considering the EELS data and the molecular packing densities obtained from Auger spectroscopy, along with the theoretical model based upon covalent and van der Waals radii, the detailed surface configurations of adsorbed NA are illustrated in Figure 6 for different conditions of pH and electrode potential. Analogous to pyridine, NA is tilted with an angle of 75° between the plane of the ring and the Pt surface (Figure 6A), leaving the carboxyl moiety free. As illustrated in Figure 6B, the disappearance of the O-H stretching band and the significant weakening of the C=O band with increasing positive potential is interpreted as a slight change in the angle of tilt and rotation, allowing the carboxylate group to form a coordinate covalent bond with the Pt surface.

The pH-dependent adsorption behavior for NA on Pt(111) has also been explored. The strong O-H stretching band (3566 cm^{-1}, Figure 5A) also vanishes when the adsorbed NA layer is rinsed with basic solution (0.1 mM KOH) (Figure 5C) and reappears after the surface is rinsed with acid (2 mM HF). Significant retention of K^+ ions is indicated by Auger spectroscopy, suggesting that an ion-exchange reaction, replacement of the protons of the carboxylate acid moiety of NA by K^+, occurs (Figure 6A and 6C, respectively). Not surprisingly, a distinct potential-dependence of K^+ retention for adsorbed NA on Pt(111) from basic solution is observed. Since this is unique to pyridine derivatives containing carboxylic acid group(s) and is not observed for pyridine, it implicates the coordination of a carboxylate moiety to the Pt surface (Figure 6C-D). Studies of NA and other pyridine carboxylate compounds prove that the ring substituents can act as surface anchoring groups providing additional binding sites to the surface under certain conditions.

It is found that chemisorption of the electroactive center to a surface causes loss of electroactivity, whereas attachment to the surface through some other part or functional group of the molecule preserves the electroactivity [1,2,14]. In contrast to the

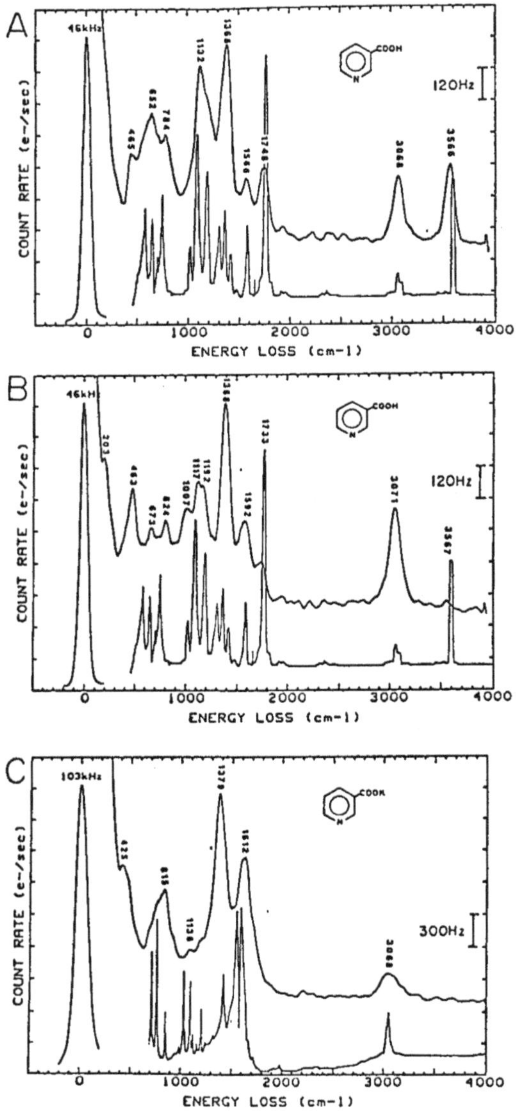

Figure 5. (A) EELS spectrum of NA adsorbed at Pt(111), -0.2 V, pH 3; lower curve in A and B is mid-IR spectrum of NA vapor; (B) As in (A), but +0.6 v; (C) EELS spectrum of NA at Pt(111), -0.3 V, rinsed at pH 10; lower curve is mid-IR spectrum of solid KNA. Experimental conditions: adsorption from 1 mM in 10 mM KF, pH 3, followed by rinsing with 2 mM HF (A and B, pH 3) or 0.1 mM KOH (C, pH 10).

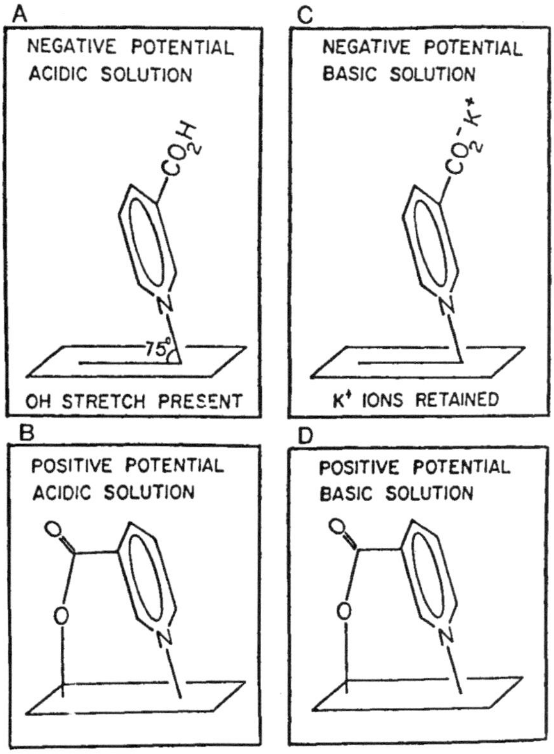

Figure 6. Structural variations of NA on Pt(111) surface vs. pH and electrode potential.

electrochemical behavior of adsorbed hydroquinone, which undergoes a complete oxidation to CO_2 due to a horizontal orientation on the Pt surface [14], 3-(pyridyl)-hydroquinone (3PHQ) forms an adsorbed layer at Pt(111) which displays a reversible two-electron redox process characteristic of the hydroquinone moiety. Shown in Figure 7 is the cyclic voltammogram obtained from a freshly adsorbed 3PHQ layer (solid curve). Notice that the anodic and cathodic peaks for adsorbed 3PHQ are slightly broadened in comparison with the corresponding peaks for solution species. This phenomenon is commonly observed for chemisorbed species, and is probably due to adsorbate-surface interactions. Also shown in Figure 7 for comparison is the cyclic voltammogram obtained after the coated surface was placed in vacuum for 1 hour during surface characterization (dotted curve). The two curves are almost identical, showing that the 3PHQ layer on Pt(111) is stable in vacuum. The surface coverage of 3PHQ is estimated by integrating the i-E curve and using Faraday's law (n_{ox}=2), yielding a value of 0.27 nmol/cm^2, consistent with the value of 0.28 nmol/cm^2 obtained from Auger data. The excellent agreement of surface coverage measurements by the two independent methods (Auger spectroscopy and coulometry) indicates that the entire 3PHQ layer is electroactive, suggesting that the hydroquinone moiety is pendant and not directly bound to the surface [10]. Analogous to PYR and NA, the surface configuration of 3PHQ is derived with an angle of 74° between the plane of the pyridine ring and the Pt surface.

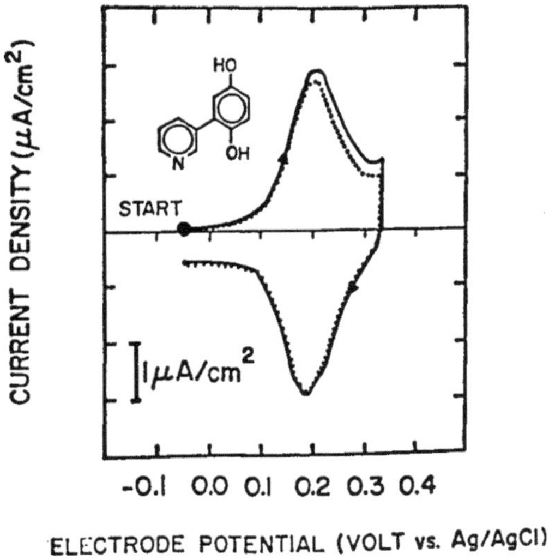

Figure 7. Cyclic voltammogram of adsorbed 3PHQ at Pt(111): solid curve (——), immersion into 0.5 mM 3PHQ, followed by rinsing with 1 mM HF; dotted curve (.....), 3PHQ layer (as above) after 1 hr in vacuum.

Angular Distribution Auger Microscopy (ADAM)

Measurements of Auger electron angular distributions from single-crystal surfaces have led to the discovery that the distributions consist primarily of the silhouettes of near-surface atoms backlit by Auger emission originating from atoms deeper in the metal, yielding direct structural information about the surface [3]. That is, the angular distribution of Auger electron emission from each element yields an image of the structure of the surface. A schematic diagram of the ADAM experiment is illustrated in Figure 8. Measurement and display of the complete angular distribution of Auger electrons greatly facilitate recognition of the features of the Auger electron distribution particularly when the data are compared with a complete theoretical simulation of the distribution.

Shown in Figure 9 is the experimental angular distribution of Auger electrons emitted from a bare Pt(111) surface with a kinetic energy of 65 eV [3]. Regions of highest intensity appear white and lowest are black (less than 10% of full scale). The data display striking variations in signal with angle of detection, and threefold symmetry with many fine features. Detailed investigation revealed that the angular distribution is created by Pt atoms behaving as spherical scatterers illuminated by other Pt atoms located deeper within the sample, acting as point sources of Auger electrons. That is, the distribution from bare Pt surface is recognizable as hexagonal layers of atomic scatterers backlit by the Auger electrons emitted from the atoms located below each layer.

Theoretical angular distributions show that the ADAM pattern for a Pt(111) surface results from the spatial relationships between Auger emitters and scatterers as illustrated in Figures 10A-D. The theoretical angular distributions illustrate how silhouettes result (layer-by-layer) when Auger electrons from atomic point emitters are scattered by atoms represented by spheres having uniform cross-section. The theoretical ADAM pattern for Pt(111) has been simulated by considering all Auger emission and scattering events occurring in the outermost 16 atomic layers, Figure 10E. The best quantitative agreement is achieved between theory and experiment when each individual scatterer is taken to be 40% transparent, with a radius equal to 90% of its atomic radius [3]. Notice that the Y-shaped distribution of Auger electron intensity (Figure 10E) is a striking feature in agreement with the observed ADAM pattern (Figure 9). The overall feature of the ADAM pattern indicates that the image is largely determined by contributions from the top 4 layers (as shown in Figure 10 A-D).

In order to further analyze the fine features of ADAM images such as Figure 9, atomic vibration has been introduced into the simulation. A simulation that includes vibrational motion equal to 9 % of the atomic radius is depicted in Figure 10F [3]. The effect of considering uniform thermal vibrations of all atoms in the first 16 layers can be seen by comparing Figure 10F with 10E. Excellent agreement resulted between theory (Figure 10F) and experiment (Figure 9).

Usefulness of ADAM for imaging the structure of surface adsorbed layers is illustrated by results obtained for a silver monatomic layer having an overlayer of iodine atoms, all resting atop the Pt(111) surface. Clearly visible at the center of the Ag Auger

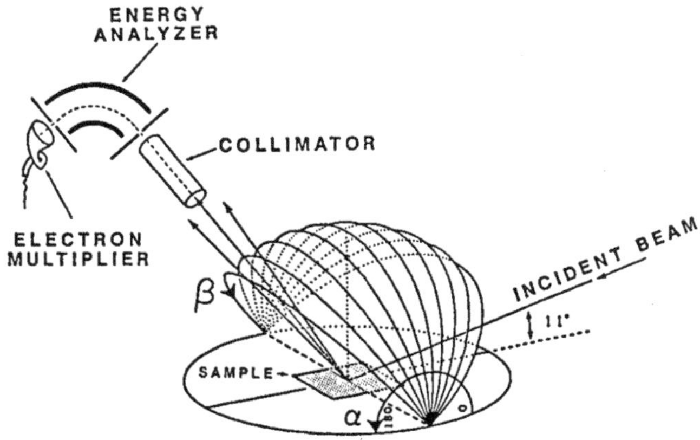

Figure 8. ADAM angular distributions are measured versus the angular coordinates (α and β) of the detector. Auger electrons emission is stimulated by an incident electron beam impinging on the sample at a fixed angle (11° from the surface plane). An Auger electron traveling along the selected trajectory (α, β) must pass through the angle-resolving collimator and an energy resolving analyzer before being amplified and counted.

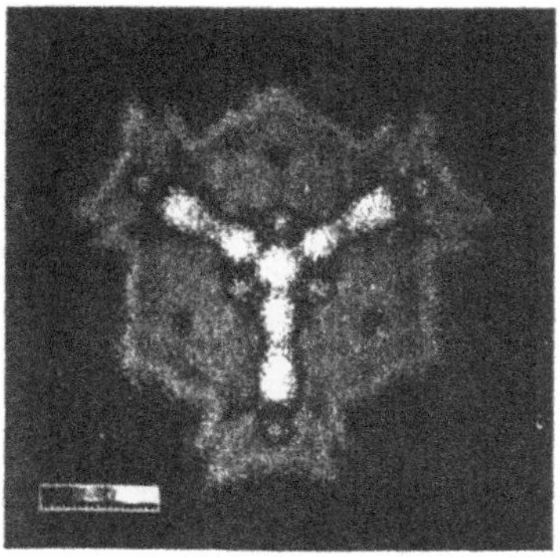

Figure 9. The measured angular distribution of Auger electrons (65 eV kinetic energy) emitted from a bare Pt(111) surface.

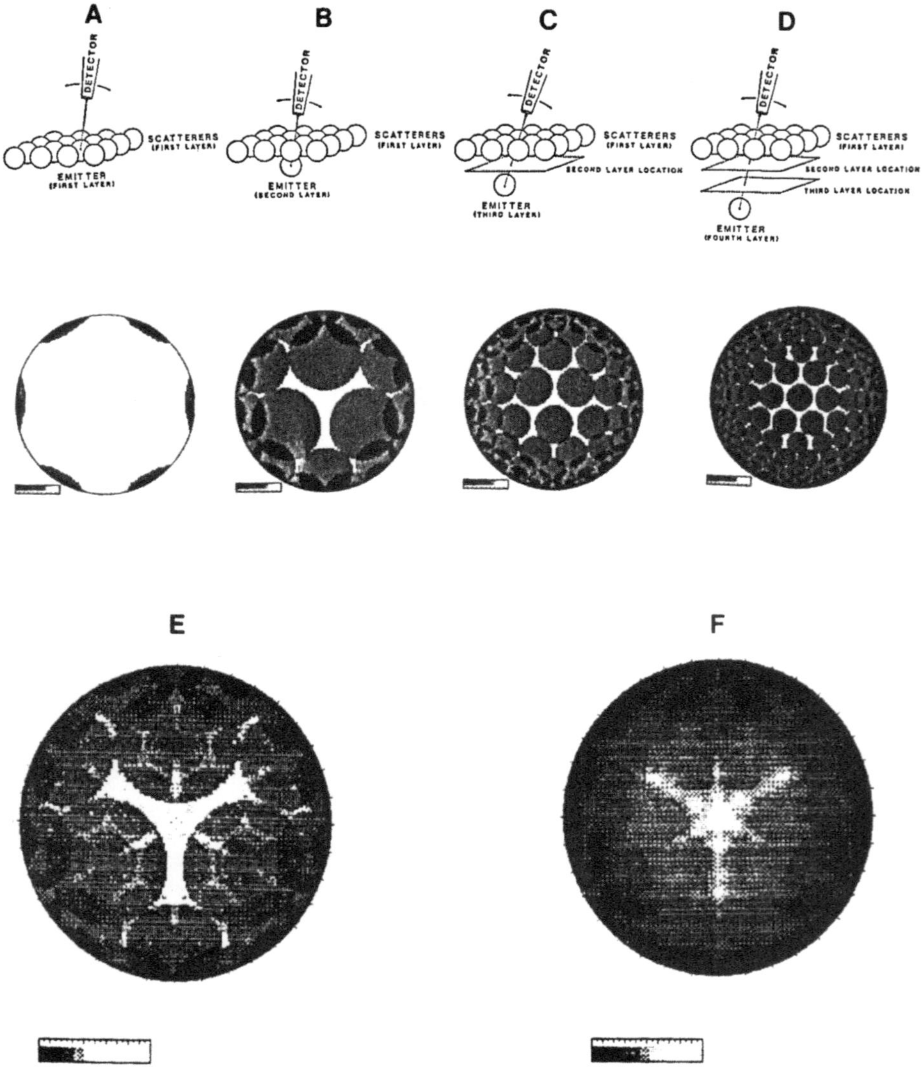

Figure 10. (A) Emitters and scatterers all in the same layer; (B)Emitters and scatterers in adjacent layers; (C and D) Emitters and scatterers separated by two and three layers, respectively; (E) Theoretical Pt(111) Auger electron angular distribution by considering all Auger emission and scattering events occurring in the outermost 16 layers; (F) As in E, but including the consideration of atomic vibrations.

distribution (Figure 11A) is the round silhouette of the overlying I atom and the silhouettes of the neighboring I atoms from the surrounding hexagonal pattern. The location, dimension, and density of the central silhouette in Figure 11A indicate that each I atom is located atop an Ag atom, as illustrated in Figure 11C. Note that the I Auger angular distribution (Figure 11B) is essentially featureless, as expected, since iodine forms the outermost layer. The ADAM data also indicate that the Ag-Ag interatomic distance is 7 % larger than the Pt-Pt distance, and that the I atoms are located atop one-third of the Ag atoms in a "($\sqrt{3}$x$\sqrt{3}$)R30°" arrangement (Figure 11C). The I-I interatomic distance is 20% greater than the van der Waals radius of closest approach [3].

Concluding Remarks

An objective of this article is to illustrate, by examples from recent research in surface electrochemistry, the power of a multitechnique approach to elucidation of the complex chemistry occurring at well-defined electrode surfaces. Although the emphasis has been on stable adsorbates at Pt(111), progress in the field has been rapid and numerous classes of molecules have been studied as well as various metal surfaces [1-2]. Viewed in perspective, this is a potentially large field that is presently at an early stage of its development. Only a few of the virtually unlimited variety of substrates and fluids have been studied, and even in those cases the evidence is still somewhat fragmentary. The experiments are difficult but becoming easier as a result of refinements in technique. The excitement of this field is growing as the pace of completion of proper experiments quickens and as the various findings begin to fit together to reveal general principles.

It has already become apparent that well-defined surface electrochemistry experiments are important to the growth of understanding in a number of surface related fundamental scientific areas, such as the surface structure of materials and of solid-liquid interfaces, surface electronic structure and chemical bonding, surface thermodynamics, surface reaction kinetics and mechanisms, surface acid-base chemistry, surface oxidation-reduction processes, surface reaction product distributions, electrical double-layer phenomena, and solvation of surfaces. Numerous practical advances follow from the fundamental developments, due to strong involvement of surface electrochemistry in areas including the behavior of catalysis and electrocatalysis; chemical and clinical analysis based upon electrochemistry; electrochemical synthesis; modification of the surface chemical properties of metals, oxides, semiconductors, and polymers; manufacture of electronic microcircuits and detectors; surface phenomena of adhesion and lubrication; corrosion; electrochemical machining; surface passivation and other surface related issues.

Acknowledgement

Acknowledgement is made to the National Science Foundation, Air Force Office

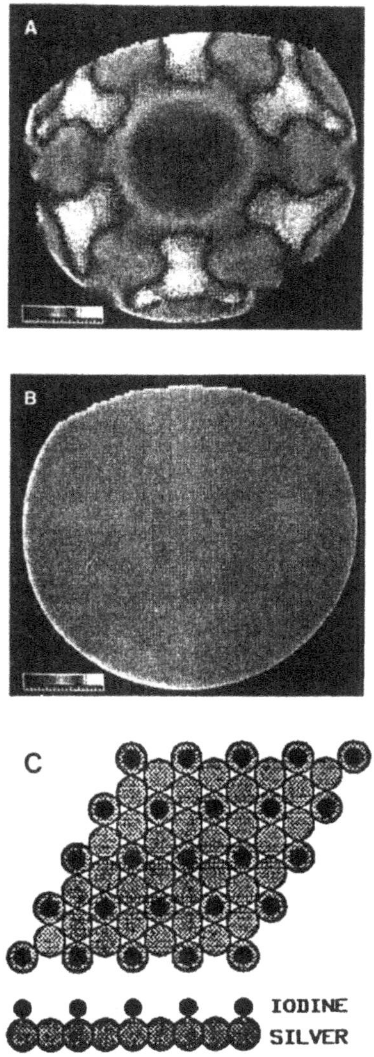

Figure 11. Measured angular distribution of (A) 355-eV Ag Auger electron and (B) 507-eV and 518-eV I Auger electrons obtained from Ag and I atomic layers on a Pt(111) surface; (C) Structural model of the I-Ag-Pt surface: A monolayer of Ag is electrodeposited on Pt(111) and the I atoms are located atop one-third of the Ag atoms in the "($\sqrt{3}$x$\sqrt{3}$)R30°" arrangement shown.

of Scientific Research, Gas Research Institute, and National Institute of Health for support of this research.

References

1. A.T. Hubbard, *Chem. Revs.*, **88** (1988) 633.
2. A.T. Hubbard, *Langmuir*, **6** (1990) 97.
3. D.G. Frank, N. Batina, T. Golden, F. Lu and A.T. Hubbard, *Science*, **247** (1990) 182.
4. G.N. Salaita, D.A. Stern, F. Lu, H. Baltruschat, B.C. Shardt, J.L. Stickney, M.P. Soriaga, D.G. Frank and A.T. Hubbard, *Langmuir*, **2** (1986) 828.
5. F. Lu, G.N. Salaita, H. Baltruschat and A.T. Hubbard, *J. Electroanal. Chem.*, **222** (1987) 305.
6. G.A. Garwood, Jr. and A.T. Hubbard, *Surf. Sci.*, **92** (1980) 617.
7. D.A. Stern, H. Baltruschat, M. Martinez, J.L. Stickney, D. Song, S.K. Lewis, D.G. Frank and A.T. Hubbard, *J. Electroanal. Chem.*, **217** (1987) 101.
8. D.G. Frank, J. Y. Katekaru, S.D. Rosasco, G.N. Salaita, B.C. Schardt, M.P. Soriaga, D.A. Stern, J.L. Stickney and A.T. Hubbard, *Langmuir*, **1** (1985) 587.
9. G.N. Salaita, F. Lu, L.L.-Davidson and A.T. Hubbard, *J. Electroanal. Chem.*, **229** (1987) 1.
10. D.A. Stern, L.L.-Davidson, D.G. Frank, J.Y. Gui, C.-H. Lin, F. Lu, G.N. Salaita, N. Walton, D.C. Zapien and A.T. Hubbard, *J. Amer. Chem. Soc.*, **111** (1989) 877.
11. L.C. Pauling, *The Nature of the Chemical Bond*, 3rd Ed., (Cornell University press, Ithaca, NY, 1960).
12. J.H.S. Green, *Spectrochim Acta*, **33A** (1977) 575.
13. J.E. Demuth, K. Christmann and P.N. Sanda, *Chem. Phys. Lett.*, **76** (1980) 201.
14. F. Lu, G.N. Salaita, L.L.-Davidson, D.A. Stern, E. Wellner, D.G. Frank, N. Batina, D.C. Zapien, N. Walton and A.T. Hubbard, *Langmuir*, **4** (1988) 637.

Adsorption of Water on Metal Surfaces:

What do we know from high vacuum measurements and quantum chemistry calculations?

Gerhard Pirug
Institut für Grenzflächenforschung und Vakuumphysik
Forschungszentrum Jülich, D-5170 Jülich, Germany

1. Extended Abstract

The adsorption of water on metal surfaces is of fundamental interest with respect to current problems in electrochemistry, corrosion or heterogeneous catalysis. Almost all surface sensitive techniques have been applied in order to characterize the chemical interaction between H_2O molecules and metal surfaces[1]. Most of the metals react readily with H_2O under the formation of hydroxide or oxide. Transition metals, e.g. Ni[2], Pt[3] or Ru[4-6], which adsorb H_2O also molecularly, attract the special attention of surface science. Here the strength of the H_2O-substrate bond is relatively weak and comparable to the hydrogen bridge bonding, which controls the crystallography of ice. The surface structure can generally be described by modified Bernal Fowler Rules[6]. Hydrogen bonding is also responsible for the cluster formation which occurs even at low H_2O coverage. Monomers have hardly been spectroscopically identified and only at temperature as low as 10 K[7]. On the contrary evidence for the formation of well developed bilayers exists[4-6]. The reactivity depends on the crystallographic structure. More open faces like the fcc [110] surface or surface defects such as monoatomic steps lead to an increased dissociation probability[1]. Also additives like alkali metals promote the dissociation of H_2O if they exceed a critical coverage [3,8-10].

Consistent with the experimental findings it has been shown by quantum chemistry calculations that molecular water is generally bonded via its oxygen lone pair orbital to the surface[11,12]. Charge transfer between the H_2O molecule and the surface does not play any significant role due to the filled orbital structure of H_2O similar to the filled shells of noble gases. Instead only a small charge transfer from the H_2O molecule to the unfilled orbital of the metal under a simultaneous polarization of charge has been postulated on the basis of cluster calculations[11,12].

The orientation of the water molecule can be changed in the presence of ionically coadsorbed alkali metal atoms[3,8-10]. This reorientation can be theoretically described by the electrostatic interaction between the permanent dipole moment of the water molecule and the electric field of the alkali metal atom[12].

2. References

1. P.A. Thiel and T.E. Madey, Surface Sci. Rep. **7** (1987) 211.
2. B.W. Callen, K. Griffiths, U. Memmert, D.A. Harrington, S.J. Bushby and P.R. Norton, Surface Sci. **230** (1990) 159.
3. M. Kiskinova, G. Pirug and H.P. Bonzel, Surface Sci. **150** (1985) 315.
4. G. Pirug, C. Ritke and H.P. Bonzel, Surface Sci., in press.
5. P.A. Thiel, F.M. Hofmann and W.H. Weinberg, J. Chem. Phys. **75** (1981) 5556.
6. D.L. Doering and T.E. Madey, Surface Sci. **123** (1982) 305.
7. S. Andersson, C. Nyberg and C.G. Tengstal, Chem. Phys. Lett. **104** (1984) 305.
8. D.L. Doering, S. Semancik and T.E. Madey, Surface Sci. **133** (1983) 49.
9. E.M. Stuve, R. Döhl-Oelze, K. Bange and J.K. Sass, J. Vac. Sci. Technol. **A4** (1986) 1307.
10. G. Pirug, C. Ritke and H.P. Bonzel, Surface Sci. submitted.
11. J.E. Müller and J. Harris, Phys. Rev. Lett. **53** (1984) 2493.
12. H.P. Bonzel, G. Pirug and J.E. Müller, Phys. Rev. Lett. **58** (1987) 2138.

L. STOLBERG and J. LIPKOWSKI
Department of Chemistry and Biochemistry
University of Guelph, Guelph, Ontario N1G 2W1
CANADA

D.E. IRISH
Department of Chemistry
University of Waterloo, Waterloo, Ontario N1L 3G1
CANADA

F. HENGLEIN and D.M. KOLB
Fritz-Haber-Institüt
Max-Planck-Gesellschaft, Faradayweg 4-6
1000 Berlin 33
GERMANY

ABSTRACT

Molecular Adsorption at Gold Electrodes

Four techniques have been applied to probe pyridine molecules adsorbed at the gold-solution interface. Adsorption isotherms, free energies of adsorption and electrosorption valencies for pyridine adsorbed at the (111), (100), (110) and (311) surfaces of gold have been determined. At the (111) and (100) planes of gold pyridine behaves as an ambidentate ligand which is coordinated to the metal surface either through the π band system of the aromatic ring or through the nonbonding orbital localized on the nitrogen atom. Consequently, two orientations for the absorbed pyridine molecules were observed; a flat orientation at the negatively charged surface and the vertical at the positively charged interface. Reorientation from the flat to the vertical position takes place at the potential of zero charge and has the character of a first order two-dimensional phase transition. At the stepped surfaces (110) and (311), the pyridine molecule behaves as a monodentate ligand and assumes only the vertical orientation.

The free energies of adsorption for pyridine at gold surfaces are on the order of -40 kJ mol^{-1} and vary for the different crystallographic orientations. Consequently, a strong effect of surface crystallography on pyridine adsorption at gold electrodes was observed [1-3].

Surface enhanced Raman Scattering (SERS) for pyridine adsorbed onto a smooth polycrystalline gold electrode surface has been investigated. It was found that, for smooth surfaces, time independent (stable within 10% of the initial value) intensities of the totally symmetric ring breathing mode of pyridine (1010 cm^{-1} band) could be measured. The potential and concentration dependence of the intensity of the 1010 cm^{-1} band was therefore determined. Independently, the isotherms for pyridine adsorption at smooth polycrystalline gold surfaces were obtained using chronocoulometry. The SERS data were compared to the surface concentration data. The SERS intensities were directly proportional to the surface concentration at low coverages, but deviated from the linear dependence and even passed through a maximum at high pyridine coverages. It has been shown that the dipole-depolarization effect can explain the observed dependence of the SERS intensities on the coverage of the electrode surface by the pyridine molecules [4].

The effect of pyridine adsorption on the electroreflectance spectra for Au(100) in the visible range of radiation has been investigated using electromodulation reflectance spectroscopy. The electromodulated reflectance spectrum of the Au(100) surface displays (in the absence of adsorbates) a characteristic bipolar feature due to the electronic transitions from the Fermi level to unoccupied p-surface states [5,6]. We have observed that, in presence of pyridine, the surface state bands are only energy shifted by flat adsorbed molecules and are totally quenched by vertically adsorbed N bonded molecules. Consequently, electroreflectance spectroscopy proved to be a powerful tool to determine the molecular orientation at the interface.

Finally, by using light from the Berlin storage ring BESSY, we have investigated the optical properties of pyridine adsorbed on gold in the photon energy range 3.5 to 6.5 eV. Electroreflectance spectra for Au were recorded under various modulation conditions resulting either in a coverage or in an electric field modulation. Under such conditions ($\Delta R/R$) represents in essence the difference between the spectra for pyridine on the surface and in the bulk. The optical transitions, $^1A_1 \rightarrow {}^1B_2$ and $^1A_1 \rightarrow {}^1A_1$ for pyridine, which occur at 257 and 197 nm for the molecule in solution, are clearly seen as bipolar - 2nd derivative - like structures in the normalized reflectance spectrum centered around 250nm and 195nm. This can be explained by a substantial line broadening of the spectrum for the adsorbed species. A series of ER spectra for pyridine on Au(110) were recorded under constant coverage conditions with the electric field vector E [111]. A very clear change of the optical response with potential was observed. Both the spectral position and the intensity varied. The origin of these changes was attributed to a variation of the tilt angle for adsorbed molecules with potential and/or a line shift broadening due to the Stark effect.

In summary, the results obtained by four techniques applied in these studies indicate that: (a) pyridine molecules are weakly chemisorbed at gold surfaces, (b) the flat or vertical orientation at the surface can be assumed, (c) the transition from flat to the vertical orientation takes place at the potential of zero charge and (d) the tilt angle for the vertical orientation changes with potential.

References

1. L. Stolberg, J. Lipkowski and D.E. Irish. *J. Electroanal. Chem.*, **238** (1987) 333.
2. L. Stolberg, J. Lipkowski and D.E. Irish. *J. Electroanal. Chem.*, XX (1990) XXX.
3. L. Stolberg, S. Morin, J. Lipkowski and D.E. Irish. *J. Electroanal. Chem.*, 1990 submitted.
4. L. Stolberg, J. Lipkowski and D.E. Irish. *J. Electroanal. Chem.*, **300** (1991) XXX.
5. S.H. Liu, C. Hinnen, C.N. Van Huong, N.R. de Tacconi, K.M. Ho. *J. Electroanal. Chem.*, **76** (1989) 325.
6. D.M. Kolb and C. Franke. *Appl. Phys.* **A49** (1989) 379.

USE OF ISOTOPES IN ELECTROCHEMISTRY

STATE OF ART

V. E. KAZARINOV
Frumkin Institute of Electrochemistry
Academy of Sciences of the USSR
Leninsky prospect 31, Moscow 117071, USSR

ABSTRACT
The present paper describes the state and the tendency in the development of isotopic methods used in the electrochemical investigations.

The state of art in the use of isotope techniques in electrochemical studies, as of the end of 1980, is described in detail in our review published in Volume 9 of The Comprehensive Treatise of Electrochemistry [1]. The number of times this review has been quoted and the character of references show that it fully characterized the status of the problem at that time.

I'd like now to speak about the changes in the use of radioactive and stable isotopes which occurred during the period since 1980 up to the present time and the trends in the development and refinement of the techniques employed.

It can be safely stated today that no fundamentally new radioactive tracer techniques for investigation of electrochemical processes have appeared in the past decade. During this period mainly the existing techniques have been refined and attempts have been made to create new complex equipment which would permit joint use of the known radio tracer technique and modern electrochemical methods. The greatest progress has been witnessed in the development and refinement of the "electrode immersion" method.

A team of scientists working in the USA at the University of Illinois under Professor A. Wieckowski devised a complex equipment [2,3], which combined the achievements in radiometry by glass scintillation sensors in electrochemical measurements with the "electrode immersion" method proposed by V.E.Kazarinov at the Institute of Electrochemistry, Academy of Sciences of the USSR. In this equipment a glass oscillator served as the bottom of an electrochemical cell. It had been previously subjected to a special polishing treatment to produce a radioactive background in the case when the electrode was lowered onto the cell bottom. Some other technical means were used to enhance the sensitivity of the radiometric apparatus. This equipment is described in detail in the work of A.Wieckowski et al[3].

The main advantage of the equipment under consideration is the possibility of making adsorption measurements on smooth metals with the roughness factor less than 5. So far only preliminary data are available on the adsorption of formic acid on smooth platinum which have been obtained by the new equipment.

Fig.1

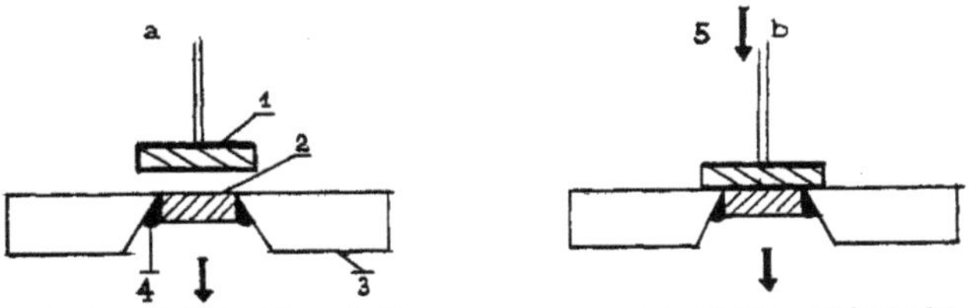

Diagram of electrode in (a) adsorption position and (b) "squeezed" position for adsorbate measurement:

1 - smooth polycrystalline platinum electrode

2 - polished glass scintillator detector

3 - Macor ceramic disk, which forms cell bottom,

4 - resin, 5 - 350-g weight applied to electrode shaft during pressing[2].

The work on the refinement of the "electrode immersion" method was also continued at our Institute. A complex equipment was devised which allowed the "electrode immersion" method to be used in conjunction with the most advanced modern methods - cyclic voltammetry and potentiodynamic pulse method [4]. The aim of this new equipment was to provide a more perfect tool for investigation of electrochemical systems containing organic substances and employing chemically modified electrodes.

For joint use of radiotracer and potentiodynamic methods in the same experiment some changes in the cell design were made which enabled measurements to be made in the case of rather large currents (up to 200 mA). Such currents passed through the cell during electrode preactivation or upon application of the measuring pulse, the working electrode being in a raised position.

The radioactivity measurements were performed with the electrode being lowered onto the membrane serving as the cell bottom; potential was kept constant and adsorbate was in the solution. Due to this fact the experimental conditions could be made to approach the steady-state conditions as closely as possible.

By pulse electrochemical measurements it is possible to determine from the current - time curves the amount of electricity Q_H^{dehydr} passing through the electrode during adsorption, and from the cathodic potentiodynamic curves - ΔQ_H - the decrease in the hydrogen adsorption on the electrode due to the organic substance adsorption and from the anodic potentiodynamic curves one could determine Q_a - the amount of electricity consumed in the oxidation of the chemisorbed substance to products removed from the electrode surface into the solution bulk. Using organic substance samples containing the radioactive isotope C^{14}, we determined at each moment of time during the adsorption process the absolute amounts of chemisorbed carbon-containing $N_{C^{14}}$. By comparing all these quantities, it is possible to obtain the following information on the nature of chemisorbed particles and on the chemisorption process:

1. The charge of surface platinum atoms bound by a chemisorption bond with adsorbed organic particles in the total number of surface platinum atoms is

$$\theta'_R = \frac{\Delta Q_H}{\overset{\circ}{Q}_H} = \frac{N_{Pt-C}}{\overset{\circ}{N}_{Pt}} \qquad (1)$$

2. The number of electrons passing through the surface when one organic molecule undergoing adsorption is transferred to a chemisorbed particle is

$$n_H = \frac{Q_H^{degidr} \cdot 1,31 \cdot 10^{15}}{0,21\ N_{C^{14}}} \qquad (2)$$

3. The number of surface platinum atoms bound by a strong chemisorption bond with one chemisorbed particle is

$$n_{Pt} = \frac{\Delta Q_H \cdot 1{,}31 \cdot 10^{15}}{0{,}21 \; N_{C^{14}}} \qquad (3)$$

It should be noted that the quantity n_{Pt} does not represent the total number of sites occupied by an adsorbed molecule because a part of surface platinum atoms, even with a complete coverage of the surface with organic .substance, remains accessible to hydrogen adsorption, and inaccessible to the given organic substance adsorption, i.e. it proves to be blocked due to steric hindrances.

If one kind of chemisorbed particles is adsorbed on the surface, the total number of occupied adsorption sites can be found by the equation

$$n_{\Sigma} = \frac{N^{\circ}_{Pt}}{N^{max}_{C^{14}}} = \frac{Q^{\circ}_H \cdot 1{,}31 \cdot 10^{15}}{0{,}21 \; N^{max}_{C^{14}}} \qquad (4)$$

In this case the total number of occupied adsorption sites and the number of surface platinum atoms bound by a strong chemisorption bond with organic particles are linked by the following relation

$$n_{\Sigma} = \frac{I}{Q^{max}_R} \cdot n_{Pt} \qquad (5)$$

4. The number of electrons consumed in the oxidation of one chemisorbed particle to products removed from the solution bulk is

$$n_e = \frac{Q_a \cdot 1{,}31 \cdot 10^{15}}{0{,}21 \cdot N_{C^{14}}} \qquad (6)$$

From electrochemical pulse measurements one usually determines only the number of electrons consumed in the oxidation of chemisorbed particles per one platinum site bound by a chemisorption bond with the adsorbed particle

$$n_a = \frac{Q_a}{Q_H} = \frac{n_e}{n_{Pt}} \qquad (7)$$

On the basis of this quantity one cannot say to what products the chemisorbed organic particle is oxidized during the anodic pulse. In the general case, the number of electrons consumed in the oxidation of one organic molecule with allowance for the prechemisorption stage is

$$\Sigma n_e = n_H + n_e \qquad (8)$$

Thus, joint use of pulse potentiodynamic and radiotracer methods affords more complete information on the processes occurring during adsorption, on the composition and nature of chemisorbed products on the surface without making any assumptions that cannot be experimentally checked.

SOME EXPERIMENTAL DATA

Fig.2 shows the dependence of the amount of substance adsorbed from a 10^{-1} M methanol solution on the adsorption time obtained by the radiotracer technique, and the typical cathodic and anodic potentiodynamic curves for the same adsorption times, measured on the same electrode, from which were determined the amounts of electricity corresponding to the adsorbed substance oxidation and the platinum surface area covered.

Table 1 lists the data on the composition of chemisorbed methanol particles for different adsorption times and potentials. It is clear from the Table that, irrespective of the adsorption time and potentials (from 0.1 to 0.6 V), the mean number of platinum adsorption sites bound with the chemisorbed methanol particle is 3 ± 0.3, the chemisorbed particle loses the same number of electrons upon its oxidation.

$N^{max}_{C^{14}}$ for methanol adsorption on a platinized platinum electrode is equal to 3.4×10^{14} particles per 1 cm^2 [5], and $Q^{max}_R = 0.75$ [6]. Hence the total number of occupied and blocked adsorption sites $n_\Sigma = 4$. As for methanol adsorption $G^{dehydr}_H = \Delta Q_H = Q_a$, the main chemisorbed product on the platinized platinum electrode from methanol solutions is the COH particle, which is completely oxidized to CO_2.

Table 1

Composition of chemisorbed product from a 10^{-2}M methanol solution**

τ min ads	E_{ads} B	θ'_R	N 10^{-14} part.cm^{-2}	n_{Pt}	n_e	n_a
3	0,4	0,17	0,82	2,8	3,1	1,1
10	0,4	0,40	1,73	3,2	2,4	0,8
35	0,4	0,51	2,3	3,0	2,9	1
60	0,4	0,60	2,7	3,0	3,1	1
35	0,4	0,5	2,3	2,8	2,4	0,9
			Hydrogenation			
15	0,05	0,5	2,3	2,9	2,5	0,9
30	0,1	0,11	0,46	3,2	2,9	0,9
30	0,2	0,30	1,38	3,0	3,1	1
30	0,4	0,50	2,4	2,8	3,0	1,1
30	0,6	0,05*	0,16	3,2	2,7	0,8

* This value is at the experimental error level and therefore the corresponding values of n_{Pt} and there fore the corresponding values of n_{Pt} and n_e are approximate.

** The correction for the electric double layer charging is taken into account in the same manner as in /5/.

Fig.2

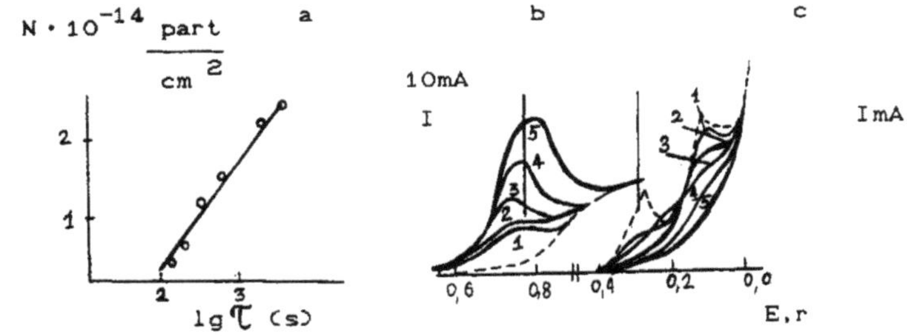

The adsorption kinetics of methanol on a platinized electrode measured by the radio tracer technique is a 0.5 M H_2SO_4 + 10^{-2} M CH_3OH

solution at the potential of 0.4 V (a), and the anodic (b) and cathodic (c) potentiodynamic curves measured under the same conditions at different adsorption times, min: 1 - 1; 2 - 3; 3 - 5; 4 - 10; 5 - 35. Dashed line - the curve in the supporting electrolyte solution (0.5 M H_2SO_4).

The development of the "foil" method proceeded in the following manner. Some technical improvements were introduced which permitted deposits of platinum group metals modified by polymer films to be used as electrodes. These studies were carried out under the direction of Professor D. Horanyi in Budapest [7-10].

Significant changes took place in the use of stable isotopes in electrochemistry. Therefore, I'd like to discuss this problem at greater length.

When stable isotopes are used in electrochemical studies, they are recorded by mass spectrometry.

Intensive use of stable isotopes in electrochemical studies began late in the 50-ies. The most appreciable results were obtained in the studies on hydrogen evolution processes [11], in the elucidation of the role of chemisorbed oxygen, anions and surface oxides in hydrogen evolution processes[12,13] and formation of peroxide compounds [14,15]. The experimental equipment used in these studies provided for periodic sampling of gaseous reaction products, followed by their analysis with a mass spectrometer for the content of certain stable isotope previously introduced into the electrochemical system. This equipment could not be operated in situ and the quality of the information obtained did not satisfy electrochemists. For this reason, by the beginning of the 80-ies the interest in the use of stable isotopes diminished considerably and, naturally, the number of publications on this subject decreased both in the USSR and abroad.

The use of stable isotopes in electrochemistry entered into a new stage when after 1980 new complex equipment was devised at Professor Vielstich's laboratory in Bonn, which permitted joint use of cyclic voltammetry and mass-spectroscopic measurements in situ[16-21]. The method became known as differential electrochemical mass spectroscopy (DEMS). It consists essentially in the following. A porous platinum electrode in the form of a platinum deposit on an inert porous base is used as the bottom of a three-electrode electrochemical cell. From the outside this porous electrode is directly connected to the vacuum system of a mass spectrometer. An electrode such as this enables the gaseous products of the electrochemical reaction occurring on it to find their way at once into the mass spectrometer and be analyzed for the content in them of certain stable isotopes previously introduced into the electrochemical system.

This method can be employed for determination of the stoichiometric composition of the adsorption products of organic substances and of the influence on it of the adsorption conditions: potential and nature of the electrode material, solution composition and pH, temperature, and also for establishing the interrelationship between the conditions of electrochemical process and the composition of reaction products. The method also permits investigation of the influence of adatoms and other modifiers on the electrocatalytic reaction.

The chief advantage of the method under consideration is that it affords instantaneous and continuous recording of the electrochemical parameters of the electrode - solution system and of the products of the reaction occurring on the electrode.

Fig.3

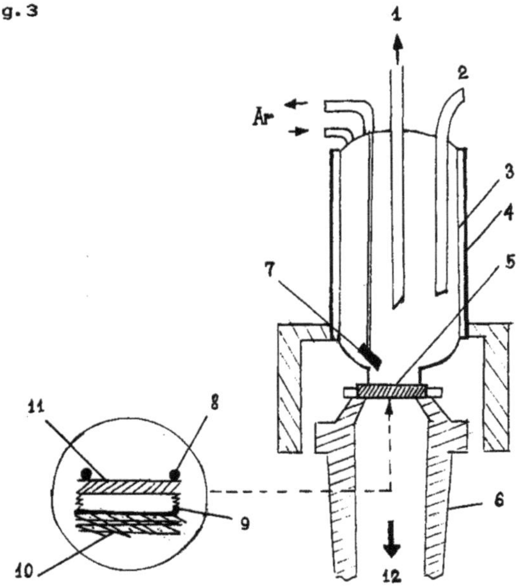

Electrochemical cell for on-line mass-spectroscopy of volatile products using porous metal layer electrodes[21].

Fig. 3

1 - to reference electrode

2 - Electrolyte inlet

3 - Glass insert

4 - Cell body

5 - Working electrode

6 - Cell support

7 - Counter electrode

8 - PTFE - ring

9 - PTFE - membrane

10 - Frit

11 - Platinum

12 - to Mass - spectrometer

The information on the energy state of adsorbate on the electrode surface can be obtained by electrochemical thermodesorption mass spectroscopy [20]. But, unfortunately, in this case it is necessary to interrupt the electrochemical process. The electrode is moved inside the vacuum system, subjected to heating. Then the desorption products are analyzed with a mass spectrometer. The method permits recording the influence of the desorption temperature on the qualitative composition and properties of desorption and adsorption products. Thus, it is possible to obtain information on the energy state of desorbate ex situ, to be more precise, on the strength of the adsorbate-adsorbent bond.

A disadvantage of the methods of investigation described above is that in their case mainly platinum group metals can be used as working electrodes (a porous metal layer about 0.1 mm thick is needed). This markedly narrows down their range of application.

This drawback is eliminated in the equipment in which a rotating disk electrode is used as a working electrode, and thus an oriented flow of gaseous reaction products is set up from the electrode through the porous diaphragm to the vacuum system of the mass spectrometer [21].

Fig. 4

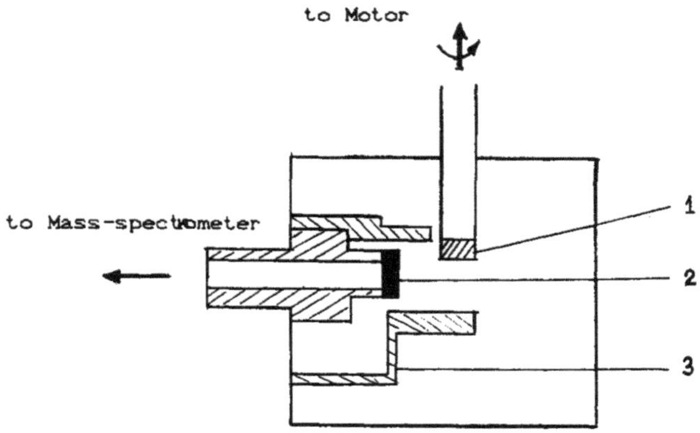

Electrochemical cell for on line mass spectroscopy of volatile products using a rotating cylinder electrode, applied in this study [21].

1 - Working electrode

2 - PTFE membrane

3 - Thread

REFERENCES

1. V.E.Kazarinov and V.N.Andreev, in *Comprehensive Treatise of Electrochemistry* 9 (Plenum Press, New York, 1984), p.393.

2. E.K.Kranskoph, K.Chan, A.Weickowski, *J.Phys.Chem.* 91(1987) p.2327.

3. A.Weickowski in *Modern Aspects of Electrochemistry* 21, ed.R.White, J.O'M.Bockris, B.E.Conway, (Plenum Press, New York-London, 1990).

4. O.A.Hazova, V.N.Andreev, Yu.B.Vasiljev, V.E.Kazarinov, *Electrokhimia*, 23, 9 (1987) 1155.

5. V.E.Kazarinov, G.Ya.Tysyachnaya, V.N.Andreev, *J.Electroanal. Chem.*, 65 (1975) 391.

6. O.A.Petry, B.I.Podlovchenko, A.N.Frumkin, Hira Lal, *J.Electroanal.Chem.* 10 (1965) 253.

7. G.Inzelt, G.Horanyi, *J.Electroanal.Chem.* **200** (1986) 405.

8. G.Inzelt, G.Horanyi, J.Q.Chambers, E.W.Day, *J.Electroanal.Chem.* **218** (1987) 297.

9. G.Inzelt, G.Horanyi, J.Q.Chambers, *Electrochim. Acta* **32**, 5 (1987) 757.

10. G.Horanyi, G.Inzelt, *J.Electroanal.Chem.* **257** (1988) 311.

11. M.Enyo , in *Modern Aspects of Electrochemistry*, ed.B.E.Conway, J.O'M.Bockris, **10**, (Plenum Press ,1975), p.161.

12. M.O.Davies, M.Clark, E.Yeager, E.Novorka, *J.Electrochem.Soc.* **106** , 1 (1959) 562.

13. E.V.Kasatkin, K.I.Rozentalj, V.I.Veselovskyi,*Elektrokhimia* **4** , 12 (1968) 1402.

14. L.A.Hanova, E.V.Kasatkin, V.I.Veselovskyi, *Electrokhimia* **9** , 4 (1973) 562.

15. M.R.Tarasevich, G.I.Zaharkin , R.M.Smirnova, *Electrokhimia* **7** , 9 (1971) 1298.

16. J.Willsau, J.Heitbaum, *J.Electroanal.Chem.* **185** (1985) 181.

17. J.Wolter, J.Willsau, J.Heitbaum, *J.Electrochem.Soc.* **132** , 7 (1985) 1635.

18. N.A.Anastasijevie, H.Baltruschat, J.Heitbaum, *J.Electroanal. Chem.* **272** (1989) 89.

19. J.Willsau, J.Heitbaum, *Electrochem..Acta* **31** , 8 (1986) 943.

20. S.Wilhelm, W.Vielstich, I.Iwasita, *J.Electoral.Them.* **229** (1987) 377.

21. S.Wasmus, E.Cataneo, W.Vielstich, *Electrochim. Acta* **35** , 4 (1990) 771.

THEORY OF SPECIFIC IONIC ADSORPTION AT A NON-TRANSITIONAL METAL SURFACE

M. A. VOROTYNTSEV
A.N.Frumkin Institute of Electrochemistry, Academy of Sciences of the USSR
Moscow 117071, Leninsky prospekt 31, USSR

ABSTRACT

Modern status of the theory of ionic adsorption at metallic electrodes has been analyzed. A brief overview of experimental techniques is given as well as results and shortcomings of traditional approaches to the description of this phenomenon. Problems on the way to create a consistent theory of specific ionic adsorption are formulated. Results obtained for thermodynamic and correlation properties of a two-dimensional ionic ensemble as well as for ionic adsorption isotherms are discussed. Possibility of the new theory to interpret experimental data on ionic adsorption is studied. Prospects for the development of the theory are outlined.

1. Introduction

Any contact of the metal and solution surfaces results in a concentration redistribution of all components of the media in contact. This phenomenon from the thermodynamic point of view is usually interpreted in terms of positive or negative adsorption of the corresponding species. The process is strongly dependent not only on the solution composition but also on the electric variable, the interfacial potential drop, E, or the electrode surface charge density, σ. Essential difference between ionic adsorption within the diffuse and compact layers has been substantiated as a result of studies by Stern, Frumkin, Grahame *et al.* Characteristics of purely electrostatic adsorption within the diffuse layer of the solution may be calculated quite satisfactory within the Gouy-Chapman theory (at least for single-charge ions), see reviews[1,2] for discussion of this question for various metals and solvents. For illustration the Parsons-Zobel plots (i.e. dependence of the inverse interfacial

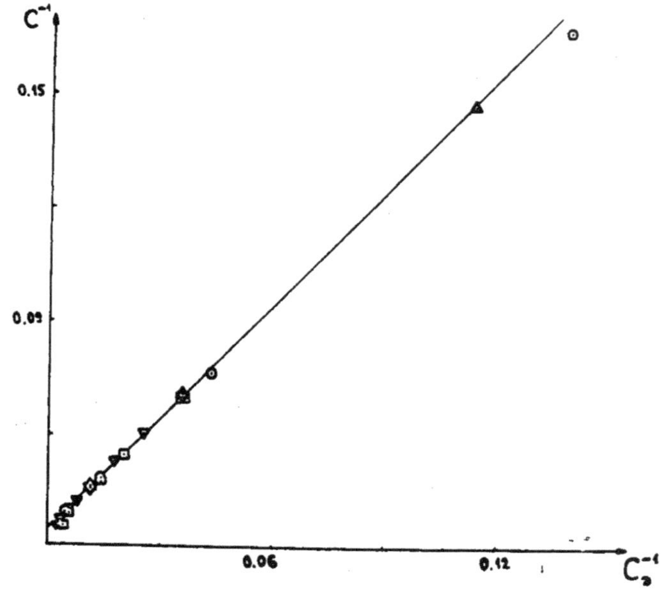

Fig. 1. Inverse capacitance, C^{-1}, of the interface between mercury and various surface-inactive electrolyte solutions in water, NaF (circles), NaH_2PO_4 (squares) and Na_2SO_4 (triangles) at PZC as a function of the diffuse-layer one, C_D^{-1}, calculated within the Gouy-Chapman model, see Ref.[1], both capacitances being in microfarades/cm^2.

capacitance, C^{-1}, on the inverse diffuse-layer capacitance *calculated within the Gouy-Chapman model*) are gathered in Fig.1 for contacts of mercury with aqueous solutions of various surface-inactive electrolytes, NaF, NaH_2PO_4 and Na_2SO_4, at PZC (potential of zero charge). Close proximity of experimental points for all these systems to a single straight line with a unit slope within a wide concentration range should be stressed. Besides confirmation of the Grahame theory for surface-inactive electrolytes, this fact testifies in favor of the hypothesis on independence of compact-layer properties of such systems on the specific choice of an electrolyte.

Quite different regularities are typical for solutions of ions adsorbing at an electrode within the compact layer. Then, the interfacial characteristics, in particular its capacitance, are essentially dependent on the specificity of interaction of a particular ion with the metal/electrolyte interphase. Such specific surface activity resulting in a considerable accumulation of ions (first of all anions) within the compact layer has been observed for a dominant majority of electrochemical systems.

Up till now, the most detailed experimental information about this phenomena has been obtained by means of capacitance measurements. Methods of binary and mixed electrolyte solutions are usually employed for this purpose. In the former case the capacitance curves, $C(E)$, are measured for various concentrations of a binary electrolyte, c, with choosing a counterion which is surface-inactive within the whole potential range studied. Double integration of capacitance curves at each solution composition, with using the corresponding constants found for surface-inactive electrolyte solutions, gives dependencies of the interfacial energy, γ, on the potential, E, and the concentration, c. The same curves can also be measured immediately for liquid metal electrodes but usually with a lower preciseness. Numerical differentiation of these data with respect to the chemical potential of the electrolyte, μ (i.e. at varying the electrolyte concentration) enables one to determine the values of Gibbs adsorption for the ion, Γ, i.e. its total adsorption within the diffuse and compact layers with respect to Gibbs reference plane (in diluted solutions $d\mu=RT\,d\ln c$). To find the amount of specifically adsorbed ions, $\Gamma_1(E,c)$, one has to subtract this ion adsorption within the diffuse layer, $\Gamma_2(E,c)$ (calculated within the framework of Gouy-Chapman theory) as well as the solvent adsorption contribution, $\Gamma^{(s)}(E,c)$, from the overall adsorption, Γ.

Thus found "experimental" dependence, $\Gamma_1(E,c)$, is related to a theoretical isotherm. In numerous papers such interpretation was carried out with the use of various isotherms which did not take into account the potential drop across the diffuse layer though this was obviously to result in an essential change of the isotherm parameters found, like the energy of individual ion adsorption or the limiting surface coverage, Γ_∞. Another important shortcoming of this analysis is a too complicated form of theoretical adsorption isotherms containing too many fitting parameters. For example, adsorption of simple inorganic ions at mercury electrode was frequently treated with use of the three-parameter Frumkin isotherm though a maximum surface coverage by ions in these systems was low enough to neglect the Langmuir factor.

The mixed electrolyte method by Hurwitz and Parsons has some important advantages with respect to measurements in the binary solutions. Capacitance measurements are carried out for a set of mixed electrolytes with a constant ionic strength and the varying ratio of concentrations of the surface-active, $c_1=mc$, and inactive, $c_2=(1-m)c$) ionic components. Both electrolytes include the same surface-inactive counter ion. Dielectric properties of both the compact and the diffuse layers do not change at replacing some part of indifferent ions 2 within the diffuse layer by the same amount of surface-active ions 1, if keeping Γ_1 constant. Then, the electrocapillary equation,

$$d\gamma = -\sigma dE - \sum \Gamma^{(i)} d\mu^{(i)}$$

results in a simple relation,

$$\Gamma_1 = -(\partial\gamma/\partial\mu)_{E,c} \simeq -(RT)^{-1} \ (\partial\gamma/\partial \ln m)_{E,c},$$

which enables one to find immediately the surface concentration of specifically adsorbed ions, Γ_1, from experimental data for the interfacial energy, γ, or capacitance, C. This peculiarity of the mixed electrolyte method not only simplifies the calculation but also gives a way to avoid the necessity to apply the Gouy-Chapman theory for calculation of Γ_2. This advantage becomes especially important far from PZC in view of insufficient verification of this theory under such conditions.

These two methods have been used to study experimentally the specific adsorption of numerous inorganic and organic ions at various metal surfaces from aqueous and non aqueous solutions (see reviews[1-11]).

Radiotracer technique[12] can be used for direct measurements of adsorption isotherm in systems with a large amount of adsorbed species per a unit visible surface area, e.g. at multilayer adsorption or owing to high electrode surface roughness. This method is also applicable in the case of a submonolayer adsorption at a flat surface if the exchange of adsorbed ions with the bulk solution is slow enough since the adsorption layer can be studied after the radioactive ions have been removed from the bulk solution. A way for direct measurement of the derivative, $(\partial\Gamma/\partial E)_c$, by optical methods has also been suggested[13].

2. Traditional theories of ionic adsorption

Numerous publications have been devoted to theoretical description of specific ionic adsorption and relative microscopic characteristics, uniary and binary interaction potentials for ions near the electrode surface. Unfortunately, their analysis of the phenomenon was usually based on inappropriate hypotheses and approximations. Binary interaction potential for adsorbed ions, $U(R)$, was usually postulated as corresponding to the metal/uniform local dielectric interface, either with no diffuse-layer ions or at extremely high concentrations. Adsorbed ions ensemble was considered either as an almost ideal gas or within the approach of "cut-off-disk" or as a regular two-dimensional lattice within a wide range of surface coverages (due to their mutual electrostatic repulsion). However, a special analysis of this question has revealed that all these assumptions concerning both the binary interaction potential and the statistical mechanical properties of the adsorbed ionic ensemble are *not* correct, see below.

3. Formulation of the problem

A consistent theoretical treatment of the phenomenon is to include a thorough study of thermodynamic (like the chemical potential of adsorbed ions, μ) and correlation (e.g. the mean force potential, $W(R)$) properties of a system of charged species within the interfacial region. For a non-transitional metal the delocalized s,p-electrons give the main contribution to the adsorption energy of an individual ion. As a result, this energy changes weakly at shifting the ion along the surface. At the same time ionic motion in the normal direction is confined to a narrow region near the equilibrium bond length. Therefore one may consider ions as a *two-dimensional* system of species at a *uniform* substrate.

The necessity for proper account of "discreteness-of-charge" effects has been established long ago by Esin, Shikhov and Ershler. This means that the chemical potential of specifically adsorbed ions may not be found as a work of ion transfer within an external one-dimensional electric field created by a uniformly "smeared-out" ionic charge along the adsorption plane (as assumed in the Stern theory). Thus, one must analyze statistical-mechanical properties of a set of *discrete* species.

The form of the binary interaction potential, $U(R)$, as a function of the distance between adsorbed ions, *R*, has been established in [15-19] for diluted electrolyte solutions. A crucially different behaviour of *U(R)*, in comparison with expressions postulated previously (see the preceding section) was revealed[15-19]. Its remarkable features are[15-19]:

-much greater intensity (if comparing with the systems like metal/vacuum, metal/uniform medium with a macroscopic dielectric constant or metal/concentrated electrolyte solution),

-very slow decrease of the interaction energy, $U \sim R^{-1}$, within an intermediate range of distances, from a few Å to several ten Å, which is replaced by a steeper decrease at larger distances,

-a marked dependence of the interionic interaction potential on the screening properties of the diffuse layer, varying as a function of the electrode charge and the solution ionic strength,

-a strong dependence on individual characteristics of the ion, in particular the degree of charge transfer at adsorption, ionic charge distribution in the adsorbed state and the distance between the ion and the metal surface which may also be electrode charge dependent.

Since all these predictions are qualitatively different from those postulated in previous publications it is of obvious interest *to compare the corresponding statistical-mechanical results both between each other and with the experimental data* for specific ionic adsorption.

As mentioned above, adsorption characteristics at flat surfaces of non-transitional metals are usually studied by capacitance and electrocapillary methods so that the adsorption isotherm cannot be measured immediately but must be calculated with the use of graphical differentiation. The latter leads to a drastic increase of dispersion of the isotherm found. In addition, necessity to carry out this operation increases considerably the experimental work since the whole interval of the adsorbing ion concentrations under study is to be divided into several equal (in a logarithmic scale) parts which should be small enough. As a result, even measuring at 8-10 concentrations is usually enough to obtain roughly approximate isotherms only. Therefore *one of our main aims was to derive a non-traditional approach* to check the theoretical conclusions which would enable one *to carry out comparison with experimental data without a procedure of graphical differentiation.*

4. Statistical-mechanical theory. Ionic adsorption isotherm[20-25,14]

Our analysis of the thermodynamic and correlation properties of a two-dimensional ensemble of discrete species has been based on the specific properties of the binary interaction potential between adsorbed ions, $U(R)$. Varying the density of the system, i.e. the surface concentration of ions adsorbed within the compact layer (this quantity is denoted below as Γ, instead of Γ_1) leads to a gradual change of the phase state.

At very low concentrations the system behaves like a two-dimensional diluted gas with a weak interaction between the species. The virial expansion may be used to determine all energetic and correlation characteristics. But the coefficients of this expansion turned out to be very large so that a considerable repulsion between adsorbed ions took place even at very low coverage,

$\Gamma \sim \Gamma_* \sim (0.5-5)*10^{-3}$ Å^{-2}.

Approaching the region of a "transitional" density, $\Gamma \sim \Gamma_*$, leads to a gradual change of the correlation properties of the system. At low coverage, $\Gamma << \Gamma_*$, the main contribution is due to pair "collisions" of adsorbed ions whereas at higher densities, $\Gamma \gg \Gamma_*$, an effective screening of the interactions due to collective plasma-like excitations arises at large interionic distances, $R \gg R_\Gamma$, the two-dimensional screening length, R_Γ, diminishing at increasing the coverage as Γ^{-1}.

At further increase of the density, $\Gamma \sim \Gamma_B \sim (0.5-2)*10^{-2}$ Å^{-2}, if the average interaction energy for an ion pair becomes about the thermal one, kT, the system is transforming into a two-dimensional irregular condensed phase. Ultimately, at approaching the monolayer coverage, a two-dimensional variant of the crystal lattice having a quasi long-range ordering is established.

Existence of plasma-like and "liquid-like" phases has not revealed in earlier publications at all. Values of virial coefficients for the gas phase were strongly underestimated and, correspondingly, the region of gas-phase densities was drastically widened. Existence of a two-dimensional crystal phase due to Coulombic forces of interionic repulsion was also supposed within a wide range of densities whereas our study has shown that such a phase cannot be realized in electrochemical systems at all.

Preceding statistical-mechanical theories of ionic adsorption resulted in complicated expressions for the chemical potential of the adsorbed state, μ, containing numerous phenomenological coefficients as well as detailed characteristics of the interfacial region. Checking the theoretical predictions by means of their comparison with experimental data was complex and inconvincing in view of the large number of these fitting parameters. Moreover, most of these theories was unable to give even a qualitative explanation for the dependence of energetics of interionic interaction on the screening properties of the diffuse layer. This failure is due to neglect of the latter factor at the choice of the binary interaction potential, $U(R)$.

The new theory has enabled us to show that because of the specific properties of the interaction potential, $U(R)$, the contribution to the chemical potential due to interionic repulsion, μ_{int}, is determined for diluted solutions by ionic interactions at large distances, about several ten Å, both at low and high surface coverages, Γ. As a result, a simple analytical approximation for the chemical potential of the adsorbed ion has been proposed for a wide range of the density, Γ,

$$\mu = \text{const} + kT \ln \Gamma + \mu_{int}, \quad \mu_{int} = kT\, 2\upsilon\Gamma$$

Here, υ is an analog (with minus) of the attraction constant determined by the long-range part of the binary interionic potential as

$$\upsilon \approx \pi \int RdR\, [1 - \exp(-U(R)/kT)]$$

Adsorption isotherm is obtained after equalizing the values of the chemical potential of adsorbed and bulk solution states. It has the form for diluted solutions,

$$\ln mc/\Gamma = W_{ads}/kT + 2\upsilon\Gamma$$

W_{ads}, one-particle adsorption energy, with no influence of other ions. Isotherm parameters, W_{ads} and υ, are essentially dependent on the ionic strength of the solution as well as an electrical variable.

5. Choice of electrical variable. Isotherm parameters[14,24,25]

The question about an appropriate choice of the electric variable has given rise to a discussion between Parsons, on the one hand, and Frumkin and Damaskin, on the other[28-32]. Parsons' opinion was in favor of the electrode charge, σ, whereas his opponents proposed argumentation to choose the electrode potential, E, their reasoning being based on experimental data for adsorption of neutral organic molecules.

The electrode charge is undoubtedly preferable if neglecting the discreteness-of-charge effects of adsorbed ions, i.e. within the model using the uniform "smearing-out" of their total charge, $ze\Gamma$, along the adsorption plane. Then, the electrical part of the chemical potential of the ionic ensemble, μ, depends on the diffuse-layer charge, σ_D, equal to $-\sigma-ze\Gamma$, only. However the necessity to account of the discrete nature of specifically adsorbed ionic charges has been proven long ago (Esin-Markov effect) so that this reasoning is not quite convincing.

To my opinion, there are two quite different criteria for a "proper" choice of the electric variable:

1) to provide the functional form of the adsorption isotherm, $\Gamma(c)$ or $\Gamma(m)$, as simple as possible if this variable is kept constant,

2) to make clear the physical meaning of the adsorption parameters determined from experimental data for $\Gamma(c)$ or $\Gamma(m)$.

This is the former way which has been usually followed earlier. The latter criterion is used below.

If the surface coverage by specifically adsorbed ions is not especially high one must take into account a discrete nature of not only their charge, $ze\Gamma$, but also of the metal and diffuse-layer charges, σ and σ_D, correspondingly. In particular, the metal charge density, σ, is equal to the sum of a uniformly distributed component, σ_0, (which has existed at the surface before specific ionic adsorption) and the induced countercharges of all these ions at the metal surface. So, the total density, σ, at a fixed distribution of all adsorbed ions is a function of the lateral coordinate, $\mathbf{R}$. Because of an analogous effect for the diffuse-layer charge, σ_D, the distribution of the total potential within this region is essentially three-dimensional, either. In particular, the boundary between the diffuse and compact layers is not equipotential. If the form of adsorption isotherm, $\Gamma(c)$ or $\Gamma(m)$, is analyzed at σ=const, the variation of the specifically adsorbed ion density, Γ, is accompanied by changes of not only the induced countercharges, $\delta\sigma$ and $\delta\sigma_D$, but also σ_0 (in view of their connection, $\sigma=\sigma_0+\delta\sigma$). This change of σ_0 results in a change of conditions of an individual ion adsorption. In particular, its one-particle adsorption energy, W_{ads}, is depending on this charge, σ_0, so that this energy turns out to be effectively dependent of not only σ but also of Γ. Analogously, one has to reconsider a contribution to μ due to interionic interaction within the adsorbed layer, either. Indeed, though this characteristics has originally been a two-particle one it is inserted into the one-particle adsorption energy, W_{ads}, within the Grahame-Parsons treatment, accounted by the contribution due to variation of the double-layer field.

Another approach seems to be justified for a clearer physical understanding at not very high surface coverages, by considering adsorbed ion characteristics at a fixed value of the uniformly distributed part of the charge density, σ_0. Variation of the coverage, Γ, at σ_0=const, means a change of the bulk concentration of the adsorbing ions. This is usually accompanied (e.g. for a binary electrolyte solution) by the change of both E and σ, because of variation of the screening properties of the diffuse layer depending on the ionic strength of the solution. As a result, the adsorption isotherm has a complicated form at a fixed potential, E. However for a mixed electrolyte a constancy of both σ_0 and c means no change of the electrode potential, E, if usual assumptions of this method are fulfilled. Then, their relation, i.e. the function, $\sigma_0(E,c)$, may be found from capacitance data for surface-inactive electrolyte solutions.

Thus, for a mixed electrolyte solution of a constant ionic strength, c, the adsorption isotherm at sufficiently low surface coverages is of a "virial" form at E=const, i.e. σ_0=const. This conclusion is valid if nonlinear effects in the response of all other components may be neglected.

Calculation of the "repulsion" constant, υ, with the use of the interaction potential, $U(R)$, found earlier[18], results in a simple analytical expression,

$$\upsilon=\bar{\upsilon}\,\Delta^2,\quad \bar{\upsilon}=z^2e^2(2kT)^{-1}\left[C_H(\sigma_0)+C_D(\sigma_0,c)\right]^{-1}$$

The parameter, Δ, represents a specific characteristic of the adsorbing ion[14,18]. It depends on the degree of charge transfer at adsorption, on the charge distribution in the adsorbed state as well as on the dielectric properties of the transition layer at the metal/solvent boundary. On the contrary, another multiplier, $\bar{\upsilon}$, may be calculated entirely with the use of experimental data for the corresponding solution of a surface-inactive electrolyte. In this formula, ze is the ion charge in the bulk solution, $C_H(\sigma_0)$, the compact layer capacitance, $C_D(\sigma_0,c)$, the differential capacitance of the diffuse layer in the Gouy-Chapman model, σ_0, the electrode charge of this system corresponding to the potential, E.

The adsorption energy, W_{ads}, depends on the total electrolyte concentration, c, and the charge, σ_0, i.e. the electrode potential, E, due to the reasons:

1)because of the average field change within the interfacial region, $\varphi(x)$,

2) via the dependence of the image potential, W_{im},

3) because of the change of characteristics of non-electrostatic interactions between the adsorbed ion, metal and solvent within the compact layer,

$$W_{ads}=W_{dl}+W^{C}_{im}+W^{O}_{ads}$$

The electrical part of the work on ionic transfer in the average field of the double layer (i.e. without account of the field relaxation at introducing an additional ion) can be expressed through the same parameter, Δ, the electrode potential with respect to PZC, $\varphi=E-E_{\sigma=0}$, as well as the potential drop, $\Psi_0(\sigma_0,c)$, calculated within the Gouy-Chapman model,

$$W_{dl}=ze[\varphi(1-\Delta)+\Psi_0\Delta]$$

The energy of image forces for specifically adsorbed ions, W_{im}, includes a considerable contribution due to short-range forces of a non-electrostatic origin. Therefore it is convenient to separate only some long-range part of this contribution to W_{ads}. This quantity, W^{C}_{im}, is the change of the image potential with the variation of the ionic strength from zero to c, at a fixed value of the charge, σ_0. An analytical expression for W^{C}_{im} is rather complicated, see[14,23,24], but it contains the same parameter, Δ, and some characteristics of the surface-inactive electrolyte, only.

The latter component, W^{O}_{ads} ,characterizes a short-range contribution to the adsorption energy of an individual ion. Generally it may be dependent of the charge, σ_0, due to the change of the dielectric properties of the interface as well as the interaction energies, ion-boundary and solvent-boundary.

Thus, the adsorption isotherm derived has a highly simple form and contains two microscopic characteristics of the ion only, Δ and W^{O}_{ads}, which cannot be found with use of the data for corresponding surface-inactive electrolyte solution. Expressions determining the former parameter, Δ, show that its value is between 0 and 1 and is increasing (if other conditions are fixed) at weakening the bond of the ion with the boundary and at diminishing the degree of charge transfer at adsorption.

6. A new method for determining the isotherm parameters from experimental data

Electrocapillary curves, $\gamma(E,c)$, can be calculated within an adequate preciseness by twofold integration of capacitance curves, $C(E,m,c)$, for each solution composition, with the use of additional data for a surface-inactive electrolyte. If the isotherm of specific adsorption is found by subsequent graphical differentiation, $\Gamma=-(\partial\gamma/\partial\ln m)_{E,c}$, then, the isotherm parameters, W_{ads} and υ, and consequently, W^{o}_{ads} and Δ, may be found with use of coordinates $\ln mc/T;\Gamma$. One should keep in mind that a set of straight lines is to be obtained, according to the theory, the slope of which is maximum in the vicinity of PZC.

However realization of the operation of graphical differentiation requires thorough measurements of the capacitance curves for a large number of molar fractions of an adsorbing ion, m, and results in a drastic increase of dispersion of the "experimental" isotherms obtained. An alternative method proposed in Ref.[25] enables one to find the isotherm parameters even if having the experimental data of a much worse quality. After using the theoretical expression for the adsorption isotherm and the thermodynamic identity for Γ, one can derive a relation between the molar fraction, m, and the shift of the capillary curve for the mixed electrolyte with respect to that for a surface-inactive electrolyte solution, $\delta\gamma$,

$$\ln m=W_{ads}/kT-\ln 2\upsilon c+V+\ln V,\quad V=\mu_{int}/kT=(1-4\upsilon\delta\gamma/kT)^{1/2}-1 \qquad (1)$$

Therefore one can find the isotherm parameters, W_{ads} and υ, for each potential, E, and ionic strength, c, from the minimization condition of the function,

$$\Phi(\upsilon)=\Sigma_i f_i^{\,2}-(\Sigma_i f_i)^2/N,$$

$$f_i=V_i+\ln V_i-\ln 2\upsilon c m_i,\quad W_{ads}/kT=-\Sigma_i f_i/N,$$

calculated with use of the experimental values of m_i and $\delta\gamma_i$.

A modification of the method is to calculate values of the adsorption energy according to equation (1) for a set of experimental values, $(\delta\gamma_i,m_i)$, with a preliminary choice of some value of the parameter, υ. After repeating this procedure for other values of υ, one can attain a practical constancy of the W_{ads} values for various molar fractions, m_i. The variety of such υ values may be considerably diminished if the function, $\bar{\upsilon}(E,c)$, is preliminary calculated with use of the data for a surface-inactive electrolyte and variation of the Δ value is carried out step-by-step in the interval between 0 and 1. It should be noted that according to the theory the values of υ and $\bar{\upsilon}$ are to be drastically changing as functions of the electrode potential, E, especially near PZC, whereas variation of Δ is expected to be much weaker. As can be seen from the discussion below, these conclusions are in conformity with the data for the real electrochemical systems studied.

7. Comparison with experimental data

The new method has been used for analyzing capacitance data in the following systems (for a mixed electrolyte solution, everywhere), see Refs.[14,20,26,27,33], 1)Bi/H_2O-J^-; 2)InGa/H_2O-J^-; 3)InGa/H_2O-Br^-; 4)Hg/ethylene glycol-Cs^+; 5)Hg/propylene glycol-Cs^+; 6)Hg/H_2O-Cl^-. Experimental data for systems (1)-(5) were taken from Refs.[34-36]. System (6) was studied experimentally[27] after creation of the above discussed theory in order to carry out its thorough verification. These systems were chosen for checking the theoretical predictions concerning the form of the adsorption isotherm and values of its parameters as functions of the metal, solvent and adsorbing ion properties (both cations and anions) as well of the ionic strength of the solution. Besides, tables of experimental capacitance data, $C(E,m,c)$, for systems (2)-(6) were available. The latter system provides the widest interval of the ideal polarizability of the contact, the highest preciseness of measurements (liquid metal, aqueous solution), the least effect of solvent contaminations. Preliminary data for this system (6) testified that the adsorption energy of chloride anion is rather small so that one was to expect a considerable fraction of its first solvation shell to be remained at adsorption as well as its localization near the outer boundary of the compact layer, i.e. the proximity of the Δ parameter to unity.

Quite reasonable agreement between the theoretical conclusions and experimental data has been found for all these systems. Results of such comparison for system (1) are discussed below as an illustration. Its interpretation was of especially high interest since these experimental data turned out to be at variance drastically with the predictions of the Grahame-Parsons theory[34] and this failure of the theory become a starting point for a hypothesis [4,38] that the counter ions (cations) were pulled

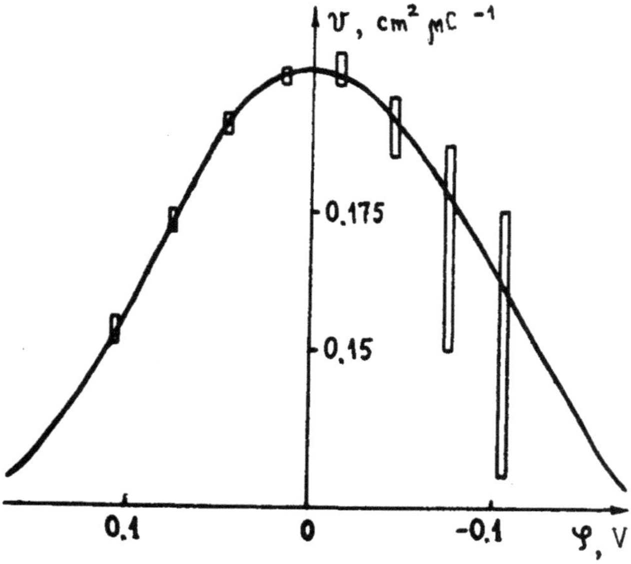

Fig.2. Comparison of experimental values of υ (their dispersion is shown) at various electrode potentials, $\varphi=E-E_{\sigma=0}$, with the theoretical values calculated at $\Delta^2=0.5$, both values being taken in $cm^2/\mu C$.

inside the compact layer under the action of the electric field of adsorbed iodide ions.

Since the deviation from the Grahame-Parsons theory was increasing at diminishing the ionic strength of the mixed electrolyte solution the same experimental data[34] for the most diluted solution, c=0.01M, were treated in Ref.[26] for verification of the new theory.

The above procedure to find the isotherm parameters was used. It has shown that the Δ parameter is practically constant, $\Delta^2 \simeq 0.51 \pm 0.01$, within the whole potential interval more positive than PZC, though the functions, υ and $\bar{\upsilon}$, are changing by several ten per cents (Fig.2). A maximum at the experimental curve, $\upsilon(E)$, observed near PZC is also in conformity with the theory. The minimum of the $\Phi(\upsilon)$ function became flat in cathodic region resulting in drastic increase of the υ values dispersion. This is due to considerable diminution of the average interaction energy between adsorbed ions, $\mu_{int}=2\upsilon\Gamma kT$, because of decreasing their adsorption, Γ. Nevertheless, the value, Δ^2=0.4-0.6, is remaining constant even in this region, within the experimental dispersion.

The total one-particle adsorption energy, W_{ads}, in isotherm (1) calculated at Δ^2=0.5, was varying as a function of the electrode potential according to an approximately linear law (Fig.3). A major contribution turned out to be given by the work term of ion transfer in the double-layer

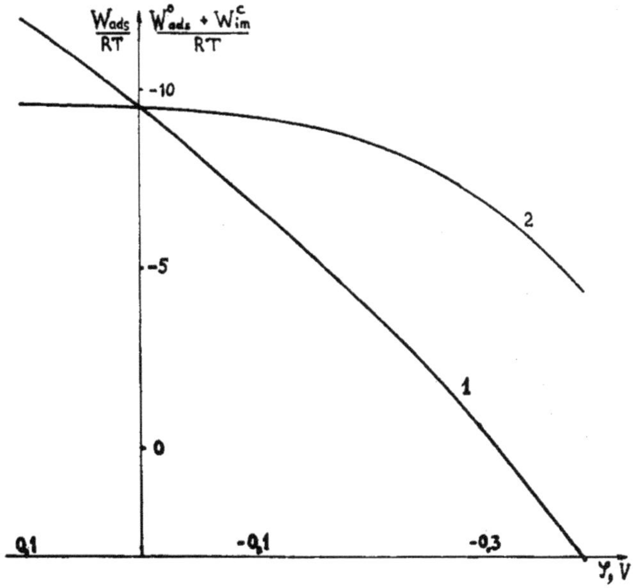

Fig. 3. Dependencies of the total single-particle adsorption energy, W_{ads}/RT, (curve 1) and of the sum of "specific" and image contributions, $(W^{o}_{ads}+W^{c}_{im})/RT$, (curve 2) on the electrode potential, $\varphi=E-E_{\sigma=0}$.

field into the adsorption plane, W_{dl}, because after its subtraction the energy was practically constant in the vicinity of PZC. Since the image force contribution, W^c_{im}, according to the theoretical formula, is sufficiently small near PZC one can make a conclusion on approximate constancy of the "specific" contribution to the adsorption energy, W^o_{ads}, within this potential interval. At large negative potentials a marked decrease of W^o_{ads} was observed.

8. Discussion

Similar results have also been obtained[27,33] for other systems (2)-(6), see the preceding section. All these experimental data can be properly described within isotherm (1). Thus found experimental values of the "repulsion" constant, υ, in agreement with the theoretical predictions, were markedly dependent of the electrode potential and the ionic strength as well as of the properties of the metal, solvent and adsorbing ion. Here are the intervals of υ variation for each of these systems, (1)-(6), 1) 250-325; 2) 60-80; 3) 90-105; 4) 85-135 (c=0.01M) and 75-120 (c=0.1M); 5) 85-130 (c=0.01M) and 75-115 (c=0.1M); 6) 130-310 (values of υ are listed in Å^2/ion). A large difference between the υ values for these systems (more than 5 times) should be noted. At the same time the values of a really microscopic parameter, Δ, turned out to be surprisingly stable for the same systems, 0.71; 0.63; 0.55; 0.84; 0.87; 1.00, correspondingly.

It must be stressed that the Δ values belong to the interval, $0.5<\Delta\leqslant1$, for all these systems, in accordance with the theoretical expression. At the same time, if one used traditional concepts on the binary interaction potential between adsorbed ions to find the υ function the values about several tens would be obtained for Δ^2 which contradicted to the physical meaning of this quantity. Thus experimental data testify directly in favor of the above given picture of interactions between charged species localized inside the compact layer. In particular this agreement shows that a dominant role in the case of adsorption at a s,p-metal is played by interaction along the lines of the electric field which pass through the diffuse layer rather than along the compact layer or via the metal surface layer.

An additional confirmation of this conclusion is given by practical independence of the Δ value for each of these systems on either the electrode potential, E, or the ionic strength of the solution, c. In other words, all these experimental dependencies, $v(E,c)$, are completely reproduced by the theoretical expression for $\bar{v}(E,c)$, the latter function containing *no* fitting parameter.

According to the theory the Δ value is to depend on the dielectric properties of the interfacial region and characteristics of an individual adsorbing ion, see above. It should be expected to increase at weakening the bonding strength of the ion with the interfacial region because of both its shift from the metal surface and the rise of its charge (diminution of the charge loss degree at adsorption). This conclusion is also confirmed by the above results. Indeed, in conformity with the relative values of the specific contribution to the adsorption energy, W^{o}_{ads}, one can see that

1) the Δ value for adsorbing iodide ion is larger at Bi than at InGa alloy,

2) for the same contact, InGa/H_2O, it is larger for the bromide anion than for iodide one,

3) the properties of the solvent have a small influence on the Δ value in the case of a weak adsorption of cesium cation.

Finally, in agreement with our preliminary expectations, the value, $\Delta \simeq 1.0$, was found experimentally for chloride adsorption at Hg from aqueous solutions which means a practically complete conservation of the ionic charge at adsorption and its localization within the outer part of the compact layer.

For all these systems the whole dependence, or its considerable fraction, of the one-particle adsorption energy, W_{ads}, on the electrode potential, E, is due to the double-layer component, W_{dl}, see above, whereas the concentration-dependent contribution, W^{c}_{im}, is usually less important. It was found in the analysis of the system (6) that at anodic potentials a major contribution of the remaining dependence on E is due to the dependence of image energy *at zero ionic strength* on the electrode charge whereas the specific contribution to the adsorption energy was practically constant. An appreciable variation of this specific component was observed at the cathodic branch, similar to the other systems. Such a dependence may be a consequence of weakening of the bond strength between the ion and metal at high charging of the electrode surface. Another possible reason might be a variation of the dielectric properties of the compact layer with charging of the electrode which is to be accompanied by some change of the short-range interaction energy between the adsorbed ion and the solvent. An observed parallelism between variations of this contribution to the adsorption energy and of the compact layer capacitance for the Hg/water contact is testifying in favor of this explanation. However, a broader set of data for various systems should be analyzed in order to make an ultimate conclusion.

9. Conclusion

A remarkable peculiarity of the theory under discussion is the possibility to derive very simple analytical formulae for most of necessary dependencies, W^c_{im}, $U(R)$, $\mu_{int}(\Gamma)$, $W(R)$, $\upsilon(E,c)$, $W_{dl}(E,c)$, $\Gamma(m,E,c)$, being based on a highly general description of the interfacial region (within the general scheme of the nonlocal electrostatics, without addressing any particular model for the interfacial structure). It is practically enough for these final formulae to be valid that a compact layer does exist, i.e. a thin region of an atomic scale having a drastically reduced value of the effective dielectric constant which results in localization of a considerable fraction of the overall potential drop there. Existence of such a region is a direct consequence of experimental capacitance data for a surface-inactive electrolyte. Only two micro-characteristics of an individual adsorbed ion, Δ and W^o_{ads}, are contained in the final expressions, owing to their consistent derivation, whereas much greater numbers of fitting parameters were introduced in previous approaches (based, furthermore, on particular and rather crude models of the interfacial region and using some inappropriate postulates on the metal/solvent+adsorbing ion structure). The new method to treat experimental data has enabled us to do it without operation of their graphical differentiation. This has given a possibility for drastic reduction of both the dispersion of individual ion characteristics under determination and the necessary amount of required experimental information (the number of different solution compositions). Despite of diminishing the number of fitting parameters, theoretical equations turned out to be in quite satisfactory agreement with adsorption data for all systems which have been analyzed. Thus found values of the Δ parameter are practically independent of the electrode potential or ionic strength of the solution and demonstrate an expected correlation with the properties of adsorbing ion, metal and solvent.

Therefore one may assume that the theory gives an adequate interpretation of the physical picture of this important phenomenon. It seems to be actual to widen the set of real systems treated to create a "collection" of the isotherm parameters, Δ and W^{o}_{ads}, which would give an "experimental" basis for a future quantum-mechanical theory of a single ion at the metal/solvent interface.

REFERENCES

1. M.A.Vorotyntsev, in Adv.Science & Technology. Electrochemistry, ed. Yu.M.Polukarov (VINITI,Moscow) 21 (1984)3-77.
2. M.A.Vorotyntsev, in Modern Aspects of Electrochemistry, ed. J.O'M.Bockris, B.E.Conway & R.E.White, (Plenum Press,New York), 17 (1986)131-222.
3. U.V.Palm,E.V.Past, Uspekhi Khimii (Adv.in Chemistry) 44 (1975)2035-2047.
4. U.V.Palm,B.B.Damaskin, in Adv.Science & Technology. Electrochemistry, ed. Yu.M.Polukarov (VINITI,Moscow) 12 (1977)99-143.
5. B.B.Damaskin,R.V.Ivanova, Uspekhi Khimii (Adv.in Chemistry) 48 (1979) 1747-1772.
6. I.A.Bagotskaya, in Adv.Science & Technology. Electrochemistry, ed. Yu.M.Polukarov (VINITI,Moscow) 23 (1986)60-100.
7. R.Payne, in Adv.Electrochem.Electrochem.Eng., ed. P.Delahay, (Interscience,New York) 7 (1970)1-76.
8. R.Payne, J.Electroanal.Chem. 41 (1973)277-309.
9. R.Parsons, ibid 150 (1983)51-57.
10. A.Hamelin,T.Vitanov,A.Popov,E.Sevastyanov, ibid 145 (1983)225-264.
11. M.D.Levi,A.V.Shlepakov,B.B.Damaskin,I.A.Bagotskaya, ibid 138 (1982)1-27.
12. V.E.Kazarinov,V.N.Andreev, in Comprehensive Treatise of Electrochemistry, ed. E.Yeager,J.O'M.Bockris,B.E.Conway & S.Sarangapani (Plenum Press,New York-London) 9 (1984)393-443.
13. A.M.Funtikov, Soviet Electrochemistry 19 (1983)1378-1385.
14. M.A.Vorotyntsev, in Adv.Science & Technology. Electrochemistry, ed. Yu.M.Polukarov (VINITI,Moscow) 26 (1988)3-93.
15. M.A.Vorotyntsev,A.A.Kornyshev, JETP-Soviet Physics 78 (1980)1008-1019.
16. A.A.Kornyshev,M.A.Vorotyntsev, Surface Sci. 101 (1980)23-48.
17. M.A.Vorotyntsev, Soviet Electrochemistry, 16 (1980)1350-1358.
18. M.A.Vorotyntsev,V.Yu.Izotov,A.A.Kornyshev, Surface-Soviet Physics (1983)N 7, 97-110.

19. M.A.Vorotyntsev, ibid (1989)N 2, 39-46.
20. M.A.Vorotyntsev, J.Research Inst.Catalysis. Hokkaido Univ. 30 (1982) 167-178.
21. M.A.Vorotyntsev,V.Yu.Izotov, Soviet Electrochemistry 19 (1983)831-833.
22. M.A.Vorotyntsev,S.N.Ivanov, JETP-Soviet Physics 88 (1985)1729-1737.
23. M.A.Vorotyntsev,K.Holub, Soviet Electrochemistry 20 (1984)256-259.
24. M.A.Vorotyntsev, ibid 21 (1985)278-281.
25. M.A.Vorotyntsev, ibid 21 (1985)354-357.
26. M.A.Vorotyntsev, ibid 21 (1985)513-516.
27. A.V.Shcheglov,M.A.Vorotyntsev,B.B.Damaskin,R.V.Ivanova, ibid 21 (1985) 1126-1129.
28. R.Parsons, Trans.Faraday Soc. 51 (1955)1518-1529.
29. R.Parsons, J.Electroanal.Chem. 7 (1964)136-152.
30. A.N.Frumkin, ibid 7 (1964)152-155.
31. B.B.Damaskin, ibid 7 (1964)155-159.
32. R.Parsons, ibid 8 (1964)93-98.
33. V.Yu.Izotov, Thesis (Inst.General Inorganic Chemistry,Kiev,1983).
34. U.Palm,M.Paarnou,M.Salve, J.Electroanal.Chem. 86 (1978)35-51.
35. L.M.Dubova, Thesis (Inst.Electrochemistry,Moscow,1978).
36. Dz.I.Dzaparidze,Sh.S.Dzaparidze, in Adsorption and Double Layer in Electrochemistry (Nauka,Moscow,1972)68-71.
37. M.A.Salve,U.V.Palm, Proc.Tartu State Univ. 332 (1974)N 9, 71-75.
38. B.Damaskin,U.Palm,M.Paarnou, J.Electroanal.Chem. 70 (1976)103-115.

SOLVENT ADSORPTION AT THE ELECTRODE / SOLUTION INTERFACE. THE ROLE OF METAL.

ZOFIA BORKOWSKA and GRAŻYNA JARZĄBEK
Institute of Physical Chemistry, Polish Academy of Sciences, Warsaw 01-224, Poland

ABSTRACT
In this paper we discuss the properties of electrode / solution interface in organic solvents on various metals. The available electrochemical data, such as double layer capacity and zero charge potential are used to evaluate influence of the metal and of the solvent on the potential drop across the interface in terms of Trasatti's correlation between the work function of electrons in the metal and in the solution.

Introduction

The electrochemical measurements performed at the ideally polarized electrode are carrying an information about whole interface including metal and solution side. It is now a well established fact, that the properties of the interface depend strongly on the nature of the solvent[1,2] and of the electrode material[3-5].

The aim of this work is to discussed the influence of the metal and of the solvent on the electrochemical properties of electrode / electrolyte interface.

Addressing this question we have collected the double layer data such as: differential capacity, C, potential of zero charge, pzc, and adsorption parameters of a given solvent from aqueous solutions on various metals.

The electrode / electrolyte interface in dimethylsulphoxide, DMSO, and n,n-dimethylformamide, DMF, have been investigated in our laboratory. The data for other solvents have been collected from the literature.

According to Trasatti[6] the potential drop across the interface, $\Delta\Phi^{M-S}$, is connected with the work function of the metal, Φ^M, and the surface potential, χ^S:

$$\Delta\Phi^{M-S} = \underbrace{\left(\Phi^M/e + \delta\chi^{M}_{(S)}\right)}_{g^M_{(S)}(\text{dipole})} - \underbrace{\left(\chi^S + \delta\chi^{S}_{(M)}\right)}_{g^S_{(M)}(\text{dipole})} + g^M_S(\text{ion}) \tag{1}$$

where $\delta\chi^{M}_{(S)}$ and $\delta\chi^{S}_{(M)}$ are the modifications of the metal and solvent surface potentials respectively when brought into contact, $g^M_{(S)}(\text{dipole})$ is the potential drop due to the dipole orientation on the surface of the metal in contact with liquid phase and

$g^{S}_{(M)}$(dipole) is the potential drop due to the dipole orientation on the surface of the liquid phase when brought into contact with with the metal. The term g^{M}_{S}(ion) counterbalance the charge of the electrode, σ. At the pzc g^{M}_{S}(ion) is equal to zero and the potential drop across the interface is as follows:

$$\text{pzc} = \Phi/e + \left[\delta\chi^{M}_{(S)} - g^{S}_{(M)}(\text{dipole})\right] + E_{(\text{ref})} \tag{2}$$

where $E_{(\text{ref})}$ is the absolute potential of the reference electrode. The term in brackets represents metal - solvent interactions.

If we consider the interface in various solvents on one metal, and $E_{(\text{ref})}$ is constant or the difference in reference potential in various solvents can be accounted for then:

$$^{S_1}_{S_2}\Delta\text{pzc} - {}^{S_1}_{S_2}\Delta\left[\delta\chi^{M}_{(S)} - g^{S}_{(M)}(\text{dipole})\right] = \Delta X \tag{3}$$

In this case the influence of the solvent on the metal - solvent interactions can be compared.

If we consider the interface between various metals and one solvent then the difference of pzc and of the work function is equal to:

$$^{M_1}_{M_2}\Delta\text{pzc} - {}^{M_1}_{M_2}\Delta\Phi/e = {}^{M_1}_{M_2}\Delta\left[\delta\chi^{M}_{(S)} - \delta\chi^{S}_{(M)}\right] = \Delta X \tag{4}$$

where ΔX is so called adsorption potential[3] and represents also metal - solvent interactions. In this case the influence of the metal on metal - solvent interactions can be compared, as it was done previously[3] for Hg, Ga, In+Ga and Bi in DMSO and acetonitrile, AN. However in this work instead of the work function difference, the potential difference at high negative charge densities, $^{M_1}_{M_2}\Delta E_{\sigma<<0}$ was used.

EXPERIMENTAL

All measurements were made in three electrode cell. The working and the counter electrode were made from the same metal. Gold and platinum electrodes were cleaned in reducing flame and cooled in a stream of argon. The reference electrode was 0.1 M NaCl calomel electrode, 0.1 NCE. Additionally the half wave potential of bisbiphenylchromium (1/0), $E_{1/2}$BBC in the solvents was determined[7].

Organic solvents were initially dried over 0.4 nm molecular sieves, then distilled trice at reduced pressure.

Inner layer capacities for solid electrodes has been obtained from measured double layer capacities using the roughness factor, r_f, needed to obtain monotonic $C(\sigma)$ curves[8].

RESULTS AND DISCUSSION

Some physico chemical properties of the investigated solvents and metals are given in Table 1 and Table 2 respectively.

Table 1. Some physico chemical properties of the solvents[1].

Solvent	Dielectric constant, ε	Dipole moment, μ/D	Polarizability $10^3\alpha/nm^3$
DMSO	46.7	3.69	10.5
DMF	36.7	7.85	7.81
PC	64.4	4.98	9.0
AN	36.0	3.92	
H_2O	78.3	1.85	1.5

Table 2. Some physical properties of metals studied. Φ is the work function and n_+ is the charge density of uniform positive background ($n_+ = 0.6022*10^{24}Z\rho M/A$ where Z is charge number, ρ metal density and A molecular weight).

METAL	$(\phi/e)/V$[6]	n^+/a.u.	Z[9]
Bi	4.35	0.012	5
Au	4.80	0.0091	1.11[x]
Pt	5.03	0.0124	1.26
Hg	4.50	0.0027	2
Ga	4.25	0.0225	3
In	4.16	0.0172	3
Pd	5.0	0.0100	1.04[x]

[x] Z given by the difference between the number of valency electrons and d-band electrons[10].

The double layer data for investigated systems are given in Table 3. Inner layer capacity curves, $C_i(\sigma)$ are presented in Figs 1 - 4.

It can be seen from Figs 1 - 4 and Table 3 that at sufficiently high negative polarization the capacity becomes metal independent as it was suggested previously in the literature[3]. So, at these conditions properties of the electrode / electrolyte interface are dependent on the solvent only. When the electrical field is relaxed the influence of the metal is visible which is reflected in the values of pzc coordinates.

In all investigated solvents, except AN, $C_i^{\sigma=0}$ at the interface with Hg is much higher as compared with Au and Pt. We suggest, that interactions between uncharged Au and Pt and investigated aprotic solvents are strong and similar to those at high negative polarizations. So it is possible that at pzc the solvent dipoles are oriented with the positive pole towards the electrode.

Table 3. Double layer data for electrode/solution interface.

Solution	Electrode	PZC/V *vs* 0.1 NCE	PZC/V *vs* BBC	$C_i^{\sigma=0}$	$C_i^{\sigma<<0}$	Ref.
DMSO + $LiClO_4$	Hg	-0.41	0.52	0.26	0.078	11
	Bi	-0.57	0.36	0.28	0.080	3,12
	In + Ga	-1.03	-0.10	0.54	0.079	3
	Au	0.06	0.99	0.12	0.08 ($r_f = 1.2$)	13
	Pt	-0.20	0.73	0.10	0.08 ($r_f = 1.3$)	14
	Pd	-0.09	0.84			15
DMF + $LiClO_4$	Hg	-0.30_5	0.54	0.22	0.070	1,16
	Bi	-0.47	0.37_5	0.26	0.072	17
	Au	0.18	1.02_5	0.08	0.07 ($r_f = 1.3$)	18
PC + $NaClO_4$	Hg	-0.38	0.62	0.20	0.08	16,19
	Au	0.16	1.16	0.09	0.07_5	20
AN + $LiClO_4$	Hg	-0.29	0.63	0.16	0.10	3,19
	Bi	-0.53	0.39	0.16	0.10	3
	In + Ga	-0.78	0.14	0.28_5	0.10	3
	Au	0.56	1.48	0.18	0.10 ($r_f = 0.75$)	21
	Pt	-0.17	0.75			22
	Pd	0.01	0.93			22
MeOH + $NaClO_4$	Hg	-0.39_5	0.50	0.21	0.09_5	1,23
+ KF	Hg	-0.39	0.50_5			24
+ KF	Au	-0.11	0.78_5			25
H_2O + $NaClO_4$	Hg	-0.50	0.48	0.39	0.17	26
+ NaF	Bi	-0.72	0.26			27
+ $NaClO_4$	In + Ga	-1.01	-0.03			28
+ $LiClO_4$	Au	-0.16	0.82	0.18	0.10 ($r_f = 2.23$)	29,30

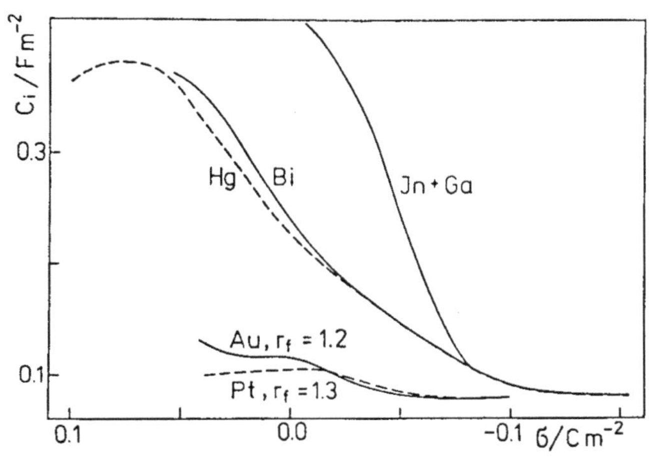

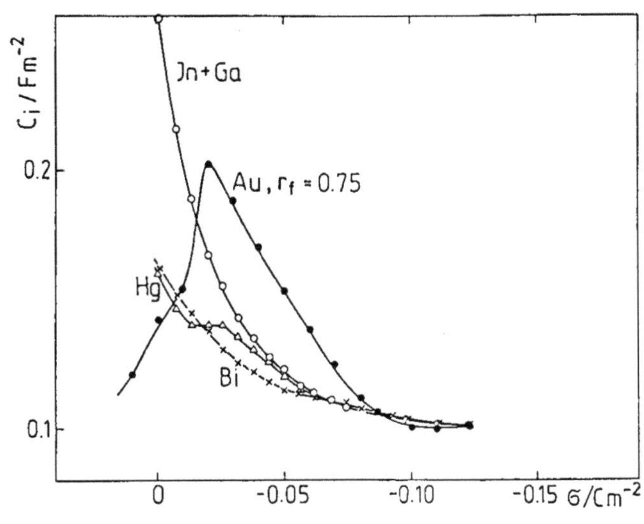

Fig. 1. Inner layer capacity curves in dimethylsulphoxide for In + Ga (3), Bi (3), Hg (11), Au (13), and Pt (14) with $LiClO_4$ as supporting electrolyte. Roughness factor, r_f, given in the figure if different from 1.

Fig. 2. Inner layer capacity curves in acetonitrile for In + Ga (3), Hg (3), Bi (3) and Au (21) with $LiClO_4$ as supporting electrolyte. Roughness factor, r_f, given in the Figure if different from 1.

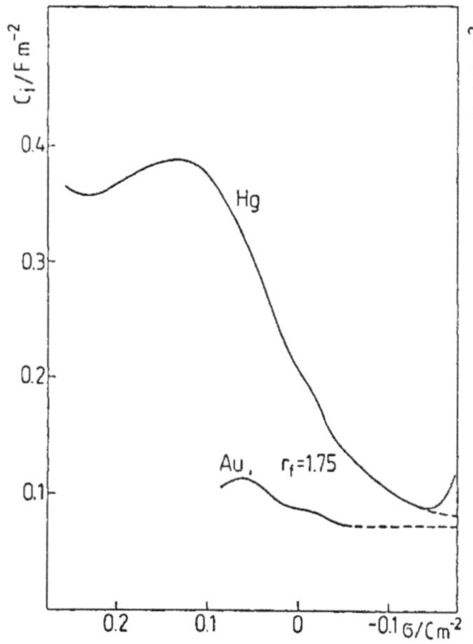

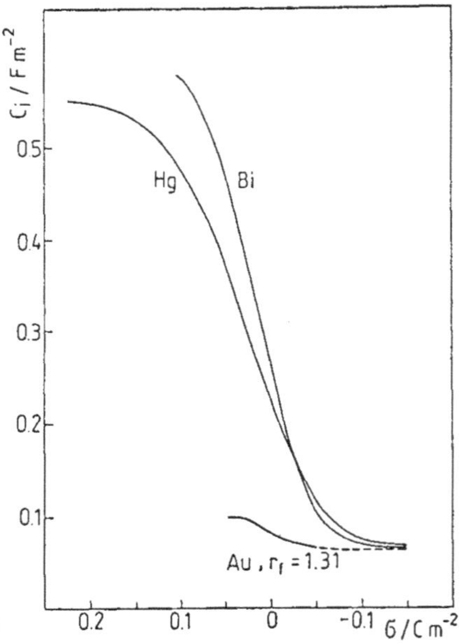

Fig. 3. Inner layer capacity curves in propylene carbonate for Hg (16) and Au (20) with $NaClO_4$ as supporting electrolyte. Roughness factor, r_f, given in the Figure if different from 1.

Fig. 4. Inner layer capacity curves in n,n-dimethylformamide for Hg (16), Bi (12) and Au (18) with $LiClO_4$ as supporting electrolyte. Roughness factor, r_f, given in the Figure if different from 1.

This suggestion is supported by much stronger DMSO adsorption from aqueous solutions on Au as compared with Hg (Table 4).

Table 4. The parameters for DMSO adsorption on various metals from H_2O + DMSO mixtures.

Electrode	Free energy, ΔG/KJ M^{-1}	Interaction parameter, a	Half coverage conc., $X_{O=0.5}$
Hg[31]	-8.4	0.69	0.01[32]
Au[29]	-26.0	-0.22	0.0003
In + Ga[33]			0.044

If we consider different solvents at the interface with one metal, the pzc values *vs* BBC changes as follows:

$$AN > PC > DMF > DMSO > MeOH > H_2O$$

The position of MeOH is not clear because the data are available only for Hg (as in unequolity given above) and for Au where the order would be reversed: H_2O > MeOH. In the same order changes $\delta\chi^{M}_{(S)} - g^{S}_{(M)}$(dipole) = X' and ΔX' related to water (Table 5). ΔX' is the most positive or the less negative for AN and the most negative or the less positive for MeOH.

Table 5. Adsorption potential, $\Delta X'$ (eq. 3), on a given metal in various solvents S_1 referred to $S_2 = H_2O$.

S_1	Metal			
	Hg	Bi	In+Ga	Au
DMSO	0.04	0.10	-0.07	0.17
AN	0.15	0.13	-0.11	0.66
DMF	0.065	0.115		0.205
PC	0.14			0.34
MeOH	0.02			-0.035

On the other hand, if we consider different metals in one solvent, pzc values changes as follows:

$$Au > Pd > Pt > Hg > Bi > In+Ga$$

In the same order changes $\Phi/e + \left(\delta\chi^{M}_{(S)} - \delta\chi^{S}_{(M)}\right)$. For sp metals, in the same order changes also Φ/e. It would suggest that for these metals the surface modification term given in brackets is either similar or negligible as compared with the work function term.

Table 6. Adsorption potential, ΔX (eq. 4), in a given solvent for various metals M_1 referred to $M_2 = Hg$.

M_1	Solvent					
	H_2O	DMSO	AN	DMF	MeOH	PC
In+Ga	-0.17	-0.28	-0.15			
Bi	-0.07	-0.01	-0.09	-0.02		
Au	0.04	0.17	0.55	0.18	0.00	0.24
Pt		-0.32	-0.41			
Pd		-0.18	-0.20			

The adsorption potential related to Hg, ΔX, (Table 6) changes as follows:

$$Au > Bi > In+Ga > Pd > Pt$$

or

$$Au > Bi > Pd > In+Ga > Pt$$

In the same order changes the surface modification term $\delta\chi^{M}_{(S)} - \delta\chi^{S}_{(M)}$. ΔX is the most positive or the less negative for Au and the less positive or the most negative for Pt.

Acknowledgements

This work was carried out within Research Project CPBP 01.15.

References

1. Z. Borkowska, *J. Electroanal. Chem.*, **244** (1988) 1.
2. S. Trasatti, *Electrochim. Acta*, **32** (1987) 843.
3. M. D. Levi, A. V. Shlepakov, B. B. Damaskin and I. A. Bagotskaya, *J. Electroanal. Chem.*, **138** (1982) 1.
4. S. Trasatti, *ibid*, **123** (1982) 121.
5. W. Schmickler and D. Henderson, *J. Phys. Chem.*, **80** (1984) 3381.
6. S. Trasatti, in *Advances in Electrochemistry and Electrochemical Engineering*, ed. H. Gerisher and C. W. Tobias, (A. Wiley, New York, 1977), vol.10.
7. Z. Borkowska and G. Jarząbek, *J. Electroanal. Chem.*, submitted.
8. A. Hamelin and A. Le Lan, *C. R. Acad. Sci. Ser. II*, **295** (1982) 161.
9. N. W. Ashcroft and N. D. Mermin, *Solid State Physic*, (Holt, Rinehost & Winston, New York, 1976), p. 227.
10. M. Brejnak and P. Modrak, *Surface Sci.*, in press.
11. Z. Borkowska and J. Stafiej, *J. Electroanal. Chem.*, **226** (1987) 283.
12. E. K. Petyarv, U. V. Palm, *Elektrokhimya*, **12** (1976) 806.
13. G. Jarząbek and Z. Borkowska, *J. Electroanal. Chem.*, **226** (1987) 295.
14. G. Jarząbek and Z. Borkowska, *Electrocatalysis*, ed. P. Nowak (Drukarnia Narodowa, Crakow, 1988), p.159.
15. K. Yu. Alekseeva, W. A. Safonov and O. A. Petrii, *Elektrokhimya*, **20** (1984) 945.
16. Z. Borkowska and W. R. Fawcett, *ibid*, **16** (1980) 131.
17. E. K. Petyarv and U. V. Palm, *ibid*, **9** (1973) 1836.
18. G. Jarząbek, unpublished results.
19. R. Payne, *J. Phys. Chem.*, **71** (1967) 1548.
20. C. Nguyen Van Huong, *J. Electroanal. Chem.*, **194** (985) 155.
21. A. Hamelin and L. M. Dubova, *ibid*, **220** (1987) 277.
22. O. A. Petrii and I. G. Khomchenko, *idib*, **106** (1980) 277.
23. Z. Borkowska and W. R. Fawcett, *Can. J. Chem.*, **59** (1981) 710.
24. D. C. Grahame, *J. Am. Chem. Soc.*, **76** (1954) 4819.
25. M. Brzostowska-Smolska and S. Minc, *Pol. J. Chem.*, **57** (1983) 1005.
26. Z. Borkowska, unpublished results.
27. A. Frumkin, *Potencjaly Nulevovo Zariada*, (Nauka, Moskwa, 1979).
28. N. B. Grigiriev, S. A. Fateev and I. A. Bagotskaya, *Elektrokhimya*, **20** (1984) 945.
29. G. Jarząbek and Z. Borkowska, *J. Electroanal. Chem.*, **248** (1988) 389.
30. J. Clavillier et C. Nguen Van Huong, *ibid*, **41** (1973) 193.
31. Z. Borkowska, *ibid*, **146** (1983) 385.
32. J. Taraszewska, *ibid*, **121** (1981) 215.
33. I. A. Bagotskaya and L. M. Dobova, *Elektrokhimya*, **14** (1978) 1264.

ELECTRODYNAMICS, SPECTROELECTROCHEMISTRY, SCANNING TUNNELING MICROSCOPY

In-situ SHG and Raman Spectroscopy at Electrodes

Bruno Pettinger and Andreas Friedrich

Fritz-Haber-Institut der Max-Planck-Gesellschaft
Faradayweg 4-6, D-1000 Berlin 33, Germany

Abstract

Second harmonic generation and surface Raman spectroscopy at metal electrodes are *in-situ* spectroscopies for the study of the metallic and electrolyte side of the electric double layer at the interface, and in this sense both spectroscopies are supplemental to some extent: Surface (un)enhanced Raman spectroscopy yields via vibronic spectra insight in electrochemical processes such as adsorption or reaction pathways.
With second harmonic generation, however, one can monitor *in-situ* at a Au(111) electrode not only the formation or lifting of the (1× 23) surface reconstruction, but also (indirectly) the specific adsorption of ions as they alter the non-linear in-plane and perpendicular polarizability of metal electrons at the interface in a distinct way. Hence, chemisorption processes can be studied in a novel way.

Introduction

The research of electrochemical processes requires not only a deeper insight into the composition of the phase boundary but also of the bonding, orientation and structure of the adsorbate. Somewhat underestimated in the past is the importance of the actual surface orientation and its possible reconstruction for electrochemical processes and kinetics. Thus, the study of the electric double layer and metal surfaces demand specific spectroscopic tools such as *in-situ* surface (un)enhanced Raman spectroscopy and *in-situ* Second Harmonic Generation.

In-situ surface Raman spectroscopy

With the advent of **S**urface **E**nhanced **R**aman **S**pectroscopy (SERS), *in-situ* vibronic spectroscopy of the interfacial layer begann to thrive [1-6], but at the same time it was obstructed by the small number of SERS displaying metals and the necessity to activate the electrodes, e.g. to roughen them to a great extent [3-6]. Such a procedure leads to an unusual electrochemical behaviour [7] in addition to the fact that such surfaces are not well defined.

Concepts to circumvent limitations characteristic of SERS were presented in recent years, such as
(i) the approach of diminished SERS, e.g. to reduce the number of so-called SERS active sites after the activation assuming that then SERS for adsorbates at more usual sites dominate the vibronic spectra [8,9];
(ii) the preparation of SERS-active gold electrodes in a separate cell which afterwards can be immersed into the electrolyte of interest [10,11];
(iii) the deposition of metal overlayers on a SERS-activated surface; in this way the surface enhancement is transferred to the overlayer. This approach extends SERS to transition metal surfaces [12-15];
(iv) the use of moderate roughening in order to retain the usual electrochemical behaviour of the electrode (e.g. only one ORC cycle at a gold electrode instead of 15 [7]).
(v) the employment of surface unenhanced Raman spectroscopy for *in-situ* studies [16,17].

Whereas at the beginning of SERS much effort had been spent in the model discussion, a shift towards to the use of SERS for *in-situ* investigations has been noticed in recent years [7-70]. On account of this development a short overview of recent activities in this respect will be presented; the intention is to show that SERS is indeed a particular interesting and valuable spectroscopic tool for *in-situ* investigations at electrodes. In this context one should note that in order to understand and interpret SERS data one can refer to the large body of knowledge in vibronic spectroscopy. Hence, for a spectrocopic analysis we do not need to know in detail the surface enhancement mechanisms.

In-situ Second Harmonic Generation

The rotational anisotropy of the Second Harmonic Generation (SHG) from single crystals can represent a sensitive *in-situ* analysis of surface reconstructions. Provided the variation of electrochemical parameters causes a change of the rotational symmetry of the overall surface (i.e. $C_s \rightarrow C_{3v}$ due to the transition of Au(111)-(1×23) → Au(111)-(1×1) [73,74]), it is possible to follow *in-situ* via SHG measurements the development or lifting of surface reconstruction. Such measurements are of particular interest as not only the actual orientation of single crystal electrode but also the presence or absence of surface reconstruction has a specific effect on the pattern of cyclic voltamograms, capacity-voltage curves, and the current value of the pzc point [75]. For a Au(111) electrode, immersed into a 0.01 M $HClO_4$ electrolyte the potential range of a stable reconstructed surface (C_s: U< 0.1 V) and of its relative wide transition regime in this electrolyte (C_s: +0.1 V ↔ +0.9 V: C_{3v}) could be determined [73]. In a more recent paper a significant influence of specific adsorbed halide ions on the surface structure was reported [76]. Contrary to the case of halide free solution, their adsorption (by scanning the potential anodically) leads to a lifting of the reconstruction in a narrow potential region. This region is determined by the type of anion and its solution concentration. Similary, their desorption during a cathodic potential scan induces the reconstruction immediately [76]. In a recent publication an analogous influence of Cd-deposition and -dissolution on the surface reconstruction was reported [77]. A further study of the SHG-potential dependence deals with the separation of various contributions to the SHG-response such as the isotropic and anisotropic parts of SHG [78]. In this context it should be noted that the non-linear susceptibility, $\chi^{(2)}$ describing the SHG response is a tensorial quantity, which can contain up to 18 non-zero elements. Its structure and the actual number of non-zero elements is determined by the surface symmetry. In order to use SHG as a spectroscopic tool one has to get a complete picture of the non-linear interaction of light with the metal surface. Thus, it is necessary to determine the important elements of the $\chi^{(2)}$ tensor and their potential dependence [78].

In the second part of this communication preliminary results of this study are presented, e.g. the potential dependence of the isotropic and anisotropic part of the SHG response.

Results and Discussion

In-situ surface Raman Spectroscopy

In-situ SERS investigations can yield valuable informations about the structure and composition of the electric double layer, the identity of adsorbates and intermediates and their interactions with the metal electrons depending on electrochemical parameters such as electrode potential, solution pH, anion or cation concentrations of the supporting electrolyte. A necessarily incomplete overview of recent activities in this direction is given in ref. [18] and is reproduced here by a table1. It concerns electrochemically motivated SERS studies over the last three years (studies on colloidal systems and investigations mainly dedicated to evaluation of the surface - enhancement mechanisms are not included): For a quick overview table 1 is divided into four parts: Studies on adsorption, thin oxide/hydroxide films, redox-processes and double layer structure. Only a few reports can be addressed in this paper.

SERS studies concerning adsorption processes [19-34] cover a great variety of topics such as correlation of adsorption isotherms with SERS-intensity (Stolberg et al. [7]), the formation of surface complexes (Feilchenfeld [20]), the vibrational coupling measured for isotopic mixtures of $^{12}CN^-/^{13}CN^-$ and benzene using C_6H_6/C_6D_6 (Gao et al. [21]), the acidity of the surface (Guzonas et al. [23]), evidence for distinct SERS-active sites (Temperini et al. [26,27]), or the orientation of aniline on gold electrodes (Holze [29]).

The next group of SERS studies in table 1 concerns Raman spectroscopy on thin oxide films [35-37]. Although in the oxidative electrochemistry on noble metals the oxidation of the electrode itself plays a prominent role, there are only comparatively few SERS studies on this subject. The two examples reported in the following indicates the great potential of such studies.

Desilvestro et al. studied gold electrodes in contact with acidic and alkaline media. Two Raman lines were noted and labeled as ν_1 (~400 cm^{-1}) and ν_2 (~550 cm^{-1}) [36]. The ν_1 mode emerges and disappears in the potential range of -0.2 V and +0.1 V, whereas the ν_2 vibration grows at around +0.3 V. These modes were assigned to M-OH vibrations as an exchange of the electrolyte with D_2O led to frequency shifts for both bands with the appropriate ratio of $\nu(H_2O)/\nu(D_2O)$ ~ 0.97. In the authors opinion, among the two modes only the ν_2 vibration indicate clearly an early stage of oxide formation in the potential range of 0.3-0.5 V. In acid media the ν_1 vibration is completely absent, whereas the ν_2 mode appears to be broadened [36].

Iwasaki et al. investigated the oxidation and reduction of an oxidized silver electrode in a sodium-hydroxide containing electrolyte observing the Ag-O stretching mode at 470 cm^{-1} and a Ag-OH^- bending mode at 860 cm^{-1} [37]. Of particular interest is that new Raman bands appear when the potential is moved cathodically to -1.4 V which were assigned to librational and scissor modes of hydrated water molecules [37].

The next group of SERS studies in table 1 refers to the study of redox-reactions with the intention to identify intermediates and evaluate reaction pathways [38-53]. One report will be addressed here, given by Weaver et al., who explored the electro-oxidation of benzidine by voltammetry, SERS and IRRAS. The Raman spectra were recorded in real time within a single potential sweep using an optical multichannel analyser [47]. In a following study on the electroreduction of nitrobenzene and azobenzene on gold electrodes the authors presented spectral evidence for the formation of nitrosobenzene as an intermediate existing only at the interface [50].

The forth group in table 1 represent SERS studies on interfacial water and hydroxide which are particular interesting examples for the potential power of SER spectroscopy, but also for its limitation. Whereas at the beginning of SERS no surface enhancement for interfacial water had been reported [54-63], later investigations revealed intense SERS for water only when particular conditions were meet [54-61], such as the use of highly concentrated (≥1 M) halide or pseudo-halide solutions.

As is well known, halide ions are specifically adsorbed at silver, gold and Cu electrodes. Their role for SERS of water is twofold as they (i) stabilize the "SERS- active sites" (which might be disordered surface sites or very small metal clusters) and (ii) reduce significantly the hydrogen bonding of the interfacial water compared to that determining the water network of the bulk electrolyte [54-70]. Most interestingly is that not only anions, but also cations influence the SER spectra for interfacial water, in particular in the H_2O stretch regime [58,62,63]. In fact, relative intensities, frequencies, halfwidth, band shapes and band-splittings, even the temporal evolution of SERS depend on the nature of the cations [65]. These findings point to the formation of hydrated water with energetically distinct sites at the interface. Unfortunately, in recent years only a few papers were devoted to this topic [64-70]: A study of

Table 1: SER spectroscopy of electrochemical processes

Topic	Metal	Electrolyte	Ref
Adsorption			
Hexamethyltetramine	Ag	0.1 M KNO_3	19
Alkynes	Ag; Au; Ag/Au	0.1 M KCl or $KClO_4$	20
CN^-, arom. comp.	Au	0.5 M K_2SO_4	21
pyridazine	Ag	0.1 M KCl	22
diazabicyclo-octane	Ag	0.1 M KCL	23
Adenosine triphosphate	Ag	0.1 M KCl	24
Thioura	Ag	0.01 M $HClO_4$	25
Pyridine	Ag	0.1 M KCl	26
Pyridine	Ag	0.1 M NaOH	27
4-pyridinecarboxaldeh.	Ag	0.1 M KCl, KBr, KI	28
Aninline	Au	0.1 M $HClO_4$, H_2SO_4	29, 30
Pyridine	Au	0.1 M $KClO_4$	31
CO_3	Ag	0.5 M KCO_3	32
Surfactants	Ag	0.01 M KBr	33
Pyridine and derivates	Ag	0.1 M KCl	34
Thin films / Oxide layers			
MnO_2 + MnOH	Au	1 M KOH	35
$AuOH^-$, AuOH, AuX	Au	1 M KOH	36
AgO, Ag_2O, Ag-H_2O	Ag	0.5 M NaOH	37
Redox processes			
$(Fe(CN)_4NO)^=$ + CN^-	Ag	0.1 M KCl	38
S_2O_8 reduction	Au/OH or $Ni(OH)_2$	1 M KOH	39
Glucose oxid. by glucose oxidase	Ag	0.1 M Na_2SO_4	40
Subst. viologens	Ag	0.1 M KCl	41
Nile blue	GC, Ag	phosphate buffer(4.3)	42
Methylenviologen	Ag	0.1 M KCl	43
4,4'-bipyiridine	Ag	0.1 M KCl, KI, K_2SO_4	44
Cystine ⇔ cysteine	Ag	2 M HCl	45
Phenancine	Ag	aqu./ methanolic sol.	46
Benzidine reduction	Au	0.1 M Na-acetate	47, 48
Me - CO ⇒ Me - O ,..	Rh/Au; Ru/Au	0.1 M $HClO_4$	49
Nitrobenzene	Au	0.1 M H_2SO_4, NaOH	50
$Co(sep)^{3+}$/ $Co(sep)^{2+}$	Ag, Cu	0.1 M KBr	51
Au–$Ni(OH)_2$ oxidation	Au	1 M KOH	52, 53
Double layer			
H_2O	Ag	0.1 M KCl	64
H_2O	Ag	0.1 M NaCl	65
O^{2-}, OH^-, H_2O	Ag	0.01 M NaCl + KMnO	66, 67
OH^-	Ag	0.01 M MgCl + tr.KMnO	68
H_2O, D_2O	Ag	1 M KCl	69
H_2O	Ag	8 M $NaClO_4$+.1MKCl	70

the anion effect by Kung and Chen [64], and three publications by Dorain on SERS for water or OH^- using very dilute solutions of alkali metal halides which contain in addition 10^{-5} M $KMnO_4$ or Cl_2 [66–68]. The later results indicate that trace quantities of oxidizing agents can produce SERS active sites [68]. Finally, there is a report on intense SERS for water in concentrated $NaClO_4$ solution by Tian et al. [70]. This is observed only when the Ag electrode is immersed into an 8 M $NaClO_4$ solution containing 0.1 M halide or pseudohalide ions. For each anion there is a characteristic SERS intensity-potential profile indicating the particular, anion depending stability range of SERS-active sites at Ag.

In recent years a few attempts were made to employ **S**urface **U**nenhanced **R**aman **S**pectroscopy (SURS) as an *in-situ* spectroscopy. Campion at al. have shown that SURS can be used routinely under UHV conditions [71,72]. One basic requirement in this respect is to increase at first the overall sensitivity of the detection system and secondly to reduce bulk Raman scattering. Two reports in this field are now available, SURS for cyano-pyridine on a rhodium electrode [16], and SURS for ClO_4^- and SO_4^{2-} ions at a Pt electrode [17]. Obviously, these approaches require a detection sensitivity which is usually not given; however, it should be noted that a combination of all modern developments in Raman spectrometer systems (concerning an increase in light collection efficiency, spectrometer throughput and photon detection quantum yield) would lead to a about 10 times larger photon detection efficiency than given in ref. [17].

From the above discussion it is obvious that at present SERS/SURS provides only a fragmentary insight in the structure and composition of the Helmholtz layer at a metal electrode, in particular due to the specific prerequisites for a SERS active surface such as roughness and the need of highly concentrated electrolytes (for the study of interfacial water). It is, therefore, quite questionable whether and to what extent, under conditions, suitable for intense SERS, the vibronic spectra are representative for the majority of adsorbates. Since SERS and in particular SURS can be performed in more dilute, or halide free solutions and at (relatively) smooth interfaces, recent results in this direction might be encouraging enough to stimulate further detailed studies.

In-situ Second Harmonic Generation at Au(111) Electrodes

As already mentioned, the SHG response from a Au-electrode is mainly determined by the non-linear susceptibility of the metallic interface. Adsorbates such as halide ions or small organic molecules exhibit here an indirect influence only by changing the density profile of surface metal electrons. That means that the localization of metal electrons into metal adsorbate bondings must alter the SHG response as well as a change of surface structure due to formation or lifting of surface reconstruction [78].

Concerning the latter phenomenum, it is well known that a Au(111) surface reconstructs to the so-called Au(111)-(1×23) structure: The hexagonal top layer is compressed by 4% into the $(1\bar{1}0)$ direction, so that every 23rd surface atom is in registry with the underlying bulk. In addition a part of surface atoms are displaced laterally by maximal 0.9 Å in $(11\bar{2})$ direction which indicates a transition between fcc and hcp stacking [79-87]. Considering the symmetry of this reconstruction it is evident that it exhibits two symmetry elements for an identity operation: (i) one vertical mirror plane and (ii) an azimuthal rotation over 360^0. Hence, the Au(111)-(1×23) reconstruction has to be attributed with a C_s symmetry which is superimposed onto the underlying C_{3v} symmetry of the Au(111) surface. Thus, in the presense of this reconstruction the overall symmetry is C_s, in its absence it is C_{3v}.

As addressed in earlier papers, SHG shows only then a rotational anisotropy when the surface symmetry contains a n-fold rotational axis with $n \leq 3$, e.g. when the symmetry is of the type C_1, C_s up to C_{3v} [88]. For symmetry reasons higher rotational axes can not be resolved by SHG. The corresponding SHG anisotropy stems from an oscillation of the induced non-linear polarization, $\mathbf{P}(2\omega)$, between maxima and minima during the azimuthal rotation of the sample. $\mathbf{P}(2\omega)$ represents a vector sum of all dipole momenta induced in a unit cell, which can add constructively or destructively, depending on the relation of the EM field vector and the spatial electron density structure; and thus, depending on the actual surface symmetry (including possible reconstruction) and on the polarization of the incident ω- and emitted 2ω-beams, very distinct anisotropy pattern can emerge.

The formal, more phenomenological description of SHG correlates the induced polarization, $\mathbf{P}(2\omega)$, with the incident EM-field, $\mathbf{E}(\omega)$, via the non-linear susceptibility tensor of third rank $\boldsymbol{\chi}^{(2)} = (\chi^{(2)}_{ijk})$:

$$\mathbf{P}(2\omega) = \boldsymbol{\chi}^{(2)} : \mathbf{E}(\omega)\mathbf{E}(\omega) \tag{1}$$

where each index i,j,k scans over x,y,z and, hence, the set of ijk includes all its permutations of ijk. For SHG, the reduced form of the non-linear susceptibility tensor and of the dyadic product of the field terms is applied (Kleinmann symmetry), because $\chi^{(2)}_{ijk} = \chi^{(2)}_{ikj}$ and $(E_jE_k) = (E_kE_j)$; thus, the reduced tensor contains 18 elements within 6 rows and the vector (E_iE_j) 6 columns. The structure of the $(\chi^{(2)}_{ijk})$ tensor and the number of non-vanishing elements is determined by the current surface symmetry. In the mathematial description of the SHG process, the rotational tensor transformation of $\boldsymbol{\chi}^{(2)}$, given in standard notation, corresponds to the azimuthal rotation of the sample in the experiment:

$$\chi^{(2)}_{ijk}(\varphi) = \sum_{rst}(S_{ir}\,S_{js}\,S_{kt}\,\chi^{(2)}_{rst}) \tag{2}$$

where S_{ir}, S_{js}, S_{kt} represent rotation matrices. Since this formalism is given elsewhere [73,88–90], we reproduce only the final results being of interest. The case of s/s polariation is the most simple one as is evident form the tensor equation (1,2): the only contribution arises from the tensor element $\chi^{(2)}_{yyy}(\varphi)$; all the others are not relevant. Hence, the SH intensity, depending on the current symmetry, is described as:

$$C_s : I^{SH}_{ss} \sim \left[\, d^{(1)}_{11}\sin\varphi - d^{(3)}_{11}\sin 3\varphi \,\right]^2 \tag{3a}$$

$$C_{3v} : I^{SH}_{ss} \sim \left[\, -d^{(3)}_{11}\sin 3\varphi \,\right]^2 \tag{3b}$$

For p/p polarisation several tensor elements contribute and we have instead:

$$C_s : I^{SH}_{pp} \sim \left[\, a^{(\infty)} + b^{(1)}\cos\varphi + c^{(2)}\cos 2\varphi + d^{(3)}\cos 3\varphi \,\right]^2 \tag{3c}$$

$$C_{3v} : I^{SH}_{pp} \sim \left[\, a^{(\infty)} + d^{(3)}\cos 3\varphi \,\right]^2 \tag{3d}$$

with
$a^{(\infty)} = d^{(\infty)}_{31} + d^{(\infty)}_{33} + d^{(\infty)}_{15}$; $b^{(1)} = d^{(1)}_{11} + d^{(1)}_{13} + d^{(1)}_{34}$; $c^{(2)} = d^{(2)}_{31} + d^{(2)}_{14}$; $d^{(3)} = d^{(3)}_{11}$; where, according to eq. 2, each amplitudes, $d^{(1)}_{ijk}$, $d^{(2)}_{ijk}$, $d^{(3)}_{ijk}$, of the tensor elements $\chi^{(2)}_{ijk}$ (note: i = 1 … 3, jk = 1 … 6) represents actually a sum of several $\chi^{(2)}_{rst}$ elements. In addition, each term $d^{(x)}_{ijk}$ contains the Fresnel factors for the transmission of incident and generated fields: $d^{(x)}_{ijk} = T_i\, d^{(x)}_{ijk}\, t_j\, t_k$ [78,88].

Fig. 1a,b illustrates the distinct potential dependences of the SHG signal for s/s and p/p polarisation. The significant influence of halide adsorption is seen in each part of the figure in addition. Fig. 1a shows SHG potential dependences when the azimuthal angle is set to 90^0, e.g. ($I_{ss}^{SH}(\varphi=90^0)$ vs. U), where the contribution of surface reconstruction is at maximum. The curves in fig. 1a indicate therefore basically the formation or lifting of surface reconstruction due to the symmetry transition $C_s \leftrightarrow C_{3v}$. One of the reasons for this sensitivity is, that one observes here the interference of two non-linear SHG sources (as evident in eq. 3a) which is remarkable even when $|d_{11}^{(1)}| \ll |d_{11}^{(3)}|$. A Fourier analysis of SHG anisotropy patterns (I_{ss}^{SH} vs. φ for a set of potentials) recorded in addition to the above curves yields a minor potential dependence of $d_{11}^{(3)}$, e.g. the potential dependences seen in fig. 1a stem mainly from the $d_{11}^{(1)}$ term which is indicative for the degree of surface reconstruction being present (see fig. 1b).

In contrast to the above case, the SHG response for p/p polarisation is composed by several contributions as eq. 3c,d show. Inspite of their significant potential dependence the anisotropy pattern has a typical form: In the anodic regime, at U > +0.9 V, a sinusoidal curve displaying three equal maxima every 120^0 is superimposed onto a strong isotropic background. In the cathodic regime (U=-0.4 V) the SHG intensity is generally reduced. In addition, one maximum is reduced less than the other two; also the isotropic part is lowered. Equation 2 indicates that there could also be a twofold contribution, however, the visual inspection is only sufficient to notice a one-fold symmetry superimposed to a three-fold one, whereas the Fourier analysis yields a quite small value for the two-fold amplitude, $b^{(2)}$. It should be mentioned, however, that in the curves recorded for s/p polarization, where no isotropic susceptibilities arise, the wofold amplitude is clearly detectable. On account of the various distinct contributions, an interpretation of I_{pp}^{SH} (U) curves is not straightforward. More adequate is to record several I_{pp}^{SH} vs. φ curves for a constant potential U which is incremented by +0.1 V after each scan. A Fourier analysis permits then the separation of various contributions and, hence, the evaluation of potential dependence of the isotropic and three anisotropic contributions to SHG for p/p polarisation.

Fig. 1b displays the ratios of all important coefficients of the $\chi^{(2)}$ tensor for a Au(111) electrode undergoing a symmetry transition $C_s \rightarrow C_{3v}$. Open symbols denote the curves for the standard electrolyte (0.1 M $HClO_4$), filled ones that for a halide containig solution (10^{-3} M Br^- in 0.1 M $HClO_4$). The sum of the directly generated isotropic amplitudes $d_{33}^{(\infty)}$ + $d_{15}^{(\infty)}$, the $d_{31}^{(\infty)}$ -coefficient and the threefold amplitude $d_{11}^{(3)}$ are plotted for a pure and bromide containing electrolyte. The reference value is always the $d_{11}^{(3)}$ coefficient at the most cathodic potential. The ratio for the $d_{11}^{(1)}$ coefficient is shown as well, in order to illustrate the region of the (1×23) → (1×1) transition in this particular electrolyte. Because of the interference

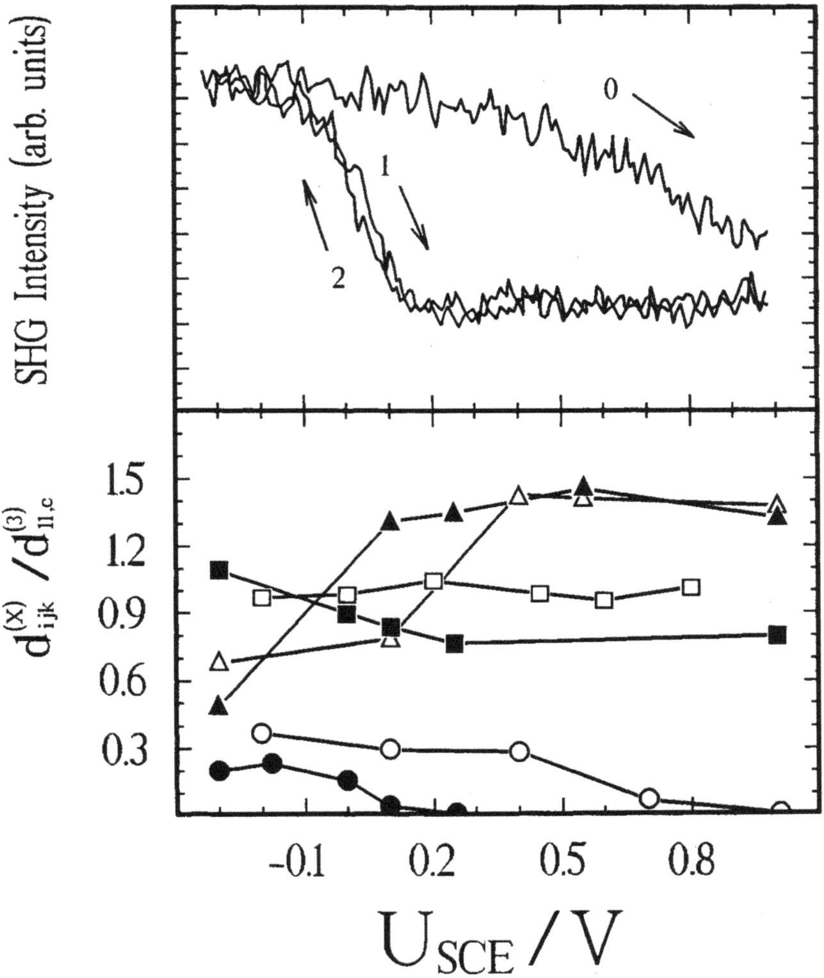

Fig. 1: *In-situ* Second Harmonic Generationat a Au(111) electrode.
(a) I_{ss}^{SH} vs electrode potential (scan rate: 2 mV/s). The azimuthal angle is 90°. The curves (0,1,2) show the lifting/formation of surface reconstruction(1×23) $\rightleftarrows$ (1×1) during the potential scan in pure 0.1 M $HClO_4$, or 0.1 M $HClO_4$ + 10^{-3} M X with X = Cl^- , Br^- or I^-.
(b) Amplitude ratios for some important coefficients versus electrode potential in pure 0.1 M $HClO_4$ (open symbols) and with 10^{-3} M Br^- (filled symbols). Note that these coefficients include the Fresnel factors of the type $d_{ijk}^{(x)} = T_i d_{ijk}^{(x)} t_j t_k$ [78]. Reference value is the $d_{11}^{(3)}(U_c) = d_{11,c}^{(3)}$ coefficient at U_c = -0.2 V or U_c = -0.3 V, respectively. Symbol definition: ○ : $d_{11}^{(1)}/d_{11,c}^{(3)}$; □ : $d_{11}^{(3)}/d_{11,c}^{(3)}$; △ : $(d_{33}^{(\infty)} + d_{15}^{(\infty)})/d_{11,c}^{(3)}$;
The solid lines are gides to the eye.

of the different coefficients (the intensity is the square of the susceptibility) small changes of the coefficients can have a pronounced influence on the anisotropy pattern. Whereas the $d_{11}^{(3)}$-coefficient is found to be constant in the double-layer region in pure $HClO_4$ solution, the adsorption of bromide leads to a decrease of the threefold amplitude. On the contrary the isotropic susceptibilities show a remarkable increase at the potential range of bromide adsorption, which takes place much more cathodically compared to the case of a pure $HClO_4$ solution. These effects can be explained by the action of contact adsorbed halide ions on the perpendicular and parallel non-linear polarizabilities. The specific adsorption of Br^- ions means a (partly) localization of metal electrons in a metal-ion bond, thus increasing the vertical polarizability by reducing the number of free electrons polarizable in the plane of the surface. Details of this study will be published elsewhere [78].

Conclusion

Raman and SHG spectroscopies represent valuable *in-situ* spectroscopic tools for monitoring the composition of the interfacial region and changes of the surface structure. The use of Raman spectroscopy for the study of electrochemical processes such as adsorption and reaction pathways has been discussed. Recent developments having a great potential include (i) the use of mild roughening procedures which do not alter the electrochemical behaviour of the electrodes but still give sufficient surface enhanced Raman activity for the study of the double-layer region; and (ii) the employment of surface unenhanced Raman spectroscopy for instance at platinum and rhodium electrodes.

With second harmonic generation at metal electrodes a sensitive tool is at hand to study not only the formation and lifting of surface reconstruction, but also the distinct change of the in-plane and perpendicular non-linear polarizability of the metal surface as a function of the potential and contact ion adsorption. This opens a new avenue for the study of ad- and chemisorption processes.

Acknowledgement

The authors would like to thank Prof. G. Ertl for continuing support and stimulating interest in these studies.
This work was partly supported by the Deutsche Forschungsgemeinschaft (research project SFB 6).

References

1) Jeanmaire, D.; van Duyne, R.; *J. Electroanal. Chem.* **84** (1977) 1.
2) Albrecht, M.; Creighton, J.; *J. Am. Chem. Soc.* **9** (1977) 5215.
3) *Surface enhanced Raman scattering* , Chang, R.; Furtak, T.; Eds.; Plenum Press: New York and London, (1982).
4) Otto, A.; in: *Light scattering in Solids*, Cardona, M.; Günterodt, G.; Eds.; Springer-Verlag: Berlin, (1984) Vol. 4 , pp. 289.
5) Metiu, H.; *Surface Enhanced Spectroscopy*, *Progr. Surf. Sci.* **17** (1984) 153.
6) Pettinger, B.; Wetzel, H.; *Surf. Sci.* **78** (1981) 398.
7) Stolberg, L.; Lipkowski, J.; Irish, D.; *J. Electroanal. Chem.* in press(1990).
8) Fleischmann, M.; Graves, P.; Robinson, J.; *J. Electroanal. Chem.* **182** (1987) 73.
9) Fleischmann, M.; Tian, Z.; *J. Electroanal. Chem.* **217** (1987) 385.
10) Weaver, M.; Corrigan, D.; Gao, P.; Gosztola, D.; Leung, L.; *J. Electron Spectrosc. and Rel. Phenom.* **45** (1987) 291.
11) Gao, P.; Gosztola, D.; Leung, L.; Weaver, M.; *J. Electroanal. Chem.* **233** (1987) 211.
12) Leung, L.; Weaver, M.; *J. Electroanal. Chem.* **217** (1987) 367.
13) Kellog, D.; Pemperton, J.; *J. Phys. Chem.* **91** (1987) 1120.
14) Pemperton, J.; Coria-Garcia, J.; Hoff, R.; *Langmuir* **3** (1987) 150.
15) Mengoli, G.; Musiani, M.; Fleischmann, M.; Mao, B.; Tian, Z.; *Electrochim. Acta* **32** (1987) 1239.
16) Shannon, C.; Campion, A.; *J. Phys. Chem.* **92** (1988) 1385.
17) Pettinger, B.; Friedrich, A.; Tiedemann, U.; *J. Electroanal. Chem.* **280** (1990) 49.
18) Pettinger, B.; in: *Adsorption of molecules at Metal electrodes*, eds. Lipkowsky, J.; Ross, P.; in press (1991)
19) Bertolino, J.R.; Temperini, M.L.A.; Sala, O.; *J. Molec. Struct.* **178** (1988) 113.
20) Feilchenfeld, H .; Weaver, M.J.; *J. Phys. Chem.* **93** (1989) 4276.
21) Gao, P.; Weaver, M .J.; *J. Phys. Chem.* **93** (1989) 6205.
22) Takahashi, M .; Niwa, M .; Ito, M .; *J. Phys. Chem.* **91** (1987) 11.
23) Guzonas, D.A.; Irish, D.E.; Aktinson, G.F.; *Langmuir* **5** (1988) 787.
24) Chen, T.T.; Kuo, C.S.; Chou, Y.C.; Liang, N.T.; *Langmuir* **5** (1989) 887.
25) Tian, Z.Q.; Lian, Y.Z.; Fleischmann, M.; *Electrochim. Acta* **35**(1990).
26) Sobocinski, R.L.; Pemperton, J.E.; *Langmuir* **4** (1988) 836.
27) Temperini, M.L.A.; Sala, D.; Locconi, G.I.; Gioda, A.S.; Macagno, V.A.; Arvia, A.J.; *Langmuir* **4** (1988) 1032.
28) Anderson, M .R.; Evans, D.H .; *J. Am. Chem. Soc.* **110** (1988) 6612.
29) Holze, R.; *J. Electroanal. Chem.* **250** (1988) 143.

31) Gao, P.; Gosztola, D.; Weaver, M.J.; *J. Phys. Chem.* **93** (1989) 3753.
31) Stolberg, L.; Lipkowski, J.; Irish, D.E.; *J. Electroanal. Chem.* **282** (1990).
32) Sasaki, Y.; Iwasaki, N.; Nishina, Y.; *Surf. Sci.* **198** (1988) 541.
33) Sun, S.; Birk, R.L.; Lombardi, J.R.; *J. Phys. Chem.* **94** (1990) 2005.
34) Beer, K.D.; Tanner, W.; Garrel, R.L.; *J. Electroanal. Chem.* **258** (1989) 313-315.
35) Gosztola, D.; Weaver, M.J.; *J. Electroanal. Chem.* **271** (1989) 1412.
36) Desilvestro, J.; Weaver, M.J.; *J. Electroanal. Chem.* **209** (1986) 377.
37) Iwasaki, N.; Sasaki, Y.; Nishina, Y.; *Surf. Sci.* **198** (1988) 524.
38) Bertolino, J.R.; Della Vedona, C.O.; Sala, O.; *Polyhedron* **8** (1989) 361.
39) Desilvestro, J.; Weaver, M.J.; *J. Electroanal. Chem.* **234** (1987) 237.
40) Holt, E.; Cotton, T.; *J. Am. Chem. Soc.* **111** (1989) 2815-2821.
41) Lu, T.; Cotton, T.M.; Hurst, J.K.; Thompson, D.H.P.; *J. Phys. Chem.* **92** (1988) 6978-6985.
42) Ni, F.; Feng, H.; Gorton, L.; Cotton, T.M.; *Langmuir* **6** (1990) 66.
43) Feng, Q.; Yue, W.; Cotton, T.M.; *J. Phys. Chem.* **94** (1990) 2082.
44) Lu, T.; Cotton, T.M.; Birk, R.L., Lombardi, J.R.; *Langmuir* **5** (1989) 406.
45) Watanabe, T.; Maeda, H.; *J. Phys. Chem.* **93** (1989) 3528-3260.
46) Takahashi, M.; Goto M.; Ito, M.; *J. Electroanal. Chem.* **51** (1989) 177.
47) Weaver, M.J.; Corrigan, D.S.; Gao, P.; Goszola, D.; Leung, L.W.H.; *J. Electron Spectrosc. and Rel. Phenom.* **45** (1987) 291.
48) Gao, P.; Gosztola, D.; Weaver, M.J.; *Anal. Chimica Acta* **212** (1988) 201.
49) Leung, L.W.H.; Weaver, M.J.; *Langmuir* **4** (1988) 1076.
50) Gao, P.; Gosztola, D. Weaver, M.J., *J. Phys. Chem.* **92** (1988) 7122.
51) Bertolino, J.R.; Temperini, M.L.A.; Sala, O.; *J. Raman Spectrosc.* **20** (1989) 725.
52) Desilvestro, J.; Corrigan, D.A.; Weaver, M.J.; *J. Electroch. Soc.* **135** (1988) 885.
53) Gosztola, D.; Weaver, M.J.; *Langmuir* **5** (1989) 776.
54) Fleischmann, M.; Hendra, P.J.; Hill, I.R.; Pemble, M.E.; *J. Electroanal. Chem.* **117** (1981) 243.
55) Pettinger, B.; Philpott, M.R.; Gordon II, J.G.; *J. Chem. Phys.* **74** (1981) 934.
56) Pettinger, B.; Philpott, M.R.; Gordon II, J.G.; *Surf. Sci.* **105** (1981) 469.
57) Pettinger, B.; Wetzel, H.; *Ber. Bunsenges. Phys. Chem.* **85** (1981) 473.
58) Macomber, S.H.; Furtak, T.E.; Devine, T.M.; *Surf. Sci.* **122** (1982) 556.
59) Chen, T.T.; Owen, J.F., Chang, R.K.; Laube, B.L.; *Chem. Phys. Lett.* **89** (1982) 356.

60) Owen, J.F.; Chen, T.T.; Chang, R.K.; Laube, B.L.; *Surf. Sci.* **125** (1983) 679.
61) Owen, J.F.; Chen, T.T.; Chang, R.K.; Laube, B.L.; *J. Electroanal. Chem.* **150** (1983) 389.
62) Fleischmann, M.; Hill, I.R.; *J. Electroanal. Chem.* **146** (1983) 367.
63) Pettinger, B.; Mörl, L.; *J. Electroanal. Chem.* **150** (1983) 415.
64) Kung, H .P.; Chen, T.T.; *Chem. Phys. Lett.* **130** (1986) 311.
65) Chen, T.T.; Owen, J.F.; Chang, R.K.; in *Dynamics on Surfaces*, Pullman, B.; et al., Eds., Reidel Publ. Comp. (1984) 365.
66) Dorain, P.B.; *J. Phys. Chem.* **90** (1986) 5808.
67) Dorain, P.B.; *J. Phys. Chem.* **90** (1986) 5812.
68) Dorain, P.B.; *J. Phys. Chem.* **92** (1988) 2546.
69) Fleischmann, M.; Hendra, P.J.; Hill, I.R.; Pemble, M.E.; *J. Electroanal. Chem.* **117** (1987) 243.
70) Tian, Z.Q.; Lian, Y.Z.; Lin, T.Q.; *J. Electroanal. Chem.* **265** (1989) 277.
71) Campion, A.; Mullins, D.; *Chem. Phys. Lett.* **54** (1983) 576.
72) Campion, A.; "*Raman spectroscopy* " in: "Vibrational spectroscopy of molecules on surfaces", eds. Yates, J.; Madey, T.; Plenum Publishing Corporation, (1987) Ch. , pp. 345.
73) Friedrich, A.; Pettinger, B.; Kolb, D.; Lüpke, G.; Steinhoff, R.; Marowsky, G.; *Chem. Phys. Lett.* **163** (1989) 123.
74) Lüpke, G.; Steinhoff, R.; Marowsky, G.; Friedrich, A.; Pettinger, B.; Kolb, D.M.; *Phys. Rev. B* **41** (1990) 6913.
75) Kolb, D.M.; Schneider, J.; *Electrochim. Acta* **31** (1986) 929.
76) Friedrich, A.; Shannon, C.; Pettinger, B.; *Surf. Sci.* in press (1991).
77) B. Pettinger, A. Friedrich, C. Shannon, subm. *J. Electroanal Chem.* (1990).
78) Pettinger, B.; Friedrich, A.; to be published.
79) Perdereau, J.; Biberian, J.; Rhead, G.; *J Phys.* **F4** (1974) 798.
80) van Hove, M.; Koestner, R.; Stair, P.; Biberian, J.; Kesmodel, L.; Bartos, I.; Somorjai, G.; *Surf. Sci.* **103** (1981) 189.
81) Melle, H.; Menzel, E.; *Z. Naturforsch.* **33a** (1978) 282.
82) Heyraud, J.; Metois, J.; *Surf. Sci.* **100** (1980) 519.
83) Tanishiro, Y.; Kanamori, H.; Takayanagi, K.; Yagi, K.; Honjo, G.; *Surf. Sci.* **111** (1981) 395.
84) Takayanagi, K.; Yagi, K.; *Trans. Jpn. Inst. Met.* **24** (1983) 337.
85) Harten, H.; Lahee, A.; Toennies, J.; Wöll, Ch.; *Phys. Rev. Lett.* **54** (1985) 2619.
86) Wöll, Ch.; Chiang, S.; Wilson, R.; Lippel, P.; *Phys. Rev. B* **39** (1989) 7988.
87) Barth, J.; Brune, H.; Ertl, G.; Behm, R.; *Phys. Rev. B* , in press.
88) Sipe, J.; Moss, D.; van Driel, H.; *Phys. Rev. B* **35** (1987) 1129.
89) Shen, Y.; *The principles of Nonlinear Optics*, Wiley, New York (1984).
90) Richmond, G.; Robinson, J.; Shannon, V.; *Progr. Surf. Sci.* **28** (1988) 1.

Second–Harmonic Generation from a Metal–Electrolyte Interface

A. Liebsch
KFA–IFF, 5170 Jülich, Germany

Abstract

The potential and angular dependence of second–harmonic generation from a metal–electrolyte interface is calculated within the time–dependent density functional approach. The perpendicular surface contribution to the harmonic polarization is found to be significant and to vary in a characteristic manner with the surface charge induced by the electrolyte. Both the overall magnitude of the surface polarization and its variation with applied potential are in qualitative agreement with recent measurements for Ag in contact with a $KClO_4$ solution.

1. Introduction

There has recently been considerable interest in applying Second–Harmonic Generation (SHG) to metal–electrolyte interfaces.[1] The motivation of these studies lies in the fact that SHG is dipole forbidden in bulk systems with inversion symmetry,[2] so that the leading polarization sources are determined by the two–dimensional electronic properties at the surface. Thus, in striking contrast to the linear light reflection coefficient which is overwhelmingly dominated by the bulk dielectric properties, the second–harmonic intensity may be used to probe the electronic states at the metal–electrolyte interface and, in particular, to study how these states are modified by varying the electric potential.[3]

Among the various terms that contribute to the total second–harmonic polarization,[4] the one that most strongly depends on the interface properties is the perpendicular component of the surface polarization. It is this term which also depends most sensitively on the net surface charge induced by the electrolyte. For negatively charged electrodes, the profile of the electronic density is slightly more diffuse than for the neutral surface.[5] This leads to an increase of the dipole moment and, accordingly, to a larger second–harmonic surface polarization. Positively charged electrodes, on the other hand, show a slighty stiffer density profile which is less polarizable than in the neutral case. Thus, positive charging implies a reduction of the surface polarization. In addition, the second–harmonic polarization exhibits a characteristic frequency dependence[6] which also undergoes changes as the metal surface is charged. For example, since the effective work function depends on the net surface charge, the available states that take part in the optical transitions induced by the laser light, are modified by the electric potential. We will see below that this effect also leads to significant changes of the second-harmonic surface polarization.[3]

There is now a wealth of theoretical as well as experimental information on the static and dynamic response properties of clean simple metals and of chemisorbed alkali-metal overlayers. The local fields and the associated induced surface charge density distributions have been calculated for a variety of systems[7,8] with the aim of analyzing physical quantities that are relevant for several surface spectroscopies. In particular, one now has a very complete and consistent understanding of the local field enhancement[9] that occurs at all simple metal surfaces and that can be observed in surface photoemission,[10] electron loss spectroscopy[11] (surface plasmon dispersion and multipole surface plasmon dispersion), and in the nonlocal corrections to the Fresnel formula for light reflection.[12]

In addition to these linear response properties, the nonlinear electronic surface response has been calculated for clean simple metals[6,13] and alkali overlayers[14] and was found to agree remarkably well with experimental data using SHG.[15,16] Until now, most calculations have made use of the so-called jellium model in which the positive ionic charges of the metal are represented by a semi-infinite uniform background. In this approximation, which is thought to be reasonable for simple metal surfaces, only the details of the charge density distribution normal to the surface and its polarization properties are taken into consideration. More sophisticated calculations that include also the band structure of realistic alkali metals, are now beginning to appear.[17,18] Such extensions will be particularly important for systems where azimuthal anisotropies in SHG have been observed that provide information on the two-dimensional crystal structure at the surface.[1]

The theoretical approach in all these works relies greatly on the density functional method[19] and its time-dependent extension.[20] This scheme treats the ground state electronic properties of the semi-infinite metal rather accurately and allows for an adequate description of the self-consistent screening response of the surface electrons to the time varying external electromagnetic fields. In the present work we extend these calculations and simulate the influence of the electrolyte on the metal surface by applying a static electric field oriented perpendicular to the surface.[5] Such a field induces a net surface charge and at the same time deforms the ground state density distribution, thereby making it considerably more or less polarizable. As pointed out above, these electrolyte-induced changes of the surface polarization properties can be readily detected by using second-harmonic generation.

2. Theory

In the p-in p-out configuration the SHG field from the metal-electrolyte interface can be written as,[21,22] $(\Omega \equiv 2\omega)$

$$E_p(\Omega) = \frac{16\pi i\Omega}{c} \frac{\varepsilon_1(\Omega)\varepsilon_1(\omega)\varepsilon_2(\omega)k_{1x}k_{1z}}{(\varepsilon_2(\Omega)K_{1z} + \varepsilon_1(\Omega)K_{2z})(\varepsilon_2(\omega)k_{1z} + \varepsilon_1(\omega)k_{2z})^2} \times \\ \frac{e(\varepsilon_2(\omega) - 1)}{16\pi m\omega^2} \Big(\frac{\varepsilon_2(\Omega)k_{1x}^2}{\varepsilon_2(\omega)} a(\omega) - 2\frac{k_{2z}K_{2z}}{\varepsilon_2(\omega)} b + \frac{d}{2} \Big) E_p^2(\omega) \tag{1}$$

where the index 1 refers to the liquid and index 2 to the metal. The axes are: z perpendicular to the surface pointing to the metal and x in the plane along the incident beam. The quantities k and K are the normalized wavevectors such that $k_{ix} = K_{ix} = \sqrt{\varepsilon_1(\omega)}sin\theta$, $k_{ix}^2 + k_{iz}^2 = \varepsilon_i(\omega)$, and $K_{ix}^2 + K_{iz}^2 = \varepsilon_i(\Omega)$. In the current theory of the nonlinear metal response, one has $b = -1$, $d = 1$, and $a(\omega)$ depends on the details of the electronic distribution at the metal surface.[6] The term d accounts for the bulk magnetic dipole contribution and the term b is due to the electronic quadrupole response to the field gradient at the interface. Because of the large dielectric constants of silver, the terms involving the coefficients b and d have a very similar angular dependence and can therefore not be separated by measurements of the variation of the SH intensity with incidence angle. However, the term involving the coefficient $a(\omega)$, which is the one of interest, gives a very different angular dependence so that it can be extracted from the SHG data.

As a first step towards an interpretation of SHG from metal–electrolyte interfaces, we extend our previous calculations of SHG from clean metals[6] by taking the influence of the electrolyte into account via a uniform static electric field oriented perpendicular to the surface. The same procedure was employed in Ref. 5 to evaluate the purely static electronic response at a metal surface. The calculations proceed in two steps: First, the ground–state properties of the metal surface in the presence of an applied static electric potential are determined by direct solution of the Schrödinger equation and iteration to self–consistency. Second, from these properties the first– and second–order nonlocal susceptibility functions[6] are constructed and the linear and nonlinear response equations are solved by the same procedure as described in Ref. 7. As before, we use the jellium model to represent the positive ionic charges. This model neglects interband transitions but provides a realistic description of the density distribution in the surface region. The use of the time–dependent density functional approach allows an accurate evaluation of the self–consistently screened surface charge densities that are induced by the laser radiation. We denote by $n_1(z,\omega)$ that part which oscillates at the fundamental frequency and by $n_2(z,\omega)$ the density that oscillates at the harmonic frequency.

From these induced surface charge densities, we obtain the linear and nonlinear dipole moments

$$d(\omega) = \int dz\ z\ n_1(z,\omega)/\sigma(\omega) \tag{2}$$

and

$$a(\omega) = -4\bar{n} \int dz\ z\ n_2(z,\omega)/\sigma(\omega)^2 \tag{3}$$

where $\bar{n}$ is the bulk density and $\sigma(\omega)$ is the integrated surface charge:

$$\sigma(\omega) = \int dz\ n_1(z,\omega) = \frac{\epsilon(\omega)-1}{\epsilon(\omega)+1}. \tag{4}$$

Here, $\epsilon(\omega)$ is the bulk dielectric function. The function $d(\omega)$ is of central importance for the description of several surface spectroscopies that involve the linear response of the surface electrons to the applied fields.[7–12] For example, the surface photoabsorption cross–section is directly proportional to $Im\ d(\omega)$,[9,10] and the linear coefficient of the surface plasmon dispersion relation at small momentum transfers is determined by $d(\omega)$.[9,11] The analogous quantity $a(\omega)$ that characterizes the nonlinear surface response is the one of interest here since it is directly proportional to the perpendicular second–harmonic surface polarization.[4,6]

Fig. 1 shows the effect of the applied electric field on the ground state electronic potential and charge density near the surface. These results are for a

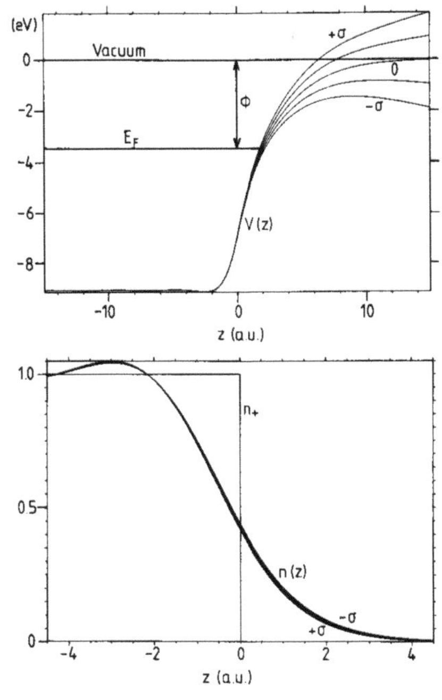

Fig. 1. Ground state potential V(z) and electronic density n(z) for neutral jellium surface ($\sigma = 0$) and for charged surfaces: $\sigma = \pm\ 0.0002$ a.u. and $\sigma = \pm$ 0.0004 a.u. ($1.8\ 10^{-4} a.u. \equiv 1\ \mu C/cm^2$).

semi-infinite jellium system whose bulk density corresponds to the free-electron density of Ag ($r_s = 3$). Although the change of the electron distribution might appear small, the dynamical response calculations show that the harmonic surface polarization depends very sensitively on the applied electric field. This is illustrated in Fig. 2 where the frequency dependence of $a(\omega)$ is plotted for various net surface charges.

In the case of the neutral surface, $Im\ a(\omega)$ exhibits two spectral features:[6] one related to the threshold for electron emission ($2\omega \approx$ workfunction Φ) and a second related to the local-field enhancement[7-11] close to $2\omega \approx 0.8\omega_p$ where ω_p is the plasma frequency. In the presence of a static field both features shift to different frequencies. In particular, because of the lower workfunction of a negatively charged surface, the threshold peak shifts downwards, causing a strong increase in both real and imaginary parts of $a(\omega)$ at 1064 nm. For positively charged surfaces, the electron distribution is pushed into the metal and becomes much stiffer. This leads to an overall decrease of the nonlinear polarizability at low frequencies.

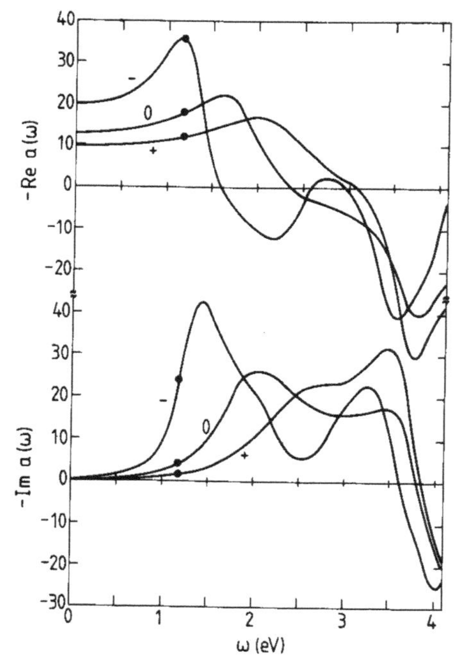

Fig. 2. Calculated frequency dependence of $a(\omega)$ for neutral surface (symbol 0), and for charged surfaces (symbols + and −) corresponding to charges ± 0.0004 a.u., respectively. The dots indicate the values at 1064 nm = 1.13 eV.

We note here that the present theoretical model allows for negative charging of the surface only up to a certain finite field value. Since the surface potential decreases linearly outside the metal (see Fig. 1), field emission will eventually set in. In practice this limit is not reached because of the interaction of ions with the metal surface. In the opposite limit of large positive surface charges, the stiffening of the density profile saturates and the electron distribution simply gets pushed rigidly inwards. A rigid shift of the density profile, however, does not influence the magnitude of the nonlinear dipole moment. Indeed, using the sum rules derived in Ref. 6, one can show that

$$\int dz\ z\ n_2(z,\omega=0) = 0.5/\bar{n} \tag{5}$$

for large positive fields. From the definition of $a(\omega)$ we therefore find that $a(0)$ reaches the limiting value -2 regardless of the bulk density of the jellium system.

3. Comparison with Experiment

In the experiment by Guyot–Sionnest and Tadjeddine,[3] Ag(111) electrodes were used that were obtained by evaporation of silver on mica. The electrolyte was an aqueous solution of 0.1M $KClO_4$ made from triply distilled water and Suprapure (Merck) salt. This system presents the advantage of having a wide region of potential accessible where the system acts as a simple capacitor, between the dissolution of silver at 0.25 V/SCE and the evolution of hydrogen at -1.35 V/SCE. ClO_4^- is an anion that is weakly adsorbing on silver. It permits the use of charge measurements that are normally taken at much lower concentration by the differential capacitance technique. The potentials are referenced to a calomel electrode (SCE).

The SHG measurements were performed using a Nd–YAG laser generating pulses of 300 ps at 1064 nm. The pulse energy at the sample was about 1 mJ over 0.15 cm^2 and the repetition rate was 30 Hz. The sample could be easily rotated to allow for the different angles of incidence between 15° and 85°. At each angle of incidence, an absolute calibration was made with a z–cut Quartz crystal set at the first Maker fringe in the incident beam. After a slight realignment to compensate for the water dispersion, the SHG signal from the sample was recorded between -1.35 V/SCE and 0.1 V/SCE in the positive direction at a scan rate of 2 mV/s and the data were accumulated every 100 mV interval. The measurements were taken on several samples evaporated at different times and with thicknesses of 1000 Å and 2000 Å. In all cases, the data were reproducible to within 10% for the SH field amplitude.

Fig. 3 shows angular scans of the square root of the ratio $R = I(\Omega)/I(\omega)^2$ measured at four different potentials. The intensities are taken outside the cell and $I(\omega)$ is corrected for the water transmission at 1064 nm (60% in our configuration). Also plotted are fits of these data to the expression for R resulting from Eq.

(1). For these fits the following values were used:[22] $\varepsilon_1(\omega) = 1.77$, $\varepsilon_1(\Omega) = 1.8$, $\varepsilon_2(\omega) = -67 + i2.44$, $\varepsilon_2(\Omega) = -11.9 + i0.33$. From the absolute calibration to the quartz reference, it was found that, at the peak, $R = 13 \pm 1.5 \times 10^{-20}\,\mathrm{cm}^2/\mathrm{W}$. From Eq. (1) and the fit at 0 V/SCE with $a = 0.7$, one should have $R = 11.2 \times 10^{-20}$ cm^2/W. The agreement is surprisingly good. This close agreement had already been observed at the silver–glass interface[22] at the same wavelength.

The fits could be improved in the small angle region by taking a with a small imaginary part while the fit at the maximum is essentially determined by the real part of a. As shown in the inset of Fig. 3, the fitted a coefficient varies over a very large range as a function of the applied potential, from +0.7 to about −10. The variation is linear with a slight steepening towards the most negative potentials. These experimental results illustrate the remarkable sensitivity of the normal surface polarization to the state of the surface.

It should be noted that, as a result of the interference of the various harmonic source polarizations, widely different values of a can nevertheless yield similar intensities (e.g. for $a = -10$ and $a = +0.7$ at 75°). It may therefore be misleading to measure the intensity at only one incident angle and to take its variation with applied potential as representative of the variation of the surface polarization. For example, at angles $\leq 45^o$ the intensity is seen to decrease monotonically towards more negative potentials even though the magnitude of a increases.

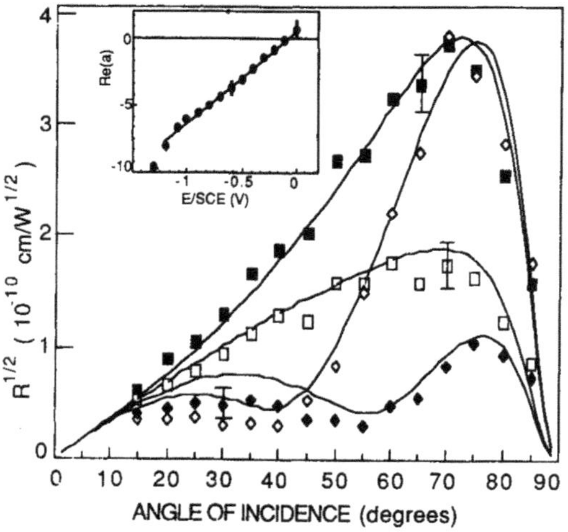

Fig. 3. Measured $R^{1/2} = I(2\omega)^{1/2}/I(\omega)$ as a function of the angle of incidence for Ag(111) at 1064 nm for various potentials. The lines are fits to the data using Eq. (1) as described in the text. Filled squares: 0 V/SCE; open squares: −0.4 V/SCE; filled diamonds: −1 V/SCE, open diamonds: −1.3 V/SCE. The inset shows the variation of the fitted $Re\ a(\omega)$ with applied potential.

Fig. 4 shows the comparison of the calculated and measured values of *Re* $a(\omega)$ at 1064 nm as a function of the surface charge. As can be seen from Fig. 2, the imaginary part of $a(\omega)$ is rather small at this frequency, except for negative surface charges. The experimental points are obtained from the results given in the inset of Fig. 3 by using measurements of the surface charge versus applied potential which have been performed on the same system. The theoretical curves (shown at 1064 nm and in the static limit) are obtained from the results shown in Fig. 2 by accounting also for the static screening of the surface charges due to the water molecules. In our model this screening leads to a reduction of the effective applied potential by the dielectric constant of water at the interface ($\epsilon \approx 6$). Also, whereas the jellium model gives a quantitative agreement with SHG measurements from Al,[15] in the case of Ag it has been shown to overestimate the surface polarization: For not too large fields, $a(0)$ is about 0.3 times smaller than the jellium value as result of band structure effects.[23]

This reduction has mainly two sources: First, the jellium model tends to underestimate the work function Φ. Whereas for Ag one finds about $\Phi = 4.5$ eV, the $r_s = 3$ jellium system has only $\Phi = 3.5$ eV. A larger work function gives a stiffer and less polarizable ground state density profile. We have estimated the importance of this effect by simply scaling the jellium surface potential upwards by a factor of about 11% in order to reproduce the measured work function. This

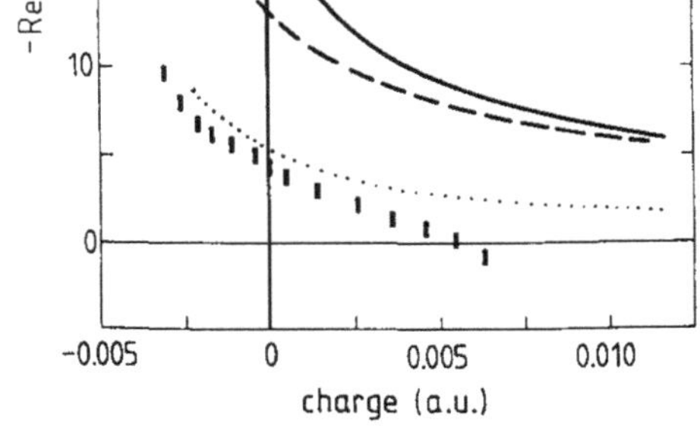

Fig. 4. Variation of $-Re\ a(\omega)$ as a function of surface charge. Filled rectangles: experimental values; solid (dashed) curve: density functional calculation at 1064 nm (in the static limit); dotted curve: solid curve reduced by factor 0.3 to account for band structure effects (see text).

leads to a reduction of $a(=0)$ by about 30%. The remaining decrease of $a(0)$ by 40% found in Ref. 23 is therefore most likely caused by the second effect, namely, the hybridization between s and d states in Ag. Apparently, the three-dimensional crystal structure, i.e., the lateral corrugation of the charge density along the surface, makes the s electrons in the surface region considerably less polarizable than in the equivalent semi-infinite free-electron gas.

Since the lowest interband transition in Ag occurs at about 4 eV, we assume the same reduction of about 70% found in Ref. 23 in the static limit to take place also at the laser frequency of 1064 nm = 1.17 eV as indicated by the dotted curve in Fig. 4. The important features of the experimental results are then seen to be reasonably well reproduced by the theory: The overall variation of a which becomes more negative with negative charging, is similar in both cases. The physical reason for this variation is that the electron density becomes more polarizable as it is pulled out of the metal. Since a is proportional to the dipole moment of the harmonic surface charge, the increase in $|a|$ reflects this enhanced polarizability. Moreover, both theory and experiment show that a is large and negative over most of the potential range. This is in striking contrast to the value of $a = +0.9$ which had been found for a silver-glass interface.[22] It is interesting to note that this value can be obtained here only for the most positive potential where the electronic distribution is displaced towards the metal. This might suggest a similar electronic distribution at the metal-glass interface.

For negative and small positive surface charges, the theoretical results for $a(\omega)$ nearly coincide with the experimental data if the reduction of the polarizability due to band structure effects is taken into account. On the other hand, as argued above, for large positive fields, the jellium calculations predict saturation of $a(\omega = 0)$ at the value -2 independently of the bulk density. The data show instead a linear variation and a change of sign to positive $Re\ a(\omega)$ when the surface charge is about 0.005 a.u. This behavior could be caused by the interaction of negative ions, in particular OH^-, with the metal surface.

From the above discussion we can conclude that SHG can provide rather detailed and useful information on the nature of the electronic states at the metal-electrolyte interface. The electric potential generated by the ions near the electrode has a significant and characteristic influence on the surface electrons which is qualitatively well described by the theory. These encouraging results suggest that one may now attempt more sophisticated theoretical studies which include the band structure of the metal and the electronic structure of the ions. Such investigations, in conjunction with the experimental tools now available, could lead to a very much more detailed understanding of the interaction between metal and electrolyte.

References

1. G.L. Richmond, J.M. Robinson, and V.L. Shannon, Progr. in Surface Science

28, 1 (1989).
2. Y.R. Shen, *The Principles of Nonlinear Optics*, Wiley, New York (1984); Nature **337**, 519 (1989).
3. P. Guyot-Sionnest, A. Tadjeddine, and A. Liebsch, Phys. Rev. Lett. **64**, 1678 (1990).
4. N. Bloembergen, R.K. Chang, S.S. Jha, and C.H. Lee, Phys. Rev. **174**, 813 (1968); J. Rudnick and E.A. Stern, Phys. Rev. B **4**, 4272 (1971); J.E. Sipe and G.I. Stegeman, in *Surface Polaritons*, edited by V.M. Agranovich and D.L. Mills (North-Holland, New York, 1982), p. 661.
5. P. Gies and R.R. Gerhardts, Phys. Rev. B **33**, 982 (1986); Europhys. Lett. **1**, 513 (1986).
6. A. Liebsch, Phys. Rev. Lett. **61**, 1233 (1988); A. Liebsch and W.L. Schaich, Phys. Rev. B **40**, 5401 (1989).
7. A. Liebsch, Phys. Rev. B**36**, 7378 (1987).
8. K. Kempa and W.L. Schaich, Phys. Rev. B**37**, 6711 (1988).
9. P.J. Feibelman, Progr. Surface Science **12**, 287 (1982).
10. H.J. Levinson, E.W. Plummer, and P.J. Feibelman, Phys. Rev. Lett. **43**, 952 (1979).
11. K.D. Tsuei, E.W. Plummer, A. Liebsch, K. Kempa, and P. Bakshi, Phys. Rev. Lett. **64**, 44 (1990); K.D. Tsuei, E.W. Plummer, A. Liebsch, E. Pehlke, K. Kempa, and P. Bakshi, Surface Science, xxx (1991).
12. A. Liebsch, G. Hincelin, and T. Lopez-Rios, Phys. Rev. B **41**, 10463 (1990).
13. M. Weber and A. Liebsch, Phys. Rev. B**35**, 4711 (1987).
14. A. Liebsch, Phys. Rev. B**40**, 3421 (1989); M. Weber and A. Liebsch, Phys. Rev. B**36**, 6411 (1987).
15. R. Murphy, M. Yeganeh, K.J. Song, and E.W. Plummer, Phys. Rev. Lett. **63**, 318 (1989).
16. H.W.K. Tom, C.M. Mate, X.D. Zhu, J.E. Cromwell, Y.R. Shen, and G.A. Somorjai, Surface Science **172**, 466 (1986); K.J. Song, D. Heskett, H.L. Dai, A. Liebsch, and E.W. Plummer, Phys. Rev. Lett. **61**, 1380 (1988).
17. H. Ishida and A. Liebsch, Phys. Rev. B**42**, 5505 (1990).
18. J.T. Lee and W.L. Schaich, Phys. Rev. B**43**, xxx (1991).
19. N.D. Lang and W. Kohn, Phys. Rev. B1, 4555 (1970); ibid. **3**, 1215 (1971); ibid. **4**, 4234 (1971).
20. A. Zangwill and P. Soven, Phys. Rev. A**21**, 1561 (1980); M.J. Stott and E. Zaremba, ibid. **21**, 12 (1980); G.D. Mahan, ibid. **22**, 1780 (1980).
21. J.E. Sipe, V.C.Y. So, M. Fukui, and G.I. Stegeman, Phys. Rev. B **21**, 4389 (1980).
22. J.C. Quail and H.J. Simon, Phys. Rev. B **31**, 4900 (1985).
23. G.C. Aers and J.E. Inglesfield, Surface Science **217**, 367 (1989).

RESONANT LOCAL MODES AND ENHANCEMENT OF PHOTOPROSESSES ON A ROUGH ELECTRODE SURFACE

A.M.Funtikov

A.N.Frumkin Institute of Electrochemistry, Academy of Sciences of the USSR Moscow 117071, Leninsky prospekt 31, USSR

Abstract

A summary is given to the estimates of the surface averaged and local electromagnetic field enhancement on a rough silver electrode. The role of resonant local modes in the enhancement of local photoemission currents, light absorption and Raman scattering is discussed.

1. Introduction

Optical properties of smooth and rough metallic surfaces differ significantly. This it has especially been proved in Raman scattering experiments [1-3]. The cross section of this process for molecules adsorbed on a specially prepared rough surface (especially for silver, gold, copper and alkali metals) increase by a factor 10^5-10^6 in comparison with a smooth one. This effect was called Surface Enhanced Raman Scattering (SERS). Sufficiently high enhancement was also obtained later on these rough surfaces for linear and nonlinear optical processes such as photoemission [4,5], luminescence [6,7], light absorption [6,8], second harmonic generation[9], etc. The electromagnetic field (EM) enhancement on rough metal surfaces is often considered as a main reason for appearence of these effects [3,10-14]. The nature of EM field enhancement is determined by excitation of Surface Plasmons (SP) on a rough surface with surface inhomogeneties of the order of 100-1000 Å. Nevertheless, the other mechanisms determined by specificity of each optical processe may also contribute to its enhancenent connected with the rougheness at atomic scale . It is difficult, for instance, to separate the contribution of EM field enhancement from SERS experiments due to appearence both large surface inhomogeneties and atomic scale roughnesses [15,16] after a roughening procedure.

On the other hand theoretical considerations give the value of EM field enhancement only for special cases: excitation of the delocalized SP propagated along a slightly rough surface[10] and resonant local modes (LM) localized near surface structural defects with simple geometries [3]. It is difficult to evaluate the EM field on a real metal surface where structural geometry is more complicated and both DSP and resonant LM can be excited.

The electrochemical interface is very suitable for the evaluation of surface EM field strength via electron photoemission into the electrolyte solution [4,5,16-18]. On the other hand, it is possible to generate electrochemically surface inhomogeneties of different geomenry and change their stability in time by changing the adsorption of ions or moleculles.

We will consider an application of the photoemission effect at a metal electrode-electrolyte solution interface for an estimate of the surface averaged and local EM field enhancement (section 2), the role of the resonant LM in light absorption (section 3). In section 4 the connection between SERS and the excitation of LM will be considered.

2. Photoemission and SP excitation

The photoemission at the metal-electrolyte solution interface can be considered to involve two stages; firstly, the photoexcitation of electrons in the metal followed by injection of hot electrons into the electrolyte solution, and secondly, thermalization and hydration of hot electrones followed by subsequent reactions of solvated electrons e^-_{aq} with scavengers. The first stage is described by the so called "5/2 law" for the photoemission current [19,20]:

$$J_o \propto M_{s,p} \, (\hbar\omega-\hbar\omega_o-e\varphi)^{5/2} \, \theta(\hbar\omega-\hbar\omega_o-e\varphi) \, E^2_{surf} \qquad (1)$$

where $M_{s,p}$ is the matrix element of optical transitions in metal for s- or p-polarized light, $\hbar\omega_o$=3,4 eV is the photoelectric threshold at the potential of zero charge, E_{surf} is EM field strength near the surface. In comparison with metal-vacuum interface, the threshold of photoemission depends on the electrode potential φ and does not depend on the nature of metal or its surface structure. If an efficient scavenger is present in the solution, i.e the first order reaction of solvated electrons take place, the photocurrent measured is proportional to the current of emitted electrons.

Assumping the matrix elements $M_{s,p}$ in Eq.1 only slightly depend on the surface geometry and $|M_s| \cong |M_p|$, it is possible to estimate the surface averaged EM field intensity, $\overline{E}^2_{surf}$. An enhancement factor $\overline{G}$ is defined as

$$\bar{G} = \bar{E}^2_{surf} / E^2_{inc}, \tag{2}$$

where E^2_{inc} is the EM field intensity of the incident light. Thus, the enhancement factor on a rough surface is given by

$$\bar{G}_{rough} = J_{rough} \, G_{smouth} / J_{smooth} , \tag{3}$$

where J_{rough} and J_{smooth} are photocurrens measured on rough and smooth surfaces. Since on an ideally smooth surface SP can not be excited by an EM wave incident from the ambient, the enhancement factor on a smooth surface, G_{smooth} can be discribed by Fresnel formulae.

A rough metal surface can be made by oxidation-reduction cycle (ORC) consisting in anodic dissolution or oxidation of some amount of silver followed by its cathodic deposition or reduction.

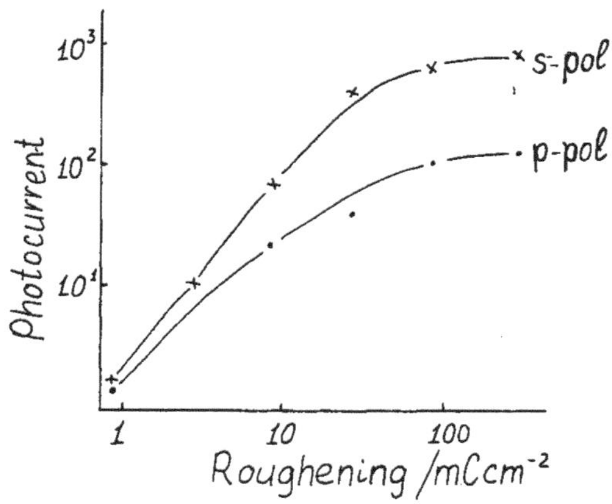

FIGURE 1. Relative increase of the photoemission current on an electropolished silver single crystal (face 100) electrode in 0.1 M Na_2SO_4 + 1 M acetone solution as a function of the charge density passed during ORC. Wavelength λ = 488 nm; angle of incidence $\theta = 65^0$; φ_{SCE} = - 1,5 V. After Ref.18.

Photoemission currents on a silver single crystal electrodes depend strongly on the degree of the surface roughening characterized by the charge Q passed during ORC (Fig.1). The estimate (3) yields nearly the same maximal values $\bar{G} \cong 200$ both for s- and p-polarized light.

Data of Fig.1 have been obtained by using stepped ORC, where silver was dissolved at the anodic potential and then reduced at the potential negative then the potential of zero charge of the polycristalline electrode, $\varphi < \varphi_{pzc} = -0.95$ V. Below we will consider results obtained by this procedure. A prepolarisation of the electrode after ORC at $\varphi > \varphi_{pzc}$ leads to a smaller quantity $\bar{G}$. The ORC typically employed in SERS experiments with the potential sweep from the initial potential to the dissolution potential and back yields also smaller quantities $\bar{G}$ (3 - 5 fold diminished). The pretreatment procedure will also play a crucial role in the local photoemission, SERS and light absorption experiments (see below).

By using the first order reaction of solvated electrones it is possible to determine only the surface averaged quantity $\bar{G}$. The local EM field at some places of the surface may be stronge enhanced due to an excitation of resonant LM localized near sharp structural defects [3,12,15] (bumps, cavities, pores, etc.). As a result the local photocurrent density and, thereby the local concentration of solvated electrons, c_{e^-} may be sufficient for an effective bimolecular reaction

$$e^-_{aq} + e^-_{aq} + 2H_2O = H_2 + 2OH^- \qquad (4)$$

It is essential that the rate of this reaction is proportional to squared concentration of solvated electrons. A simple estimate yields the nessesery $c_{e^-} \sim 10^{-2}$ M/L, correspondingly the photocurrent density $j_0 \sim 1$ kA/cm^2 for the effective rate of the reaction (4).

The first order reaction of e^-_{aq} is practically impossible due to an extremly small rate constant[21]. If no efficient scavanger is present in the solution and the current density of photoemitted electrons is not sufficient to provide a high value of c_{e^-}, solvated electrons are able only to return to the electrode at a very rapid rate (lifetime $\sim$ 1 ns). As a result the photocurrent measured will be nearly zero.

Fig.2 shows the photocurrent measured in absence of specially added scavengers in the solution on a rough silver surface created by the above mentioned stepped roughening procedure. The relatively small photocurrents at low laser powers may be ascribed to the presence of acceptor impurities in the solution. At higher powers the value of the photocurrent measured tend to that of emitted electrons (curve A). It may be related to the high efficiency of the reaction (4) due to abnormally enhanced local concentration c_{e^-} and local photocurren dencities $j_0 \sim 1$ kA/cm^2. A comparison of this value with that obtained at the same experimental conditions on a smooth single crystal surface yields six orders enhancement of the local photocurrent dencity. Solution of the three dimensional diffusion problem with quadratic term describing reaction (4) presents great difficulties even for an isolated structural defect. Therefore it is possible to make only a crude estimation. Assuming, the local photocurrent enhancement to be determined by that of the EM field near a surface, in accordance with Eq.1, we obtaine a local enhancement factor

$$\bar{G}_{loc} = E^2_{loc} / E^2_{inc} \sim 10^6. \tag{5}$$

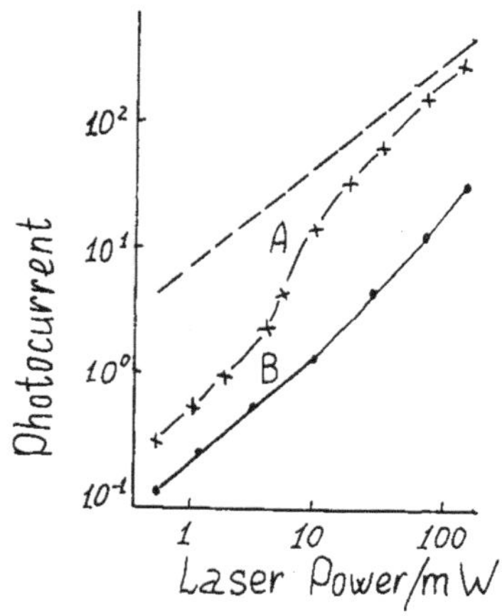

FIGURE 2. Dependence of the photocurrent on incident radiation power at an electropolished silver single crystal (face 100) electrode after ORC with Q = 10 mK/cm^2 in 0.2 M Na_2SO_4 solution at E = - 1.5 V. Normal incidence; λ = 488 nm; light spot diameter of the order of 100 μm. Dotted line is related to the photocurrent of emitted electrons measured in the presence of a scavenger in the solution and with defocused light spot (diameter of the order of 3 mm). Curve A obtained immediately after ORC. Curve B - after the prepolarization at φ = - 0.2 V during 100 s.

Taking into account the value $\bar{G}$ = 200 obtained above, we can conclude that the EM field is nonuniformly distributed along the surface due to an excitation of LM localized near structural defects. It is obvious that the above estimate is averaged over structural defects, where the excitation of LM is possible and much higher EM field enhancement may be expected in some places of the surface. The quantity $\bar{G}$ is very sensitive to the pretreatment used. Sweeped ORC or a prepolarization of a silver electrode at $\varphi > \varphi_{pzc}$ after stepped ORC lead to much smaller photocurrens. This effect manifests itself especially on single cryctal surfaces after ORC with not very large charge Q (Fig.2, curve B).

It would be noted that homogeneous reactions of radicals generated by solvated electrons can be also used for G_{loc} estimations. In this case the heterogeneous reactions of radicals have to be additionally taken into account. Other arguments in support of so high value of quantity G_{loc} may be found in Refs.[16-18].

3. Electroreflectance and resonant LM

Electroreflectance (ER) spectra of a silver single crystal electrole after an ORC with reduction of silver at $\varphi > \varphi_{PZC}$ have no any sharp structures in visible[22,23]. In ultraviolet the derivative - like structure appears near the cut off frequency of delocalized SP, $\hbar\omega_s = 3.5$ eV. This structure arises from the fact that ω_s is potential dependent and hence the minimum in static reflectance due to SP excitation on a rough surface is shifted along ω_s-axis with potential. Additional derivative -like structures appear in ER spectrum in visible, when silver is redeposited at $\varphi < \varphi_{pzc}$ (Fig.3). These structures may be assigned to the excitation of resonant LM . The derivative like structures in ER spectra appear due to the potential dependence of eigen frequences of LM on the surface charge of structural defects. It is essential that these structures appear only on roughened single crystal. There are structural defects with different shape on a rough polycristalline electrode, correspondingly a set of LM-eigen frequencies covering the whole spectral range below the cut off frequency of DSP forms a broad absortion maximum. It would be reasonable to expect that structural defects created on a single cryctal surface have more narrow distribution in shape. As a result, the nonuniform broadering of the absorption band will not be very high.

The derivative like structures in ER spectra are disappeared after a sufficiently long prepolarization of the electrode at $\varphi > \varphi_{pzc}$. This result correlates with the irreversible decreasing of the local EM field enhancement on a rough silver electrode after the same preexposure. In may be supposed that at positive potentials structural defects with a small radius of curvature relax to a more stable geometry due to a decrease in the activation energy of silver atoms self-diffusion . This smoothing effect results in the destruction of resonances of LM. In contrast the derivative - like structure in ER spectra at $\hbar\omega_s$ = 3.5 eV is changed but not disappeared after the prepolarization. This structure is related to the excitation of DSP which is sensitive to the averaged characteristics of the surface. Electroscattering spectra[24] on a rough silver surface related to the excitation of the DSP and its radiative decay are also in,sensitive to the prepolarisation treatment. Thus, the effect of the prepolarization may be related rather to transformations of the surface inhomogeneties with a small radius of curvature then to "smooth" roughnesses.

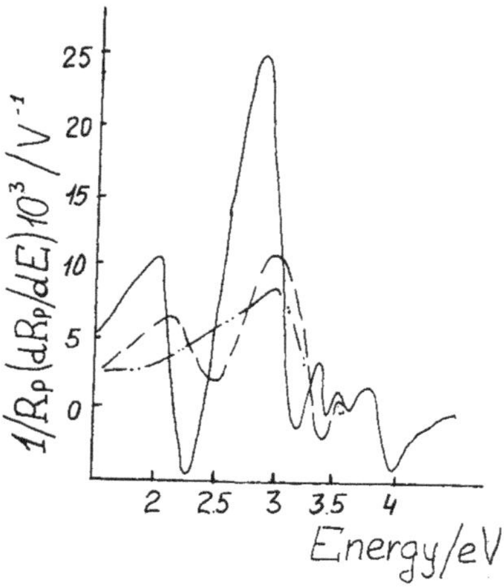

FIGURE 3. ER spectra of Ag (100) - 0,5 M NaF after ORC with Q = 10 mC/cm^2 measured at φ = - 1.3 V. θ = 60^0; p-polarisation. (——) immediately after ORC; (----) after prepolarization at φ = - 0.3 V during 3 s; (-..-..) after prepolarization at φ = - 0.3 V during 100 s. After Ref.[24].

4. SERS and LM excitation

Below, we will consider SERS spectra on rough silver electrodes obtained under the same ORC that was used in photoemission and ER experiments. Fig 4 shows SERS spectra of adsorbed water and ions SO_4^{2-}.The line at 1644 cm^{-1} correspondes to the scissor vibrations of water and it is shifted to smaller frequencies for HDO and D_2O. No lines due to stretching vibrations of adsorbed water were found. The SERS spectrum of SO_4^{2-} have been obtained during a short accumulation time (1 s) by using the Optical Multichannel Analyzer because of the irreversible quenching of SERS at $\varphi > \varphi_{pzc}$. At more negative potentials the adsorprion of SO_4^{2-} tends to zero.

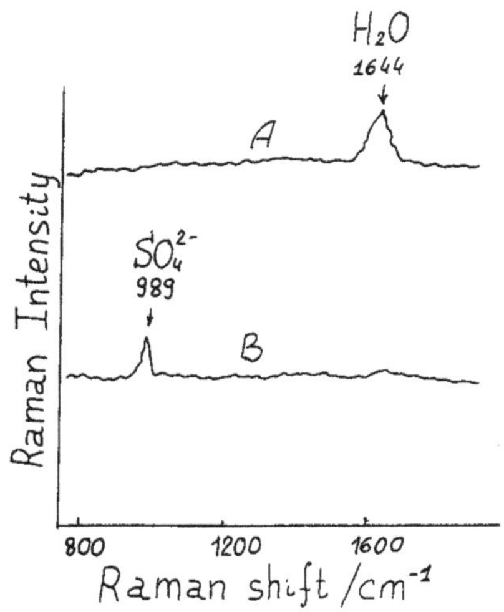

FIGURE 4. SERS spectra of adsorbed water and SO_4^{2-} on an electropolished silver single crystal electrode (face 100) after ORC with Q = 30 mC/cm^2 in 0.1 M Na_2SO_4 + 10^{-3} M H_2SO_4. λ = 514 nm. A - 1.4 V; B - 0.7 V.

It would be noted that SERS spectra of adsorbed water [25,26] may be obtained after ORC with redepositon of about 1000 monolayers of silver at $\varphi > \varphi_{pzc}$ only in a narrow range of φ near the silver dissolution ponentials . It is possible to extend the region of the existence of SERS to negative potentials if the halides coadsorbed from concentrated solutions is present on the surface. A short exposure of the electrode at sufficiently negative potentials leads to the disappearance of SERS spectra. In contrast to that, SERS spectra of Fig.4 disappear after a prepolarization at sufficiently positive potentials. The rate of quenching of SERS correlates with that of photocuurents (Fig 5) and derivative - like structures in ER spectra.

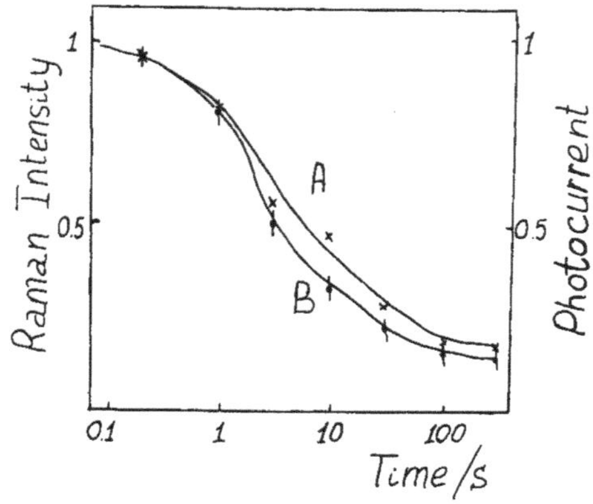

FIGURE 5. Dependences of the SERS (line 1644 cm^{-1} of scissor vibrations of water) and the photocurrent on the prepolarization time at φ = - 0.5 V plotted simultaneously. Silver single crystal (face 100) electrode after ORC with Q = 10 mQ/cm^2. 0.1 M Na_2SO_4 solution; λ = 514 nm; laser power 200 mW; light spot diameter of the order of 100 μm. After Ref.18.

The rate of quenching of these photoprosesses is strongly influenced by the adsorption of halide ions. The prepolarization of a silver electrode under the adsorption of halide with the coverage about 0.1 leads to the 10-fold increase of the rate of quenching. Such behavior can be explained by the influence of the adsorption of halides on the activation energy of Ag atoms self-diffusion. This results in the disappearance of sharp structural defects, where the resonant LM excitation is possible.

Thus, the LM excitation play a crucial role in the SERS under conditions discussed. Nevertheless, the SERS is short-range effect, i.e it is determined by the first layer of adsorbed water. It is clear , e.g. from the potential dependence of the SERS. The intensity of line 1644 cm^{-1} is reversibly diminished from the maximal value at φ = -1.4 V to zero at φ = -1.0 V. In spite of the high value of G_{loc} estimated in section 2 the contribution of the pure EM mechanism into SERS signal is therefore considered to be small in comparison with the Raman scattering from the bulk solution.

The mechanism of SERS should account for both the resonant LM exitation and the interaction of the adsorbed molecules with the surface. The inelastic Mie scattering mechanism considered in Refs. [13,27] may be used. It is assumed that intramolecular vibrations of an adsorbed molecule modulate the polarizability of LM due to charge injection and withdrawal by adsobate on the surface. This create sum and difference bands on the Mie scattering light corresponding to the Raman effect. If we assume that the adsorbed molecules vibrate in phase, 9 orders Raman scattering enhancement may be estimated for an isolated 200-Å-diam silver sphere . For uncoupled adsorbate molecules the Raman scattering enhancement is only 5 orders of magnitude.

SERS of SO_4^{2-} ions (Fig 5) were obtained with coverage smaller than 0.1. At so low coverage the adsorbate molecules is expected to vibrate incoherently. Nevertheless, the high magnitude of the Raman enhancement may be estimated due to sharper resonances excited on a rough silver electrode. The quantity G_{loc} estimated in section 2 for a real silver surface is 4 orders higher than that for isolated silver sphere. The polarizability of surface structural defects is therefore expected to be more sensitive to the modulation of the surface charge.

References

1. M.Fleischman, P.G Hendra and A.J McQuillian, Chem Phys. Lett. 26 (1974) 1631.
2. (a) D.L.Jeanmaire and R.P.Van Duyne, J. Electroanal. Chem. 84 (1977) 1; (b) M.G.Albrecht and J.A.Creighton, J. Am. Chem. Soc. 99 (1977) 5215; (c) V.V.Marinyuk, R.M.Lazorenko-Manevich and Ya.M.Kolotyrkin, Elektrokimiya, 14 (1978) 452.
3. See, e.g. (a) R.K.Chang and T.E.Furtak (Eds.), Surface Enhanced Raman Scattering, Plenum Press, N-Y and London, 1982; (b) M. Moskovits, Rev. Mod Phys. 57 (1985) 783.
4. J.K Sass, R.K.Sen, E.Meyer and H.Herischer, Surf. Sci. 44 (1974) 515.
5. H.Neff, J.K.Sass and H.J.Lewerenz, J. Phys. Chem. 84 (1980) 1135.
6. A.M. Glass, A.Wokaun, J.P.Heritage, J.G.Bergman, P.F.Liao and D.H.Olson, Phys. Rev. B 24 (1981) 4906.
7. G.Ritchie, E.Burstein, Phys. Rev, B 24 (1981) 4843.

8. S.Garoff, D.A.Weitz, T.J.Gramila, and S.D.Hanson, Opt. Lett. 6 (1981) 245.
9. C.Y.Chen, A.R.B. de Castro and Y.R.Shen, Phys. Rev. Lett. 46 (1981) 145.
10. H.Raether, in: Surface Polaritons (V.M.Agranovich and D.L.Mils, eds, 1982), North Holland, Amsterdam, p 331.
11. J.I.Gersten A.Nitzan, J. Chem. Phys. 73 (1980) 3023.
12. M.Moskovits, J.Chem Phys. 69 (1978) 4159.
13. J.A.Creighton, C.G.Blatchford and M.G.Albrecht, J. Chem. Soc., Faraday Trans. II, 75 (1979) 790.
14. H.Chew, M.Kerker and P.I. McNulty, Phys. Rev. A 13 (1976) 396.
15. P.K.Aravind, A.Nitzan and H.Metiu, Surf. Sci. 110 (1981) 189.
16. A.M.Funtikov, S.Sigalaev and V.E.Kazarinov, Elektrokhimiya, 24 (1988) 419.
17. A.M.Funtikov, S.Sigalaev and V.E.Kazarinov, Elektrokhimiya, 25, (1989) 762.
18. A.M.Funtikov, S.Sigalaev and V.E.Kazarinov, J.Electroanal. Chem. 228 (1987) 197.
19. A.M.Brodsky and Yu.Ya.Gurevich, Sov. Phys. JETP, 27 (1968) 114.
20. L.I.Korshunov, Ya.M.Zolotovisky and V.A.Bendersky, Elektrokhimiya, 4 (1968) 499.
21. A.K.Pikaev, Solvatirovanniy Elektron v Radiacionnoy Khimii, Nauka. Moscow, 1973.
22. D.M.Kolb and P.Koetz, Surface Sci. 198 (1977) 96.
23. D.M.Kolb, in: Spectroelectrochemistry, ed. R.J.Gale (Plenum, New-York, 1988), p 87.
24. A.M.Funtikov, V.E.Kazarinov, M.I.Urbakh and T.Tadjeddine, Surface Sci. 239 (1990) 59.
25. M.Fleischman, P.T.Hendra, I.R.Hill and M.E.Remble, J. Electroanal.Chem. 117 (1981) 243.
26. B.Pettinger, M.R.Philpott and J.G. Gordon, J. Chem. Phys. 74 (1981) 934.
27. H.Abe, K.Manzel and W.Schulze, J. Chem. Phys. 74 (1981) 792.

OPTICAL DIAGNOSTICS OF THE STRUCTURE OF ROUGH ELECTRODE SURFACES

M. URBAKH
School of Chemistry, Tel-Aviv University, Ramat Aviv, 69 978 Tel-Aviv, Israel

ABSTRACT

Characterization of surface structure in vacuum has progressed rapidly in recent years. In electrochemistry, electrode surfaces are of fundamental importance, however very little is known about their structure. Recent advances in optical experiment permit the detailed analysis of the microscopic properties of electrochemical interfaces and their changes under strictly controllable state of the surface, in particular, on application of external fields, under adsorption and in the reconstruction of a surface. The aim of the present review is to discuss the latest achievements in the optics of rough metal-electrolyte interface. Optical properties of slightly and highly rough electrode surfaces are considered. It is shown what information about the geometrical structure of adsorbed layers can be extracted from optical data. The possibility of optical diagnostic of the initial stages of electrocrystallization is considered.

1. Introduction

Real interfaces are always characterized by a certain degree of roughness which depends on the actual material, the method of surface treatment and the presence of adsorbed particles on the surface. A major part of electrochemical measurements has been carried out using polycrystalline electrodes. The fact should not be forgotten that polycrystalline samples are used in most applications of practical importance. If the properties of the different crystal faces are considerably different, then such polycrystalline surfaces are highly inhomogeneous. The contacts between faces cause an additional inhomogeneity. Experimental data show[1] that properties of a polycrystalline electrode cannot be deduced, in the general case by simple averaging of characteristics of the individual faces. It may be due to local defects of the surface structure or to specific potential distribution near the contacts between faces that such electrodes exhibit high catalytic activity.

Mostly the samples of practical interest consist of microcrystals with dimensions smaller than or comparable to the wavelength of the light; the samples are thus heterogeneous. Composites, island films and microdeposits on different supports are becoming widely used as electrocatalysists[2-4]. It was demonstrated[2-4] that these materials exhibit different electrocatalytic properties depending on their geometry compositional structure and characteristic sizes of separate elements.

Surprisingly, however, the geometrical structure of in situ electrode surfaces, although of fundamental importance in electrochemistry, are poorly understood. The only measurable parameter in electrochemistry that characterizes the distinction between rough and smooth electrodes has so far been the roughness factor, i.e., the true-to-apparent surface ratio. The results of experiments unambiguously indicate that the roughness factor by no means characterizes completely the surface inhomogeneity. At the same value of the roughness factor, the corrosion rate, electrocatalytic activity and photoelectrochemical effects may differ significantly[1,2-4,7-9]. The use of optical methods gives a much better insight into the microscopic properties of heterogeneous electrodes than traditional electrochemical methods.

Of the very many experimental techniques used to study the morphology of surface, the optical methods are of considerable importance, viz., surface plasmon spectroscopy (SPS),[10] spectroscopy of diffuse light scattering,[11] electroreflectance (ER),[12] ellipsometry,[13] etc. In contrast to the overwhelming majority of other experimental techniques, employed only in the case of solid-vacuum interface, the optical methods can be used to study clean and adsorbate-covered metal surfaces under equilibrium conditions in contact with air, solution and other media. In this connection the optical methods are widely used in the fundamental researches and in applications as well.

Under UHV conditions, temperature is the only thermodynamic variable. In an electrochemical environment, the electric field or the charge density at the surface can be controlled by changing the potential across the double layer. The surface morphology can be effectively changed by potential cycling. It is also essential that the use of the modulation technique based on the potential modulation allows the accuracy of optical measurement to be considerably improved.

In what way do rough surfaces differ optically from ideally smooth ones? First, light reflection from rough surfaces gives rise not only to a specular reflected wave but also to diffusely reflected waves, leaving the surface at arbitrary angles. Second, the presence of roughness makes possible the generation of surface plasmons (SP) by the electromagnetic wave incident on the metal, and also scattering of SP, which results in their conversion into bulk excitations or other states of SP. Finally, one more feature of rough surfaces is important from the point of view of the enhancement effects. Namely, electromagnetic modes can be excited which are practically localized in the vicinity of inhomogeneities with a small radius of curvature.

The above mentioned effects result in considerable changes in the distribution of the electromagnetic field intensity near the surface, as well as in the SP-dispersion law, the light reflection coefficients and ellipsometric parameters[7-9,14-17].

To gain information on the structure of rough surfaces from experimental data, a theory is necessary, in which the observable quantities are related to the geometrical and electronic characteristics of the interfaces. In theoretical description of the optical properties of rough interfaces one usually proceeds from the Maxwell equations in which the surface properties are taken into account in terms of the dielectric function of the system[7,8,16]. It is impossible at present to provide a united theory which described the interaction of light with an arbitrary roughness. The main problem, therefore lies in choosing an approximation taking into account the physical features of a particular problem.

In this paper we will discuss the properties of the so-called microrough surfaces for which the characteristic sizes of the inhomogeneities are no larger than hundreds of Å and thus are much smaller than the wavelength of light. Enhancement effects of optical processes were discovered on such surfaces and they are of the greatest interest from the standpoint of electrochemical applications.

It was shown[8,14,18,19] that on the basis of the results of optical experiments, microrough surfaces may be divided into two groups: (i) Those for which the characteristic sizes of inhomogeneities in the direction perpendicular to the interface (height) are less than those in the direction along the surface. These are surfaces with "smooth" roughness, sometimes called "slightly rough surfaces". These roughnesses are always found on real electrode surfaces and are additionally formed after slight loosening. (ii) Surfaces for which the sizes of the inhomogeneities in the direction perpendicular to the interface and along it are of the same order, i.e., surfaces containing the inhomogeneities with small radii of curvature. These surfaces are sometimes called "highly rough".

It has been theoretically proven[14,19] that on slightly rough metal surfaces only delocalized surface modes propagating along the interface can exist. On highly rough surfaces resonance electromagnetic modes strictly or approximately localized in all directions may also arise.[8,9,15] As a consequence, the optical properties of slightly and highly rough surfaces differ qualitatively.

The division of surfaces into two groups may be based, e.g., on measurements of the light reflection coefficients and ellipsometric parameters.[18] When a surface is loosened, minima appear, at first in the spectra of the light reflection coefficients corresponding to delocalized SP, and thus we have a case of "smooth" roughness. As loosening progresses, additional minima appear, lying in a longer wavelength spectral region and corresponding to localized surface excitations.

2. Slightly Rough Surfaces

In describing the optical properties of slightly rough surfaces, one usually resorts[7,8,16] to the perturbation theory of the interaction of light with the roughnesses, which takes into account only the first non-disappearing term of the perturbation theory series (single light scattering). In the framework of these approximation the changes in the optical quantities (ellipsometric parameters, light reflection coefficients, SP dispersion relation,...) caused by the influence of roughnesses prove to be proportional to the mean square height of roughness, δ^2, and the Fourier component of correlation function, g(k). For instance, the expressions for the angular distribution of the differential intensity of diffuse light scattering can be written as:

$$\frac{1}{I_o}\left[\frac{dI_{\alpha\beta}}{d\Omega}\right] = \left(\frac{\omega}{c}\right)^4\delta^2 g(\Delta k)F_{\alpha\beta}(\varepsilon_m,\theta_i,\theta_f\varphi) \qquad (1)$$

Here $I_{\alpha\beta}$ is the intensity of α-polarized light scattered in the solid angle $d\Omega = \sin\theta_f d\theta_f d\varphi$; I_o is the intensity of the incident β-polarized light,(α,β = s,p); θ_i and θ_f are the angles of incidence and scattering, respectively, which lie in the plane of incidence, φ is the azimuth angle; Δk is the change of the wave vector due to scattering and $F_{\alpha\beta}$ are optical factors depending on the dielectric constants of the electrode and electrolyte and on the angles θ_i, θ_f and φ.

Numerous studies carried out so far at the metal-vacuum interface lead to the conclusion that the light scattering method and SPS allows the root mean square height of roughness to be determined very accurately (within 1-3 Å). The resolution along the surface, however, is much lower (no better than 200 Å). In-situ scanning tunneling microscopy (STM) can be used to obtain information about surface structure at smaller tangential scales.[20]

The presence of inhomogeneities in the electrode bulk also leads to appearance of diffusely scattered light. In a first approximation, the intensity of the scattering from bulk and surface inhomogeneities can be simply summed. These two effects can be separated[8,10] by measuring the angular dependence of scattered light. If at a fixed angle θ_f, the light intensity is determined as a function of the angle ψ, between the plane of polarization and the plane of observation, then at $\phi = 90^0$ this function should be proportional to $\cos^2(\psi - \theta_f)$ for surface roughness and to $\cos^2\psi$ for bulk inhomogeneities. The alteration of the surface structure by potential cycling gives additional possibility to analyze the surface contribution to the integral signals.

At the metal-electrolyte interface, in addition to measurements of the intensities of diffusely scattered light, it is possible to measure[7] the derivatives of these intensities with respect to the potential with the use of modulation techniques. The latter method became to be known as electroscattering (ES).

Geometrical Structure of Adsorbtion Layers

The major interest for electrochemistry is the investigation of the structural properties of adsorption layers. In this case the situation is more complex than for bare metal surfaces. We have, as a rule, a rough surface covered by a non-uniform adsorption layer. To describe the phenomena occurring in these systems one has to examine the geometric structure on two scales: the atomic scale and the "macroscopic" scale (of the order of hundreds of Å). The most effective methods of obtaining information about the behavior of a system on the atomic scale is STM. For investigation of the large-scale geometric structure, one can use ellipsometry, electroreflectance and electroscattering methods.[21,22] Thus, STM and optical methods are complimentary for the study of the structure of heterogeneous electrode surfaces.

From the results of the experiments on the ER and the ES one can conclude whether the adsorption layer formed on rough surface is uniform, i.e., the adsorbate follows the surface relief and uniformly covers the substrate, or non-uniform and the adsorbate mostly covers some areas on the surface and is adsorbed to a much less extent on others. For example, in the case of a polycrystalline substrate, adsorption may be different on different faces. When the characteristic sizes of such areas with different adsorption properties are greater than 100 Å this phenomenon can be detected by optical methods.[21]

The ER and ES methods have been used for the investigations of the adsorption of hydrogen and Cl^- and Br^- anions on polycrystalline platinum electrode.[21] The Pt surface is a classical example of the adsorption-nonuniform electrode surface. On the base of potential dependencies of ER and ES signals it has been shown that weakly-bound hydrogen and anions are distributed randomly along the Pt surface. The authors of[21] also concluded that on smooth polycrystalline Pt surface the structure of H-adlayer changes as the coverage becomes approximately as large as 0.20 - 0.25, i.e., a transition occurs from a uniform distribution of H_{ad} along the surface to a domain structure that is correlated with the inherent inhomogeneity of the substrate. The fast cyclic treatment, which changes the surface morphology, enhances the correlation of the inherent and adsorbate inhomogeneities.

The ER and ES methods can give qualitatively new information in the investigation of the behavior of electrochemical systems in the vicinity of the point of second-kind phase transition on the surface. In this region the fluctuations of the dielectric function along the surface increase drastically and, accordingly, so does the intensity of the diffuse light scattering. It was shown in[23,24] that ES measurements give an opportunity to determine critical indexes and, correspondingly, the nature of the phase transition.

Enhancement Effect

It should be noted that in the frequency range in which SP are not exited the parameter of the expansion into a series of the perturbation theory is the greatest of the quantities δ^2/σ^2 and $\delta^2\omega^2/c^2$. In the presence of sufficiently long-lived SP, some new parameters arise in the theory and the range of the applicability of the single scattering approximation is reduced. This most significantly affects the diffuse light scattering coefficients, the angular distribution of the SP radiation and the field intensity near the surface. In order to calculate these quantities, an integral equation of the type of the Bethe-Saltpeter equation was used[19,25,26], in which the effects of multiple SP scattering were taken into account. For instance, it was demonstrated that the value of the enhancement ratio of the optical effects on slightly rough surfaces is determined by the relation between three SP decay lengths due to the absorption in the metal, $l^{(m)}_{sp}$, the radiation decay, $l^{(r)}_{sp}$, and the transitions between different SP states, $l^{(sp)}_{sp}$. In the frequency range where:

$$(l^{(sp)}_{sp})^{-1} > (l^{(m)}_{sp})^{-1} + (l^{(r)}_{sp})^{-1} \tag{2}$$

the electromagnetic field intensity near rough surface may increase (by factor of 10 in the case of silver electrode at $\hbar\omega$ = 3 eV). In the situation considered the enhancement of the incident wave results from accumulation of its electromagnetic energy in the SP field. The enhancement ratio is determined by the competition of the SP excitation and decay processes. The field enhancement mechanism described leads to experimentally observed effects, for example, to enhanced Raman scattering and absorption of adsorbed molecules, increased second harmonic generation and quantum yield of photoemission at a rough surface.[8,9,15]

2. Highly Rough Surfaces

Use of the concepts of slightly rough surfaces is insufficient for understanding of all available experimental data taken together.[8,9,15,18] It should be stressed that the slightly rough surface model gives an enhancement ratio of electric field too small by at least an order of magnitude to explain the increase of the photoemission current, and SERS observed during the anodization of the silver electrode.[9,15]

When roughness is increased inhomogeneities of a small radius of curvature are formed on the surface, in the vicinity of which localized resonance excitations can exists. In describing theoretically the optical properties of such highly rough surfaces one is faced with the necessity for a simplified model of inhomogeneities. A rough surface may be represented as a set of bumps (local centers) resting on a flat substrate. Usually bumps differ in shape and size, and this makes a precise calculation practically impossible. Further complications arise due to the necessity of taking into account the electromagnetic interactions between individual bumps. Therefore, in order to elucidate the role of inhomogeneities of a small radius of curvature in optical phenomena, it would, probably, be a good approach to try to single out general properties of simplified models which can be carried over to real situations.

Diagnostics of The Initial Stages of Electrocrystallization

The main results on the behavior of the electric field in the vicinity of bumps on a metal surface were obtained[8,9,15] using the numerical calculations for simple geometrical forms of inhomogeneities (spheres, spheroids, etc.). The ideas concerning the optical properties of separate bumps on the surface have been used to develop an optical method of diagnosing the initial stages of electrocrystallization of metals.[35,36] Experimentally, deposition of Ni and Hg onto carbon was studied by ellipsometry[35] and deposition of Cu on Pt by specular reflection of light.[36] In the case of Ni and Hg deposition the observed oscillating dependence of the ellipsometric parameter ψ upon the dimensionless ratio, $q = r\omega/c$, (where r is the radius of nucleation centers) for $q > 4$ is caused by the interference of the waves reflected from the substrate and the nucleation centers. For Ag, Au and Cu deposition the effects connected with the excitation of the local electromagnetic resonances at the nucleation centers are also essential. The former effect is especially strong for small nucleation centers ($r < 0.1$ μm), when the interference extrema are absent. The strong nonmonotonicy of the optical signals as the functions of the parameter q makes it possible to use the ellipsometry and the light reflection for diagnosing the sizes of nucleation centers and the surface geometry at the initial stages of electrocrystallization.

Interaction of Local Resonances

Two different approaches to the description of the electromagnetic interaction between bumps on a surface have been proposed. The first[14,17,22,27] is based on the application of kinetic equations relating to one another the number of photons (in unit volume) in the incident flux, the number of SP and local modes per unit surface area.The second approach[17,28-31] to the description of the optical properties of highly rough surfaces is based on the simulation of inhomogeneities by dipole or multipole centers, randomly distributed in the surface region and studied using the multiple scattering theory. In these cases the problem of statistical averaging over the roughness is usually solved using one of the version of the mean field theory.[7,15,32]

In the framework of these approaches the effect of interaction of local modes with SP and with one another on the spectrum of surface electromagnetic excitations, light reflection coefficients and on behavior of the field near the surface was examined. It was shown[8,9,15] that the local increase of the field intensity in the vicinity of the bump at Ag surface can be as large as 4-5 orders. At the same time, the surface averaged field enhancement ratio may reach the values of the order of 10^2-10^3, but no greater, since with increasing concentration of the centers, a considerable interaction arises between individual local modes which decreases their life time and, accordingly, the enhancement ratio.

The frequencies and intensities of the absorption bands corresponding to local modes depend on the shape and size of inhomogeneities and therefore it is possible from optical data to asses the characteristics of inhomogeneities and to investigate the potential dependence of their formation and relaxation.

Measurements of the specular reflection and diffuse light scattering can be also used for the examination of the bulk structure of thick (thickness of a few hundred Å) nonuniform surface films, for instance oxide or polymer films. In this case the effective dielectric constant of the film should depend on its thickness due to electromagnetic interaction of the local fields induced at film inhomogeneities with the substrate.[37] This fact is very important for the interpretation of data of numerous optical experiments in electrochemical conditions but was not taken into account previously.

Kinetic equations and multiple scattering approaches were also used to describe the optical properties of island films.[27,31] Separate island can be considered as a clusters consisting of closely spaced, mutually interacting local centers. In these approaches, it is possible to study the behavior of the optical characteristics of island films in the vicinity of the percolation threshold for the metal phase. According to experimental data[33,34], the reflection and transmission spectra of island films change greatly near the percolation threshold. At a low metal-phase concentration, the spectra have extrema, which correspond to the local modes of separate metal particles. When the metal-phase concentration becomes greater than the percolation threshold, these extrema vanish.

References

1. M.A. Vorotyntsev, in *Modern Aspects of Electrochemistry*, eds. J. O'M. Bockris, B.E. Conway and R.E. White, (Wiley, New York, 1986), vol **17**, p. 131.
2. *Electrocatalysis*, eds. W.E. O'Grady, P.N. Ross and F.G. Wills, (Pennington, New Jersey, 1982).
3. V.S. Bagotzky and A.M. Skundin, *Electrochim. Acta* **29** (1984) 757.
4. V.S. Bagotsky and A.M. Skundin, *Electrochim. Acta* **30** (1985) 899.
5. M. Bachtler, W.J. Lorenz, W. Schindler and G. Saemann-Ischenko, J. *Electrochem. Soc.* **135** (1988) 2284.
6. D.R. Riley and J. T. McDevitt, *J. Electroanal Chem.* **295** (1990) 373.
7. A.M. Brodsky and M.I. Urbakh, *Electrodynamics of Metal-Electrolyte Interface*, (Nauka, Moscow, 1989) in Russian.
8. A.M. Brodsky and M.I. Urbakh, *Progress in Surface Sci.* **33** (1990) 91.
9. A.G. Malshukov, in *Chemical Physics of Solvation*, eds. R.R. Dogonadze, E. Kalman, A.A. Kornyshev and J. Ulstrup, (Elsevier, Amsterdam, 1988), Part C, p. 433.

10. H. Raether, *Surface Plasmons on Smooth and Rough Surfaces and on Gratings*, in *Springer Tracts in Modern Physics*, (Springer, Berlin, 1988), Vol. **111**.
11. H. Reather, in *Surface Polaritons*, eds. V.M. Agranovich and D.L. Mills, (North-Holland, Amsterdam, 1982), Chap. 9.
12. A.M.Brodsky, L.I.Daikhin and M.I.Urbakh, *J. Electroanal. Chem.* **171** (1984) 1.
13. R.M.A. Azzam and N.M. Bashara, *Ellipsometry and Polarized Light*, (North-Holland, Amsterdam, 1976).
14. M.I. Urbakh, *Electrochim Acta* **34** (1989) 1777.
15. *Surface Enhanced Raman Scattering*, eds. R.K. Chang and T.E. Furtak, (Plenum, New York, 1982).
16. A.A. Maradudin, *in Surface Polaritons*, eds. V.M. Agranovich and D.L. Mills, (North-Holland, Amsterdam, 1982), Chap. 10.
17. A.M. Brodsky and M.I. Urbakh, *J. Electroanal. Chem.* **228** (1987) 179.
18. T. Lopez-Rios, Y. Borenstein and G. Vuye, *Phys. Rev.* **B30** (1984) 659.
19. A.M. Brodsky and M.I. Urbakh, *Surface Sci.* **167** (1986) 231.
20. R. Christoph, M. Siegenthaler, H. Rohrer and M. Wiese, *Electrochim. Acta* **34** (1989), 1011.
21. A.M. Brodsky, L.I. Daikhin, A.M. Foontikov, V.E. Kazarinov, G.A Tsirlina and M.I. Urbakh, *Surface Sci.* **226** (1990) 137.
22. A.M. Foontikov, V.E. Kazarinov, M.I. Urbakh and A. Tadjeddine, *Surface Sci.* **239** (1990) 50.
23. L.I. Daikhin and M.I. Urbakh, *Elektrokhimiya* **24** (1988) 925.
25. A.M. Brodsky, L.I. Daikhin and M.I. Urbakh, *Elektrokhimiya* **25** (1989) 827.
25. K. Arya, R. Zeyer and A.A. Maradudin, *Solid St. Comm.* **42** (1982) 461.
26. G. Brown, V. Celli, M. Haller, A.A. Maradudin and A. Marvin, *Phys. Rev.* **B31** (1985) 4993.
27. L.I. Daikhin and M.I. Urbakh, *Surface Sci.* **236** (1990) 187.
28. *Acoustic, Electromagnetic and Elastic Wave Scattering - Focus on the T - Matrix Approach*, eds. V.K. Varadan and V.V. Varadan, (Pergamon, New York, 1980).
29. A. Wachniewski and H.B. McClung, *Phys. Rev.* **B33** (1986) 8053.
30. L.I. Daikhin and M.I. Urbakh, *Poverkhnost* **12** (1989) 36.
31. L.I. Daikhin and M.I. Urbakh, *Surface Sci.* **241** (1991) 395.
32. O. Hunderi, *Surface Sci.* **96** (1980) 1.
33. E. Dobierzewska-Mozrzymas and P. Biegausk, *J. Phys.* **F18** (1988) 2061.
34. P. Gadenne, Y.Yagil and G.Deutscher, *Physica* **A157** (1989) 279.
35. M.Y. Abyanch, W. Visscher and E. Barendercht, *Electrochim. Acta.* **28** (1983) 285.
36. G.A. Tsirlina, A.V. Smolin, A.I. Danilov, A.M. Foontikov, Yu. M. Polukarov and V.E. Kazarinov, *Elektrokhimiya* **26** (1990) 360.
37. L.I. Daikhin and M.I. Urbakh, *in press*.

Fundamental Aspects of Scanning Tunneling Microscopy in Polar Solvents

J. K. Gimzewski

IBM Zurich Research Laboratory
CH-8803 Rüschlikon

and
J. K. Sass

Fritz-Haber-Institut der Max-Planck-Gesellschaft,
Faradayweg 4-6, D-1000 Berlin 33

Abstract

The general concepts and interpretational difficulties of STM are briefly outlined. Attention is drawn to the low apparent barrier heights found in aqueous environments. Examples of electrochemical applications of STM, with recent achievements of atomic resolution, are presented. The complex electron-molecule interactions for tunneling through liquids are considered in detail. Recent results for the relaxation times of electron hydration in water and for the transmission times through potential barriers provide the basis for proposing a novel concept of electron transfer in polar solvents.

1. Introduction

The application of scanning tunneling microscopy [1] to the study of the solid-electrolyte interface is undoubtedly still in its infant stages. Several recent investigations [2-4] have indicated, for example, that an STM-junction filled with an aqueous electrolyte solution may exhibit an unexpected variance of its physical properties. In addition, there are indications that, although atomic resolution in water [5,6] and other polar liquids [7] has been reported recently, STM operators face considerable difficulties in achieving even somewhat lower resolution routinely on a day-to-day basis [8]. A particularly intriguing aspect of general interest is the low range of apparent tunnel barrier heights [9], which are deduced from the variation of the current when the gap dimension is changed. Compared to most typical values of 2-3 eV found in vacuum [1] only a few tenths of an electron volt, on average, have been reported for the apparent barrier in a recent careful study [2] where aqueous solutions were employed.

In this short review article we shall focus our attention on these unresolved problems and shall try to at least provide suggestions for explaining the striking difference between electron transfer across a vacuum gap and that through a polar liquid. Our conceptual approach will also lead to an extension of the analogy between interfacial redox reactions and *in-situ* charge transfer in a STM, previously considered by us in a preliminary report [10]. The basis for this extension is provided by the recent microscopic information about the coupling between electrons and the motions of the solvent nuclei, as well as the progress in characterizing the tunneling times of electrons through a classically impenetrable potential barrier. As a result of this emphasis in the present article it is not our intention to comprehensively discuss the great variety of images which by now have been accumulated with this local-probe technique in liquid electrochemical environments. The recent review [11] by Siegenthaler and Christoph is recommended as a source of such type of information. A description of various experimental details has also recently been provided [12].

We have organized the content of this chapter as follows. In the next section the basic concepts of an STM, both operational and interpretational, are briefly reviewed and the importance of the apparent barrier height as a junction parameter is emphasized. Section 3 contains selected examples of images measured *in-situ* and a description of the peculiarities which have been observed with polar liquids, i.e. predominantly H_2O, in the gap. The physical considerations relevant to electron transfer in a medium with substantial orientational polarizability are outlined in section 4 and proposals for electron propagation pathways are discussed there. Finally, in section 5 we conclude by considering future developments.

2. Basic Considerations

In its basic form STM involves positioning a sharp tip in close proximity of a surface and measuring the electrical current I that flows under the influence of an applied voltage V. The detected variation in current as a function of the spatial coordinates of the tip support system x, y and z (z defined as the surface normal) and applied bias voltage represents the experimental information. The different modes of operation[1], such as topographic, spectroscopic and barrier height measurement, involve a judicious selection of one or more parameters to be varied and the remaining parameters to be kept more or less constant.

In the topographic mode, for example, the current is measured and maintained constant by a feed-back adjustment of the piezo translator in the z-direction while an image is obtained by rastering in x and y and recording the voltage applied to that z-translator. The other modes are variations on this basic concept and have been described in detail elsewhere [1]. The philosophy of what one learns from all this activity is a matter of taste and imagination, but is eventually governed by the degree to which microscopic theories of tunneling and electron injection are developed.

Ideally, one may wish to imagine a picture of a single atom apex, supported by a trimer, from which electrons tunnel according to nonpertubative theories involving weak tip-surface interactions, with the overlap of the tip and surface wave functions probing contours of constant charge density, reflecting in turn surface atomic structure. Indeed for some phenomena such as the (1x2) reconstruction of the Au(110) surface [13], for example, simple theories can provide a reasonable description of experimental data [14]. Similarly, the qualitative aspects of spectroscopic data obtained on semiconductor surfaces can be modelled satisfactorily using fairly simple concepts [14]. In contrast, the nature of the tunnel barrier and how it fluctuates in time, thus changing its topology, is only poorly understood. Here one must go on to consider the tip and surface as a coupled system and determine its joint characteristics [15].

One of the first STM measurements by Binnig, Rohrer and coworkers[16], to prove that it was indeed a tunnel junction as opposed to a point contact , was the investigation of the dependence of the tunnel current on the z-position of the tip, which can be used to define a quantity called the apparent barrier height φ_a

$$\varphi_a = (d\ln I/dz)^2 \qquad (1)$$

(note here that z is not strictly speaking the variation of gap separation; see below). If we consider φ_a as being loosely related [1] to the inverse tunneling decay length κ

$$I \propto \exp(-2\kappa s) \qquad (2)$$

(where s is now the actual gap spacing) or even more loosely to the mean of the work functions of the tip $\varnothing_t$ and the sample $\varnothing_s$

$$I \propto \exp(-2[(2m/\hbar^2)(\varnothing_t+\varnothing_s)/2]^{1/2}s) \qquad (3)$$

then a range of values of $\varphi_a \sim$ 1-5 eV are typical for metal-vacuum-metal (MVM) tunneling. Even in UHV, and especially during the early development of STM, values of φ_a of the order of meV were occasionally measured. Simple theories based upon an idealized MVM junction [17] cannot account for such low values of the barrier. Other explanations of these and related observations, such as various forms of elastic deformation which would cause differences between the actual apex atom motion and the motion inferred from the voltage applied to the z-piezo translator [18] or collapse of the barrier at small distances leading to ballistic electron injection [19], have been proposed from the early stages.

Fig. 1a contains a clear manifestation of the processes occurring in the STM as the tip is first brought towards and then driven into a metal surface in a controlled manner in UHV. For low values of Δz, corresponding to large tip-sample separations, an approximately exponential relationship is found corresponding to a value of φ_a of ~ 2-3 eV. However, at very close distances φ_a is observed to essentially collapse. This behaviour has been interpreted as the transition from tunneling to ballistic electrical transport across a point contact [20]. A detailed theoretical analysis by Lang [21], in which the tip was modelled as a Na atom on a jellium surface and the sample as a second jellium surface, has shown good agreement with the experimental observation (see Fig. 1b).

In view of the numerous reports of atomically resolved STM images such attention to the basic phenomena outlined above may seem obvious and appear pedantic. However, emphasis of this point is very necessary since almost all STM experiments conducted in aqueous media including biological samples, humid air environments and hydrophobic types of surfaces and tips, in addition to electrochemical environments, have yielded anomalously low values of φ_a.

An example [2] of such measurements is shown in Fig. 2. Two representative current I vs. tip translation Δz results are shown in Fig. 2a, with the tip moved from the initial position first towards the sample and then away from it. During the approach the expected logarithmic dependence is observed over a limited range of Δz values, but upon removing the tip backwards considerable curvature appears in the log I vs. Δz relationships and a different current is measured for nominally similar values of Δz. The

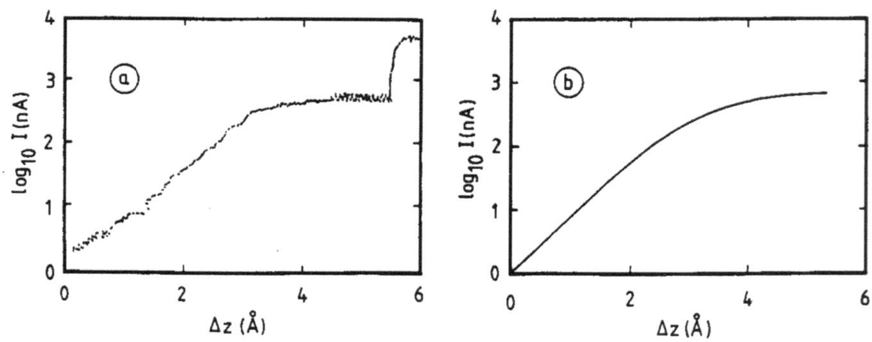

Fig. 1: Dependence of the current I on the translation Δz of the tip, with increasing Δz corresponding to an approach towards the sample; a) experimental curve (V_t = 20 mV); b) calculated result; (after Lang [21]).

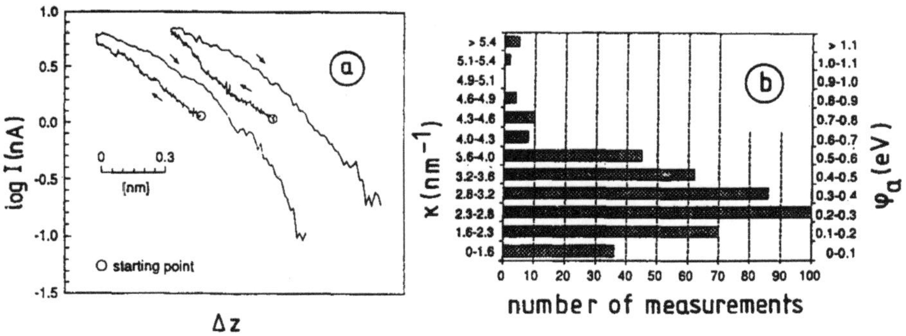

Fig. 2: a) Examples of the current I vs. translation Δz relationship obtained on Au(111) in 0.5M $NaClO_4$; note the hysteresis-like behaviour when the tip first approaches the sample from the initial position (o) and is then retracted; b) Statistics of a large sample of apparent barrier height measurements (as shown in Fig. 2a), indicating a surprisingly wide spread of φ_a-values; (after Binggeli *et al.* [2]).

apparent barrier heights deduced from a large body of such measurements (cf. Eq. (1)) exhibit considerable scatter, as is shown by the distribution of values in Fig. 2b. Most often it is found that φ_a is a few tenths of an eV, but the origin of the large spread of barrier heights without an apparent change of experimental parameters is not understood [2].

Such observations immediately provoke the fundamental question which phenomena control the tunnel barrier in spatially confined polar liquids. For applied purposes, the ability to obtain topographic data under a variety of environments has resulted in much STM activity, despite the remaining uncertainties about the imaging mechanism involved. In the following section some representative examples of the *in-situ* use of STM for electrochemical applications are presented that definitely provide an encouraging outlook for *in-situ* applications. The fundamental question of electron transport in an STM operating in aqueous media will be taken up again in section 4.

3. Selected Examples of *In Situ* Imaging

The microtopography of solid surfaces plays an important role in many electrochemical processes. Much work has therefore gone into incorporating the STM into an electrochemical cell, to thereby permit *in-situ* investigations of a variety of processes such as deposition and dissolution. Technically, the construction of such a system is complex and solutions to various problems, such as high faradaic current densities at the tip and independent control of the tip and sample potential, have to be solved [11,12]. With the appropriate system and low lateral drift rate of the image window, local electrochemical processes may be evaluated from a sequence of images recorded with different electrochemical parameters. Fig. 3 shows an example in which a Ag(100) surface was imaged under electrochemical conditions corresponding to repetitive cycles of deposition and dissolution of Pb [11]. It is clear from inspection of the images that distinct changes occur upon cycling the electrochemical parameters. There is also a sufficient similarity between all the images (see circular dots to guide the eye) that one can conclude that steps are preferentially formed only at certain parts of the surface upon cycling.

Although one obtains an overall impression of the shape of the surface in these and many other images in an electrochemical environment there is a lack of quantization of topography into units of atomic height (or fractional atomic height for certain dislocations) which can be obtained with STM in an UHV environment. To illustrate this, Fig. 4a shows the topography of a highly stepped region on a Cu(111) surface [22] imaged in UHV . Although the surface is rough it is built from discrete layers of monoatomic height (~ 2.5 Å). This is a surface physicist's ideal conception of roughness. However, if we simulate a decrease in resolution by applying a recursive filter Fig. 4b is obtained. The surface now appears to have 'wavy' form and much of the details which can be interpreted in terms of terraces, steps, or dislocations, are lost. The image in fact now has qualitative resemblance to those pictures shown in Fig. 3. Interpretation is further complicated by the role that the tip shape plays when tunneling on rough surfaces. As illustrated in Fig. 5 the role of tunneling from different points on the shank of the tip may distort the image recorded [23]. When there are topographical features on the sample contour which the the end of the tip, because of its own dimensions, cannot trace, the shape of the tip is instead imaged. One should be aware, in general, that there are also other interpretational difficulties in STM. One important aspect is the role played by interatomic forces between the tip and surface. Their influence is discussed elsewhere in detail [24] and additional problems have also been reviewed [14,25].

Achieving the kind of atomic resolution which has been more or less standard in UHV studies for quite some time has been the major goal in electrochemical studies for several years, but the first successful reports have appeared only recently. One example [5] is shown in Fig. 6, where individual copper atoms in an adlayer on a Au(111) substrate can be distinguished. The interesting result of this study is the marked influence of the electrode potential on the type of long-range order observed in the copper adsorbate. Whereas in Fig. 6a only a strongly corrugated ($\sqrt{3}$x$\sqrt{3}$)R30° phase is visible, lowering the potential by only 20mV leads to the appearance of a weakly corrugated (5x5)superstructure, coexisting with the other pattern (Fig. 6b). The

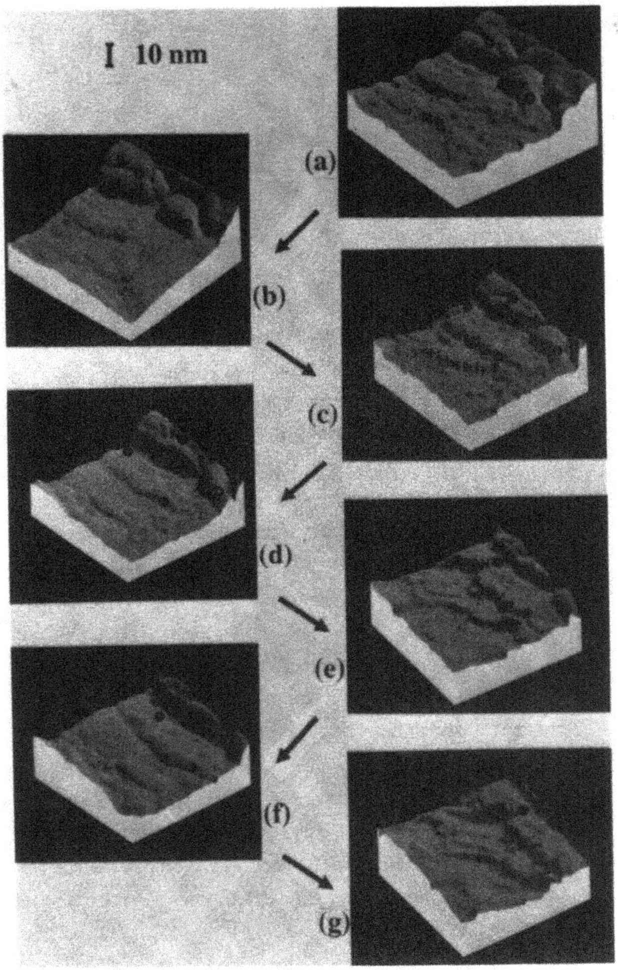

Fig. 3: STM images (window: 20 x 20 nm) of repetitive lead (ca. 10-15 ML) deposition (Figs. 3a,c,e,g; E_{SCE} = -155 mV) and removal (Figs. 3b,d,f; E_{SCE} = -955 mV) on Ag(100) in 0.5M $NaClO_4$ + 0.001M Pb^{2+}; Vt = -5 mV, I = 80 pA; (after Siegenthaler and Christoph [11]).

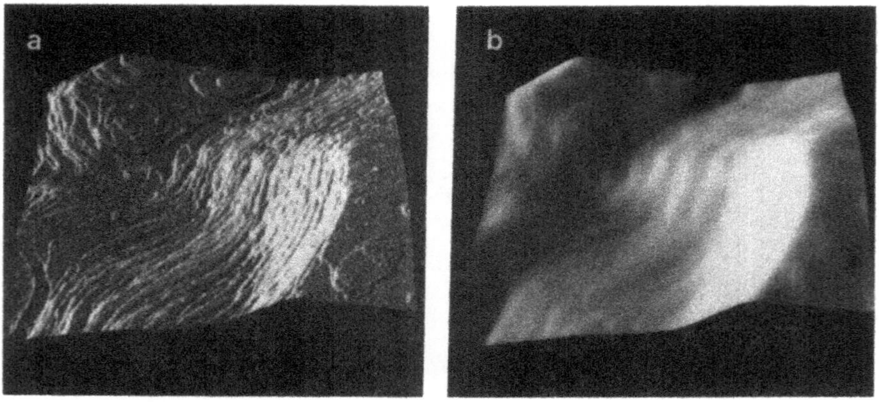

Fig. 4: a) Large area (75 x75 nm) STM topograph of a highly stepped region on Cu(111), measured in UHV; b) same image after recursive filtering to simulate loss of lateral resolution; both figures are represented as pseudo-illuminated, three-dimensional images; (after Berndt *et al.* [22]).

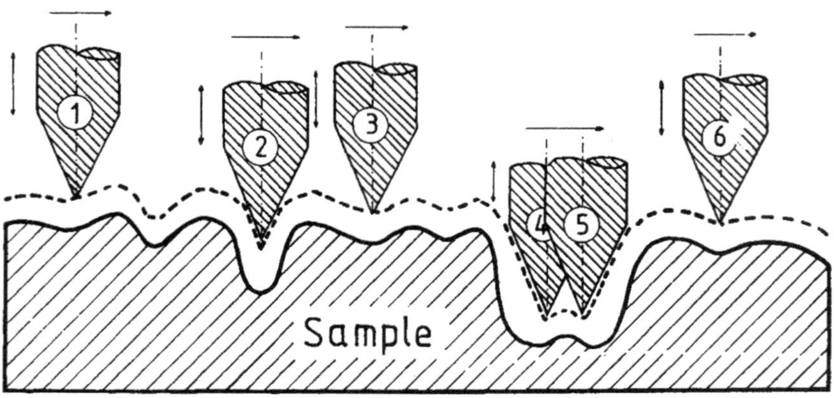

Fig. 5: Schematic illustration of the influence of the tip shape on the measured STM trace, indicating the loss of vertical and lateral resolution at narrow or deep channels of the sample contour; (after Gimzewski *et al.* [23]).

difference between these observations and pseudomorphic Cu adlayers on Au(111), formed in ultrahigh vacuum, are attributed to the structure determining role of anions (i.e. sulphate) coadsorbed from the electrolyte solution [5]. Another impressive example of atomic resolution achieved *in situ* is the recent imaging [6] of the (√7x√7)R19.1° and the (√3x√3)R30° iodine adlattices on Pt(111), where again the electrode potential can be used to transform the former phase, created by vapour deposition in the gas phase, into the latter one.

An intriguing question with regards to the electron transmission mechanism across the STM gap is the dependence of the resolution on the specific properties of different solvents. Preliminary results, obtained by us in collaboration with K.H. Besocke [7], on evaporated and annealed Au(111) films immersed in both propylene carbonate and triethylene glycol (without a conducting salt) seem to indicate that either solvent is more suitable than water for imaging a metal electrode surface. Examples of the high-resolution images which were obtained in these liquids are shown in Fig. 7. In triethylene glycol the (1x23) reconstruction of Au(111) [26] could be imaged with considerable detail at presumably close proximity of the tip and the sample (Fig. 7a). At somewhat larger gap spacing (Fig. 7b) features appear which, because of their dimensions, are tempting to attribute to individual solvent molecules, more strongly bound and therefore immobilized during the STM scan to specific sites on the surface. In Fig. 7c a similar example of very good resolution of atomic steps on a highly structured area of the Au(111) surface in propylene carbonate is shown. These images definitely provide an incentive for extending electrochemical studies, at present almost exclusively confined to aqueous solutions, to other protic and aprotic solvents. Unfortunately, in these experiments it was not possible to obtain lnI vs. Δz relationships, i.e. values of the apparent barrier height.

4. Tunneling Trajectories in Polar Liquids

Several major experimental [27,28] and theoretical [29,30] advances in characterizing the interaction of electrons with polar liquids, in particular at very short times ($t < 10^{-12}$ sec), underlie the considerations in this chapter. The other aspect, which we propose to be very relevant to charge transfer in liquids, is the duration of electron tunneling through a classically impenetrable potential barrier. For many years there has been considerable confusion regarding this problem [31], but it appears that some clarification has been achieved recently [32].

A previously somewhat neglected issue for analyzing the difference of electron transfer across the STM-gap, when there is a polar liquid present, is the construction of an appropriate potential energy surface in the region between the tip and the sample. When considering this task, it becomes immediately obvious, however, that even in water as the most extensively studied polar solvent only very approximate descriptions of the electron energy level spectrum are at present available. The traditional view which evolved in radiation chemistry [33] and *in-situ* photoemission studies [34] distinguished between a quasi-free energy range above the V_0-level (with the latter equivalent to the vacuum level, i.e. representing the lowest non-bonding level) and the hydrated electron state e^-_a where the solvent dipoles are rearranged around the excess charge to yield the lowest energy configuration.

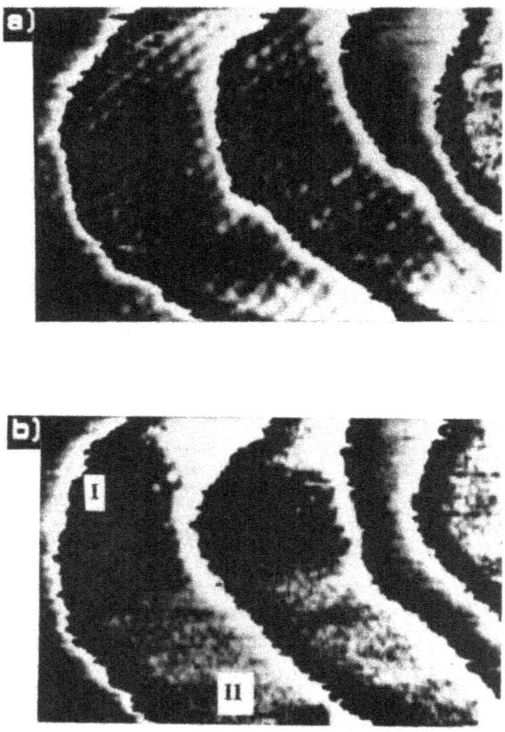

Fig. 6: STM images (8 x 12 nm) of copper adlayers on the same area of a Au(111) substrate in 0.05 M H_2SO_4+5mM $CuSO_4$; a) E_{SCE} = 0.215 V; b) E_{SCE} = 0.200 V ; whereas in Fig. 6a only a ($\sqrt{3}$x$\sqrt{3}$) R30° superstructure (phase I) is observed, a (5x5) structure (phase II) is seen to coexist with phase I in Fig. 6b; V_t = -0.17V, I = 2.5 nA; (after Magnussen *et al.* [5]).

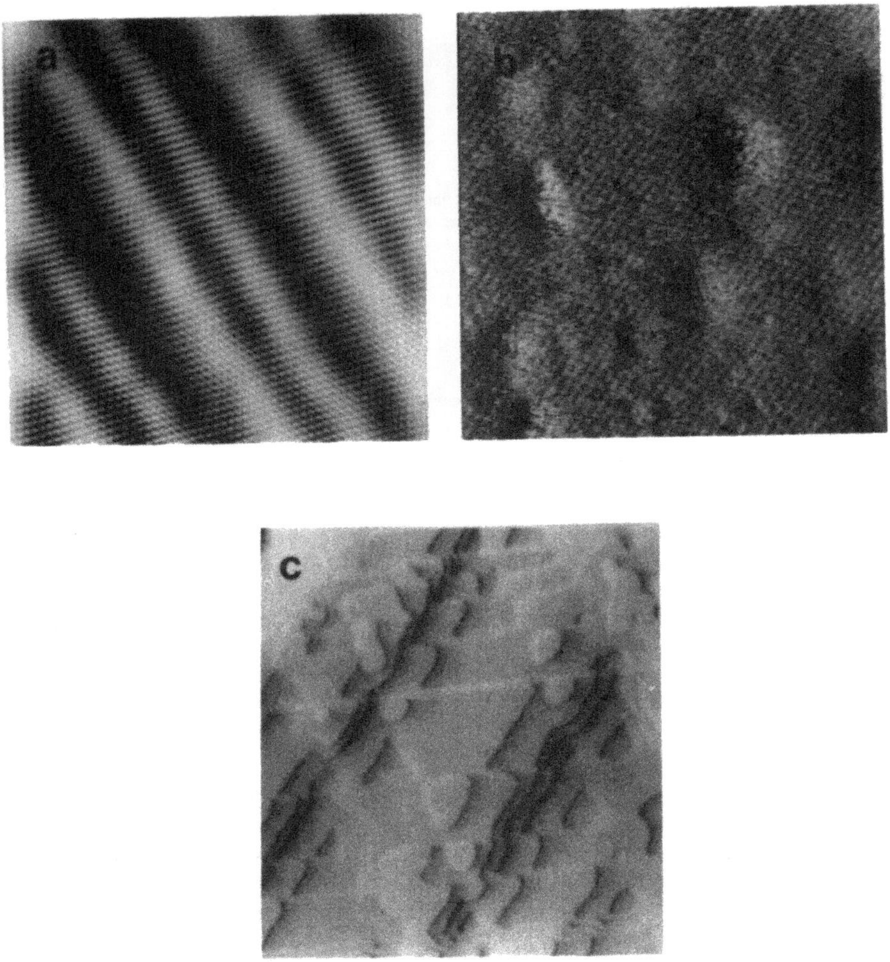

Fig. 7: STM images of evaporated Au(111) films (flame-annealed at ca. 1000 K) in organic solvents; a) area with (1x23) reconstruction in triethylene glycol, viewed at close proximity of tip and sample; V_t = 0.02 V, I = 20 nA, (13.8 x13.8 nm); b) unreconstructed region in triethylene glycol at larger gap spacing, compared to Fig. 7a; V_t = 0.14 V, I = 18 nA, (13.8 x13.8 nm); c) highly stepped area in propylene carbonate; V_t = 0.03 V, I = 20 nA, (220 x220 nm); (after Haiss *et al.* [7])

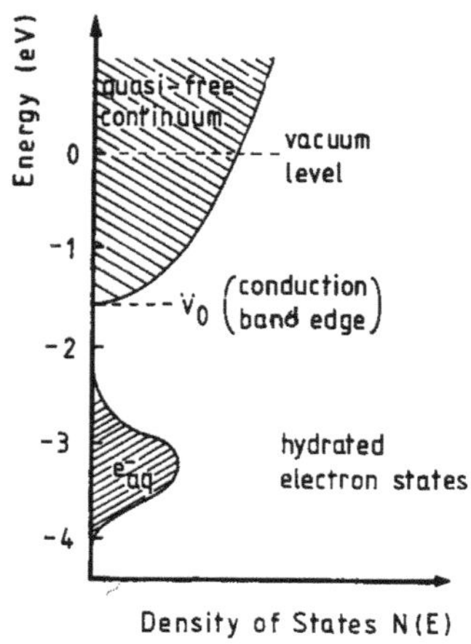

Fig. 8: Schematic density-of-states diagram of electron energy levels in liquid water.

Approximate density-of-states distributions of these energy levels are shown in Fig. 8, with the energetic positions of the V_0-level and of the hydrated electron chosen as a compromise between slightly conflicting views about the appropriate interpretation of different spectroscopic results and thermodynamic considerations [33-35]. The distribution for the quasi-free states is based upon the assumption of a conduction band , with an increase of the density of states as the square root of the energy. As will be discussed subsequently, this picture is probably too simplistic and a conclusive evaluation of these states must await further theoretical and experimental developments.

The orientation of the water molecules in the first hydration shell, deduced some years ago from electron spin echo measurements in aqueous glasses [36], is schematically depicted in Fig. 9a. One of the positive hydrogen arms in each of the six water molecules is pointing towards the negative electronic charge, which constitutes an arrangement not unlike that found in neutron scattering for negative ions such as the halides [37]. This simple picture may be contrasted to the result of a recent molecular dynamics (MD) simulation of electron hydration in a cluster of $(H_2O)_{128}$ [29], which is shown in Fig. 9b. Although it appears that in the latter case the electron charge is not as localized in the centre of the first hydration shell, it has in fact been suggested [30] that the somewhat simplified situation shown in Fig. 9a is quite representative of the majority of configurations occurring in the MD simulations.

In a one-dimensional picture for electron transfer through an aqueous environment in the STM we should therefore like to suggest, based upon the above description of the electronic energy levels, two stages of modification of the potential barrier, compared to vacuum, as shown in Fig. 10. The lowering of the barrier due to the electronic interaction with the solvent (the "fast" response) which has recently been addressed by Schmickler and Henderson [38] is illustrated in Fig. 10a. This reduction (cf. Fig. 6) is not sufficient, however, to explain the values of the apparent barrier height which have been described in section 2. In Fig. 10b a suggestion for a further, very substantial reduction is presented, justification of which seems to us to require a fairly demanding discussion of two time scales. One of these two time domains is related to the fastest orientational response of the solvent molecules, whereas the other, which is probably the more difficult one to assess, encompasses the time span available to a tunneling electron for interaction with the environment, i.e. the solvent dipoles, in the classically forbidden region of the potential barrier. Hypothetically speaking (with some qualitative justification provided below), if the tunneling electron resides in the STM-gap sufficiently long to allow the solvent molecules to reorient, or if there are spontaneously occurring configurations which are similar to Fig. 9, unoccupied electronic levels with much lower energy than V_0 are available to the propagating electron and therefore the measured barrier height will

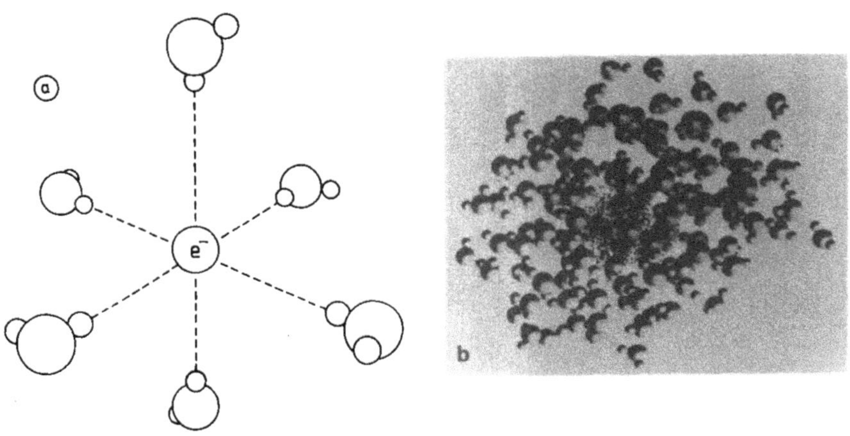

Fig. 9: Structural models of the hydrated electron; a) first hydration shell from electron spin echo measurements in aqueous glasses (after Kevan [36]); b) molecular dynamics simulation for a $(H_2O)^-_{128}$ cluster; (after Barnett *et al.* [29]).

also be much lower. In the remainder of this chapter we shall present theoretical and experimental evidence which motivated us to consider this novel concept of electron transfer/tunneling in polar solvents. We also note in passing that the investigations of dynamical polarization effects in tunneling through semiconductor heterojunctions described by Gueret [39] and subsequently calculated by Persson and Baratoff [40], in which the temporal evolution of the image potential and its relation to the plasma frequency of the conducting electrodes were studied, are conceptually very similar to the processes that are explored subsequently.

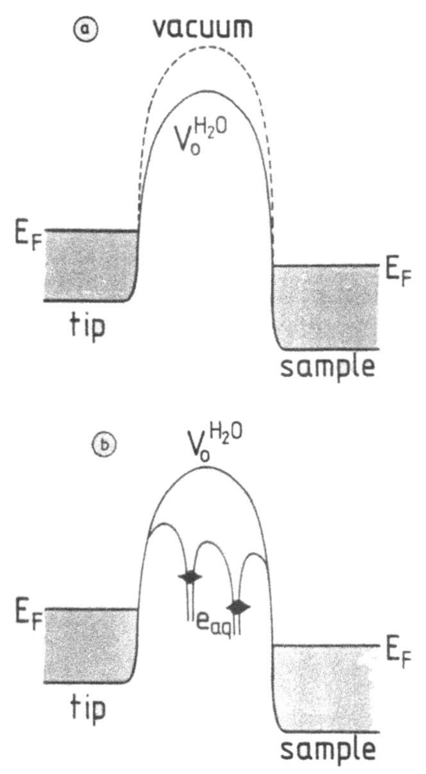

Fig.10: a) Lowering of the STM barrier height in water, compared to vacuum, due to the energetic location of the V_0-level (cf. Fig. 8); b) additional reduction of the barrier by either spontaneously occurring or induced solvent configurations e^-_{aq} with attractive electrostatic potential (schematic representation).

The molecular-dynamics "experiments" [29,30,41-43] provide a wealth of microscopic, time-resolved information about the electron-solvent interaction, even though the numerical results of, for example, the binding energy of a hydrated electron [44] may deviate by several tenths of an eV from the experimental findings [35] for a particular H_2O-cluster size or some other physical parameter. We shall mention here only a few aspects of these studies which are particularly relevant to electron transfer in the STM and relate them to the very interesting experimental results of hydrated electron formation in both the gas phase [35] and in solution [28]. In Fig. 11 the calculated evolution and shift of optical absorption features after the "sudden" creation of quasi-free electrons (cf. Fig. 8), indicate that already after 10^{-14} seconds a substantial lowering of the energy, probably of the order of 1 eV, has occurred [30]. After about 30-40 femtoseconds the first reorientation stage of the water molecules is more or less completed and is then followed by a substantially longer process towards the fully relaxed hydration cage. This latter solvent reorganization is characterized by the growth of an absorption band at a fixed wavelength (cf. Fig. 11), which can be identified with the well-known optical absorption of hydrated electrons observed experimentally [33]. The dynamical features obtained by this MD simulation essentially confirm earlier results of spectroscopic studies in the femtosecond range [28], although a somewhat larger relaxation time was estimated in the experiment.

The nanometer proximity of the tip and the sample in the STM also necessitate the consideration of the spatial range of the electron-solvent interaction. At least in one dimension the solvent extension may vary, depending on gap spacing, between one and several monolayers. Although the aggregation geometry, which is present in the STM, has not been investigated it seems possible to gain some qualitative insight into this problem by considering the theoretical and experimental results for the electronic levels in hydrated electron clusters with different numbers, n of water molecules. In Fig. 12 we compare theoretical data [44] of the vertical detachment energies of an electron, from gas phase $(H_2O)^-_n$ clusters, with corresponding experimental findings [35] for a wide range of cluster sizes. Both sets of results indicate a decrease of the detachment energy with increasing cluster size where the $n^{-1/3}$ dependence may be derived from a long-range electrostatic continuum model. For n between 6 and ~ 25, which is probably the range relevant to the usual gap spacings of ~ 0.3 - 1.0 nm, the electron energy is seen to vary by the significant amount of ~ 1eV. This trend, if applicable to the STM-geometry, would imply that the potential-energy surfaces in the liquid depend on the separation of the tip from the sample and also that there are substantial anisotropies in the directions parallel and perpendicular to the sample surface, because of the different extensions of the dipolar medium. With this information in mind we wish to consider now the problem of tunneling times and return to the electron-water interaction later in this section.

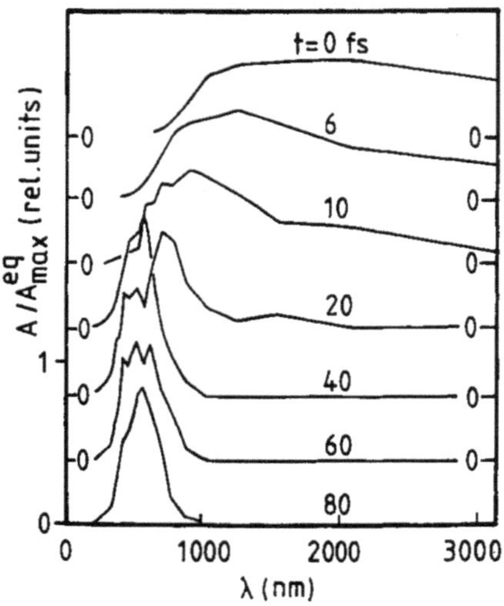

Fig.11: Molecular dynamics simulation of the femtosecond evolution of optical absorption features during the hydration of electrons in water at 300 K; (after Rossky and Schnitker [30]).

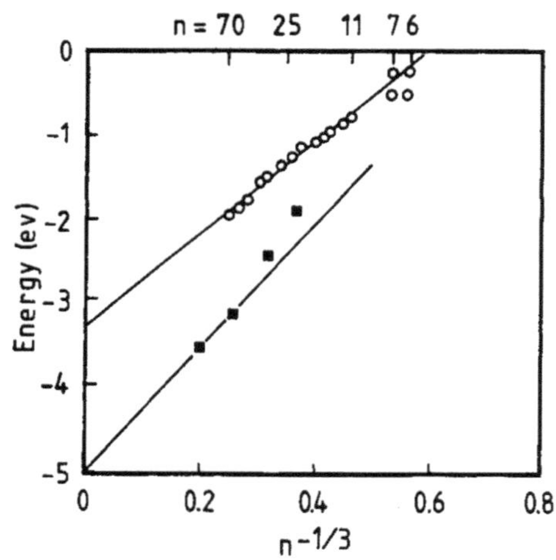

Fig.12: Comparison of calculated (■) and measured (o) vertical electron detachment energies as a function of the number n of water molecules in a $(H_2O)^-_n$ cluster; (after Barnett *et al.* [44] and Coe *et al.* [35]).

In the last years there has been an extensive debate about the proper definition of the average time spent by a transmitted tunneling electron in the region of a classically impenetrable one-dimensional potential barrier [31]. The controversies associated with the identification of this average transmission (or traversal) time arise because time is not regarded in quantum mechanics as an observable which can be represented by a Hermitian operator. Leavens [32] has most recently given a lucid analysis of this problem and has suggested a novel theoretical approach which is based on Bohm's causal interpretation of quantum mechanics. Subsequently, we shall rely on his theoretical concepts and use the orders of magnitude of his calculated transmission times to illustrate our ideas, while being aware that a rigourous evaluation of this fundamental physical aspect is perhaps not yet available. However, it must also be emphasized that our concept does not depend on Bohm's interpretation being the correct one.

The height and range of widths of the rectangular potential barrier, which were used by Leavens [32], roughly represent the STM-situation of electron tunneling between two metals in nanometer proximity at low bias (see insert in Fig. 13). The zero of the energy scale may be equated to the bottom of the conduction band, with the Fermi level located at ~ 5 - 6 eV and the work function amounting to 4 - 5 eV as the remainder of the total barrier height of 10 eV. In addition to specifying the kinetic energy of the incident electron, the minimum-uncertainty-product wavepacket description of the electron used in the Bohm trajectory calculation also requires a value of either Δz or Δk, the spread in either real or momentum space, respectively. Unfortunately, this latter parameter is difficult to estimate for a given experimental situation and introduces a certain measure of uncertainty into the following discussion which should hence only be regarded as qualitative. The average transmission times that resulted in this calculation are shown in Fig. 13 for a mean incident kinetic energy of 5 eV and two values of $\Delta z(\Delta k)$, as a function of barrier thickness. For reference, the times in which free electrons with kinetic energy of 5 eV traverse the distances d, in the absence of a barrier, are also shown.

There is an additional aspect in Leaven's treatment of the problem which is related to Bohm's interpretation of quantum mechanics and the statistical nature of the tunneling event. As far as we know, the Bohm interpretation is the only one that has led to calculated distributions of real-valued transmission times and the prediction of trajectories, although admittedly very rare ones, which spend a considerably longer time in the barrier compared to the average values shown in Fig. 13. This is illustrated in Fig. 14 for a barrier thickness of 5 Å by the probability distribution of the transmission time for the typically 10^4 trajectories used in the calculation for one set of barrier parameters. The significance of these rare events is perhaps best illustrated by comparison to chemical reactions in solution, where also the very unlikely, highly activated processes are dominant [45]. To demonstrate

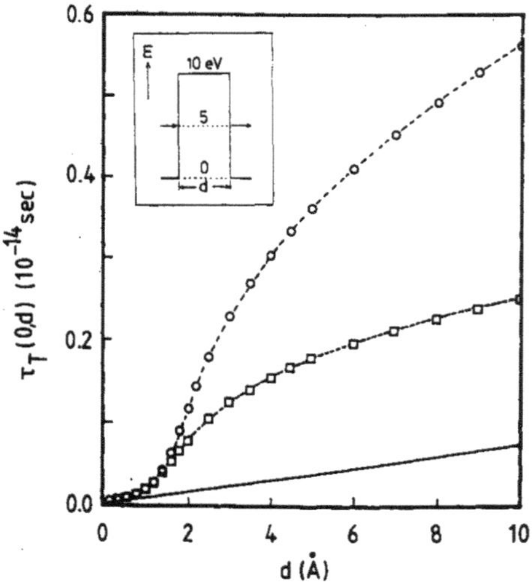

Fig.13: Average transmission times of electrons tunneling through a rectangular potential barrier (see insert) with mean incident kinetic energy of 5 eV; width of minimum-uncertainty-product wavepackets are Δz=12.5 Å (o) and Δz= 6.25 Å (□) The straight line represents the free-particle (plane-wave) traversal times; (after Leavens [32]).

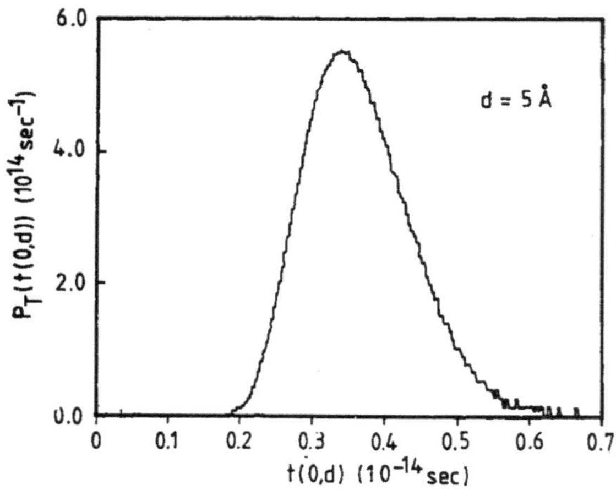

Fig.14: Distribution of real-valued transmission times of electron wavepackets with E = 5 eV and Δz = 12.5 Å, obtained from Bohm trajectory calculation; (after Leavens [32]).

the mechanism which we propose it is also useful to consider first a static barrier which is determined purely by the electronic response of the liquid (cf. Fig. 10a). According to Leavens and Aers [46] those trajectories which penetrate most deeply into the barrier region and "almost" result in transmission, but are ultimately reflected, are the ones which will be most affected by introducing the spontaneous and induced polarization fluctuations, which very crudely may be thought of as the time-modulated modification of a barrier previously used in theoretical analyses [31]. Since these electrons reside a longer time in the liquid gap they may induce the highest degree of polarization and thus have the largest chance to be transmitted, instead of being reflected, as a result of the self-consistent barrier lowering.

Comparing Figs. 11 and 13 it is evident that the time scales for the response of the water molecules to an excess charge and for the transmission through a barrier of dimensions which represent about one to three diameters of the H_2O molecule could very well be of the same order of magnitude if the actual range of Δz is of order 10 Å.We propose on the basis of the considerations outlined above the following scenario for tunneling through molecular layers of water, or perhaps more generally of polar solvents. Electrons from the source (e.g. the tip in a STM or the donor in a redox reaction) penetrate the barrier region quantum-mechanically and initiate an orientational polarization of the H_2O dipoles. This polarization increases with the time spent by the electron in a particular region of the gap. Since the potential is lowered as a result of these interaction processes the electron is capable of moving more freely through the environment in its immediate vicinity.

To these considerations we must now add two probably very important aspects. As a result of the thermal fluctuations of the solvent molecules the potential barrier which is probed by the tunneling electron is also time dependent and is expected to influence the further evolution of the trajectory within the gap. This may be particularly relevant since even in vacuum the electron transfer between the tip and the sample must be considered in three dimensions [9]. However, whereas in vacuum the spatial distribution of the tunneling current density is closely related to the electric field lines within the gap, the random fluctuations of the three-dimensional potential in the presence of the solvent dipoles give rise to a much more complicated situation. In particular, one may envision substantial propagation components parallel to the sample surface as the fluctuations of the potential exceed the potential gradient imposed by the external STM voltage. A schematic and certainly somewhat speculative representation of the more diffuse electron trajectories, expected from this model, is shown in Fig. 15. Electrons can now emanate from a wider region on the end of the tip, because of the occurrence of prehydration cages which are formed either spontaneously or are induced by electrons leaking out quantum-mechanically from the tip. Subsequently, the three-dimensional tunneling trajectory may similarly follow a fairly randomized, three-dimensional path of, loosely

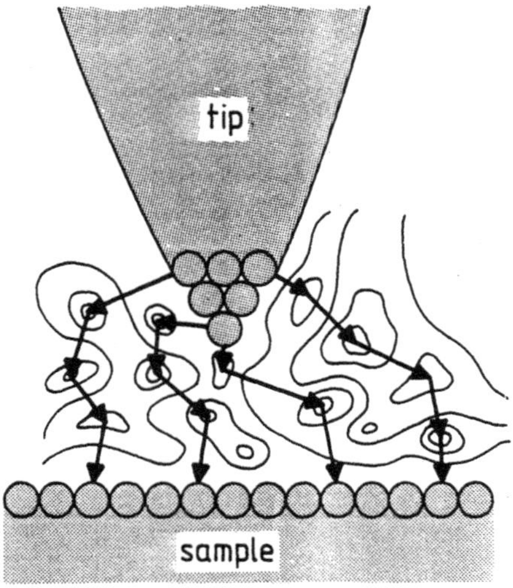

Fig.15: Schematic illustration of more diffusive STM trajectories in an aqueous environment, compared to tunneling in vacuum, due to spontaneous and induced polarization fluctuations of solvent dipoles, represented by qualitative contour lines of the electrostatic potential.

speaking, the least resistance with regards to the height of the potential barrier and only on average in the direction of the sample because of the superimposed external voltage. Such tunneling trajectories may actually explain the greater difficulties in obtaining atomic resolution in an aqueous environment, because of the more diffusive nature of the electron pathways.

The amount by which the apparent barrier height would be lowered due to prehydration of the tunneling electrons, as proposed in this section, is difficult to predict, but may clearly be of the order of 1 eV or more. There are, however, additional effects which influence the potential energy distribution between the tip and the sample and contribute to the complexity of the electron trajectory in the liquid, compared to vacuum tunneling. For example, the size dependence of the electron binding to water clusters (cf. Fig. 12) can be expected to induce an increase of the barrier reduction, which we predict as a result of very fast dipole relaxation, with increasing gap spacing, as the extension of the polarizable medium in the direction normal to the sample surface is enlarged. It is not unlikely that the curvature in the log I vs. s measurements in Fig. 2a is caused by this mechanism, but further studies are required to confirm this contention.

It is also conceivable that the geometry of the STM junction, with its large extension of liquid parallel to the sample, is associated with an anisotropy in the three-dimensional potential energy surfaces, such that the dynamical barrier is higher in the z-direction. This would tend to make propagation of electrons parallel to the sample more favourable and lead to an even larger spread of the current distribution on the sample, thus further degrading (cf. Fig. 15) the lateral resolution in topographic measurements. The evolution of any particular electron trajectory, which as a very rare event [45] among the many attempts surmounts the gap, depends upon the microscopic details of the temporal fluctuations of the solvent orientation and spatial arrangement. This should be represented in Fig. 8 by a distribution of shallow prehydration traps [47] below the conduction band edge, but again there is not yet sufficient information about these states.

To conclude this section, we briefly reconsider[10] possible similarities between electron traversal across a liquid STM gap and electron transfer in homogeneous or heterogeneous redox reactions. Fleming and Wolynes[45] have recently highlighted the impact that advances during the last decade in ultrafast spectroscopy and in large-scale computer simulations have had on our thinking about chemical reactions in solution. Solvent dynamics, i. e. the time scales associated with different modes of nuclei reorganization, have also increasingly been recognized as a key ingredient for the more formal, traditional theoretical concepts in interfacial electrochemistry and have been successfully used to explain trends in experimental data for different solvents [48]. To the extent that the above ideas of dynamical polarization induced by tunneling electrons stand the test of future, more rigourous analysis of the two competing time scales, application to electron transfer from one redox species to another, or to a metal electrode, suggests itself immediately. It appears, however, that the STM geometry with its ability to systematically vary parameters such as distance and tunnel voltage is more suitable initially to study the relevance of these effects in different solvents. In general, it is encouraging that the ultrafast relaxation studies in liquid water [28] seem to have identified a relaxation time, although associated with the slower one of two reorganization processes detected, which compares well with the longitudinal relaxation time derived in phenomenological models [48] and this may indicate a future unification of both approaches.

5. Future Developments

The major purpose of this contribution has been to stimulate interest in exploiting to a larger extent the well-defined experimental parameters of the STM for fundamental studies of the physical processes which are associated with electron transfer across a potential barrier composed of polar solvent molecules. Too much emphasis is perhaps placed upon the production of good images in current electrochemical studies, with not enough concern about the much more complex electron-molecule interaction mechanisms, compared to tunneling in vacuum. This point has also recently been stressed by Siegenthaler and Christoph [11]. In the course of such more fundamental studies we expect insight into molecular details which may be very relevant to dynamical aspects of chemistry in solution [45]. A systematic variation of the solvent, similar to the intriguing studies described by Fawcett and Foss [48], should be very helpful for this purpose.

We anticipate also that spectroscopic studies of the STM junction, involving either emission [10,49] or absorption [50] of photons in the gap region, will be applied extensively to the solid-electrolyte interface. The prevalence of the inelastic radiative tunneling mechanism [51], recently confirmed by theoretical studies [49,52], may be of even greater significance in solution [53]. Time-resolved measurements of such processes are a possibility which should be investigated.

Another goal in future electrochemical studies, which has been addressed previously by Siegenthaler and Christoph [11], is the determination of structural molecular properties of the solution layer immediately adjacent to the electrode surface. We have proposed [10] that even in UHV such studies can be meaningfully performed and, if the difficulties of *in situ* imaging of the interfacial solvent layer persist, this simulation of the double layer [54] should probably be used initially.

Acknowledgements

It is a pleasure to acknowledge M. Büttiker, W. R. Fawcett, N. D. Lang, H. Rohrer and H. Siegenthaler for valuable discussions. We are particularly indebted to C.R. Leavens for his critical reading of the manuscript. We would also like to thank D. Lackey for his help during the writing of this manuscript. Partial support by the Deutsche Forschungsgemeinschaft through Sfb 6 is acknowledged.

References

1. H. Rohrer, in *Scanning Tunneling Microscopy and Related Methods*, ed. R. J. Behm, N. Garcia and H. Rohrer, NATO ASI Series E: Appl. Sciences, Vol. **184** (Kluwer, Dordrecht, 1990), p. 1.
2. M. Binggeli, D. Carnal, R. Nyffenegger, H. Siegenthaler, R. Christoph and H. Rohrer, J. Vac. Sci. Technol., in press.
3. S. M. Lindsay and B. Barris, J. Vac. Sci. Technol. A **6** (1988) 544.
4. A. Stemmer, A. Hefti, U. Aebi and A. Engel, Ultramicroscopy **30** (1989) 263.
5. O. M. Magnussen, J. Hotlos, R. J. Nichols, D. M. Kolb and R. J. Behm, Phys. Rev. Lett. **64** (1990) 2929.
6. S.- L. Yau, C. M. Vitus and B. C. Schardt, J. Am. Chem. Soc. **112** (1990) 3677.

7. W. Haiss, D. Lackey, J. K. Sass and K. H. Besocke, submitted for publication; experiments were performed on Besocke's beetle-type STM at the KFA, Jülich.
8. R. J. Nichols, private communication; H. Siegenthaler, private communication.
9. N. D. Lang, Phys.Rev. B **37** (1988) 10395.
10. J. K. Sass and J. K. Gimzewski, J. Electroanal. Chem. **251** (1988) 241.
11. H. Siegenthaler and R. Christoph, in Ref. 1, p. 315.
12. T. R. I. Cataldi, I. G. Blackham, G. A. D. Briggs, J. B. Petica and H. A. D. Hill, J. Electroanal. Chem. **290** (1990) 1.
13. J. K. Gimzewski, R. Berndt and R. R. Schlittler, Surf. Sci. , in press.
14. J. Tersoff, in Ref. 1, p. 77.
15. G. Doyen, E. Koetter, J. P. Vigneron and M. Scheffler, Appl. Phys. A **51** (1990) 281.
16. G. Binnig, H. Rohrer, Ch. Gerber and E. Weibel, Appl. Phys. Lett. **40** (1982) 178.
17. T. E. Feuchtwang and P. H. Cutler, Physica Scripta RS **8** (1987) 10.
18 J. H. Coombs and J. B. Petica, IBM J. Res. Develop. **30** (1986) 455.
19. J. Gomez-Herrero, J. M. Gomez-Rodriguez, R. Garcia and A. M. Baro, J. Vac. Sci. Technol. A **8** (1990) 445; Y. Kuk and P. J. Silverman, J. Vac. Sci. Technol. A **8** (1990) 298.
20. J. K. Gimzewski and R. Möller, Phys. Rev. B **36** (1987) 1284.
21. N. D. Lang, Phys. Rev. B **36** (1987) 8173.
22. R. Berndt, R. R. Schlittler and J. K. Gimzewski, unpublished data.
23. J. K. Gimzewski, A. Humbert, J. G. Bednorz and B. Reihl, Phys. Rev. Lett. **55** (1985) 951.
24. S. Ciraci, A. Baratoff and I. P. Batra, Phys. Rev. B **41** (1990) 2763.
25. J. P. Pelz and R. H. Koch, Phys. Rev. B **41** (1990) 1212.
26. J. V. Barth, H. Brune, G. Ertl and R. J. Behm, Phys. Rev. B **42** (1990) 9307.
27. H. Haberland, H. Langosch, H.-G. Schindler and D. R. Worsnop, J. Phys. Chem. **88** (1984) 3940.
28. A. Migus, Y. Gauduel, J. L. Martin and A. Antonetti, Phys. Rev. Lett. **58** (1987) 1559.
29. R. N. Barnett, U. Landman and A. Nitzan, J. Chem. Phys. **89** (1988) 2242.
30. P. J. Rossky and J. Schnitker, J. Phys. Chem. **92** (1988) 4277.
31. M. Büttiker and R. Landauer, Festkörperprobleme XXV (1985) 711; E. H. Hauge and J. A. Støvneng, Rev. Mod. Phys. **61** (1989) 917.
32. C. R. Leavens, Sol. Stat. Commun. **76** (1990) 253.
33. A. Henglein, Ber. Bunsenges. Phys. Chem. **78** (1974) 1078.
34. J. K. Sass and H. Gerischer, in *Photoemission and the Electronic Properties of Surfaces*, ed. B. Feuerbacher *et al.* (Wiley-Interscience, New York, 1977), p. 469.

35. J. V. Coe, G. H. Lee, J. G. Eaton, S. T. Arnold, H. W. Sarkas, K. H. Bowen, C. Ludewigt, H. Haberland and D. R. Worsnop, J. Chem. Phys. **92** (1990) 3980.
36. L. Kevan, J. Phys. Chem. **85** (1981) 1628.
37. J. E. Enderby and G. W. Neilson, Rep. Prog. Phys. **44** (1981) 593.
38. W. Schmickler and D. Henderson, J. Electroanal. Chem. **290** (1990) 283.
39. P. Gueret, in *Electronic Properties of Multilayers and Low-Dimensional Semiconductors Structures*, ed. J. M. Chamberlain *et al.* (Plenum, New York, 1990), p. 317.
40. B. N. J. Persson and A. Baratoff, Phys. Rev. B **38** (1988) 9616.
41. R. N. Barnett, U. Landman and A. Nitzan, Phys. Rev. Lett. **62** (1989) 106.
42. K. A. Motakabbir, J. Schnitker and P. J. Rossky, J. Chem. Phys. **90** (1990) 6916.
43. M. Sprik, J. Phys. Condens. Matter **2** (1990) SA161.
44. R. N. Barnett, U. Landman, C. L. Cleveland and J. Jortner, Chem. Phys. Lett. **145** (1988) 382.
45. G. R. Fleming and P. G. Wolynes, Physics Today, May 1990, p. 36.
46. C. R. Leavens and G. C. Aers, Sol. Stat. Comm., submitted for publication; C. R. Leavens, private communication.
47. R. N. Barnett, U. Landman and A. Nitzan, J. Chem. Phys. **91** (1989) 5567.
48. See, for example: W. R. Fawcett and C. A. Foss, Jr. , Electrochimica Acta, in press; and references therein.
49. R. Berndt, A. Baratoff and J. K. Gimzewski, in Ref. 1, p. 269.
50. R. J. Hamers and K. Markert, Phys. Rev. Lett. **64** (1990) 1051.
51. J. K. Gimzewski, J. K. Sass, R. R. Schlittler and J. Schott, Europhysics Lett. **8** (1989) 435.
52. P. O. Johannson, R. Monreal and P. Apell, Phys. Rev. B **42** (1990) 9210.
53. R. Mcintyre and J. K. Sass, Phys. Rev. Lett. **56** (1986) 651.
54. J. K. Sass, D. Lackey, J. Schott and B. Straehler, Surf. Sci., in press.

Invited paper for the

Proceedings of ICTP Working Party on Electrochemistry
Trieste, Italy
August 17–September 7, 1990

THEORY OF BULK AND INTERFACE CONSTANT PHASE ELEMENTS IN ELECTRODE-ELECTROLYTE SYSTEMS

S. H. LIU

SOLID STATE DIVISION
OAK RIDGE NATIONAL LABORATORY
Operated by
MARTIN MARIETTA ENERGY SYSTEMS, INC.
Under
Contract No. DE-AC05-84OR21400
With the
U.S. DEPARTMENT OF ENERGY
Oak Ridge, Tennessee 37831

January 1991

THEORY OF BULK AND INTERFACE CONSTANT PHASE ELEMENTS IN ELECTRODE-ELECTROLYTE SYSTEMS

S. H. LIU
Solid State Division, Oak Ridge National Laboratory
Oak Ridge, Tennessee 37831-6032, USA

ABSTRACT

This paper summarizes the progress gained in the last few years in our understanding of bulk and interface constant-phase-angle (CPA) behavior in electrode-electrolyte systems. It is now fairly well established that the interface constant-phase element originates from the fractal nature of the interface. The complex geometry gives rise to a fractal distribution of parallel current paths, and the competition between these paths results in the fractional power law behavior of the impedance across the interface. On the other hand, the early hope of relating the CPA exponent to the fractal dimension of the interface has been shown to be unattainable. Our understanding of the bulk CPA behavior, which is most prevalent in solid electrolytes, is only tentative. It is illustrated using a simple model that, under nonlinear dynamical laws that govern the flow of ions in the electrolyte, a current in the solid can generate a fractal distribution of vacancies which tend to impede the flow. The current is forced to negotiate a complex path through the solid, and the resulting fluctuation in path length and flow rate could be a source of the CPA behavior.

1. Introduction

Impedance spectroscopy is a simple and powerful tool for probing the physical and chemical properties of electrode-electrolyte systems. In a typical experiment an ac current is sent through the system and the voltage drop across the system is measured. From the frequency dependence of the impedance, one tries to infer the combination of known circuit elements–resistance, inductance and capacitance–which reproduces the measured behavior. The interpretation of these circuit elements gives insight into the various physical and chemical reactions taking place in the system.

Simple as the program may sound, it is difficult to carry out because the behavior of real systems cannot be modelled by a finite number of circuit elements, no matter how they are chosen. The difficulty lies in the constant-phase behavior, which necessitates the inclusion of a mysterious circuit element whose impedance has the power-law frequency dependence:

$$Z(\omega) \propto (j\omega)^{-\eta}, \tag{1}$$

where $j = \sqrt{-1}$, ω is the angular frequency, and the exponent η satisfies $0 < \eta < 1$. Fig. 1 shows a set of actual data taken at the Oak Ridge National Laboratory.[1,2]

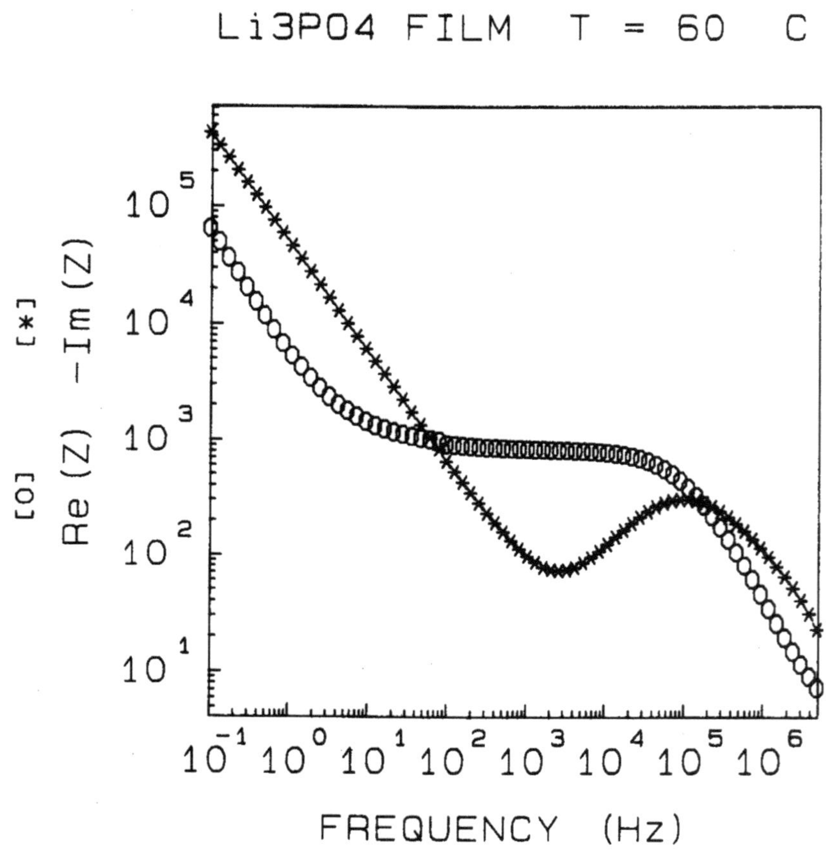

Fig. 1. The real and imaginary parts of the impedance of a 1 μm amorphous Li_3PO_4 film with gold electrodes plotted as functions of the frequency in log-log scale.

In this figure the real and imaginary parts of $Z(\omega)$ of a $1\mu m$ amorphous Li_3PO_4 film with gold electrodes are plotted against the frequency ($f = \omega/2\pi$) in log-log scale. Straight lines on the graph with negative slope indicate power-law frequency dependence. In particular, when real and imaginary parts form two parallel lines, it is an unmistakable indication of CPA behavior. One finds two regions in Fig. 1 in which CPA behavior is apparent: at the low frequency end below 3 Hz and at the high frequency end above 10^5 Hz. At the low frequency end where the CPA component of the impedance is overwhelming, one can extract the exponent η from the data in two ways: from the slope of the straightline portion of the data, or from the constant ratio

$$-ImZ(\omega)/ReZ(\omega) = \tan(\eta\pi/2). \tag{2}$$

The last property gives the name constant-phase-angle, i.e. the argument of tangent function in the above equation, to the mysterious circuit element, which is also called constant-phase element (CPE). At the high frequency end the CPA exponent must be determined from curve fitting because of the large background. To fit the data in Fig. 1 one needs two constant-phase elements, one with $\eta \simeq 0.92$ at the low frequency end and one with $\eta \simeq 0.77$ at the high frequency end.[2] Through carefully controlled experiments, it has been determined that the low frequency CPE originates from physical processes taking place at the interface while the high frequency one originates from interactions in the bulk of the electrolyte. In the complex ω plane the CPA impedance has a branch cut along the positive imaginary axis. This analytic property cannot be reproduced by any finite combination of RLC elements because the impedance of the combination can only have a finite number of poles, never a branch cut. The only known examples of branch cuts are infinite ladder networks and transmission lines, both having $\eta = 1/2$. Attempts have been made to concoct infinite ladder networks with varying RC elements or transmissions line with spatially varying RC distributions to simulate the interfacial CPE,[1-3] but one then faces the challenge of how to justify the concoction on physical grounds. For the bulk CPE one must assume a continuous distribution of activation energies for ion hopping.[4] Again, no explanation of the origin of the distribution has been offered.

In the last few years we have witnessed significant progress toward a qualitative to semi-quantitative understanding of the interfacial CPE. It is now fairly well established that the behavior has its origin in the fractal geometry of the inherently rough interface. The understanding of bulk CPE is comparatively rudimentary. It seems that a theory based on self-organized fractal geometry holds promise as a possible explanation. This paper contains a review of these recent developments, with emphasis not on data fitting but on elucidating the processes which lead to the CPA behavior.

2. Theory of Interfacial CPE

The ideal electrode-electrolyte interface as discussed in textbooks in electrochemistry consists of a planar metal electrode in contact with the electrolyte. Consider a blocking electrode where no chemical reaction takes place, a bias potential across the interface causes a surface charge to accumulate on the metal and a screening cloud of the opposite charge to form in the electrolyte. The combination of charge distributions creates an interfacial capacitance. A probing ac signal encounters an impedance which consists of the ohmic resistance in the electrolyte and the interfacial capacitance. The ohmic resistance in the metal electrode can be ignored. Therefore, the real part of the total impedance should be frequency independent and the imaginary part should be inversely proportional to the frequency. The observation that this simple theory does not apply to real interfaces was first made in 1926.[5] In recent years it has become increasingly clear that the interface, no matter how carefully it is prepared, is uneven on the submicron scale, and the CPA behavior is intimately linked to the microscopic roughness.[6-10] When the interface is made increasingly smooth, the value of η approaches unity.

A logical connection between roughness and fractional power-law impedance was first suggested by Le Méhauté who pointed out that rough surfaces are fractals, i.e. the roughness looks the same under different degrees of magnification.[11] Before we present his argument and the subsequent development of this idea by other authors, we make a brief digression to define what a fractal is in terms readily understandable by physical scientists. A good example of a fractal object is the monster curve known as the Koch island, which is constructed by the procedure illustrated in Fig. 2. One begins with an equilateral triangle. Next, one divides each side into three parts and erects an equilateral triangle using the middle part as the base. This results in the six-fold symmetric star. In the third stage the twelve sides of the star are trisected and smaller equilateral triangles are erected in the middle of each. The Koch island is constructed by repeating this procedure *ad infinitum*. The boundary of the island is obviously continuous, but is not differentiable because it makes an infinite number of zigzags between every two points on the curve. The length of the curve between any two points is infinity, but the area of the island is finite. A small part of the curve is similar to a larger part when magnified by any integral powers of 3. This property, called self-similarity under a scale transformation, is common to all fractal objects. Other examples of fractals can be found in the two books by Mandelbrot.[12,13] Fractals in nature are not as regular as the Koch island. They usually appear random, but are self-similar in a statistical sense, that is, given a sufficiently large sample of similar objects, the magnification of a small part of one object can be matched arbitrarily closely with some other object. The scale of magnification can be arbitrary. The range of length in which self similarity holds is bounded from above by the size of the object and from below by the size of the smallest building block, an atom for example.

The term "fractal dimension" describes the self-similar property under scale transformation in a quantitative manner. This concept, first proposed by Hausdorff

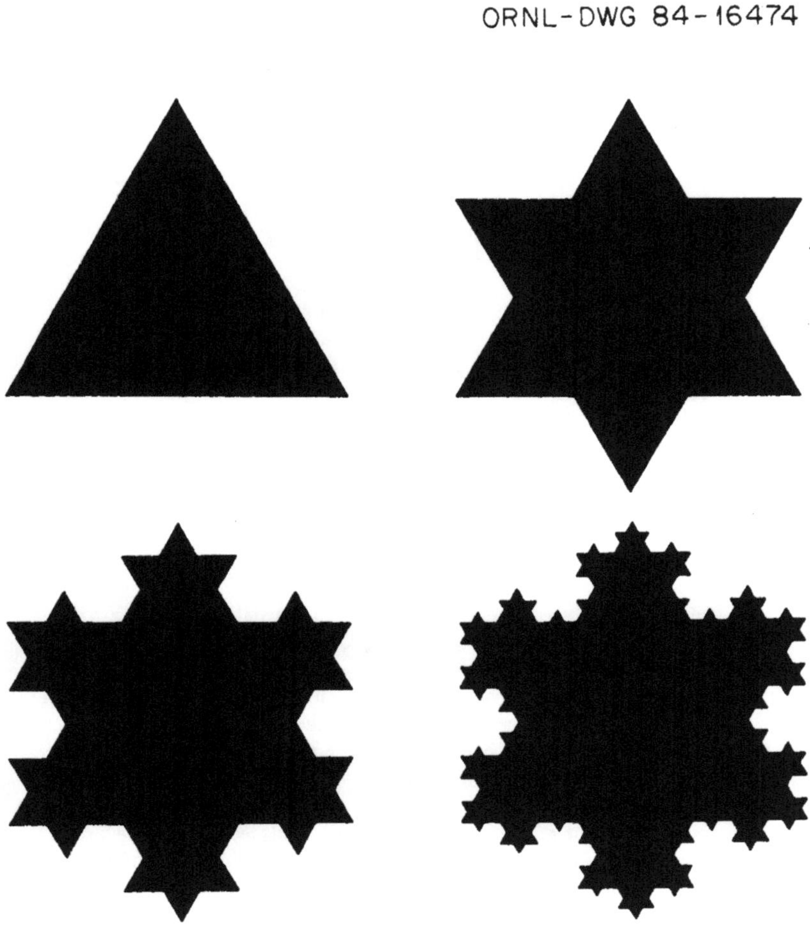

Fig. 2. The first four stages in the construction of the Koch island. In the limit of infinite stages, the boundary of the island is a fractal with fractal dimension $d = \ln 4/\ln 3 = 1.2619$.

in 1919, is a direction generalization of dimension in Euclidean geometry. When the sides of a square are magnified by a factor of 2, the area is 4 times larger. If we write $4 = 2^d$, we find that $d = 2$ is the dimension of the area. Similarly, the volume of a cube is 8 times larger under a similar magnification of its sides. This gives $d = \ln 8/\ln 2 = 3$ as the dimension of the volume. Going back to the Koch island in Fig. 2. The process of constructing the fourth stage from the third may be regarded in a slightly different way, that is, the two stages are images of the same object viewed with different degrees of magnification. In the third stage the resolution only allows us to see 48 straight segments forming the boundary but not the smaller features within each segment. In the next stage all lengths are magnified 3 times so that the same resolution allows us to resolve each segment into 4 because the middle third now appears to consist of two segments. Details beyond this level cannot be resolved. A similar definition, $4 = 3^d$ or $d = \ln 4/\ln 3 = 1.2619$ is the dimension of the Koch curve. We obtain the same result by holding the length scale constant but improving the resolution. The total length of the curve appears to be 4 times larger every time the resolution is improved threefold. In the limit of infinite resolution the length diverges, as discussed previously. The area of the Koch island remains two dimensional because it has a finite limiting value.

Fractals need not be monster curves. An example, the Cantor bar, is shown in Fig. 3. Under coarse resolution the object looks like a continuous bar. With improved resolution one sees a gap in the middle which separates the bar into two, each is $1/a$ the length of the original bar, $a > 2$. With higher resolution each one of the bars is seen to be subdivided in the same manner. The dimension of the object is $d = \ln 2/\ln a < 1$. Under infinite resolution the object seems to disappear because it has a dimension less than the dimension of the Euclidean space it is imbedded.

In Fig. 4 we show the electron micrograph of a surface of the solid electrolyte Ag β-alumina,[10] which is a good example of the electrode-electrolyte interface. There are hills and valleys of various sizes, and each hill has smaller hills and valleys of various sizes so that the picture looks the same under different magnifications. The fractal dimension of such a surface is larger than 2, which means that the observed area increases with increasing resolution. Many real surfaces have the property that self similarity is achieved by scaling the height in a ratio different from the lengths in the plane of the surface. This type of scale invariance is called self affinity. The surface depicted in Fig. 4 is actually self-affine, but the electron micrograph does not convey adequately the depth information of the surface features. Notice that although every fractal object has a well-defined dimension in a range of length scales, the fractal dimension does not contain sufficient information about the fractal. Fractals with the same dimension may look distinctly different.

The first model for rough interface was the surface traced out by shifting the Koch island in the direction perpendicular to the plane of the shape.[11] An area on the surface is bounded by Koch curves in one direction and straight lines in the perpendicular direction. Since the area is measured by the product of the lengths of the boundary curves in the two directions, the dimension of the surface is the sum of the dimensions of the boundary curves, $d = 1 + \ln 4/\ln 3 = 2.2619$. Le Méhauté

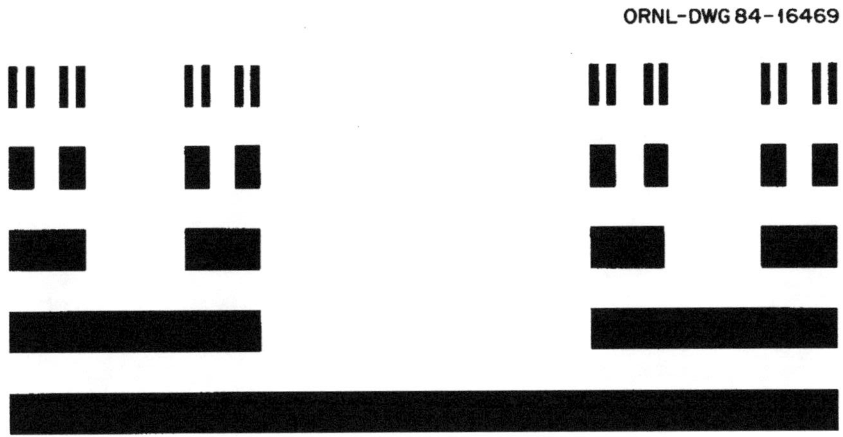

Fig. 3. The first fives stages in the construction of a Cantor bar.

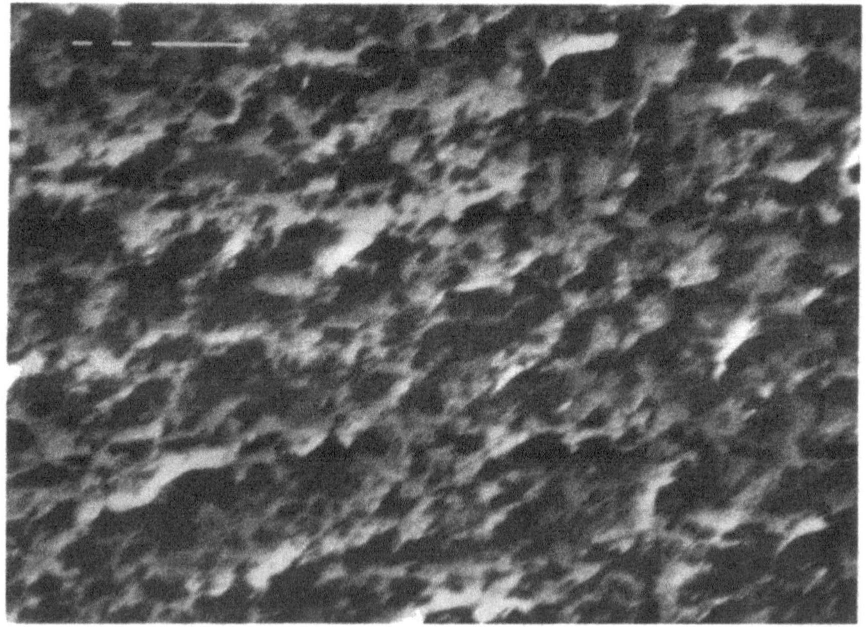

Fig. 4. The electron micrograph of a surface of the solid electrolyte Ag β-alumina.

and Crepy,[11] have used the scaling argument to find the following relation between the CPA exponent η and the fractal dimension:

$$\eta = 1/(d-1). \tag{3}$$

Since the scaling argument is not familiar to most investigators in electrochemistry, we will develop the theory based on a different model proposed by the present author.[14] The model, shown in Fig. 5, is constructed by connecting together the various stages of the Cantor bar in Fig. 3. Just like the Koch island model, the surface is assumed to be smooth in the direction perpendicular to the page, so the fractal dimension of this surface is $d = 1 + \ln 2/\ln a$. The model interface is self-affine, because self similarity is achieved when a part is magnified by an integral factor of a in directions parallel to the plane but by a factor of unity perpendicular to the plane.

The two sides of the interface can be assigned to metal and electrolyte in two ways, yielding two distinct physical models. In the first situation we assign the electrolyte to the black side to obtain a grooves within grooves model, which was inspired by De Levie's observation that polished metal surfaces exhibit grooves with jagged surfaces covered with smaller grooves.[7] When an ac signal flows from the electrolyte (black) into the electrode (white), it encounters ohmic resistance in the electrolyte and interfacial capacitance at the faces of every segment. The equivalent circuit of the interface is shown in Fig. 6. The circuit branches out at every new stage of the Cantor bar. The resistance increases by a factor of a at every branch because of the reduction in the width of the bar. The number of interfacial capacitors increases in proportion with the number of branches, but the size of the capacitors remains the same. This is an approximation which amounts to ignoring the interfacial capacitance in the dips between branches. One can see in Fig. 5 that at higher stages the areas in the dips become negligible compared with the areas of the sides. We have verified numerically that the capacitance of the dips does not affect the input impedance of the circuit at low frequencies. The common ground represents the electrode.

The input impedance of the network in Fig. 6 has the form of an infinite continued fraction:

$$Z(\omega) = R + \cfrac{1}{j\omega C + \cfrac{2}{aR + \cfrac{1}{j\omega C + \cfrac{2}{a^2 R + \ldots}}}} \,. \tag{4}$$

The function $Z(\omega)$ satisfies the frequency scaling relation:[15]

$$Z(\frac{\omega}{a}) = R + \frac{aZ(\omega)}{j\omega C Z(\omega) + 2} \,. \tag{5}$$

Fig. 5. A model of rough electrode-electrolyte interface based on the Cantor bar model.

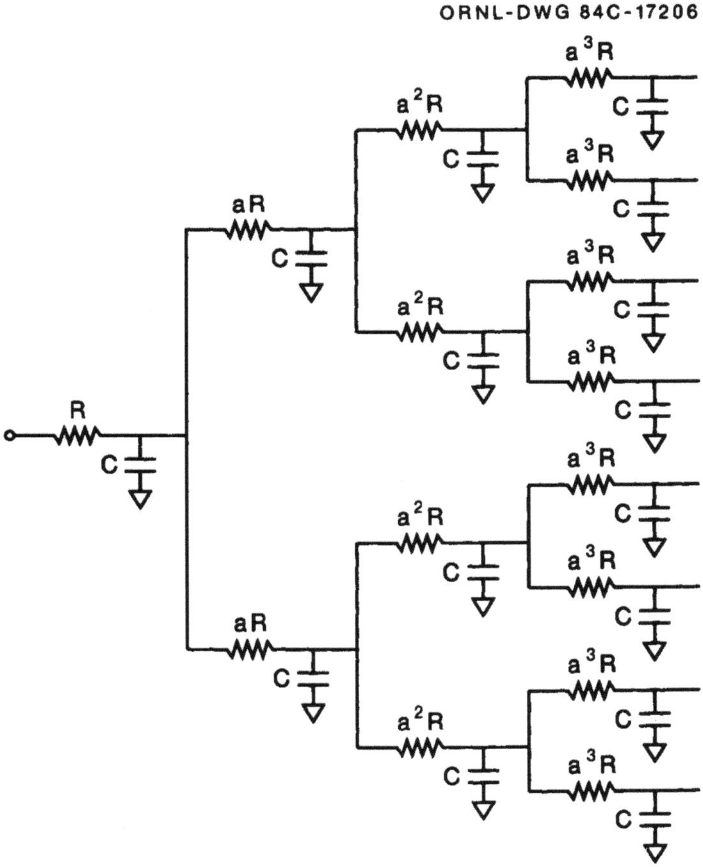

Fig. 6. The equivalent circuit of the model interface in Fig. 5.

In the low frequency limit Eq. 5 reduces to

$$Z(\frac{\omega}{a}) \simeq \frac{a}{2}Z(\omega) . \tag{6}$$

The relation in Eq. 6 is satisfied by

$$Z(\omega) = A(j\omega)^{-\eta} , \tag{7}$$

where A is a constant and

$$\eta = 1 - \ln 2/\ln a = 3 - d . \tag{8}$$

A smooth surface has $a \ll 1$ so that $\eta \to 1$ as observed.

The result in Eq. 8 is different from that in Eq. 3, and this generated some heated debate as to which one is the correct answer, because it was hoped that η was a universal function of the fractal dimension. We will show later that this is not the case because η depends on details of the surface geometry not measured by d. The insight leading to this conclusion as well as a physical understanding of the CPA behavior are gained from studying this primitive model.

As a start, we give a simple picture of the origin of the CPA behavior. We terminate the network after a finite number of stages and calculate the input impedance by using the recurrent relation method.[14] The results for the real part of $Z(\omega)$ are plotted in Fig. 7. At low frequencies the real part of Z reaches a plateau whose height increases by a factor of $a/2$ for every additional stage. At high frequencies the impedance has the limiting value R. Between these two limits the real part of the impedance exhibits the CPA behavior. The imaginary part of $Z(\omega)$ is inversely proportional to the frequency in both high and low frequency limits and has the power-law dependence in the middle range. These results show that the power-law frequency dependence is the result of competition between resistive and capacitive paths. We can look at Fig. 7 in a different way, namely that signals at lower frequencies tend to penetrate more stages of the network. This happens because a signal of lower frequency must propagate farther down the network before it finds sufficient capacitive paths to cross the interface. In doing so it also encounters higher resistance. In total it experiences a higher impedance than a signal of higher frequency. The self-similar structure ensures that the frequency dependence of the impedance follows the power-law. Real surfaces usually are self similar over a finite range of length scales, and this determines the frequency range in which the CPA behavior is evident.

The finite network calculation gives a concrete illustration of dynamical scaling. In Ref. 14 we showed that the impedance of a network with n stages has the following asymptotic expression at low frequencies $\omega RC \ll 1$ and large n:

$$Z_n(\omega) \simeq R\frac{a(a+1)}{(a-1)(a-2)(2a-1)}\left(\frac{a}{2}\right)^n + \frac{1}{2^n j\omega C} . \tag{9}$$

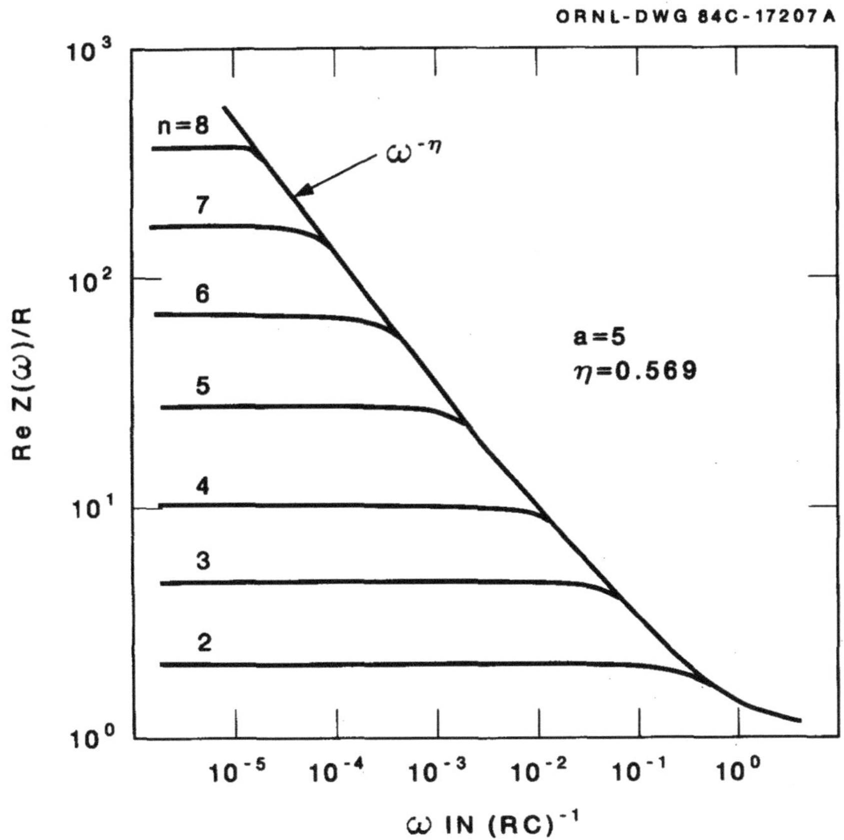

Fig. 7. The frequency dependence of the real part of the input impedance of the network in Fig. 5 after finite numbers of stages.

This equation means that for a signal frequency which penetrates n stages of the network, it experiences a resistance proportional to $(a/2)^n$ and a capacitance measured by 2^n. If we scale the frequency down by a factor of a, we deduce from Eq. 9 that, in the asymptotic regime,

$$Z_{n+1}(\frac{\omega}{a}) = \frac{a}{2} Z_n(\omega). \tag{10}$$

In the limit of large n there is negligible difference between Z_{n+1} and Z_n and the above equation reduces to the earlier result in Eq. 6. The relation in Eq. 10 can be interpreted in a different way, i.e. the signal with scaled down frequency can penetrate one more layer and thus experiences an impedance which is $a/2$ higher. Thus, the impedance is self similar, and is scaled up by a factor of $a/2$ when the frequency, which is used as the yardstick, is scaled down by a. The frequency exponent is the fractal dimension of the impedance.

We can now use the scaling argument to deduce the CPA exponent of the cylindrical surface generated by shifting the Koch island, assuming that the volume inside the cylinder is filled with electrolyte. For clarity we present the derivation put forward by Nyikos and Pajkossy.[16] We write, in analogy with Eq. 9 that for a surface with large but finite stages n

$$Z_n(\omega) \simeq R_n + 1/j\omega C_n \ , \tag{11}$$

for sufficiently small ω. The quantities R_n and C_n are difficult to calculate, but their scaling behavior can be found by simple arguments. If we increase the stage to $n+1$, the capacitance will scale according to the surface area, i.e. $C_{n+1} = \frac{4}{3} C_n$. The resistance is dominated by a newly added groove in the middle of each side. Since the width of the new groove is 1/3 of the larger groove, the resistance is scaled up by $R_{n+1} = 3R_n$. This allows us to obtain

$$Z_{n+1}(\omega) = 3R_n + 3/4j\omega C_n = 3Z_n(4\omega) \ .$$

Or for $n \to \infty$

$$Z(\omega/4) = 3Z(\omega) \ . \tag{12}$$

This gives the relation between η and the fractal dimension d in Eq. 3.

The scaling argument points to the source of difference between the formulas in Eqs. 3 and 8, both of which link η to d. The frequency exponent depends on the scaling properties of two quantities, effective R and C. Both scale according to details in the geometric properties of the fractal and indepedently of each other. The fractal dimension is but one measure of the surface geometry. It is, therefore, not surprising that a unique relation between η and d is not possible. Keddam and Takenouti reached the same conclusion by studying an interface whose cross section is a Koch curve with generalized scaling ratios.[17]

We now discuss another exactly soluble model which demonstrates that η is not a universal function of d. The model is inspired by the electron micrograph in Fig. 4 in which the irregularities are better described as bumps upon bumps. Shown in Fig. 8, the model is the product of two Cantor dusts, one associated with the x direction and one with the y direction.[18] We, therefore, named it the Cantor block model. As in the Cantor bar model, the figure shows the electrolyte protruding into the electrode. Every stage is a rectangular block, and in the most general case each block branches into $N_x N_y$ smaller blocks whose sides are scaled by $1/a_x$ and $1/a_y$ in the plane of the interface and $1/a_z$ in the perpendicular direction, $a_x > N_x$ and $a_y > N_y$. In the figure both N_x and N_y are equal to 2. The equivalent circuit of this model has the same structure as that in Fig. 2. The number of new branches at every stage is $N_x N_y$, and each branch consists of a series resistor and a capacitor connected to ground. The resistors scale up by $a_x a_y / a_z$ in each successive stage due to the reduction in cross-sectional area of the branch by $1/a_x a_y$ and the decrease in length by $1/a_z$. The capacitance is dominated by the contributions of the lateral surfaces, whose areas are reduced by factors $1/a_x a_z$ and $1/a_y a_z$ respectively in successive stages. Again, the input impedance can be expressed as an infinite continued fraction

$$Z(\omega) = R + \cfrac{1}{j\omega C + \cfrac{N_x N_y}{\frac{a_x a_y}{a_z} R + \cfrac{1}{j\omega \frac{C}{a_z}\left(\frac{1}{a_x} + \frac{1}{a_y}\right) + \cfrac{N_x N_y}{\left(\frac{a_x a_y}{a_z}\right)^2 R + \cfrac{1}{j\omega \frac{C}{a_z^2}\left(\frac{1}{a_x^2} + \frac{1}{a_y^2}\right) + ..}}}}} \tag{13}$$

The impedance has the constant-phase behavior under a set of conditions, i.e. $a_x > a_y$, $a_x > a_z$, $a_x > a_z^2$, $a_y a_z < N_x N_y$, and $N_x N_y a_z < a_x a_y$. When all these conditions are fulfilled the frequency exponent is found to be

$$\eta = 1 - \ln(N_x N_y / a_y a_z) / \ln(a_x / a_z^2) \; . \tag{14}$$

The detail of this calculation is given in Ref. 18. The determination of the fractal dimension is subtle because the surface is self-affine and contains three scaling ratios a_x, a_y and a_z. According to Mandelbrot, the dominant scaling ratio is the one that gives the most rapid length reduction in the limit of large number of stages.[19] In the Cantor block model the dominant scale is a_x, so we measure the area using yard sticks of lengths 1, a_x^{-1}, a_x^{-2}, ..., a_x^{-n+1}, etc. At every stage we ignore features smaller than the yardstick. In this way the area at the nth stage is found to have the asymptotic value

$$A_n \propto (N_x N_y a_x^2 / a_y a_z)^{n-1} \; . \tag{15}$$

The fractal dimension is defined by $A_n \propto (a_x)^{(n-1)d}$, which yields

$$d = 2 + \ln(N_x N_y / a_y a_z) / \ln a_x \; . \tag{16}$$

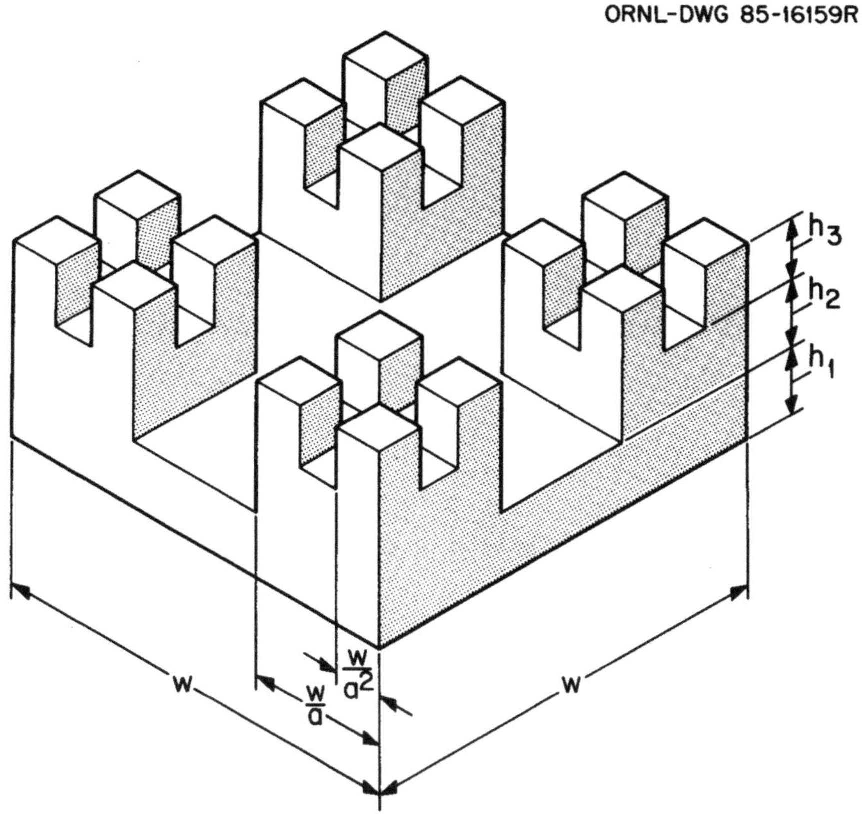

Fig. 8. The Cantor block model for rough electrode-electrolyte interface.

It is now quite apparent from Eqs. 14 and 16 that there is no univeral relation between η and d except the following inequality

$$\eta \leq 3 - d \,. \tag{17}$$

The Cantor dust models can be randomized, namely that the branching numbers and scaling ratios can be treated as random numbers with well-defined averages.[15] It has been found that randomization does not change the conclusions deduced for regular fractals, so it seems that the regular fractal models are more realistic than they may appear.

Bates and coworkers have put the various theoretical predictions to test by creating a set of rough electrodes and measuring their fractal dimensions and CPA exponents.[20,21] The electrodes were prepared from platinum and low resistance (0.007 Ω-cm) silicon with and without a thin gold coating. Pieces of platinum and silicon of $\simeq$ 1 cm^2 were polished using various grades of emery paper and alumina powder. When viewed at low magnification, the resulting surface finish had a definite lay, indicating that the roughness consisted of parallel grooves. The height profiles of the plates were measured using a profilometer equipped with an 0.5 μm diamond stylus. This instrument has a height resolution of 0.005 μm and a minimum horizontal step size of 0.04 μm. Several profiles were measured for each plate perpendicular to the lay at different positions near the center section, which was used in the electrical measurements. A typical profile is shown in Fig. 9. The profiles were analyzed by assuming that they are fractional Brownian curves, i.e. self-affine fractals whose scales in horizontal and vertical directions are related by $a_z = a_x^H$, where $H \leq 1$. To test of this hypothesis one calculates the structure function defined by

$$S(\delta) = \langle \Delta z^2(\delta) \rangle \,, \tag{18}$$

where $\Delta z(\delta) = z(x+\delta) - z(x)$ and $\langle \ \rangle$ denotes an average in the horizontal direction. If the profile is fractional Brownian, the structure function should be related to δ by[19]

$$S(\delta) \propto \delta^{2H} \,. \tag{19}$$

The data for two such profiles are shown in Fig. 10. It can be seen that the power-law relation in Eq. 19 is well satisfied for over one decade of δ in the submicron region, giving $H \simeq 0.9$ for both profiles. The range of length scale is limited by the resolution of the profilometer. The fractal dimension of the surface is related to H by $d = 3 - H$,[19] and this gives $d \simeq 2.1$. For reasons not yet fully understood, natural rough surfaces of all kinds, including the surfaces of the continents on Earth, have fractal dimensions around d = 2.1–2.2. After the profile measurements the plates were fixed onto glass substrates and wires were attached for impedance measurements.

The CPA exponents were determined by measuring the impedances with these electrodes immersed in 0.1 M H_2SO_4 solution. There is no theory for η for fractional Brownian surfaces, so we can only make qualitative comparison between experiment

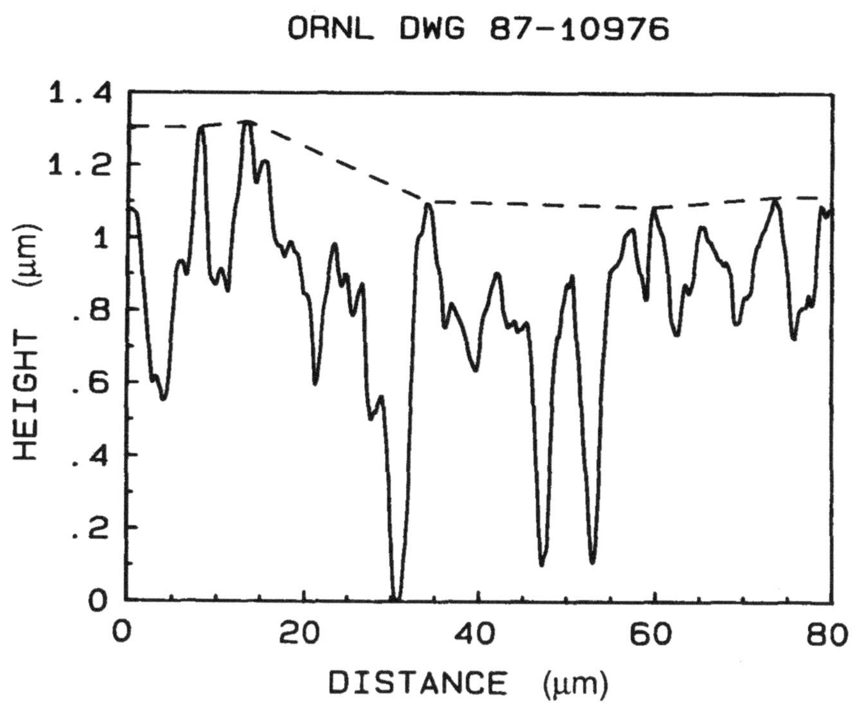

Fig. 9. Typical profile of a rough electrode created by polishing a solid surface with emery paper or alumina powder.

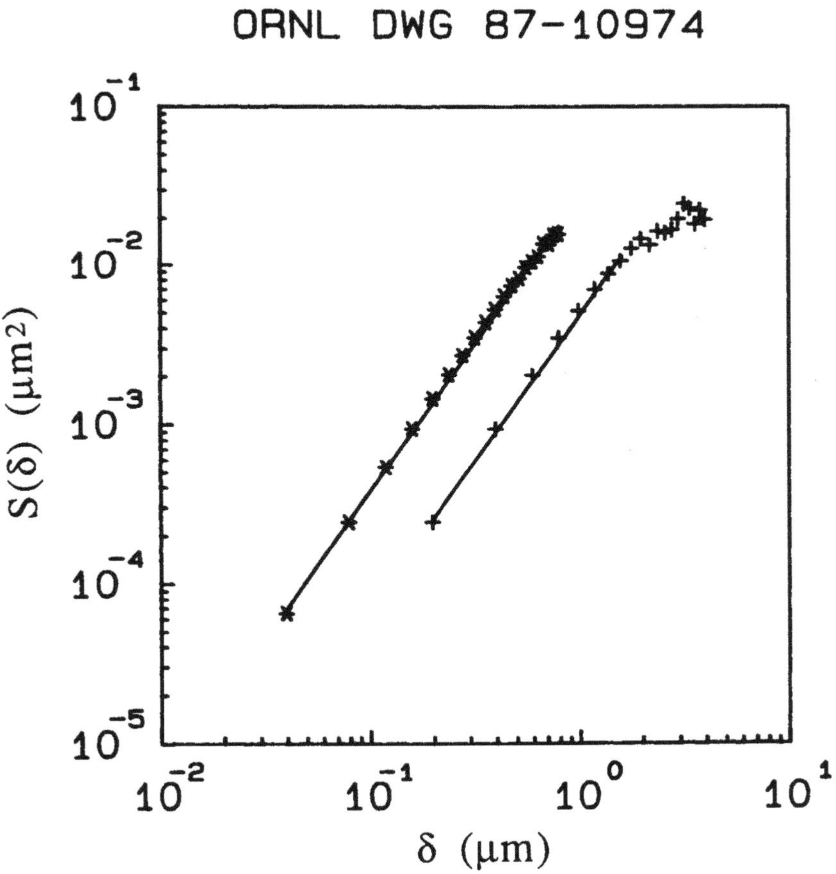

Fig. 10. The height-to-height correlation function of rough electrode profiles showing that they are well represented by fractional Brownian curves.

and theories. Since d is very close to 2, we can write $d = 2+\epsilon$, where $\epsilon = 1-H \ll 1$, both Eq. 3 and Eq. 8 give

$$\eta \simeq 1 - \epsilon = H \ . \tag{20}$$

We present the results of this study in an η versus H plot in Fig. 11. Each dot on the graph represents the result for one electrode, and the dashed line across the graph represents Eq. 20. The labels beside the points indicate how the electrode was prepared. For instance, Au/Si/600 was a gold-plated Si surface ground with 600 grit paper, and Pt/0.05 was a Pt surface polished with 0.05 μm alumina powder, etc. One can see clearly that there are points with almost equal H but very different η, and those with nearly equal η but very different H. Further examination reveals that Si and Pt surfaces form two distinct groups, i.e. the Si electrodes, whether coated with Au or not, have higher η than Pt electrodes. There is also a distinction in the shape of the protrusions in the surface, that hard Si surfaces tend to have more rounded protrusions than soft Pt surfaces. Therefore, the shape of the protrusion seems to play a strong role in determining the CPA exponent. Notice that the data are not inconsistent with Eq. 17, which can be written as $\eta \leq H$. On the other hand, Cantor dust type models are incapable of shedding light on the role of surface shape, because there is no clear way to relate the sharpness of the surface protrusions to any of the scaling parameters in the models. Further progress toward an analytical solution of the problem seems to be extremely difficult.

Bates and Chu reasoned that, under weak electric fields, the diffusion of ions in the electrolyte toward the electrode surface can be mapped onto the random walk problem, and the latter can be simulated by the Monte Carlo method. Assuming that a constant voltage is turned on across a CPE at $t = 0$, it follows that the current will have a time dependence given by $t^{-\eta}$ and the total charge given by

$$Q(t) \propto t^{1-\eta} \ . \tag{21}$$

The principle of the simulation is as follows. Random walkers are released from the counter electrode far away, and are allowed to migrate in the space between the electrodes. If a walker reaches the electrode, the migration time is recorded and another walker is released. A walker is discarded whenever it moves out of the space between the electrodes. By repeating this process a sufficient number of times, a set of data for $Q(t)$ is accumulated and the value of η can be extracted. The authors used the measured profiles for the electrode, but to save computer time, they argued that a line joining a few of the highest points of the profile, the dotted line in Fig. 9, is a good approximation for an equipotential surface so that random walkers can be released at random points on this line. While this approximation requires justification, the authors obtained values of η very close to the experimental values. The distinction between Si and Pt surfaces is clearly reflected in the results. There is one difficulty with the simulation procedure, however. As reported by the authors, they obtained $\eta \simeq 0.9$ for the cosine profile and 0.8 for the sawtooth profile. The sawtooth profile has been studied by Springer and Raistrick by numerically solving the Laplace equation for the potential and

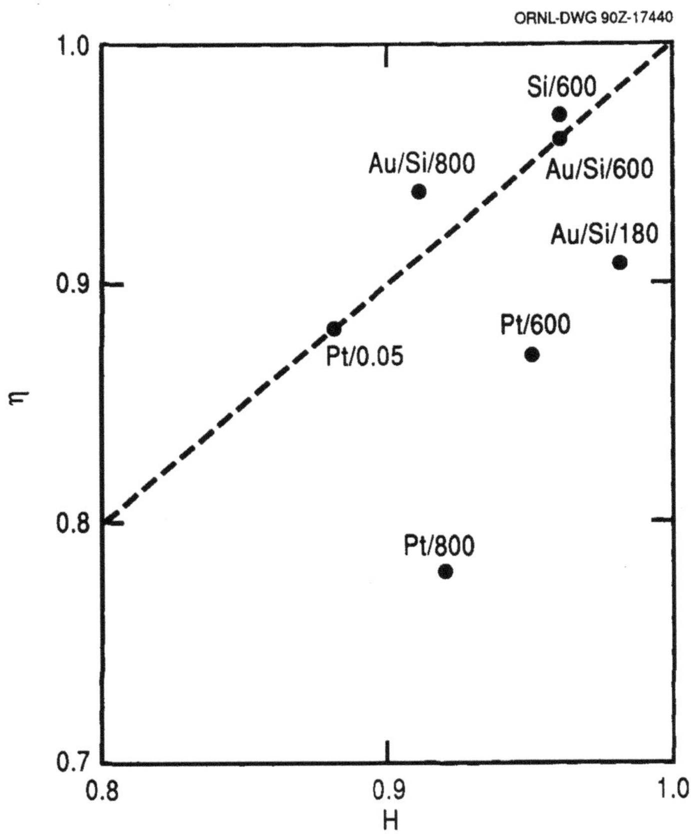

Fig. 11. The η versus H plot for several actual electrodes. The dotted line is the theoretical relation $\eta = H$.

current distributions.[22] The calculation gives $\eta = 1/2$ at low frequencies, indicating classical diffusion, and at high frequencies the real part of the impedance is constant and the imaginary part is inversely proportional to frequency. There is no nontrivial CPA behavior anywhere. It seems that the Monte Carlo simulation method should be used with caution, and more work is needed to verify the findings reported in Ref. 21.

3. Porous Electrodes

Porous electrodes are characterized by having open channels of all sizes extending into the bulk of the electrodes. One of the first models for such systems was constructed by reversing the sides of the electrode and electrolyte in the Cantor bar model in Fig. 5.[23] The black side is now the electrode, and the white side is a collection of channels which are filled with electrolyte. The electrical properties of each channel can be calculated by modeling it as a finite RC transmission line with a capacitance load at the end. The impedance of the entire system is found by combining the channels in parallel. Let the width of the widest channel be W, then in the next stage there are 2 channels of width W/a. In the third stage there are 2^2 channels of widths W/a^2, etc. It is convenient to consider a structure with $(M+1)$ stages and subsequently take the limit of infinite M. The length of the widest channel is Mh, where h is the height of each stage of the Cantor bar. Then the lengths of the next two stages are $(M-1)h$ and $(M-2)h$ respectively. Consider a channel of length $L_m = mh$. The input admittance of the channel is[24]

$$Y_m = \frac{\alpha_m}{r_m}\frac{r_m + \alpha_m Z_L(m)\tanh(\alpha_m L_m)}{r_m \tanh(\alpha_m L_m) + \alpha_m Z_L(m)}, \tag{22}$$

where $\alpha_m = (j\omega r_m c_m)^{1/2}$, r_m and c_m are the resistance and capacitance per unit length respectively, $Z_L(m)$ is the impedance of the end capacitance. For the widest channel, $m = M$, we set $r_M = r$ and $c_M = c$. Then the scaling relation between the channels gives $r_m = a^{M-m}r$, $c_m = c$, $\alpha_m = \alpha a^{(M-m)/2}$, $\alpha = (j\omega rc)^{1/2}$, $Z_L(m) = a^{M-m}Z_L$, and $Z_L = 1/j\omega C'$, with C' being the interfacial capacitance of the flat end of the widest channel. The input admittance of the entire structure is

$$Y_{M+1}(\omega) = \sum_{m=1}^{M} 2\Big(\frac{2}{a^{1/2}}\Big)^{(M-m)}\frac{\alpha}{r}\Big(\frac{ra^{-(M-m)/2} + \alpha Z_L \tanh(\alpha hma^{(M-m)/2})}{ra^{-(M-m)/2}\tanh(\alpha hma^{(M-m)/2}) + \alpha Z_L}\Big). \tag{23}$$

The quantity of interest is the low-frequency behavior of the total admittance in the limit of $M \to \infty$. A few simplifying approximations are now apparent. In the limit of large M and m the load impedance is unimportant because the line is much longer than the wavelength. Furthermore, the lengths of the mth stage and the $(m+1)$th stage are approximately equal. This allows us to deduce the following frequency scaling relation

$$Y(\frac{\omega}{a}) = \frac{2}{a}Y(\omega), \tag{24}$$

where $Y(\omega)$ is the limit of Eq. 23 when $M \to \infty$. The above equation is satisfied by $Y(\omega) \propto (j\omega)^\eta$, where

$$\eta = 1 + \ln 2/\ln a = 3 - d \ , \tag{25}$$

and d is the fractal dimension of the interface. The series in Eq. 23 converges in the limit of infinite M provided that $2 < a^{1/2}$. Recall that $Y(\omega)$ is the inverse of the impedance $Z(\omega)$, we have shown that the inverse Cantor bar is a CPE.

The inverse Cantor bar suggests another model, which consists of long channels bounded by simple curves such that there are $1, N, N^2, ...$ channels with cross sections scaled by 1, $a^{-2}, a^{-4}, ...$ One can show that as long as $a < N < a^{3/2}$, the set of transmission lines is a CPE with $\eta = 2 - \ln N/\ln a$.[23] The area of the pores is a fractal with dimension $d = 1 + \ln N/\ln a$. Thus the relation $\eta = 3 - d$ is obeyed. Randomizing the pore numbers and pore sizes does not affect the relation between η and d.

The physical origin of the CPE in the multi-channel models is substantially the same as in the branching models. High frequency signals tend to follow the wide, low impedance channels while low frequency signals are forced to seek out the more numerous, narrow, and high impedance channels. The shift of the effective current paths with frequency plus the geometric scaling of the paths lead to the power-law frequency dependence of the total impedance.

The geometry of porous solids is not limited to long and non-crossing channels. Sapoval and his coworkers have made an exhaustive study of many soluble geometries.[25] They obtained the same result for the multi-channel models as discusssed above. Their results for several sphere upon sphere models are particularly interesting. Just like the Cantor dust models, these can also be mapped onto series-parallel branching networks, and analyses of their input impedance again demonstrate that η is not simply related to d. There are fractal surfaces with no CPA behavior and CPE surfaces which are not fractal. Geometrical details are important, but it is not yet clear to what extent one can apply these understandings to naturally occuring porous materials.

In principle, there is another frequency regime of interest, at sufficiently low frequencies where the diffusion length in the electrolyte is comparable to the size of the electrode. The electric response is controled by the diffusion process, and even ideal flat electrodes have the CPA exponent $\eta = 1/2$. It was shown by Pajkossy and Nyikos[26] and by Sapoval *et al.*[25] that for a fractal electrode of dimension d the following simple relation exists:

$$\eta = (d - 1)/2 \ . \tag{26}$$

In the Faradaic regime where electrochemical reaction takes place, one expects different values of η for the same fractal interface. This problem has been discussed by Le Méhauté[27] and by Sapoval *et al.*[25]

4. Theory of Bulk CPE

To appreciate the mystery of the bulk CPA behavior in ionic conductors, we first discuss the classical theory of the ac response of such a material. Ionic conductors are a special kind of ionic solids in which one species of ions, usually the cation, can migrate within the lattice. At sufficiently high temperatures vacancies are created by thermal activation. Electrical conduction becomes possible when an electric field drives mobile ions into vacancies nearby. In doing so they leave new vacancies behind and other ions can move into them. One can equally well describe the conduction process as vacancy diffusion. Because the paths of vacancy migration tend to zigzag through the crystal, there is an ohmic resistance associated with this process. Also, the collection of ions of opposite charges can be polarized by an electric field, creating a capacitive element. The resistive and capacitive elements, R and C, are in parallel, so that the input impedance has the expression

$$Z(\omega) = \frac{R}{1 + j\omega RC} \, . \tag{27}$$

By separating the real and imaginary parts, one finds that the real part is frequency independent for $\omega \ll (RC)^{-1}$ and falls off like ω^{-2} for $\omega \gg (RC)^{-1}$, while the imaginary part behaves like ω and ω^{-1} respectively in these two limits. Therefore, the CPA behavior as observed experimentally is outside the scope of the classical theory. Attempts have been made to model the CPA behavior by invoking a continuous distribution of activation energies,[1,4] but this type of effort does nothing to advance our understanding of the origin of the CPE.

The bulk CPA behavior is but one manifestation of the universal $1/f$ fluctuation seen in many systems, such as sun spot activities, traffic flow, the flow of sand in an hour glass, and the flow of electric current through a resistor.[28] In recent years we have seen accelerated progress toward better understanding of the $1/f$ fluctuation, and there is widely held believe that the phenomenon is akin to what is taking place at the critical temperature of second order phase transition, i.e. that systems that undergo $1/f$ fluctuations have no natural length scale, and thus are self-similar under scale transformations. We will show that this so-called 'self-organized criticality' concept may also be a good starting point for elucidating the bulk CPA problem.[29–31]

Let us consider a simple, one-dimension model for ionic conductor, shown in Fig. 12. The larger ions define the lattice while the smaller ions are mobile. The classical picture for ionic conduction through vacancy migration is depicted in Fig. 12(a). In Fig. 12(b) we consider a new mode of conduction, that when driven by an electric field, it is possible to find two ions in the same interstitial space temporarily. These two repel each other so strongly that they both move out of that interstitial position and cause two nearby interstitials to be doubly occupied, as shown in Fig. 12(c). Thus, the relaxation of one doubly occupied site creates two doubly occupied sites. In the next step, the relaxation of the two doubly occupied sites results in three doubly occupied sites, etc. As the process continues more doubly occupied sites are

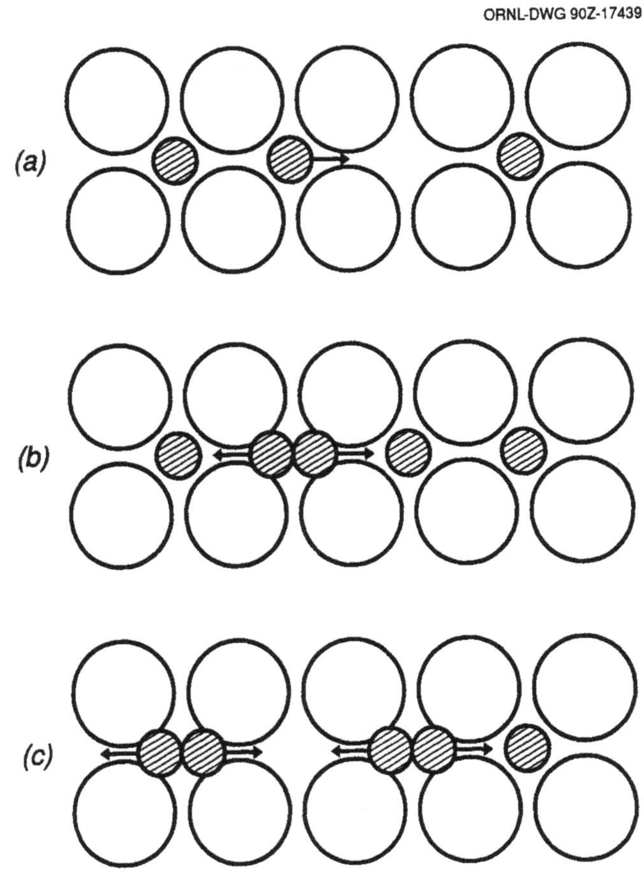

Fig. 12. Simple model for ionic conduction in fast ion conductors. (a) A schematic representation of vacancy diffusion. (b) A doubly occupied lattice position in which the two ions experience strong mutual repulsion. (c) Both ions move out of the initial position into neighboring positions on opposite sides, causing two positions to be doubly occupied.

created at every new step until a boundary site is reached, at which point one ion is ejected out of the solid. Afterwards, the system heals itself by slowly returning to singly occupied sites. This mode of conduction is governed by a set of dynamical rules. Let q_i be the charge at the mobile ion site labelled by i, then the site is stable if $q_i \leq 1$ and unstable otherwise. For an unstable site the charge redistributes according to

$$q_i \rightarrow q_i - 2,$$

$$q_{i\pm 1} \rightarrow q_{i\pm 1} + 1 . \tag{28}$$

If there are more than one unstable site at any moment, they are relaxed simultaneously. The charge is allowed to flow out at the boundary so that the entire system will eventually return to a stable state. In this way the new conduction mechanism maps exactly onto the sand pile problem first studied by Bak *et al.*[29−31]

In higher dimensions there are more than two nearest neighbors in any lattice of interstitial positions. Two ions in the same interstitial site can move in many possible ways into two neighboring positions. This problem has not yet been studied. Our present knowledge of higher dimensional systems comes from a slight variation of the model as follows. Consider a regular lattice in a D-dimensional space whose lattice points are $\mathbf{R}_l$. The lattice points represent the sublattice of the mobile ions, and $q(\mathbf{R}_l)$ is the charge at the site $\mathbf{R}_l$, which has m nearest neighbors. The site is stable if $q(\mathbf{R}_l) \leq m$ and unstable otherwise. For an unstable site the dynamical rules of charge redistribution are

$$q(\mathbf{R}_l) \rightarrow q(\mathbf{R}_l) - m,$$

$$q(\mathbf{R}_l + \delta_i) \rightarrow q(\mathbf{R}_l + \delta_i) + 1 . \tag{29}$$

In the above equation δ_i for $i = 1, 2, ..., m$ denote the set of vectors linking a lattice point to its nearest neighbors. In the event that many sites are unstable, they all relax simultaneously. The charge is allowed to flow out of the boundary.

To simulate a conduction process one should inject extra charges at one end of the system and collect charges flowing out of the other end. The time sequence of charge flow, when properly Fourier analyzed, would give information on the frequency response of the system. This has not yet been done. What has been done is a simulation of the diffusion problem in which every lattice point has a higher than stable amount of charge.[32] The square lattice ($m = 4$) enclosed in a square boundary has been studied in detail. In Fig. 13 we show the final charge distribution starting from the unstable situation where $q = 5$ at every site. Only one quadrant of the square is shown, where the center of the square is at the lower left corner. The values of q at various points are coded as follows: 1 by squares, 2 by diamonds, 3 by open circles, and 4 by open space. There is a complex pattern of nonuniform charge distribution which consists of a set of nesting heart shapes along the diagonal. This pattern is created during the process when excess charges flow out of the boundary. The ratios of the linear measures of the nearest pair of heart shapes are rational approximations of the irrational quantity $(3 - \sqrt{5})/2$, which is

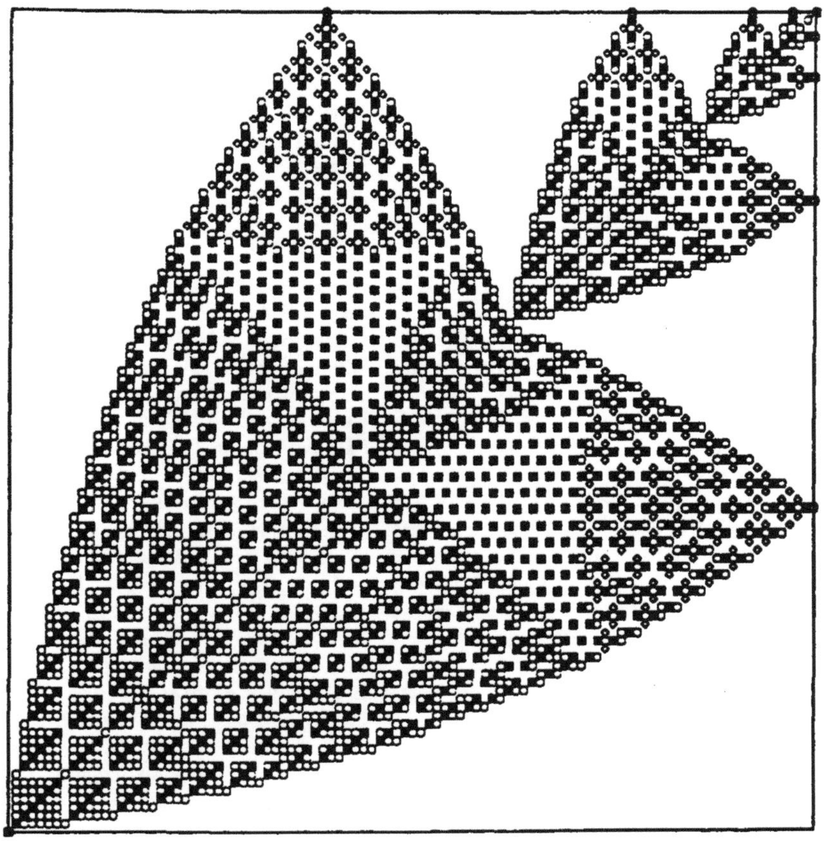

Fig. 13. Final charge distribution of a square lattice with initial charge q = 5 at every point. Only one quadrant is shown, with the center of the square at the lower left corner. The values of q are coded as follows: 1 by squares, 2 by diamonds, 3 by open circles, and 4 open space.

the classic Golden Section ratio. What is relevant to the conduction problem is the time sequence of the total outward flow, which is plotted in Fig. 14 for a 801×801 square system. The time is measured by the number of iteration steps. At the beginning the total flow follows the $t^{1/2}$ law, indicating that the flow obeys the law of classical diffusion. Later in the process the flow slows down and the total charge takes on a different power-law time dependence with the exponent 0.40, as shown in the insert of Fig. 14. This behavior is known as anomalous diffusion, and we will show that this behavior is linked to CPA behavior in the conduction process.

In classical diffusion the total charge $Q(t)$ is proportional to $t^{1/2}$. When such a system is driven by a constant external voltage, the current follows the Ohm's law and the total charge $Q(t) \propto t$. In the anomalous situation the diffusive process obeys the law $Q(t) \propto t^{\nu}$. By using the scaling argument, we infer that when driven by a constant voltage the charge will be given by $Q(t) \propto t^{2\nu}$. It follows from Eq. 21 that the CPA exponent is given by

$$\eta = 1 - 2\nu \ . \tag{30}$$

For the square system under discussion, we find $\nu = 0.4$ so that $\eta = 0.2$. We have also carried out the simulation for the cubic lattice bounded by cubic boundaries and obtained $\nu = 0.25 \pm 0.01$ and $\eta = 0.50 \pm 0.02$. It is not clear at present whether this number depends on the lattice structure. Perhaps all one can conclude from this study is that the new conduction mechanism may lead to a CPE with $\eta \simeq 0.5$. This number is small compared with $\eta = 0.77$ for amorphous Li_3PO_4. Crystalline conductors tends to give smaller values of η, typically around 0.6.[1]

Examination of the intermediate patterns during the charge relaxation process reveals an intimate connection between anomalous diffusion and the growth of the complex pattern in the final state. In Figs. 15–18 we show four intermediate patterns for the square system with starting charge $q = 5$ at every site. Very early in the relaxation process the uniform distribution quickly breaks down into plane waves with crests of $q = 6$ and troughs of $q = 4$ as shown in Fig. 15. A rudimentary heart shape develops at the corner, and it interacts with the waves and changes the crest and trough values to 5 and 3. A thin strip of $q = 4$ separates the front of the waves from the boundary. As time progresses the wave front moves toward the boundary, and upon reaching it ejects a pulse of charge. Afterwards the wave front retreats toward the center until it reaches the tip of the heart shape. At this point the motion reverses itself and the shape grows by one grid point. At a later time, shown in Fig. 16, the shape has grown in size, and smaller shapes have formed following the wake of the leading one. The growth of the nonuniform region causes the width of the wave front to shrink and the distance traversed by the wave front to lengthen. Consequently, every time the wave front returns to the boundary, it ejects a pulse that is two units less than the previous one, and the interval between successive pulses increases by two units. The total flow after t iterations is given by

$$Q(t) = \sum_n [N - 2(n-1)]\delta_{t,n^2-n+1} \simeq Nt^{1/2} \tag{31}$$

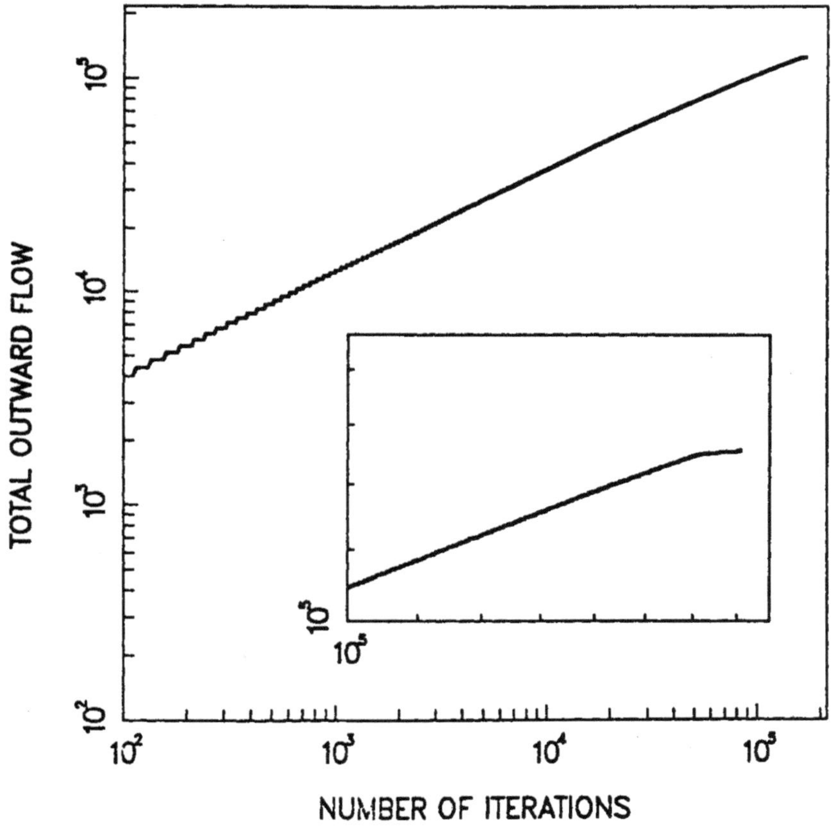

Fig. 14. The charge distribution after 160 iterations of the square grid with initial charge $q = 5$ at every point. In addition to the four codes used in Fig. 13, we present 5 by pluses, 6 by crosses, and 7 and 8 by asterisks. It requires 10,738 iterations to reach the final pattern in Fig. 13.

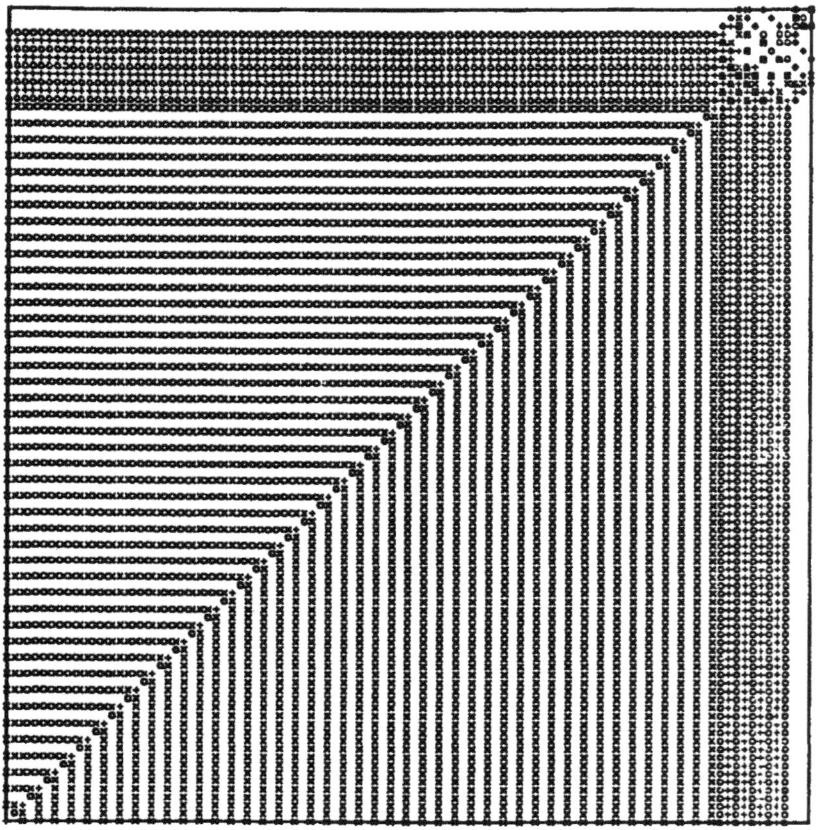

Fig. 15. Same as Fig. 14 but after 1320 iteration steps.

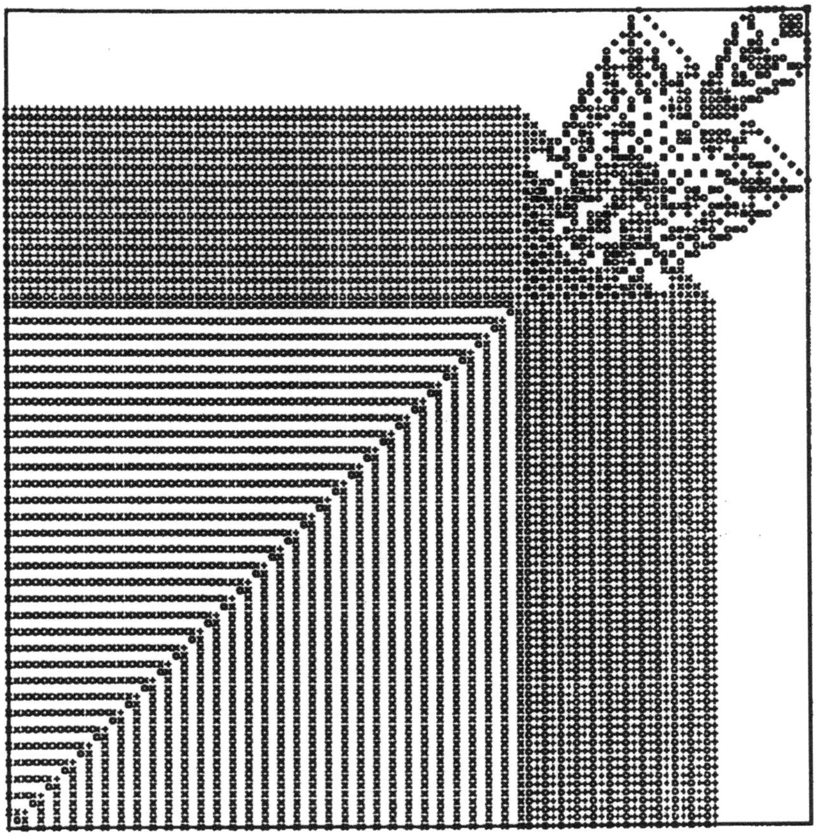

Fig. 16. Same as Fig. 14 but after 2608 iteration steps. Waves of anomalous diffusion begins to show clearly.

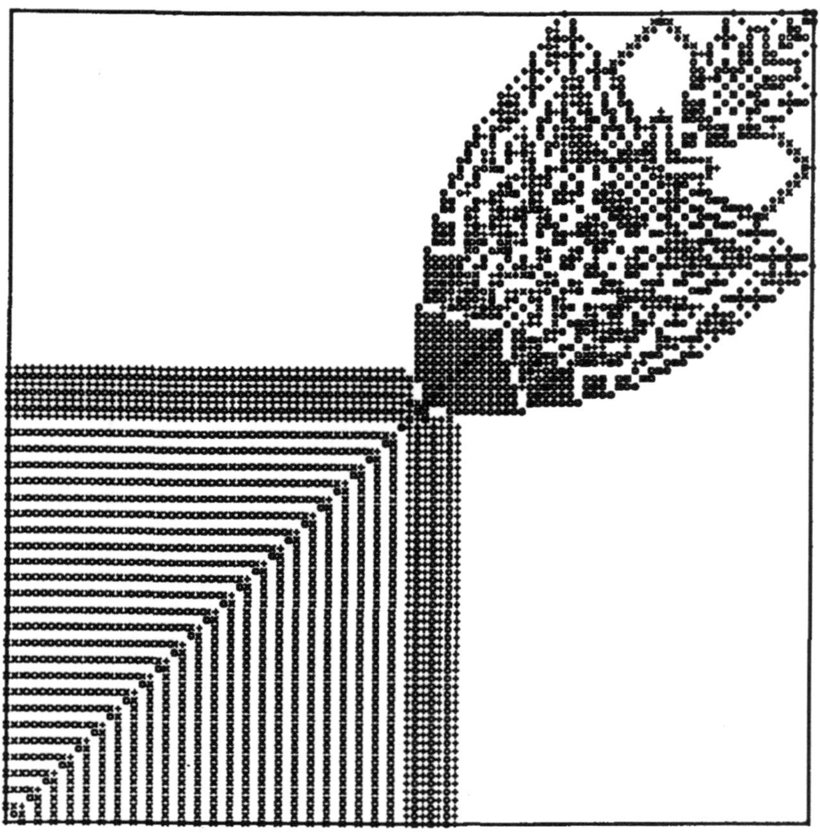

Fig. 17. Same as Fig. 14 but after 9690 iteration steps. Anomalous component of charge diffusion now dominates the flow.

Fig. 18. Total outflow of charge $Q(t)$ for a 801×801 square lattice as a function of t = iteration steps. The insert shows the flow during the last half of the relaxation process, showing the $t^{0.4}$ law of anomalous diffusion.

for $1 \ll t \ll N$. In the above equation N is the size of the boundary, n sums over the sequence of pulses, and the Kronecker δ indicates that the nth pulse emerges at $t = n^2 - n + 1$. This contribution can be identified as the classical component of the diffusion process.

As the nonuniform region grows to about one-half of its final size, shown in Fig. 17, we can clearly see waves flowing toward the boundary in the region between some of the shapes. The wave fronts make 45° angles with the boundary. The excess charge in these waves must first flow through the leading shape, and are ejected at the boundary sporadically. We identify this as the emerging anomalous diffusion component. Near the end, shown in Fig. 18, the classical diffusion component becomes insignificant and the anomalous component dominates the total flow. This sequence of events demonstrates that the anomalous behavior is due to charge flow through the nonuniform part of the intermediate distribution. The flow is self-similar in time because the nonuniform region is self-similar in space. We conclude that the anomalous diffusion behavior comes from charge flow through a self-generated self-similar distribution of vacancies.

Much more work needs to be done to clarify many important issues, chief among which is a direct simulation of conduction process based on a more realistic model whose margin of stability is $q = 2$. We believe that this line of inquiry will yield a deeper understanding of the origin of bulk CPE.

5. Acknowledgments

The author wishes to thank J. B. Bates for suggesting the series of investigation of interface and bulk CPE, to Theodore Kaplan for a critical reading of the manuscript, and to A. A. Kornyshev for his invitation to discuss our work at the Working Party on Electrochemistry held at the International Centre of Theoretical Physics in Trieste, Italy in 1990. This review grew out of the notes of that lecture. The research was supported by the U.S. Department of Energy, Division of Basic Energy Research, under Contract No. DE-AC05-84OR21400 with Martin Marietta Energy Systems, Inc.

References

1. J. B. Bates, *unpublished data.*
2. Bates, J. C. Wang, and Y. T. Chu, *J. Non-Crystalline Solids* (to appear); and references cited therein.
3. J. C. Wang and J. B. Bates, *Solid State Ionics* **18&19**, 224 (1986).
4. J. C. Wang and J. B. Bates, (to be published).
5. I. Wolfe, Phys. Rev. **27**, 755 (1926).
6. P. H. Bottelberghs, in *Solid State Electrolytes* (Academic Press, New York, 1978) pp. 145–172.

7. R. de Levie, *Electrochimica Acta* **10**, 113 (1965).
8. P. H. Bottelberghs and G. H. J. Broers, *J. Electroanal. Chem.* **67**, 155 (1976).
9. R. D. Armstrong and R. A. Burnham, *J. Electroanal. Chem.* **72**, 257 (1976).
10. J. B. Bates, J. C. Wang and Y. T. Chu, *Solid States Ionics* **18&19**, 1045 (1986).
11. A. Le Méhauté and G. Crepy, *Solid State Ionics* **9&10**, 17 (1983).
12. B. B. Mandelbrot, *Fractals: Forms, Chance and Dimension* (Freeman, San Francisco, 1977).
13. B. B. Mandelbrot, *The Fractal Geometry of Nature* (Freeman, San Francisco, 1983).
14. S. H. Liu, *Phys. Rev. Lett.* **55**, 529 (1985).
15. T. Kaplan and L. J. Gray, *Phys. Rev.* **B32**, 7360 (1985).
16. L. Nyikos and T. Pajkossy, *Electrochimica Acta* **30**, 1533 (1985).
17. M. Keddam and H. Takenouti, *Electrochimica Acta* **33**, 445 (1988).
18. T. Kaplan, L. J. Gray and S. H. Liu, *Phys. Rev.* **B35**, 5379 (1987).
19. B. B. Mandelbrot, in *Proceedings Sixth International Symposium on Fractals in Physics* (North Holland, Amsterdam, 1987).
20. J. B. Bates, Y. T. Chu and W. T. Stribling, *Phys. Rev. Lett.* **60**, 627 (1988).
21. J. B. Bates and Y. T. Chu, *Solid State Ionics* **28–30**, 1388 (1988).
22. T. E. Springer and I. D. Raistrick, *Calculation of Electrochemical Impedances by Solution of the Laplace Equation using Green's Boundary Integral Method*, abstract presented at the First International Symposium on Electrochemical Impedance Spectroscopy, Bombannes, France, 22–26 May, 1989.
23. T. Kaplan, S. H. Liu and L. J. Gray, *Phys. Rev.* **B34**, 4870 (1986).
24. R. B. Adler, L. J. Chu and R. M. Fano, *Electromagnetic Energy Transmission and Radiation* (Wiley, New York, 1965).
25. B. Sapoval and J.-N. Chazalviel and J. Peyriére, *Phys. Rev.* **A38**, 5867 (1988).
26. T. Pajkossy and L. Nyikos, *Electrochimica Acta* **34**, 171 (1989).
27. A. Le Méhauté, in *The Fractal Approach to Heterogeneous Chemistry* (Wiley, New York, 1989), pp. 311–328.
28. For a review of $1/f$ noise, see W. H. Press, *Commun. Mod. Phys.* **C7**, 103 (1978).
29. P. Bak, C. Tang and K. Wiesenfeld, *Phys. Rev. Lett.* **59**, 381 (1987).
30. C. Tang and P. Bak, *Phys. Rev. Lett.* **60**, 2347 (1988).
31. P. Bak, C. Tang and K. Wiesenfeld, *Phys. Rev.* **A38**, 364 (1988).
32. S. H. Liu, T. Kaplan and L. J. Gray, *Phys. Rev.* **A42**, 3207 (1990).

CHARGE TRANSFER

CHARGE TRANSFER IN CHEMISTRY AND BIOLOGY: A SELECTIVE OVERVIEW

Hans E.M. Christensen, Lars S. Conrad,
Jan M. Hammerstad-Pedersen and Jens Ulstrup
Chemistry Department A, Building 207
The Technical University of Denmark
2800 Lyngby, Denmark

ABSTRACT

We provide an overview of some elements of condensed phase charge transfer of importance to chemical and biological electron and atom transfer phenomena currently investigated. Issues addressed specifically include: solvent fluctuations incorporating also vibrational and spatial dielectric dispersion; molecular modes; features of electrochemical processes; the electronic factor in long-range electron transfer; relations between direct or long-range superexchange electron transfer and optical single-and multi-photon absorption and scattering processes; and finally, stochastic approaches to chemical rate theory.

We illustrate the theoretical formalism by two case examples from inorganic biochemistry, namely first, the adjacent and remote binding site electron transfer modes in native and modified forms of the blue photosynthetic copper protein plastocyanin, and secondly, stochastic effects in peptide and ester hydrolysis of the proteolytic zinc enzyme carboxypeptidase A.

1. Introduction and Some Historical Notes

Electron transfer (ET) is both the simplest conceivable chemical process and at the same time an extremely widespread phenomenon[1-17]. ET is the central molecular event in electrochemistry, in inorganic and organic redox chemistry, in the photophysics of excited molecules, and in the key steps of photosynthesis and respiration. Close physical and formal relations between ET and a range of other electronic processes can also be discerned. These include: (a) thermal and optical electronic processes in solid-state impurities; this recognition has in fact been crucial to the development of quantum mechanical ET theory. (b) Electronic relaxation in large molecules. (c) Electronic energy transfer. (d) Optical absorption and emission in molecules. (e) Impurity atom diffusion and dipolar relaxation in solids, and (f) proton and heavier atom group transfer as encountered in acid-base and nucleophilic substitution processes, hydrolytic enzyme processes etc. In still broader terms ET exhibits features common to wider classes of phenomena in larger-size systems, such as surface-modified electrochemical ET, quantum well structures, tunnel junctions including scanning tunnel microscopic systems, and hopping, band and soliton conduction in polymeric, semiconductor, and low-dimensional materials.

Charge transfer in condensed phases has now been a thriving and prosperous scientific research area for more than 40 years. Around the bicentenary of great electrochemical discoveries it is appropriate to recall a few, more recent achievements of historical importance in charge transfer science. Charge transfer in solution is much harder to detect than electrochemical charge transfer where the current directly reveals the electron flow, and charge transfer investigations for homogeneous systems began much more recently. The earliest recognition of charge transfer and tunnelling in chemical systems in the 1920's were associated with atom transfer (proton or hydrogen) and emerged out of the acid-base concept[18] and spectroscopic investigations (NH_3 inversion)[19,20]. Electron transfer as an electrochemical event was clearly recognized by Gurney in 1931[21]. Views of solution ET appeared in the 1930's[22] but the first definite outer sphere ET system report seems to have been on the process[23-25]

$$^{*}MnO_4^{-} + MnO_4^{2-} \rightleftharpoons {}^{*}MnO_4^{2-} + MnO_4^{-} \tag{1.1}$$

This report appeared in 1940 but was not disclosed until much later, and the recognition of the ubiquitous prevalence of ET as a key event in solution reactions had to wait for the more general availability of radioisotopes and the conceptual distinction between intrinsic ET and nucleophilic substitution lability which evolved out of coordination chemistry in the late 1940's and early 1950's.

At about this time a parallel development had grown out of a different area, namely electronic processes in solid materials. By the early 1950's the nature of radiative and radiationless electronic relaxation, electronic autolocalization etc. including the crucial role of electronic-vibrational coupling was understood and quantum mechanical theories for these phenomena available[26-30]. The recognition of intermolecular ET as a tunnel event, and the joint availability of accumulated redox ET data and a theoretical framework for electronic-vibrational transitions rooted in solid-state physics are what paved the way for modern ET science. This was most strikingly expressed in the first quantitative ET theories of R.A. Marcus (1956)[31] and of Levich and Dogonadze (1959)[32]. The former, rooted in classical transition state theory was the first to incorporate correctly solvent dynamics into chemical ET theory. The latter was the first theory to view chemical ET as a quantum mechanical transition between separate electronic Born-Oppenheimer states. These two theories mark the initiation of modern chemical charge transfer science, the achievements of which have become the subject of a voluminous literature. Broad groups of examples are given in refs. 5 and 10, and many more can be found in the reviews in refs. 1 - 4, 6 - 9 and 11 - 17. Noting this, we also now leave the historical view and proceed to a physical and formal approach to elements of charge transfer in solution. This overview will be selective, biased towards some areas in which recent interesting progress has been witnessed, and cover views on the continuous solvent environment, the electronic factor, and relations between optical and thermal multi-level ET processes. We shall conclude by a brief discussion of two specific case studies which illustrate important features of this overall view of charge transfer in solution.

2. The Solvent Environment and Friction-free or Damped Motion on Potential Surfaces Spanned by Collective Nuclear Coordinates

Solute molecular charge distributions are coupled, usually strongly, to a virtually infinite number of fluctuating environmental nuclear modes. This endows a fundamentally different character on the electronic processes compared with gas phase reactions. At first the reactants are infinitely remote from one another, with stationary, and widely different donor and acceptor levels. They then initiate migration towards a reactive encounter complex, overcoming on the way both diffusional barriers and inter-reactant forces. Once located in the reactive precursor complex, other solute-solvent fluctuational forces (fig.1) unimportant at infinite distances, now at the same time lead the electronic energy levels to converge (fig.2.), and the electronic orbitals to billow out spatially towards the reaction partner. This, and the perturbative interaction, are exactly what cause the nonradiative electronic process to occur. Once transferred, the electron is trapped by renewed nuclear fluctuations towards the product equilibrium configuration and ultimately, to infinite separation of the product molecules. All these events, i.e. the handling of the solvent, the nature of the fluctuational nuclear motion, the solute-solvent interaction, the environmental effects on the spatial features of the electron clouds, and

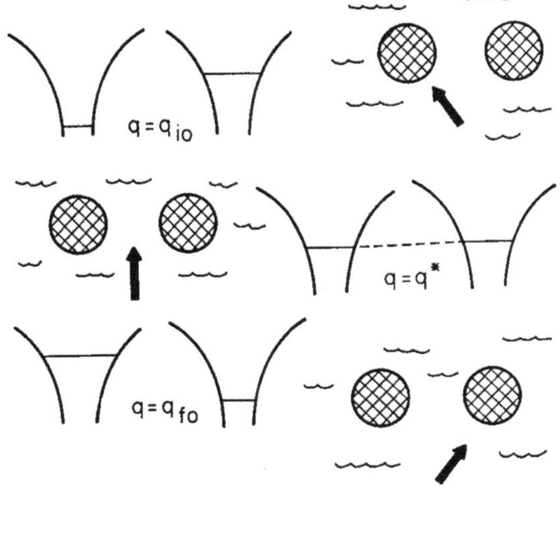

Fig.1. The sequence of events, qualitatively, in ET between two solute molecules. The energy diagrams refer to the electronic potential wells at the reactants' equilibrium nuclear configuration (top), at the transition state (middle), and at the products' equilibrium configuration (bottom) cf. fig.2. The bold arrow represents the inertial polarization from the "excess" electron transferred.

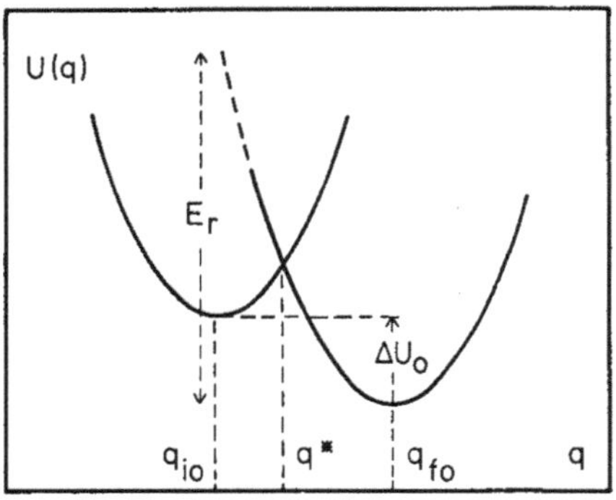

Fig.2. One-dimensional nuclear potential (free) energy diagram with indications of quantities which appear in the rate constant.

the transition probability itself, are to be covered by a comprehensive charge transfer theory, features of which we now proceed to discuss.

2.1 Transitions in one-dimensional nuclear potentials

The nature of the transition can be illustrated first by the simple one-dimensional view in fig.2. The transition probability is[1-4,7-9,12-16,33]

$$W_{fi} = \frac{2\pi}{\hbar}(V_{eff})^2 A v_v \sum_w S_{vw}\delta(E_{iv}-E_{fw}) \tag{2.1}$$

where we have invoked the Condon approximation to separate the electronic and nuclear motion. The former is contained in the electronic transition matrix element, V_{eff}, which couples the two electronic states; the latter by the Franck Condon overlap factor of the v'th and w'th nuclear wave function in the reactants' and products' state, respectively. E_{iv} and E_{fw} are the corresponding vibrational energies, and the rate constant is averaged and summed with respect to the initial and final vibrational manifolds, respectively. $2\pi\hbar$ is finally Planck's constant.

In the classical high-temperature limit W_{fi} takes the following, in many contexts quite adequate form[30]

$$W_{fi}^{cl} = \frac{2\pi}{\hbar}(V_{eff})^2 (Z_i^{cl})^{-1}\int_{-\infty}^{\infty} dq \exp\left[\frac{-U_i(q)}{k_BT}\right]\delta[U_i(q)-U_f(q)]$$

$$Z_i^{cl} = \int_{-\infty}^{\infty} dq \exp\left[\frac{-U_i(q)}{k_BT}\right] \tag{2.2}$$

What we notice is that non-vanishing contributions to the q-integral only appear at the intersection between $U_i(q)$ and $U_f(q)$ and that the nuclear motion contributes fully to the activation free energy. Also, the classical motion ensures a smooth activation energy variation when the surfaces are shifted. Eq.(2.2) can be converted to the form

$$W_{fi}^{cl} = \frac{2\pi}{\hbar}(V_{eff})^2 (Z_i^{cl})^{-1}\left[\frac{\partial}{\partial q}\left|U_i(q^*)-U_f(q^*)\right|\right]^{-1}\exp\left[\frac{-U_i(q^*)}{k_B T}\right] \tag{2.3}$$

since

$$\delta(\varphi(x)) = \sum_i \delta(x-x_i)/(\partial\varphi/\partial x)_{x=x_i}$$

where $\varphi(x)$ is any function and x_i the roots of the equation $\varphi(x) = 0$. In particular, for a pair of harmonic oscillators of frequency ω and horizontal displacement q_o

$$U_i(q) = \frac{1}{2}\hbar\omega q^2 \; ; \qquad U_f = \frac{1}{2}\hbar\omega(q-q_0)^2 + \Delta U_0 \tag{2.4}$$

eq.(2.3) takes the following form known in many contexts

$$W_{fi}^{cl} = \left(\frac{\pi}{k_B T\hbar^2 E_r}\right)^{\frac{1}{2}}(V_{eff})^2\exp\left[-\frac{(E_r+\Delta U_0)^2}{4E_r k_B T}\right]; \qquad E_r = \frac{1}{2}\hbar\omega q_0^2 \tag{2.5}$$

or

$$W_{fi}^{cl} = \kappa\frac{\omega}{2\pi}\exp\left[-\frac{(E_r+\Delta U_0)^2}{4E_r k_B T}\right]; \qquad \kappa = \sqrt{\frac{4\pi^3}{(\hbar\omega)^2 k_B T E_r}}\,(V_{eff})^2 \tag{2.6}$$

where κ is the transmission coefficient.

Difficulties arise when we attempt to use a discrete-mode view to quantum mechanical oscillator systems. The rate constant is again given by eq.(2.1), i.e. it now has a divergent form and vanishes for all energies except when $E_{iv} = E_{fw}$. The only way that this can be overcome is by assuming that even though the process is dominated by a single nuclear mode, residual coupling to a surrounding continuum nevertheless remains. Eq.(2.1) can then be integrated over a narrow energy interval of about $\hbar\omega$ to remove the divergency represented by the delta function. For the nuclear part of eq.(2.1), A_N, we thus obtain[32,34]

$$A_N = \sum_{v,w} A_N^{vw}; \qquad A_N^{vw} = \exp\left(\frac{-E_{iv}}{k_B T}\right) S_{vw}\,\delta(E_{iv}-E_{fw})$$
$$A_N^{vw} \rightarrow \int_{E_{iv}-\frac{1}{2}\hbar\omega}^{E_{fw}+\frac{1}{2}\hbar\omega} A_N^{vw}\frac{1}{\hbar\omega}d(\hbar\omega) \approx \frac{1}{\hbar\omega}\exp\left(\frac{-E_{iv}}{k_b T}\right) S_{vp(v)}; \qquad p(v) \approx v + |\Delta U_0|/\hbar\omega \tag{2.7}$$

Common usage of single-mode models rests on this "coarse graining" when applied to the temperature and reaction free energy dependence of chemical rate constants. For example, the single-harmonic oscillator approximation gives the following compact rate constant form[32,34]

$$W_{fi} = \frac{2\pi}{\hbar\omega}(V_{eff})^2 \exp\left[-\frac{E_r}{\hbar\omega}(2\bar{v}+1)\right]\left[\frac{(\bar{v}+1)}{\bar{v}}\right]^{|\Delta U_0|/2\hbar\omega} I_{|\Delta U_0|/\hbar\omega}\left\{2\frac{E_r}{\hbar\omega}\left[\bar{v}(\bar{v}+1)\right]^{\frac{1}{2}}\right\} \tag{2.8}$$

where

$$I_n(z) = I_{|n|}(z) = \left(\frac{1}{2}z\right)^n \sum_{k=0}^{\infty} \frac{(z/2)^{2k}}{k!\,(n+k)!}; \quad \bar{v} = \left[\exp\left(\frac{\hbar\omega}{k_B T}\right) - 1\right]^{-1} \tag{2.9}$$

In particular, at high T eq.(2.8) coincides with eq.(2.5). The low-T limiting form is

$$W_{fi}^{low-T} = \frac{2\pi}{\hbar^2\omega}(V_{eff})^2 \exp\left(\frac{-E_r}{\hbar\omega}\right)\frac{(E_r/\hbar\omega)^{|\Delta U_0|/\hbar\omega}}{(|\Delta U_0|/\hbar\omega)!} \tag{2.10}$$

i.e. the rate constant is independent of T. This reflects directly that nuclear motion here proceeds by tunnelling rather than activated barrier passage as at high T.

2.2 Coupling to a continuous dispersive solvent and the "effective" solute-solvent Hamiltonian

Once the importance of the solute-solvent coupling and vibrational dispersion is recognized, direct incorporation of the real solvent dynamics can be approached. This motion is collective and reflected in the appropriate susceptibility spectrum which represents the distribution of the different solvent modes as they respond to an external time dependent field. The field response relations emerge transparently for charge transfer when the field is the electric induction $\vec{D}(\vec{r})$. The solvent nuclear dynamics, coupled linearly to the field, are then constituted by the components of the inertial polarization $\vec{P}(\vec{r})$, the response of which is determined by the frequency dependent dielectric permittivity $\varepsilon(\omega)$.[1-4,13,15]

This view apparently leads to normal mode-like behaviour of the collective solvent motion although damping features are in fact included. Linearity of the solvent thus implies that for each polarization component, ν, the correlations[1-3,35]

$$\vec{P}_\nu(\vec{r}) = \frac{c_\nu}{4\pi}\vec{D}(\vec{r}); \qquad \vec{P}(\vec{r}) = \sum_\nu \vec{P}_\nu(\vec{r}) \tag{2.11}$$

are valid, where c_ν characterizes the response of the ν'th mode. Eq.(2.11) also leads to a free solvent Hamiltonian function of the form[1-3,35]

$$H_{sol} = \int d\vec{r} \sum_\nu \frac{2\pi}{c_\nu}\left\{\left[\vec{P}_\nu(\vec{r})\right]^2 + \left[\frac{1}{\omega_\nu}\dot{\vec{P}}_\nu(\vec{r})\right]^2\right\} \tag{2.12}$$

where ω_ν is the frequency of $\vec{P}_\nu(\vec{r})$. It is also conceivable that the quadratic form of eq.(2.12) enables a transformation by which the continuum Hamiltonian function can be recast into an equivalent form in which the solvent is represented as a continuous set of harmonic oscillators. This form is, for each electronic state s (= i,f)

$$H^s_{sol} = \frac{1}{2}\sum_\nu \hbar\omega_\nu\left[(q_\nu - q^s_{\nu 0})^2 + p_\nu^2\right] \tag{2.13}$$

where p_ν is the momentum of the oscillator spanned by the coordinate q_ν and $q^s_{\nu 0}$ the equilibrium value of q_ν in the electronic state s. The Hamiltonian operator, corresponding to eq.(2.13) is, finally

$$\hat{H}^s_{sol} = \frac{1}{2}\sum_\nu \hbar\omega_\nu\left[(q_\nu - q^s_{\nu 0})^2 - \frac{\partial^2}{\partial q_\nu^2}\right] \tag{2.14}$$

Eq.(2.14) is often denoted as the "effective" Hamiltonian of the solvent medium.[1-3,16,35] This appellation refers to its "operational" nature in the following sense:

(A) The procedure used to reach eqs.(2.12)-(2.14) rests on a simple view of the solvent nuclear motion, such as frictionless harmonic vibrations. This requires, however, that the solvent is approached macroscopically and collective solvent coordinate variables introduced. Individual solvent molecular motion is of a different, stochastic or dissipative, and strongly anharmonic character.

(B) Construction of the effective Hamiltonian follows a well defined and precise procedure. Its great virtue is that it leads to a set of displaced harmonic solvent coordinate-spanned potential surfaces in the reactants' and products' electronic states, with tractable nuclear wave functions etc. When these are combined with the rate constant in eq.(2.1), useful analytical rate constant forms are obtained for wide temperature and reaction free energies.

(C) A consequence of the procedure is that the c_ν- and $q_{\nu o}$-quantities are macroscopic system properties and all energies involved free energies. Moreover, we need to disclose a direct relation between these quantities and observable properties of the solute-solvent system. We do so by noting that after computational treatment the final rate constants do not contain the instantaneous coordinate values q_ν but only their equilibrium shifts, $q_{\nu o} \equiv q^f_{\nu o} - q^i_{\nu o}$. $q_{\nu o}$ appears in the solvent reorganization free energy $E_s = \sum_\nu E_{s\nu} = ½\sum_\nu \hbar\omega_\nu q^2_{\nu o}$, and always in the combination[36]

$$\sum_\nu q^2_{\nu 0} f(\omega_\nu, T) = \sum_\nu \left(\frac{E_{s\nu}}{\hbar\omega_\nu}\right) f(\omega_\nu, T)$$

where $f(\omega_\nu,T)$ is a function which characterizes the vibrational excitation of the ν'th mode at the temperature T. In the continuum limit the combination is

$$\frac{1}{2}\int_0^\infty d\omega\,\rho(\omega) f(\omega,T) \equiv \int_0^\infty d\omega \left(\frac{\mathscr{E}_s}{\hbar\omega}\right) f(\omega,T)$$

where $\rho(\omega)$ is the oscillator density. We can therefore provide a link to measurable or calculable macroscopic field and dielectric properties, maintaining the important vibrational dispersion effects, as long as we can derive relations between these properties and the quantities $q_{\nu o}$ and c_ν.

(D) This point can be illustrated in the following simple way. The overall macroscopic solvent reorganization free energy is

$$E_s = \int d\vec{r}\, \Delta\vec{P}(\vec{r})\, d\Delta\vec{D}_0(\vec{r}) = \left(\frac{1}{8\pi}\sum_\nu c_\nu\right) \int d\vec{r}\,[\Delta\vec{D}_0(\vec{r})]^2 \qquad (2.15)$$

where $\Delta\vec{D}_0(\vec{r})$ and $\Delta\vec{P}_0(\vec{r})$ are the induction and polarization changes which accompany the charge transfer process.

By the Kramers-Krönig relation[37]

$$c = \sum_\nu c_\nu = \frac{2}{\pi}\int_0^\infty \frac{d\omega}{\omega}\frac{\mathrm{Im}\,\varepsilon(\omega)}{|\varepsilon(\omega)|^2} = \varepsilon_o^{-1} - \varepsilon_s^{-1} \qquad (2.16)$$

where ε_o is the optical and ε_s the static dielectric constant, we obtain

$$E_s = \int_0^\infty d\omega\,\mathscr{E}_s(\omega) \qquad (2.17)$$

$$\mathscr{E}_s(\omega) = \frac{2E_s}{\pi\omega c}\frac{\mathrm{Im}\,\varepsilon(\omega)}{|\varepsilon(\omega)|^2} = \frac{1}{4\pi^2}\frac{1}{\omega}\frac{\mathrm{Im}\,\varepsilon(\omega)}{|\varepsilon(\omega)|^2}\int d\vec{r}[\Delta\vec{D}_0(\vec{r})]^2$$

The link between the "operational" q_o-variables and the macroscopic solvent properties can therefore be summarized by eq.(2.16) and by

$$E_s = \frac{1}{2}\sum_\nu \hbar\omega_\nu q_{\nu 0}^2 \rightarrow \frac{1}{2}\int_0^\infty \hbar\omega\,\rho(\omega) q_{\omega 0}^2 = \int_0^\infty d\omega\,\mathscr{E}_s(\omega) \qquad (2.18)$$

$$\mathscr{E}_s(\omega) = \frac{1}{2}\hbar\omega\,\rho(\omega) q_{\omega 0}^2 = \frac{1}{4\pi^2\hbar}\frac{1}{\omega^2}\frac{\mathrm{Im}\,\varepsilon(\omega)}{|\varepsilon(\omega)|^2}\int d\vec{r}\,[\Delta\vec{D}_0(\vec{r})]^2$$

2.3 The rate constant in the dynamic, "friction-free" limit

The rate constant in the vibrationally dispersive medium is[3,13,38]

$$W_{fi} = \frac{\beta}{\hbar}(V_{eff})^2\left\{\frac{1}{2\pi}\left|\Phi_s''(\Theta^*)\right|\right\}^{-\frac{1}{2}} \exp\left[-\beta\Theta^*\Delta G_0 - \Phi_s(\Theta^*)\right]; \quad \beta = (k_B T)^{-1}$$

$$\Phi_s(\Theta) = \frac{1}{2\pi^2\hbar^2}\int d\vec{r}\left[\Delta\vec{D}_0(\vec{r})\right]^2 \int_0^\infty \frac{d\omega}{\omega}\frac{\mathrm{Im}\varepsilon(\omega)}{|\varepsilon(\omega)|^2}\frac{\mathrm{sh}\frac{1}{2}\beta\hbar\omega\Theta\,\mathrm{sh}\frac{1}{2}\beta\hbar\omega(1-\Theta)}{\mathrm{sh}\frac{1}{2}\beta\hbar\omega} \tag{2.19}$$

where $\Theta^* = -k_B T d\ln W_{fi}/d\Delta G_o$ is the transfer coefficient, determined by

$$\frac{\Delta G_0}{2E_s} = \frac{1}{\pi c}\int_0^\infty \frac{d\omega}{\omega}\frac{\mathrm{Im}\varepsilon(\omega)}{|\varepsilon(\omega)|^2}\frac{\mathrm{sh}\frac{1}{2}\beta\hbar\omega(2\Theta^*-1)}{\mathrm{sh}\frac{1}{2}\beta\hbar\omega} \tag{2.20}$$

and we now use the reaction free energy as the vertical potential surface displacement.

Eqs.(2.19) and (2.20) apply to a broad range of temperatures and reaction free energies, and for rather arbitrary dielectric permittivity functions, provided that the dispersion is broad and the solute-solvent interaction strong. Dispersion functions often used practically are the Debye

$$\frac{\mathrm{Im}\varepsilon(\omega)}{|\varepsilon(\omega)|^2} = c\frac{\omega\tau_D}{1+\omega^2\tau_D^2} \tag{2.21}$$

and resonance

$$\frac{\mathrm{Im}\varepsilon(\omega)}{|\varepsilon(\omega)|^2} = c_0\frac{\Gamma_R^2+\omega_R^2\tau_R^2}{2\omega_R}\left[\frac{\Gamma_R}{\Gamma_R^2+(\omega-\omega_R)^2} - \frac{\Gamma_R}{\Gamma_R^2+(\omega+\omega_R)^2}\right] \tag{2.22}$$

functions[39], τ_D^{-1} being the Debye and ω_R the resonance frequency, while Γ_R is the damping coefficient. In particular, in the high-T limit W_{fi} from eq.(2.19) takes the same form as in eqs.(2.5)-(2.7) except that by eqs.(2.15)-(2.17), (2.21) and (2.22) the solvent dielectric dispersion is now specifically incorporated in E_s and the frequency ω replaced by an "effective" frequency, ω_{eff}, which represents all the relaxational solvent features, i.e.

$$W^{high-T} = \kappa\frac{\omega_{eff}}{2\pi}\exp\left[-\frac{(E_s+\Delta G_0)^2}{4E_s k_B T}\right]$$

$$\kappa = \sqrt{\frac{4\pi^3}{(\hbar\omega_{eff})^2 k_B T E_s}}; \quad E_s = \frac{1}{8\pi}(\varepsilon_o^{-1}-\varepsilon_s^{-1})\int d\vec{r}\left[\Delta\vec{D}_0(\vec{r})\right]^2 \tag{2.23}$$

W_{fi} is independent of T in the low-T range, reflecting that nuclear mode reorganization now proceeds by nuclear tunnelling rather than thermal activation (cf. eqs.(2.8)-(2.10).

The formalism above incorporates dielectric, thermodynamic and electronic properties of the reacting charge transfer systems. In this form it is broadly applicable to electron and light atom group transfer processes over wide ranges of temperature and reaction free energy. The formalism is representative of contemporary charge transfer theory, which now involves a variety of methods, including vibrational density matrix[1-3,15,40], trajectory[41,42] and path integral[14,43], molecular dynamics[44,45], and quantum chemical approaches[46] both to the reaction dynamics and to the various system parts. Charge transfer theory can be extended to include many important physical effects not directly considered above. Prominently among these are:

2.3.1 *Inclusion of local intramolecular nuclear modes*

A set of local molecular coordinates $\{Q_c\}$ can be included in rather arbitrary form. A straightforward way is by construction of the potential surfaces

$$U_s(\{q_v\},\{Q_c\}) = \frac{1}{2}\sum_v \hbar\omega_v (q_v - q_{v0}^s)^2 + U_{sc}(\{Q_c\}) \tag{2.24}$$

and subsequent generalization of the rate constant to the form[47-49]

$$W_{fi}(\Delta G_0) = (Z_i^{qu})^{-1}\sum_v\sum_w \exp\left(\frac{-\varepsilon_{iv}}{k_BT}\right) S_{vw} W_{fi}^{sol}(\Delta G_0 + \varepsilon_{fw} - \varepsilon_{iw})$$

$$Z_i^{qu} = \sum_v \exp\left(\frac{-\varepsilon_{iv}}{k_BT}\right) \tag{2.25}$$

U_{sc} is the local nuclear potential, ε_{iv} and ε_{fw} the vibrational spectra in the reactants' and products' state, respectively, and S_{vw} the local mode Franck Condon overlap factor coupling the v'th and w'th vibrational levels in the reactants' and products' state, respectively. W_{fi}^{sol} is finally the rate constant in the absence of local mode motion and given by eq.(2.19) or similar equations.

In this form frequency changes, anharmonicity, normal mode mixing etc. are included. Local modes represent the molecular spectroscopic and structural properties of the molecules. Nuclear tunnelling is also primarily associated with these modes as their quantum mechanical freezing begins at higher temperatures than for most solvent modes, and the behaviour of the tunnel branch of free energy relations extending into the inverted free energy range is primarily determined by these modes.

2.3.2 *Adiabatic and diabatic processes*

We consider separately the electronic factor, V_{eff}, of the rate constant (eqs.(2.5), (2.7) and (2.19)) in section 4. In the limit of strong inter-reactant reactions the transmission coefficient, κ, approaches unity, and the rate constant must here be approached differently, resting either on accurate solution of the quantum mechanical equations of motion or on summation of the perturbation series. We refer to the literature for details of this[1,2,15,50,51] but note that the adiabatic limit is important for the observation of solvent relaxational features as the rate constant here depends directly on the solvent vibrational spectrum via ω_{eff} (cf. eq.(2.23)). In section 3 we shall return to the adiabatic limit in the context of stochastic views on condensed phase chemical processes.

2.3.3 *Electrochemical processes, electrocatalysis and tunnel processes at film-covered electrodes*

As noted above, the importance of electron tunnelling in electrochemical electron transfer was recognized early. Contemporary charge transfer theory applies to electrochemical processes by combining the molecular and solvent views with the electronic and molecular structure of the interfacial electrode-electrolyte region.[1,4,52] In this way the electronic structure of the metal or semiconductor electrode and dynamic double layer effects come to constitute part of a comprehensive charge transfer theory and a frame for current-voltage relations, solvent and double layer effects in electrochemical kinetics etc.

Extensions to electrocatalysis aroused by electronic structural effects of surface adsorbates[53,54], and to elastic, resonance, and inelastic tunnel modes at surface modified metal electrodes[55,56] have been reported more recently. In addition to "normal" overpotential dependence of the nuclear activation free energy, these processes can also exhibit strong characteristic, and sometimes resonance-like overpotential variation of the electronic tunnel factor. We return to some of these features in section 4.

2.3.4 *Specific models for the system parts*

Construction of models for the various system parts, to be used along with the general frame of charge transfer theory, is a crucial part of electron and atom group transfer science. We have encountered specific representations for local nuclear modes above, and in section 4 we discuss some approaches to the electronic factor. Models for the third major system part, the solvent reorganization free energy (and the inter-reactant interactions), have rested abundantly on continuum theory which has, on the other hand reached a degree of sophistication well beyond that of structure-less continuum models. Effects included are geometric features, dielectric properties including mutual reactant polarizability[57], while solvent structural effects are covered in considerable detail by the vibrational and spatial dispersion of the dielectric permittivity function in nonlocal dielectric theory.[58]

Continuum models are likely to be of importance also in future activities in charge transfer theory due to their analytical transparency, the composite nature of the systems which can be handled, and their often striking numerical accuracy. Molecular models including fairly large assemblies of solvent molecules and based on molecular dynamics and other trajectory methods have, on the other hand also recently appeared.[44,45] They have revealed new intriguing stochastic features of the trajectories leading the system from the reactants' to the products' potential surface, and once suitably accurate molecular interactions have been found, they will represent clues also to the values of the solvent parameters of the system.

2.3.5 Extension to other physical properties

We noted in section 1 that one of the early achievements of charge transfer theory was the recognition of close formal and physical relations between radiative absorption or emission and nonradiative processes in local condensed phase "impurity" states.[30] Such correlations are illustrated in fig.3 which indicates that wherever a thermal process occurs, optical processes are in principle feasible as well. The quantitative correlations rest primarily on transition probabilities, but can be converted to the observables of the two kind of processes.[13,59] For optical processes these are the molecular absorption coefficient at the light frequency ν, $\kappa_{fi}(\nu)$, the emission spectral intensity distribution $I_{em}(\nu)$, and the excited state fluorescence lifetime τ_f, all of which are therefore covered by the comprehensive frames of charge transfer theory.

The three quantities are[13]

$$\kappa_{fi}(\nu) = \frac{2\pi}{3c\ln 10} h\nu W_{fi}^{abs}(\nu)$$

$$I_{em}(\nu) = \frac{4}{\pi c^3} \nu^4 \Delta\nu h^2 W_{fi}^{em}(\nu); \qquad \tau_f^{-1} = \frac{4h}{\pi c^3} \int_0^\infty \nu^3 W_{if}^{em}(\nu) d\nu \tag{2.26}$$

where $W_{fi}^{abs}(\nu)$ and $W_{fi}^{em}(\nu)$ are the transition probability per unit time for absorption or

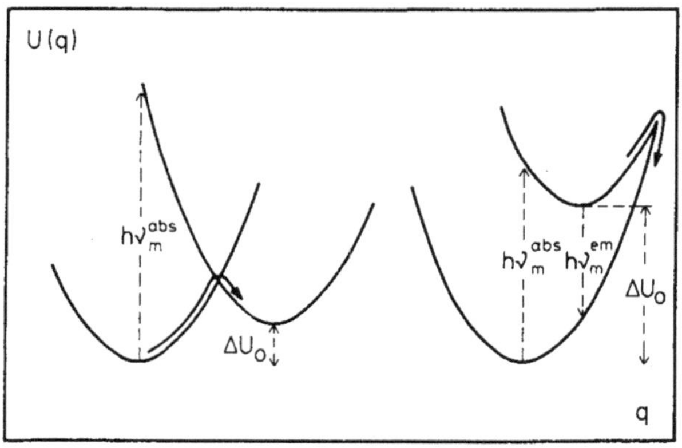

Fig.3. Schematic illustration of optical (dashed arrows) and thermal (fully drawn arrows) electronic transitions in two pairs of potential surfaces. "m" indicates the absorption or emission maximum.

emission, respectively, of a photon of frequency ν, c is the light velocity in the medium, and $\Delta\nu$ a narrow frequency interval, regarded as an instrument property.

With reference to fig.3, it is conceivable that the following simple correlations between $\kappa_{fi}(\nu)$, $I_{em}(\nu)$, and τ_f, and the thermal rate constant for the same pair of potential surfaces, $W_{fi}^{th}(\Delta G_o)$ must be broadly valid[13]

$$\kappa_{fi}(\nu) = \frac{2\pi}{3c\ln 10} h\nu \left[\frac{M_{fi}(q_{opt}^*)}{V_{eff}(q_{th}^*)}\right]^2 W_{fi}^{th}(\Delta G_0 - h\nu)$$

$$I_{em}(\nu) = \frac{4}{\pi c^3}(h\nu)^2\nu^2\Delta\nu \left[\frac{M_{fi}(q_{opt}^*)}{V_{eff}(q_{th}^*)}\right]^2 W_{fi}^{th}(\Delta G_0 + h\nu) \qquad (2.27)$$

$$\tau_f^{-1} = \frac{4}{\pi c^3}\int_0^\infty d\nu (h\nu)\nu^2 \left\{\frac{M_{fi}\left[q_{opt}^*(\nu)\right]}{V_{eff}\left[q_{th}^*(\nu)\right]}\right\}^2 W_{fi}^{th}(\Delta G_0 + h\nu)$$

where M_{fi} is the optical transition dipole. The two processes are dominated by different nuclear configurations, q_{opt}^* and q_{th}^*, namely around the equilibrium configurations in the ground and excited states for radiative absorption and emission, respectively, and around the intersection for the thermal process.

Eqs.(2.26) and (2.27) provide the basis for detailed optical bandshape analysis, competition between radiative and nonradiative excited state decay etc. In section 4 we return to similar but extended correlations for three-level instead of two-level processes.

With these observations we also conclude our general overview and proceed to two other areas of charge transfer in condensed media which have attracted particular attention in recent years. These are stochastic approaches and the role of the electronic tunnel factor specifically in ET processes.

3. Stochastic Views of Environmental Motion in Charge Transfer

Solvent dynamics is incorporated in considerable detail in the formalism above. By the correlated solvent motion inherent in the effective Hamiltonian this extends to important facets of solvent relaxation. A simple view of this appears from the specific form of ω_{eff} for the resonance oscillator distribution in eq.(2.22), for which

$$\omega_{eff}^2 = \Gamma_R^2 + \omega_R^2 \qquad (3.1)$$

In the limit of narrow, weakly damped bands $\Gamma_R^2 \ll \omega_R^2$ and $\omega_{eff} \rightarrow \omega_R$, corresponding to friction-less motion. In the opposite limit $\omega_{eff} \rightarrow \omega_R$, and the motion is now determined by the frictional properties of the medium.

The view of vibrationally equilibrated normal mode-like solvent motion inherent in this formalism, however, conceals non-equilibrium diffusional, or stochastic element of the solvent motion along the reaction coordinate, This is presently being approached by molecular, many-body simulations[43-45]. As analytical theories which are also powerful tools in approaching the solvent "collectively", stochastic methods have, however, come to stand strongly in contemporary condensed phase chemical rate theory.

In stochastic theories system motion along a molecular or collective reaction coordinate X spanning the potential U(X) is endowed with Brownian-motion, random walk character, characterized by Langevin-type equations of motion[60]

$$\ddot{X} + \Gamma\dot{X} + \frac{\partial U(X)}{\partial X} = F(t) \tag{3.2}$$

where Γ is the friction or damping coefficient and f(t) a random force, i.e. $<f(t)f(0)> = 2\pi\Gamma k_B T\delta(t)$. This feature of chemical solution reactions was first recognized in quantitative terms by Kramers whose theory is still the most important conceptual frame for stochastic rate theory. Kramers' result is[61]

$$\frac{W_{fi}^{KR}}{W_{fi}^{TST}} = \left[1+\left(\frac{\Gamma}{2\omega_b}\right)^2\right]^{\frac{1}{2}} - \left(\frac{\Gamma}{2\omega_b}\right); \quad W_{fi}^{TST} = \frac{\omega_0}{2\pi}\exp\left(\frac{-\Delta G^{\ddagger}}{k_B T}\right) \tag{3.3}$$

where ω_b is the potential curvature at the barrier top, while the undamped transition state theory result, W_{fi}^{TST}, is determined by the curvature at the well bottom of the reactants' part of the potential surface, ω_o.

By eq.(3.3) friction thus leads to a "transmission coefficient" $\kappa^{KR} < 1$. $\kappa^{KR} \rightarrow 1$ for "weak" damping, i.e. for $\Gamma \ll \omega_b$. In the opposite limit of strong damping, eq.(3.3) shows that $\kappa^{KR} \rightarrow \omega_b/\Gamma$. This limit reflects the random tumbling of the system in the transition region and is equivalent to conversion of eq.(3.2) to the Smoluchowski equation for the probability distribution $P(X,\dot{X},t)$

$$\frac{\partial P}{\partial t} = \frac{\partial}{\partial X}\left[D\left(\frac{\partial P}{\partial X} + \frac{1}{k_B T}P\frac{\partial U}{\partial X}\right)\right]; \quad D = k_B T\,\Gamma^{-1} \tag{3.4}$$

and the stationary rate constant is directly determined by the damping feature in the pre-exponential factor.

Adoption of equations such as eqs.(3.1)-(3.4) raises several fundamental questions which have been addressed in a recent increasingly voluminous literature[62]. Stochastic methods originate from large-particle motion in a medium. The first question that arises is therefore the conditions under which stochastic views apply when the size of the randomly walking particle approaches that of the solvent molecules. Other questions are related to the nature of the friction coefficient. This quantity cannot rightaway be identified with "conventional" friction coefficients, although both essentially represent collision of solute particles with solvent molecules. Also, since real solvents are vibrationally dispersive with widely different relaxation times, this must also apply to the friction coefficient itself. This will necessarily give rise both to more entangled friction dependence than represented by eq.(3.3) (cf. also eq.(3.1)) and to multi-exponential kinetics in different time ranges. Finally, since both the friction coefficient in stochastic theories and the effective frequency in dynamic theories are "collective" quantities which represent to a large extent the same physical solute-solvent properties, there must be precise isomorphic features in the two approaches in the stationary time ranges.

We attend briefly to a few of these elements. The frequency dependent friction coefficient Γ(t) is the conceptual basis of the Grote-Hynes theory[63] and for many experimental investigations[64,65]. The fundamental result is that the rate constant can be given a similar form as Kramers' result

$$\frac{W_{fi}^{GH}}{W_{fi}^{TST}} = \left\{1 + \left[\frac{\Gamma(\omega_r)}{2\omega_b}\right]^2\right\}^{\frac{1}{2}} - \left[\frac{\Gamma(\omega_r)}{2\omega_b}\right] \tag{3.5}$$

where the reactive frequency, ω_r, is determined by

$$\Gamma(\omega_r) = \int_0^\infty dt \exp(-\omega_r t)\,\Gamma(t) \tag{3.6}$$

The point is then that only time ranges around ω_b are probed and passage of sharp barriers little affected by friction if ω_b significantly exceeds the maximum of the Γ(t) dispersion.

Both the correlation between dynamic and stochastic views, and multi-exponential behaviour can be illustrated in the following way. Suppose that we add a linear coupling term to a "reactive" mode Q_c embedded in a solvent heat bath[66,67]

$$U(Q_c,\{q_\nu\}) = \frac{1}{2}\hbar\Omega_c Q_c^2 + \frac{1}{2}\sum_\nu \hbar\omega_\nu q_\nu^2 + \hbar\sum_\nu \lambda_\nu Q_c q_\nu \tag{3.7}$$

Eq.(3.7) is meant to represent the barrier-separated reactants' and products' part of an overall adiabatic potential surface. Suppose further that the system is initially "prepared" in a given configuration, say $Q_c = 0$ (or other nonequilibrium configuration). The system equations of motion then lead to a probability distribution of the form[67]

$$P(Q_c,t) = \left[\frac{2\pi k_B T}{\hbar\Omega_c}(1 - e^{-\Gamma t})\right]^{-\frac{1}{2}} \exp\left\{\frac{-E_A}{k_B T\,[1 - \exp(-2\Gamma t)]}\right\}$$
$$E_A = \frac{1}{2}\hbar\Omega_c Q_c^2; \qquad \Gamma = \pi\sum_\nu \lambda_\nu^2\,\delta(\Omega_c - \omega_c) \tag{3.8}$$

where E_A is the energy at the barrier top and Γ the damping coefficient.

Eq.(3.8) obviously is not a rate constant. On the other hand, $P(Q_c,t)$ satisfies eq.(3.4) with the diffusion coefficient $D = (k_B T/\hbar\omega)\Gamma$. Also, the stationary diffusion current in the limit of large t gives the following rate constant[67]

$$W_{fi} = \Gamma\left[\frac{(E_r + \Delta U_0)}{\sqrt{4\pi k_B T E_r}}\right] \exp\left(-\frac{E_A}{k_B T}\right) \tag{3.9}$$

where E_r and ΔU_o have the same meaning as in section 2. Eqs.(3.7)-(3.9) represent a kind of link between dynamic and stochastic views, as the frictional damping is here related to a specific coupling term in the otherwise "ballistic" potential energy function. This view can be extended to a two-dimensional system in which additional multi-exponentiality arises in different time ranges depending on the damping features of each of the two modes.[67]

Among several important reports of stochastic effects in multi-mode systems we finally attend to a very recent one by H. Sumi[68], with a view of specific application in section 5. Following earlier work by Sumi and Marcus[69], a two-mode system consisting of a relaxational mode X of relaxation time τ_{rel} and a ballistic mode was considered. Time evolution was represented by the overdamped Smoluchowski limit

$$\frac{\partial P(X,t)}{\partial t} = D\frac{\partial}{\partial X}\left[\frac{\partial P}{\partial X}+\frac{1}{k_BT}\frac{\partial U(X)}{\partial X}\right]P(X,t) - k(X)P(X,t) \tag{3.10}$$

where k(X) is the rate constant, with reorganization along the non-relaxational mode, at given value of the relaxational coordinate X. This approach is adequate when the two relaxation times are well separated and in time ranges where the fast mode is fully equilibrated.

In the model of Sumi and Marcus broad ranges of "fast" and "slow" diffusion relative to the chemical rate constant, and of the relative coupling strengths in the two modes were considered, giving in general single-exponential time evolution only in certain limits of these system properties. Only single-exponential behaviour

$$P(X,t) = P(X)\exp(-k_s t) \tag{3.11}$$

was considered in ref.68, corresponding to a slow reaction, $k_s \ll \tau_{rel}^{-1}$ and the long-time limit where P(X,t) is independent of the initial conditions.

The key result of the steady-state solution of eq.(3.10) in the form of eq.(3.11) is that k_s can be written as[68]

$$k_s = \left(k_{eq}^{-1}+k_f^{-1}\right)^{-1} \tag{3.12}$$

k_{eq} is the vibrationally equilibrated rate constant. It can take any of the forms discussed in section 2, symbolized here by

$$k_{eq} = \nu_N \exp\left(\frac{-\Delta G^{\ddagger}}{k_BT}\right) \tag{3.13}$$

Importantly, ν_N extends to diabatic reactions and incorporates both the electronic transmission coefficient and nuclear tunnel factors. k_f includes the relaxational features and is found to take the following compact form

$$k_f \approx \tau_{rel}^{-\delta} \nu_N^{1-\delta} \exp\left(\frac{-\gamma \Delta G^{\ddagger}}{k_B T}\right); \qquad \gamma = \frac{E_S}{E_R} \tag{3.14}$$

where δ is a constant quantity between zero and unity. In the limit where E_S approaches the total reorganization free energy E_R, δ is approximately 2-(E_R/E_S), while $\delta \approx 0.6$ for $E_R \approx E_S$.

Eqs.(3.12)-(3.14) cover nicely the steady-state range from the purely equilibrated case $k_s \approx k_{eq}$ when $k_f \gg k_{eq}$ to the non-equilibrated opposite diffusive limit where $k_s \approx k_f$, and in the fractional power law form often observed[64,65]. We notice that the latter limit is important, for given location of the potential surfaces and $\Delta G^{\ddagger}$ when τ_{rel} is "small" and γ significant, i.e. when the fluctuational mode relaxation is slow and its reorganization constitutes a substantial part of the overall nuclear mode reorganization. We shall use eqs.(3.12)-(3.14) in section 5 in a case of specific system data analysis.

4. The Electronic Tunnel Factor

4.1 Formal perturbation schemes

Most of our attention in sections 2 and 3 has focused on the nuclear part of the charge transfer system. In this section we attend to elements of the electronic part, V_{eff}, and the charge transfer inducing perturbation. This is a composite quantity, as apparent from the following total system Hamiltonian form[70,71]

$$H = H_0 + H_{sol} + H_{mol} + T_e + V_{DA} + V_{eD} + V_{eA} + \sum_{c_D} V_{ec_d}^{loc} + \sum_{c_A} V_{ec_A}^{loc} + V_{ep}^{sol} + V_{Dp}^{sol} + V_{Ap}^{sol} \tag{4.1}$$

H_o, H_{sol} and H_{mol} are the Hamiltonians of the donor and acceptor core electrons including the inertia-less polarization, the undistorted inertial solvent, and the undistorted local modes, respectively. The following terms are the electronic kinetic energy, the screened interaction of the "excess electron" with the donor and acceptor core, the local donor and acceptor modes, and with the solvent. V_{DA} is finally the mutual donor-acceptor core interaction, while V_{Dp}^{sol} and V_{Ap}^{sol} are donor and acceptor core interactions with the inertial solvent modes.

Decomposition of H into a zero order part and a perturbation can be effected by several schemes. An "intuitively attractive" way is to introduce the reactants' and products' channel Hamiltonians, H_R and H_P, which include all terms suitably allocated to the reactants' and products' state separately. This poses no conceptual problems, except that the inertial polarization term, V_{ep}^{sol} must be decomposed into two terms arising from the donor state polarization in the absence of the acceptor, $V_{ep,D}^{sol}$, and vice-versa, i.e. $V_{ep}^{sol} = V_{ep,D}^{sol} + V_{ep,A}^{sol}$. The reactants' channel Hamiltonian is then

$$H_R = H_0 + H_{sol} + H_{mol} + T_e + V_{DA} + \sum_{c_D} v_{ec_D}^{loc} + V_{ep,D}^{sol} + V_{Ap}^{sol} + \left[V_{eA} + \sum_{c_A} V_{ec_A}^{loc} + V_{ep,A}^{sol} \right]_d \qquad (4.2)$$

while the perturbation, V_R, is

$$V_R = \left[V_{eA} + \sum_{c_A} V_{ec_A}^{loc} + V_{ep,A}^{sol} \right]_{n-d} \qquad (4.3)$$

with a similar decomposition of the Hamiltonian H_P.

The subscripts "d" and "n-d" refer to "diagonal" and "non-diagonal" parts of the interaction terms, respectively. This distinction is necessary since we wish to characterize the system time evolution by the separate reactants' and products' stationary electronic states. Normal perturbation schemes must therefore be modified to include this feature and the associated wave function nonorthogonality. The resulting form of the perturbation matrix elements which couple the reactants' (ψ_R) and products' (ψ_P) electronic wave functions, is

$$(V_R)_{n-d}^{RP} = \frac{1}{1 - S_{RP}^2} \left(V_R^{RP} - S_{RP} V_R^{RR} \right); \; V_R^{RP} = \langle \psi_P | V_R | \psi_R \rangle; \; V_R^{RR} = \langle \psi_R | V_R | \psi_R \rangle \qquad (4.4)$$

ψ_R is thus the eigenfunctions of the Hamiltonian in eq.(4.2), H_R, with an analogous relation between ψ_P and H_P.

$(V_{RP})_{n-d}$ in eq.(4.4), combined with eqs.(4.2) and (4.3) is formally what has to be inserted for V_{eff} in the rate constants in sections 2 and 3. Mutual approach of the donor and acceptor thus distorts the channel for infinitely separated molecules by the diagonal interaction terms, while only the nondiagonal part induces the transition. Moreover, the perturbation is not merely the direct interaction between the transferring electron and the donor or acceptor cores, i.e. V_{eA} or V_{eD}, but includes also interaction with the instantaneous molecular mode field and the inertial polarization in the region to which the electron is transferred. Normally only the first term is considered but the second term is important for example in tunnel spectroscopy, while the latter is important for neutral acceptor groups or in field electron emission into solutions where no acceptor is present.

4.2 *Analytical and numerical approaches to the electronic factor*

Once the perturbation scheme is constructed, a broad selection of models and computational methods for the electronic matrix element can be adopted. Reported procedures cover both simple analytical methods and comprehensive many-electron calculations on the *ab initio* level. The following classification can serve as a crude frame:

4.2.1 *Electron tunnel views*

Historically the earliest[21] and conceptually the simplest, single-electron tunnelling views have maintained heuristic and practical value throughout the evolution of charge transfer science. Tunnel theories can rest on the tunnel probability between separate wells or stationary level splitting in suitable double-well potentials. A broadly applicable form for V_{eff} is[8,9,12-17,55,70-75]

$$V_{eff}(R) = V_{eff}^{0}\exp(-\lambda R) \tag{4.5}$$

where R is the distance between the two centres, the orbital exponent λ characterizes the spatial decay of the electronic density, while V_{eff}^{o} is a weakly R-dependent energy quantity obtained from details of the specific models used.

Eq.(4.5) has the following merits and implications:

(A) It is formally correct for longer-range ET in the sense that much more accurate many-electron calculations often substantiate the basic approximately exponential form over significant R-ranges.

(B) V_{eff}^{o} and λ can be related to physical quantities such as conduction band edge energies, energy gaps to vacant anti-bonding molecular orbitals etc.

(C) Eq.(4.5) can be extended to include formally electronic and molecular structural features in the tunnel barrier structure, random barrier height fluctuations etc.

(D) Eq.(4.5) finally represents a frame for different related tunnel system classes, say solid-electrolyte tunnelling and tunnelling in solid-state junctions.

While "operationally" useful and formally often correct, eq.(4.5) must, however, of course be applied to many-electron anisotropic molecular systems with care. When V_{eff} is brought to incorporate details of the molecular electronic structure, both a pronounced angular spatial dependence and a more entangled R-dependence can in general be expected, and preciser models and methods are needed.

4.2.2 *"Analytical" forms of wave functions and electronic coupling*

In such approaches the electronic donor and acceptor molecular electronic wave functions are approximated by suitable single-atom wave function representations of s-, p-, d- and f-symmetry or simple linear combinations of such functions. Alternatively, the molecular reactants are represented by potential wells for which the Schrödinger equation can be brought to provide electronic wave functions in principle of analytical form. In either case these wave functions obviously exhibit both distance and angular spatial dependence. When combined into the transition matrix element and electronic overlap integrals the dependencies are transferred to V_{eff} to give resulting forms such as[75-77]

$$V_{eff} = V_{eff}(R,\Theta,\varphi) \propto (\lambda R)^{m} f(\Theta,\varphi)\exp(-\lambda R) \tag{4.6}$$

where Θ and φ are the orientational angles, while m is a constant and $f(\Theta,\varphi)$ a function which can be given analytical or numerical forms depending on the complexity of the model. A key point is thus that very pronounced orientational angle dependence of V_{eff} arises once the electronic wave functions contain significant non-spherically symmetric components. Examples of such approaches can be found in ref.75 and ref.76.

4.2.3 Numerical approaches to the electronic factor in direct ET

More comprehensive numerical electronic structural computations of V_{eff} have been applied both to simple model-like systems such as the O_2/O_2^- electron exchange, to ET between transition metal complexes, and to large organic molecules.[77] Ideologically, reported approaches rest on both single-electron views in the sense of omission of electronic relaxation in all other Hartree-Fock electrons, and incorporate many-electron relaxational effects on the ET. They also extend to multi-centre ET in the sense that electron exchange is between suitably constructed LCAO's with crucial participation of say ligand-localized electron density. Electronic structure calculations on the *ab initio* level and in the context of solution ET have finally recently incorporated attempts to include explicitly also the external solvent in the continuum limit, and supplemented by individual solvent molecules added in the cavity in which the reacting molecules are accommodated.

Two groups of recent reports are illuminating. In a recent report[78] M.D.Newton investigated the ET pattern for electron exchange in several aqua, amine and metallocene complexes using eq.(4.4) in combination with INDO electronic structure calculations. This method appears to compare well with *ab initio* methods. The approach furthermore rested on many-electron views, but it was shown that single-electron views in the above sense represents well the systems considered. A range of reactant approach geometries were considered, to yield the following overall view:

(1) Direct electron exchange between metal-centred orbitals was always of minor importance compared to indirect ET via ligand groups.

(2) The whole range from significantly diabatic to pronounced adiabatic behaviour is spanned by the various conceivable inter-reactant geometries. This behaviour is, however, rather system specific and not directly related to the inter-reactant separation. For example, π-type t_{2g} ET in the $[Fe(H_2O)_6]^{2+/3+}$ couple proceeds most favourably in a face-to-face geometry due to the large number of ligand contacts. On the other hand, the δ-type e_g ET in $[Co(NH_3)_6]^{2+/3+}$ is more sensitive to geometric and distance features and proceeds most favourably in an apex-to-apex configuration.

(3) ET in the unsaturated-ligand Cp_2M metallocene systems (Cp = cyclopentadienide, M = Fe, Co) is coupled to hole transfer via the Cp-rings. This is strongly reflected in the observation that the most favourable contact orientation for the two

complexes is D_{5h} where the Cp-rings in the reactants directly face each other. Again, no direct V_{eff}/R relation is found in the R-range around suitable contact distances, Different conceivable orientations bring the ET system into either the adiabatic or diabatic ranges, so that the overall electronic system behaviour is determined both by V_{eff} and by the energies of the various ET configurations.

This investigation emphasizes the amount of electronic structural information inherent in V_{eff} and κ, and the expected intimacy between the electronic and the ET system properties.

In several other recent reports Mikkelsen and Ratner have undertaken *ab initio* investigations of a solute benzene/benzene-model system, located in a spherical cavity in an external structureless dielectric solvent[46,79]. As expected, D_{6h} is the most favourable ET orientation, and since no other molecular groups add to the ET path, a strong distance decay is found. Moreover, V_{eff} appears to depend quite insignificantly on the external solvent properties, but this is understandable in view of the cavity geometry used, where solvent polarization effects are spatially quite remote (cf. section 4.4). However, very strong, almost resonance-like dependencies of the orientation of added individual water molecules arise, parallel orientations being the most facile for H_2O-assisted ET. This configuration furthermore enable ET over distances say 5 Å or so where direct ET between benzene and benzene has long ago almost completely decayed.

Investigations such as those reported in refs.46, 78 and 79 nicely illustrate the amount of mechanistic ET information in contemporary electronic structure calculations. They also point to a new key issue, namely the crucial importance of intermediate environmental "matter" between the two ET centres and the amazing facilitation of long-range ET processes in the presence of instantaneous suitably orientationed solvent molecules or inertial solvent polarization. We proceed now to such effects in sections 4.3 and 4.4.

4.3 Bridge-group facilitated off-resonance and resonance long-range ET

The exponential decay of the electronic factor is supported by a considerable body of experimental data[77]. A major part of these rest on intramolecular ET systems, frequently of impressive molecular architecture, where the donor and acceptor groups are separated by variable-length rigid or flexible, covalently bound molecular bridge groups. A number of these systems also offer a basis for investigations of corresponding optically induced intramolecular ET processes[59,80-82]. The orbital decay factors in intramolecular ET, however, commonly displays notably smaller values than expected from accurate theoretical estimates for separate molecules, suggestive of non-trivial functions of the bridge groups. Experimental and theoretical investigations of intramolecular bridge assisted ET have now evolved into a major research area (for summaries, see refs.77, 83 and 84). Extremely elaborate preparative work is involved in synthesizing these compounds, and calculations of their electronic structure-function relations pose great theoretical challenges. The area has acquired additional perspectives as synthetic intramolecular ET systems both can possess interesting molecular "switch" properties and mimic fast electronic charge separation in organized biological electron transport. They finally represent a clue to real ET patterns of redox proteins where the detailed electronic percolation pathways to protein-buried redox centres is crucially determined by the properties of the intermediate amino acid residues.

Even the simplest tunnel view represented by eq.(4.5) holds a clue to facile bridge-assisted ET compared to "direct" ET. Most frequently the vacant bridge group orbitals are located far higher than the donor and acceptor valence orbitals. They are, on the other hand, still lower than complete ionization energies so that the effective barrier height for ET is lower through the bridge group than through vacuum or solvent, the lowest antibonding orbitals of which are much closer to the ionization energy. Bridge-assisted ET is thus both more facile and has a slower distance variation than direct ET. In preciser terms bridge-assisted ET is handled by multi-electron basis sets which include the orbitals of the bridge groups. In the diabatic limit this can be combined with higher order perturbation theory. For stronger interactions other methods such as extended Hückel theory, CNDO and INDO, local density and *ab initio* methods must be used[77,83,84], but by suitable LCAO construction of the appropriate intermediate group orbitals the actual covalently linking atoms can be brought to have a small amplitude[85] so that the frame of perturbation theory indeed has a wide range of applicability. For this reason we now restrict our considerations to the limit of higher order perturbation theory which, as noted, in practice can be little restrictive.

4.3.1 Off-resonance superexchange ET

The exact function of the bridge group emerges transparently in the off-resonance case where the orbital energies differ strongly from those of the donor and acceptor at the ET moment. This is shown in fig.4a. Expansion of the overall interaction between the zero order donor and acceptor states via the intermediate group states, using the Lippman-Schwinger, Green function, or other expansion schemes then gives for V_{eff}

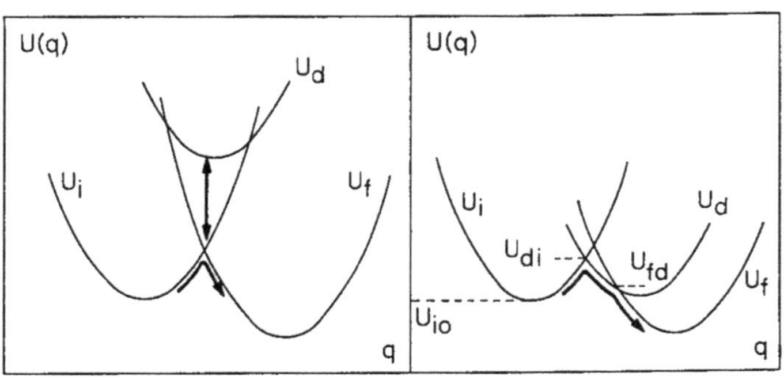

Fig.4. ET in two three-level systems with high (left) and low (right) intermediate state energy. The single-headed arrows indicate the ET path, the double-headed bold arrow the energy gap for coupling to the intermediate state at the moment of electron transfer.

$$V_{eff} = \sum_{i=1}^{N} V_{DA}^{(i)}; \quad V_D^{(i)} = \frac{V_{D(d1)}^{(1)} V_{(d1)(d2)}^{(1)} \cdots V_{(di)A}^{(1)}}{(E_D - E_{d1})(E_D - E_{d2}) \cdots (E_D - E_{di})} \tag{4.7}$$

where the superscript "(i)" indicates the order and number of bridge group states involved, E_D, E_A (= E_D) and E_{dj} (j = 1,...i) the energies of the donor, the acceptor, and the j'th bridge group at the ET moment, i.e. including all fluctuational environmental effects, while the V's are the first order transition matrix elements coupling the states indicated by the subscripts.

Equations such as eq.(4.7), broadly applied in the recent literature, have a range of implications for the molecular bridge group functionality:

(a) The equation incorporates all the information about the donor, the acceptor and the bridge group electronic structure, and in principle includes the nonorthogonality scheme in section 4.1. It can also rightaway be extended to several electronic states on particular bridge groups and to hole transfer rather than electron transfer.

(b) Long-range ET is more facile the better the nearest-neighbour overlap and the smaller the energy gap. In particular this implies that through-bond ET is more facile than through-space ET, and ET through organic material such as protein more facile than through solvent water.

(c) If several pathways are conceivable, say both electron and hole transfer, interference effects can arise.

(d) An entangled distance variation of V_{eff}, depending on the electronic properties of all the intermediate groups, can in principle be expected. However, in the simple case that all the intermediate state exchange matrix elements and energy denominators take the same values of ß and Δ, respectively, then the simple form

$$V_{DA}^{(i)} = \frac{\beta_D \beta_A}{\Delta} \left(\frac{\beta}{\Delta} \right)^{i-1} \tag{4.8}$$

is obtained, where $ß_D$ and $ß_A$ are short-hand notations for $V_{D(d1)}$ and $V_{(di)A}$, respectively. If furthermore only the N'th term is kept in eq.(4.7), then this equation can be recast as[86]

$$V_{eff} \approx V_{DA}^{(N)} = \frac{\beta_D \beta_A}{\beta} \exp\left[-\frac{1}{a} \ln\left(\frac{\Delta}{\beta} \right) R \right] \tag{4.9}$$

where a is the average geometric extension of each bridge group along the ET path. This equation represents a kind of link to the tunnel view and a relation between the tunnel barrier and the electronic properties of the bridge groups.

We notice finally that superexchange ET offers a range of extensions of correlations such as eq.(2.27) to both single- and multi-photon processes in multi-level systems. Of particular interest are relations involving Raman scattering. Such correlations have been investigated[87,88]. A representative form involves the differentially scattered light intensity into a narrow solid angle $\Delta\Omega$ and frequency range $\Delta\nu'$ centred around ν' after excitation of incoming light of frequency ν

$$\frac{\Delta^2\sigma}{\Delta\nu'\Delta\Omega} \approx \frac{1}{6\pi^2c^4}\hbar^2\nu(\nu')^3\left[\frac{M_{Dd}^{(1)}(q_{opt}^*)M_{dA}^{(1)}(q_{opt}^*)}{V_{Dd}^{(1)}(q_{th}^*)V_{dA}^{(1)}(q_{th}^*)}\right]^2\left(\frac{\Delta E}{h\nu_m^{DB}-h\nu}\right)^2 W_{fi}^{(2)}(\Delta G_0^{RP}+h\nu'-h\nu) \qquad (4.10)$$

where "d" represents a single bridge group state and $W_{fi}^{(2)}$ is the rate constant for superexchange ET via this state. Such equations are the basis for Raman spectral lineshapes, excitation profiles etc.

4.3.2 *Resonance effects in long-range ET*

Eqs.(4.7)-(4.9) represent facilitated long-range ET but are only valid far from resonance between the intermediate state levels and the crossing between the reactants' and products' potential surfaces. In the opposite limit of low intermediate state energy the process acquires an altogether quite different character since the intermediate state is now populated temporarily. Due to the partially or totally vibrationally unrelaxed of the process (fig.4b) it cannot, however, merely be regarded as two sequential independent single-ET steps.

For friction-less nuclear motion and rather general potential surfaces the high-T rate constant is thus[89,90]

$$W_{fi} = \frac{8\pi^{3/2}k_BT}{Z_i^{cl}\hbar}\gamma_{di}\gamma_{fd}\exp\left(-\frac{U_{di}}{k_BT}\right)$$

$$\gamma_{di} = \frac{(V_{eff}^{(1)di})^2}{\hbar v_{di}|\Delta U'_{di}|}; \quad \gamma_{fd} = \frac{(V_{eff}^{(1)fd})^2}{\hbar v_{fd}|\Delta U'_{fd}|} \qquad (4.11)$$

$$v_{di} = \sqrt{2k_BT\omega_i/\hbar}; \quad v_{fd} = \sqrt{2|U_{di}-U_{fd}|\omega_d/\hbar}$$

where $|\Delta U'_{di}| = |U'_d-U'_i|$ and $|\Delta U'_{fd}| = |U'_f-U'_d|$ are the gradient differences between the appropriate potential surfaces at the crossings, v_{di} and v_{fd} are the nuclear velocities at the same points, while $V_{eff}^{(1)di}$ and $V_{eff}^{(1)fd}$ are the electron exchange couplings. What we notice is then that the rate constant contains *two* rather than a single electronic factor. Eq.(4.11) further implies that first the i → d transition occurs. However, before intermediate state vibrational relaxation is completed the second ET step, d → f occurs, where the potential energy difference $|U_{di}-U_{fd}|$ is converted to nuclear kinetic energy, as reflected in the nuclear velocity at the crossing point of lowest potential energy.

Processes represented by eq.(4.11) and similar equations have far-reaching implications for fast photo-induced charge separation in bacterial and plant photosynthesis, for molecular multi-level switch systems, for fast protein ET reactions induced by reactive anion radicals, and for the opening of new inelastic resonance tunnel modes in STM surface adsorbate systems. We should here that notice current-voltage or free energy relations for such processes are fully covered in the whole range from superexchange to vibrationally unrelaxed two-step ET by the formalism from which eqs.(4.7)-(4.11) emerge. The apparent divergencies at "exact" resonance between the donor and bridge group levels or between the two crossing points are thus artefacts. They are rooted in the form of eqs.(4.7) and (4.11), respectively as limiting forms valid far from these resonances (i.e. by k_BT or so). More exact quantum mechanical handling of three-level multiphonon resonance ET with strong electronic-vibrational coupling in all the three electronic states reduces these effects to finite and rather mild resonances around the critical points[92].

4.4 Environmental effects on the electronic factor

We conclude this section by some brief observations on the effects of the environmental polarization field on the electronic factor in long-range ET. Wave functions which extend beyond the confinements of molecular geometry are exposed to such field effects. While the approximately exponential decay can perhaps be expected to persist, the decay parameters, however, come to depend on the polarization features. These effects can be several-fold. For example, in the superexchange mode the energy denominators are modified by differential solvation between the donor and intermediate groups. In addition, the polarization effects in question are not the ones prevailing at equilibrium but at the moment of ET. Since this configuration mostly differs strongly from the equilibrium values and furthermore depends on both the temperature and reaction free energy, so must the electronic factor. In the exponential view and with reservations discussed elsewhere[93-95] this dependence takes the approximate form

$$V_{eff} \approx V_{eff}^{eq} \exp(\lambda_{el}^{0} \xi \Theta^{*} R) \tag{4.12}$$

where λ_{el}^{o} is the decay parameter of the electronic wave function, at the equilibrium polarization of the reactants' electronic state, V_{eff} similarly the equilibrium value of V_{eff}^{eq}, ξ a quantity of about 0.5 and Θ^* the transfer coefficient introduced earlier. Eq.(4.12) shows that the electron clouds expand on system motion along the potential surface for $\Theta^* > 0$ (say when $\Theta^* \approx 0.5$) and that this effect can be very significant for long-range ET. Moreover, V_{eff} varies strongly over wide free energy ranges where Θ^* changes significantly. Consequences of this in biological long-range ET and extensions beyond the Condon approximation are discussed in detail elsewhere[93-94].

5. Two Case Examples of Application of Charge Transfer Concepts

A wealth of experimental investigations have been framed by electron and atom group transfer theory. They relate to many different facets including in particular: thermodynamic properties of the reacting molecules and free energy relations; spectroscopic and structural properties; dielectric environmental properties; electronic properties of the reacting molecules or electrodes, and dynamic environmental properties. We refer to several recent comprehensive reviews where overviews of such investigations are available[1-17] but conclude our presentation here rather by a short discussion of two specific and well characterized case examples. These are biological systems. The first one focuses on long-range ET in a specific redox metalloprotein in higher plant photosynthetic electron transport, the second one on conformational relaxational features in a specific hydrolytic metalloenzyme. These systems might at first sight appear slightly unfamiliar in the context of electrochemistry. Both are, however, perfectly well defined molecular systems to which charge transfer concepts and formalism adequately apply in new and intriguing fashions.

5.1 Electron transfer pathways in higher plant plastocyanins

Plastocyanin (PC) is a small (MW $\approx$ 10.500) single-copper protein (fig.5) the function of which is to transmit electrons between two large membrane-bound complexes cytb$_6$/f and photosystem I[96]. Two molecular structural features stand out prominently, mechanistically indicative of its electron transfer functionality. One is a strongly asymmetric location of the ET centre, i.e. the Cu atom, about 6 Å from a hydrophobic part of the protein surface. The other one is a protruding Tyr-83 residue amidst a group (seven) of negatively charged residues, approximately 15 Å from the Cu atom. There is

Fig.5. Structure of oxidized poplar plastocyanin. Crystallographic data from ref. 99 and Brookhaven Protein Data Bank[98]. Evans & Sutherland PS 350. The copper atom and its four tetrahedrally organized ligands are indicated in bold lines at the top. The favourable long-range ET channel extends from Cu, via ligated Cys-84 and terminates at the protruding Tyr-83 to the right. The two Phe-residues also indicated, represents another but much less favourable path.

considerable recent evidence that both these sites are involved in the electron transfer pattern of PC. Within the framework of diabatic electron transfer theory we now address the question why the much longer electron transfer path is competitive, and how the surrounding negative patch may facilitate this.

We use the following second order rate constant form

$$k_{et} = \kappa \frac{\omega_{eff}}{2\pi} \Delta V \exp\left\{ - \frac{w_r + (E_r + \Delta G_0 + w_p - w_r)^2/(4E_r)}{k_B T} \right\} \quad (5.1)$$

where E_R is the total molecular, solvent and protein reorganization free energy, w_r and w_p the inter-reactant interaction free energies in the reactants' and products' state, respectively, ΔV the volume of the reaction zone, while all other terms have been introduced above.

We consider first one particular, rather convincing indication that ET does occur near Tyr-83 and the negative patch[99]. Tyr-83 can be NO_2-substituted in the *o*-position. This increases strongly the acidity of the phenol side group, so that deprotonation

$$\underset{|}{NO_2\text{-}\varphi\text{-}OH} \rightleftharpoons \underset{|}{NO_2\text{-}\varphi\text{-}O^{\ominus}} + H^{\oplus} \quad (5.2)$$

occurs in a pH-range where the protein is stable. On deprotonation an additional negative charge is evolved and at the same time the PC redox potential drops by about 20 mv. Since $\Delta G_o + w_p - w_r \ll E_R$ these two effects can be separated by noting that

$$\frac{k_{et}^{deprot}}{k_{et}^{prot}} \approx \exp\left[\frac{-\frac{1}{2}(\Delta G_0^{deprot} - \Delta G_0^{prot})}{k_B T}\right] \exp\left[\frac{-\frac{1}{2}(w_r^{deprot} + w_p^{deprot} - w_r^{prot} - w_p^{prot})}{k_B T}\right] \quad (5.3)$$

By inserting the redox potential change and the experimental rate constants the interesting observation emerges that for the negatively charged redox partners $[Fe(CN)_6)]^{3-/4-}$ virtually all rate constant changes are located in the thermodynamic driving force. On the other hand, for the positively charged reaction partner $[Co(phen)_3]^{3+}$ the redox potential corrected rate constant increases by about 50 % on NO_2-PC deprotonation. In other words, the positively charged reaction partner is very notably affected by the additional negative charged evolved on deprotonation, whereas the negatively charged reaction partners are not. This is indicative that the ET site of the former is close to Tyr-83 and the negative patch, while the site for the latter is quite different and probably close to the Cu-atom.

The feasibility of a long-range ET path has been substantiated by quantum chemical calculations on the extended Hückel level[85]. By using the superexchange formalism in section 4 on four suitably located remote site channels, two of which are shown on fig.5, the notable outcome emerged that one particular channel, namely Cu/Cys-84/Tyr-83 (fig.5) is not only far more efficient than all other conceivable remote site channels but its efficiency even approaches that for the much shorter-range ET to Cu at the hydrophobic patch at the top of fig.5. This can be traced to the through-bond character and overall favourable energy gap and overlap features of this channel compared with other suitably located pathways. The two transmission coefficients are about 2×10^{-3} and 4×10^{-5}, i.e. that for the long-range channel is the lower by a factor of about 50. This can easily be overcome by the favourable inter-reactant work terms for positively charged reaction partners, where about 100 mv is needed, well compatible with a simple Debye-Hückel screened electrostatic potential.

For other electronic effects of NO_2-modification on the electron transfer reactivity of PC we refer to a recent report[100]. The conclusion is that quite straightforward combination of ET theory with different kinds of data can indeed provide very useful insight even into subtle details of the intriguing dual-path electron transfer pattern of higher plant PC's. It is very notable that similar considerations relating to the ET pattern of PC from the cyanobacterium *A. Variabilis* which has no negative patch, shows that short-range ET here seems to dominate entirely, also for positively charged reaction partners.[101]

5.2 *Conformational dynamics in carboxypeptidase A catalyzed peptide and ester hydrolysis*

Carboxypeptidase A (CPA) is a digestive enzyme (MW ≈ 34.000) catalytic in peptide and ester hydrolysis[102]. The catalytic events are intimately associated with active site structural elements around a zinc atom, and both the native enzyme and enzyme-inhibitor complexes simulating enzyme-substrate complexes, intermediates, and the transition state is characterized in great structural detail. The enzyme exhibits considerable structural mobility, reflected particularly in large conformational shifts of individual amino acid residues on enzyme-substrate complex formation. Representative elements of these views are shown in fig.6.

The conformational mobility can be expected to be reflected in relaxational features of the protein structure. The fluctuations of residues around the active site can also be frictionally coupled to the external solvent and therefore damped. As solvent relaxation is also physically directly related to the solvent viscosity, careful control of the latter can be expected to arouse characteristic viscosity effects on the catalytic events in the enzyme process, ultimately informative of the dynamics and mobility of the protein structure throughout the process.

In several respects CPA is a prototype-like enzyme, being one of the best characterized metalloenzymes altogether. Even then important details in the sequence of elementary steps constituting the enzyme mechanisms remain open. Key steps broadly representative of peptide and ester hydrolysis are nucleophilic attack on the scissile terminal peptide or ester bond by Zn-bound water, directed by nearby Glu-270. This is followed by tetrahedral intermediate formation and one or several proton transfers, and concluded by tetrahedral intermediate decay and liberation of the terminal residue. However, crucial details relating to the precise involvement of the metal atom, the

Fig.6. Structures of bovine carboxypeptidase A and its Gly-Tyr-inhibitor complex. Crystallographic data from ref. 103 and Brookhaven Protein Data Bank[98]. Bold lines indicate native conformations of the residues (Arg-127, Tyr-248 and Glu-270) which are strongly shifted on enzyme-inhibitor complex formation, the shifted conformations being indicated by thinner-line representations of the same groups. Two α-helices are seen at the mid-left and at the bottom. The small separate wiggle is the inhibitor, and the circle indicates the zinc atom.

molecular identity of the intermediates, and the precise function of a number of charged residues around the active site remain elusive, except that they are different for peptides and esters and exhibit a fair amount of substrate specificity inside each of these substrate classes.

In view of this it is perhaps a matter of taste to assess whether the viscosity patterns hold clues to the mechanism or add to their entanglement. We illustrate this by two recent reports[104] in which the solvent viscosity effects on the CPA catalyzed hydrolysis of the closely related peptide and ester substrate, Bz-Gly-Phe and Bz-Gly-OPhe were compared. Viscosity data for the peptide Cbz-Ala-Ala-Ala were reported previously[105]. The enzyme processes involve a considerable amount of local molecular reorganization in addition to protein and solvent reorganization. They also involve proton transfer, although the diabatic tunnel feature is not pronounced as shown by the small kinetic isotope effects of 1.3 and 1.6 for peptide and ester hydrolysis, respectively. Further, the external viscosity can be varied over a tenfold range by glycerol or sucrose addition, on which CD and thermodynamic data indicate no change in the solute CPA secondary and tertiary structure.

The interesting observations then emerge that Bz-Gly-Phe exhibits no viscosity dependence at all. This is surprising since the peptide Cbz-Ala-Ala-Ala follows eq.(3.14) with the significant value of $\delta \approx 0.76$. On the other hand, the structurally closely related ester Bz-Gly-OPhe does have a strong viscosity dependence, also following eq.(3.14), now with $\delta \approx 0.65$.

Such results, although intriguing do not offer the same molecular details of the enzyme mechanism as we could obtain for PC. They do, however, indicate the following as to the conformational protein dynamics for the various substrates. First, they substantiate other observations that even though an overall common mechanism operates its dynamic details are very substrate specific. Secondly, the viscosity behaviour is indicative of a much closer or more "rigid" "match" between enzyme and substrate for Bz-Gly-Phe than for the other two substrates. With reference to sections 2 and 3, and eqs.(3.12) and (3.14) in particular, this can be rephrased in different quantitative ways. The absence of viscosity effects thus indicates that the nuclear reorganization is dominated by local, friction-less modes, whereas solvent-damped protein motion affects the conversion of the other substrates much more strongly. For example $\delta \approx 0.7$ implies that protein/solvent reorganization constitutes about two thirds of the total reorganization free energy. A similar conclusion is in fact reached on the basis of the effective environmental Hamiltonian formalism, the frequency of which, ω_{eff} contains suitably weighted contributions from all the system modes. Although ω_{eff} does not follow the inverse fractional power law, the conclusion about the distribution among the different mode reorganization terms still coincides with the stochastic view. Still another, "dynamic" view of the structural rigidity of the enzyme-substrate complex is to note that the critical limit between the thermally equilibrated and the stochastic regime for $E_S \approx (2/3)E_R$ is approximately $\tau_{rel} \approx 10^2 \nu^{-1}$.[68] For given ν ($\approx 10^{13}$ s^{-1}), τ_{rel} would therefore appear to be smaller than about 10^{-11} s for Bz-Gly-Phe and larger than this value for the other two substrates.

All the observations are in line with a more rigid enzyme-substrate structure for Bz-Gly-Phe than for Cbz-Ala-Ala-Ala and Bz-Gly-OPhe, with better screening from the external solvent and less involvement of conformational dynamics. The viscosity data do not hold clues to *why* this is so, but a reasonable speculation could relate the different behaviour to the natural function of CPA and the preference of the enzyme for terminal aromatic residues.

6. Acknowledgement

This work is supported by The Danish Natural Science Research Council, The Carlsberg Foundation, Brødrene Hartmanns Fond and Direktør Ib Henriksens Fond.

7. References

1. R.R.Dogonadze, in *Reactions of Molecules at Electrodes*, ed. N.S.Hush, (Wiley, New York, 1971), p. 135.
2. a. R.R.Dogonadze and A.M.Kuznetsov, Physical Chemistry. Kinetics, (VINITI, Moscow, 1973). b. Kinetics and Catalysis, (VINITI, Moscow, 1978).
3. R.R.Dogonadze and A.M.Kuznetsov, *Progr.Surf.Sci.* **6** (1975) 1.
4. J.Ulstrup, *Charge Transfer Processes in Condensed Media*, (Springer, Berlin, 1979).
5. *Faraday Discussions Chem.Soc.* **74** (1982).

6. *Progr.Inorg.Chem.* **30** (1983).
7. R.R.Dogonadze and A.M.Kuznetsov, in *Comprehensive Treatise of Electrochemistry*, eds. J.O'M.Bockris, B.E.Conway and E.Yeager, (Plenum, New York, 1983), Vol.**7**, p.1.
8. M.D.Newton and N.Sutin, *Ann.Rev.Phys.Chem.* **35** (1984) 437.
9. R.A.Marcus and N.Sutin, *Biochim.Biophys.Acta* **811** (1985) 265.
10. *J.Phys.Chem.* **90** (1986) 3452-3862.
11. a. H.Frauenfelder and P.G.Wolynes, *Science* **228** (1985) 337. b. J.T.Hynes, in *Theory of Chemical Reaction Dynamics*, ed. M.Baer, (CRC Press, Boca, Boston, 1985), Vol. **4**, p.172. c. J.T.Hynes, *J.Stat.Phys.* **42** (1986) 149.
12. K.V.Mikkelsen and M.A.Ratner, *Chem.Rev.* **87** (1987) 113.
13. E.M.Itskovitch, J.Ulstrup and M.A.Vorotyntsev, in *The Chemical Physics of Solvation. Part B. Spectroscopy of Solvation*, eds. R.R.Dogonadze, E.Kálmán, A.A.Kornyshev and J.Ulstrup, (Elsevier, Amsterdam, 1986), p.223.
14. *Tunnelling*, eds. J.Jortner and B.Pullman, (Reidel, Dordrecht, 1986).
15. A.M.Kuznetsov, J.Ulstrup and M.A.Vorotyntsev, in *The Chemical Physics of Solvation. Part C. Solvation in Specific Physical, Chemical and Biological Systems*, eds. R.R.Dogonadze, E.Kálmán, A.A.Kornyshev and J.Ulstrup, (Elsevier, Amsterdam, 1988), p. 163.
16. E.D.German and A.M.Kuznetsov, *Kinetics of Solution Charge Transfer Reactions*, (VINITI, Moscow, 1990).
17. *Perspectives in Photosynthesis*, eds. J.Jortner and B.Pullman, (Kluwer, Dordrecht, 1990).
18. J.N.Brønsted and K.J.Pedersen, *Z.Phys.Chem.* **108** (1924) 185.
19. F.Hund, *Z.Phys.* **43** (1927) 805.
20. C.E.Cleeton and N.H.Williams, *Phys.Rev.* **45** (1934) 234.
21. R.W.Gurney, *Proc.Roy.Soc.* **A 137** (1931) 696.
22. a. M.J.Polissar, *J.Am.Chem.Soc.* **58** (1936) 1372. b. D.A.Schaffer, *J.Phys.Chem.* **40** (1936) 1021.
23. H.C.Hornig, G.L.Zimmerman and W.F.Libby, *J.Am.Chem.Soc.* **72** (1950) 3808.
24. W.F.Libby, in *Tunnelling in Biological Systems*, eds. B.Chance, D.DeVault, H.Frauenfelder, R.A.Marcus, J.R.Schrieffer and N.Sutin, (Academic Press, New York, 1979), p.3.
25. R.D.Cannon, *Electron Transfer Reactions* (Butterworths, London, 1980).
26. S.I.Pekar, *Untersuchungen über die Elektronentheorie der Kristalle*, (Akademie Verlag, Berlin, 1954).
27. K.Huang and A.Rhys, *Proc.Roy.Soc.* **A204** (1950) 406.
28. M.Lax, *J.Chem.Phys.* **20** (1952) 1752.
29. R.Kubo, *Phys.Rev.* **86** (1952) 929.
30. R.Kubo and Y.Toyozawa, *Progr.Theor.Phys.* **13** (1955) 160.
31. R.A.Marcus, *J.Chem.Phys.* **24** (1956) 966.
32. a. V.G.Levich and R.R.Dogonadze, *Dokl.Akad.Nauk SSSR, Ser.Fiz.Khim.* **124** (1959) 123. b. *Coll.Czech.Chem.Comm.* **26** (1961) 193.

33. L.D.Landau and E.M.Lifshitz, *Quantum Mechanics*, (Pergamon, Oxford, 1965).
34. J.Jortner, *J.Chem.Phys.* **64** (1976) 4860.
35. R.R.Dogonadze and A.M.Kuznetsov, *Elektrokhimiya* **7** (1971) 763.
36. A.M.Kuznetsov and J.Ulstrup, *Chem.Phys.* **107** (1986) 381.
37. D.N.Zubarev, *Irreversible Statistical Thermodynamics*, Plenum, New York, 1977).
38. R.R.Dogonadze, A.M.Kuznetsov, M.G.Zakaraya and M.A.Vorotyntsev, *J.Electroanal.Chem.* **75** (1977) 315.
39. H.A.Fröhlich, *Theory of Dielectrics*, 2nd ed. (Clarendon, Oxford, 1958).
40. R.R.Dogonadze, G.M.Chonishvili and T.A.Marsagishvili, *J.Chem.Soc.Faraday Trans.II* **80** (1984) 355.
41. Z.T.Chu, A.Warshel and W.W.Parson, *Photosynt.Res.* **22** (1989) 39, and references there.
42. a. A.M.Kuznetsov, *Khim.Fiz.* **1** (1982) 1496. b. *Chem.Phys.Lett.* **91** (1982) 34.
43. J.Sines, S.Allison, A.Wierbicki and J.A.McCammon, *J.Phys.Chem.* **94** (1990) 959.
44. R.A.Kuharski, J.S.Barder, D.Chandler, M.Sprik, M.L.Klein and R.W.Impey, *J.Chem.Phys.* **89** (1988) 3248.
45. D.A.Zicki, G.Ciccotti, J.T.Hynes and M.Ferrario, *J.Phys.Chem.* **93** (1989) 6261.
46. a. K.V.Mikkelsen and M.A.Ratner, *J.Phys.Chem.* **93** (1989) 1759. b. *J.Chem.Phys.* **90** (1989) 4237.
47. R.R.Dognadze and A.M.Kuznetsov, *Elektrokhimiya* **2** (1967) 1324.
48. N.R.Kestner, J.Logan and J.Jortner. *J.Phys.Chem.* **78** (1974) 2148.
49. J.Ulstrup and J.Jortner, *J.Chem.Phys.* **63** (1975) 4358.
50. R.R.Dogonadze and Z.D.Urushadze, *J.Electroanal.Chem.* **32** (1971) 235.
51. E.K.Kudinov and Yu.A.Firsov, *Zhur.Eksp.Teor.Fiz.* **49** (1965) 867.
52. a. H.Gerischer, *Z.Phys.Chem.* **26** (1960) 223. b. **26** (1960) 325.
53. a. W.Schmickler, *J.Electroanal.Chem.* **113** (1980) 159. b. **137** (1982) 189.
54. A.M.Kuznetsov and J.Ulstrup, *J.Electroanal.Chem.* **195** (1985) 1.
55. J.Ulstrup, *Surf.Sci.* **101** (1980) 564.
56. W.Schmickler and J.W.Schultze, *Mod.Asp.Electrochemistry* **17** (1986) 359.
57. For a review, see: E.D.German and A.M.Kuznetsov, *Electrochim.Acta* **26** (1981) 1595.
58. A.A.Kornyshev, in *The Chemical Physics of Solvation. Part A. Theory of Solvation*, eds. R.R.Dogonadze, E.Kálmán, A.A.Kornyshev and J.Ulstrup, (Elsevier, Amsterdam, 1985), p.77.
59. N.S.Hush, *Electrochim.Acta* **13** (1968) 1005.
60. N.Van Kampen, *Stochastic Processes in Chemistry and Physics*, (North-Holland, Amsterdam, 1985).
61. H.A.Kramers, *Physica* **7** (1940) 284.
62. Many references in refs.11, 14, and 15 above.
63. a. R.F.Grote and J.T.Hynes, *J.Chem.Phys.* **73** (1980) 4465. b. **75** (1981) 2191. c. **77** (1982) 3736.

64. a. P.Barbara and W.Jarzeba, *Acc.Chem.Res.* **21** (1988) 195. b. *Adv.Photochem.* **15** (1990) 1, and references there.
65. M.J.Weaver and G.E.McManis III, *Acc.Chem.Res.* **23** (1990) 294, and references there.
66. A.M.Kuznetsov, *Electrochim.Acta* **32** (1987) 1271.
67. A.A.Belousov, A.M.Kuznetsov and J.Ulstrup, *Chem.Phys.* **129** (1989) 311.
68. H.Sumi, *J.Phys.Chem.*, in press.
69. H.Sumi and R.A.Marcus, *J.Chem.Phys.* **84** (1986) 4894.
70. A.M.Kuznetsov and J.Ulstrup, ref.5, p.31.
71. See discussions in refs.4, 13, 15, and 16.
72. C.B.Duke *Tunneling in solids*, (Academic Press, New York, 1969)
73. C.B.Duke, in *Tunnelling in Biological Systems*, eds. B.Chance, D.DeVault, H.Frauenfelder, R.A.Marcus, J.R. Schrieffer and N.Sutin, (Academic Press, New York, 1979), p.31.
74. M.Redi and J.J.Hopfield, *J.Chem.Phys.* **72** (1980) 6651.
75. B.Brocklehurst, *J.Phys.Chem.* **83** (1979) 536.
76. P.Siders, R.J.Cave and R.A,.Marcus, *J.Chem.Phys.* **81** (1984) 5613.
77. See reviews in refs.8, 9, 12 and 16.
78. M.D.Newton, in ref.17, p.157.
79. K.V.Mikkelsen, E.Dalgaard and P.Swanstrøm, *J.Phys.Chem.* **91** (1987) 3081.
80. C.Creutz, ref.6, p.1.
81. C.A.Stein, N.A.Lewis and G.Seitz, *J.Am.Chem.Soc.* **104** (1982) 2596.
82. N.S.Hush, *Coord.Chem.Rev.* **64** (1985) 135.
83. Chapters in: *Electron Transfer in Biology*, eds. M.K.Johnson, R.B.King, D.M.Kurtz, Jr., C.Kutal, M.L.Norton and R.A.Scott, (ACS, Washington D.C., 1990), ACS Ser. **226**.
84. Chapters in: *Metal Ions in Biological Systems*, eds. H.Sigel and A.Sigel, (Marcel Dekker, New York, 1991), Vol.**27**, in press.
85. H.E.M.Christensen, L.S.Conrad, K.V.Mikkelsen, M.K.Nielsen and J.Ulstrup, *Inorg.Chem.* **29** (1990) 2808.
86. H.M.McConnell, *J.Chem.Phys.* **35** (1961) 508.
87. M.G.Zakaraya and J.Ulstrup, *Chem.Phys.* **135** (1989) 49.
88. A.M.Kuznetsov and J.Ulstrup, *Comm.Inorg.Chem.* **10** (1990) 197.
89. A.M.Kuznetsov and Yu.I.Kharkats, *Elektrokhimiya* **8** (1976) 1277.
90. Literature on long-range resonance effects such as those implied by fig.4b is overviewed in refs.2b, 4, 15 and 16.
91. Yu.I.Kharkats, A.K.Madumarov and M.A.Vorotyntsev, *J.Chem.Soc.Faraday Trans.II* **70** (1974) 1578.
92. Yu.I.Kharkats and J.Ulstrup, in preparation.
93. A.M.Kuznetsov and J.Ulstrup, ref.14, p.345, and references there.
94. A.M.Kuznetsov and J.Ulstrup, *Phys.Stat.Sol.* **b114** (1982) 673, and references there.
95. A.M.Kuznetsov and J.Ulstrup, *Bioelectrochem.Bioeng.* **21** (1989) 289.

96. For recent reviews, see: a. A.G.Sykes, *Chem.Soc.Rev.* **14** (1985) 283. b. *Struct.Bond.* (1990), in press.
97. J.M.Guss and H.C.Freeman, *J.Mol.Biol.* **169** (1983) 521.
98. a. F.C.Bernstein, T.F.Koetzle, G.J.B.Williams, E.F.Meyer, Jr., M.D.Brice, J.R.Rodgers, O.Kennard, T.Shimanouchi and M.Tasumi, *J.Mol.Biol.* **112** (1977) 535. b. E.E.Abola, F.C.Bernstein, S.H.Bryant, T.F.Koetzle and J.Weng, in *Crystallographic Databases - Information Content, Software Systems, Scientific Applications*, (Data Commission of The International Union of Crystallography, Born/Cambridge/Chester, 1987).
99. H.E.M.Christensen, J.Ulstrup and A.G.Sykes, *Biochim.Biophys.Acta* **1039** (1990) 94.
100. H.E.M.Christensen, L.S.Conrad, K.V.Mikkelsen and J.Ulstrup, in preparation.
101. H.E.M.Christensen, L.S.Conrad and J.Ulstrup, in preparation.
102. For a recent review, see: D.W.Christianson and W.N.Lipscomb, *Acc.Chem.Res.* **22** (1989) 62.
103. a. D.C.Rees, M.Lewis and W.N.Lipscomb, *J.Mol.Biol.* **168** (1983) 367. b. D.C.Rees and W.N.Lipscomb, *Proc.Nat.Acad.Sci.USA* **80** (1983) 7151.
104. a. D.E.Khoshtariya, J.M.Hammerstad-Pedersen and J.Ulstrup, *Biochim.Biophys.Acta*, in press.
105. B.Gavish and M.M.Werber, *Biochemistry* **18** (1979) 1269.

QUANTUM ELECTRODE KINETICS

ALEXANDER M. KUZNETSOV
The A.N.Frumkin Institute of Electrochemistry of the Academy of Sciences of the USSR, Leninskii prospect 31, Moscow 117071, USSR

ABSTRACT

An introduction to the quantum mechanical theory of the elementary act of electrochemical processes is presented. Electron and proton transfer reactions are considered.

1 Introduction

The aim of this paper is to give a brief introduction to the quantum mechanical theory of electrode kinetics and to demonstrate the quantum mechanical nature of phenomena of charge transfer at interfaces between an electrode and a solvent. The charge transfer through the interface is often the rate determining step which governs the whole process of the current passage between two electrodes immersed into electrolyte solution. The quantum mechanical nature of electrochemical processes was qualitatively understood already in pioneer work of Gurney[1] who suggested to consider an electrochemical reaction of electron transfer as a process of electron tunnelling from the electrode to the ion in the solution. It is recognized at present that the physical mechanism of the elementary act of electrochemical reaction is much in common with that for other processes of charge transfer in condensed media.

A modern step of the theory of charge transfer began in 1956 when Marcus[2,3] used the idea that the polarization of the medium creates a considerable Franck - Condon barrier for the process of electron transfer and calculated this barrier using method of classical thermodynamic fluctuations. The quantum mechanical approach to the theory originates from the paper by Dogonadze and Levich of 1959[4] who applied the theory of radiationless processes to the calculation of the probability of the electron transfer in such systems. Later this approach was generalized to electron transfer processes at metal and semiconductor electrodes[5,6]. Other approaches were also developed[7-9]. A great number of papers were published since that time. Some recent review articles may be found in refs.10-13. General physical ideas of the theory are presented below in Section 2 for electron transfer processes. The quantum aspects of phenomenon are most strikingly manifested in proton transfer reactions presented in Section 3. Recent tendencies in the theory and concluding remarks are summarized in Section 4.

2. Electron transfer processes

A process of electron transfer from metal electrode to an ion having unchanged intramolecular structure and located at some distance x from the electrode surface represents the simplest electrochemical reaction. We start from nonadiabatic reactions for which the total transition probability per unit time W determining the reaction rate constant and hence the electric current may be found as a sum of averaged probabilities for independent transitions of electrons from any energy level ε in the metal to the energy level ε_I in the ion. According to the Franck - Condon principle the electron transfer occurs with a remarkable probability only if the matching of the energy levels $\varepsilon=\varepsilon_I$ takes place. Due to strong interaction with fluctuating molecular surrounding in the solution the position of the energy level ε_I varies in time and the matching of the energies is transient, therefore the process has Landau - Zener character. For the reactions under consideration the main reason leading to time variation of the electron energy ε_I in the ion is its interaction with the polarization of the medium P_t.

We have to distinguish the inertialess P_e and inertial $P=P_t-P_e$ polarizations of the medium. Inertialess polarization, i.e. polarization with characteristic time of fluctuations τ_0 which is shorter than that for the electron motion τ_e, does not create Franck -Condon barrier and follows adiabatically the motion of the electron, i.e. it reacts adiabatically to the shift of instant position of the electron. It is mostly the electron polarization of solvent molecules.

In general it influences on the energy of the initial and final states and on the profile of the potential energy of the electron. In many cases it is not necessary to consider the dynamic behaviour of this part of polarization. However one has to take into account the screening of the interactions between charged particles due to this polarization[11,14].

Inertial polarization $P=P_t-P_e$ is characterized by the times of fluctuations τ which are much longer than τ_e. Therefore it can not follow the motion of the electron and creates a sort of external field for the electron which fluctuates in time around initial or final equilibrium values. They correspond to initial and final equilibrium values of inertial polarization P_{0i} and P_{0f}, which are determined by respective charge distributions in the reactants since in the Born - Oppenheimer approximation the interaction of the inertial polarization with the electron may be considered as the interaction with "electron cloud", i.e. the interaction with the distribution of the electron density. P involves orientational, librational and vibrational polarizations, and a part of electronic polarization. Most of fluctuations of inertial polarization are classical. They are responsible for the activation factor in the transition probability. A quantum part of inertial polarization along with the transferable electron determines the value of the preexponential factor[10-13].

In real media the frequency spectrum of electronic polarization spreads often to a frequency range of the transferable electron ($\tau_0 \sim \tau_e$). In this case the motion of the electron and electronic polarization must be considered self-consistently[15].

It is assumed sometimes that in this limit the electronic polarization may be in equilibrium with the distribution of the electron density at the transitional configuration (at least for the adiabatic reactions)[15]. For the adiabatic reactions it would lead to an additional contribution to the activation barrier[15]. However it should be noted that the relationship $\tau \sim \tau_e$ is usually valid for the diabatic electron states. The characteristic time τ_e^a for an adiabatic state is much longer ($\tau_e^a >> \tau_e$). Therefore in the case where $\tau_0 \sim \tau_e$ the electronic polarization of the medium (the fluctuations of whichare quantum) must be included along with the transferable electron into the quantum subsystem of diabatic states. Its behaviour in this case is similar to that of other quantum degrees of freedom (electrons, intramolecular quantum vibrations, protons etc. See Section III) and the change of its state influences mainly the preexponential factor. Its effect on the activation barrier may be due only to a distortion of the shape of the potential energy surfaces for classical subsystem caused by a modulation of the diabatic wave functions of quantum subsystem by classical fluctuations[11].

The transition probability may be calculated in the first order of the perturbation theory in the interaction of the electron with acceptor using Fermi golden rule (see Section III). If all degrees of freedom of the system may be separated to classical and quantum ones, the transition probability may be written in the form[16]

$$W(\varepsilon)=(\omega_{eff}/2\pi)æ \exp(-F_a(\varepsilon)/k_BT) \qquad (1)$$

where F_a is the activation free energy due to classical degrees of freedom, æ is the transmission coefficient describing the change of the state of quantum subsystem, ω_{eff} is an effective frequency of fluctuations for classical degrees of freedom.

In the case of Gaussian fluctuations $F_a(\varepsilon)$ has a quadratic form[7,8,16]

$$F_a(\varepsilon)=[E_r+\Delta F(\varepsilon)]^2/4E_r \qquad (2)$$

where $\Delta F(\varepsilon)$ is the free energy of transition and E_r is the reorganization energy[7,8]. The detailed discussion of this quantity may be found in refs.7,8,17-19. For linear dielectric medium the reorganization energy is equal to

$$E_r=(1/8\pi)\sum_k c(k)|\Delta D(k)|^2 \qquad (3)$$

where $c(k)=1/\varepsilon_{opt}-1/\varepsilon_s(k)$ and $\Delta D(k)$ is Fourier transform of (D_i-D_f) which is the change of the electrostatic induction in the medium due to the transition. ε_{opt} and $\varepsilon_s(k)$ are high-frequency and low-frequency limits of the dielectric function $\varepsilon(k,\omega)$.

It should be emphasized that Eq.3 is valid for both homogeneous and inhomogeneous media including the cases where the electrostatic induction depends on the dielectric constants of the media[17,18].

The effective frequency ω_{eff} is determined by the ratio of vibrational partition functions of the initial and transition states and may be calculated using the effective Hamiltonian method[20,21].

The electric current is calculated according to Eq.4 [16]

$$i_c = e\int dx\, d\varepsilon\; c_z(x)\; \rho(\varepsilon) f(\varepsilon) W_c(\varepsilon,\phi,x) \qquad (4)$$

where $c_z(x)$ is the concentration of the electron acceptor at the distance x from the electrode, $f(\varepsilon)$ and $\rho(\varepsilon)$ are the Fermi distribution function and the density of states for the electron in metal, ϕ is the electrode potential. The calculation according to Eq.4 leads to an exponential dependence of electric current on the potential(Tafel law) in some interval of the potential variation which agrees with experimentally observed dependencies. However the theory allows also more complicated i_c vs. ϕ dependencies.

3. Proton transfer

Proton is the smallest and lightest nucleus characterized by high-frequency vibrations in most chemical compounds ($\hbar\omega \sim 0.3$eV). So it exhibits quantum behaviour even at room temperature ($k_BT \sim 0.025$eV). The reaction of proton transfer at an electrode involves two steps, which, e.g. might be

$$H_3O^+ + e \Rightarrow H_{ads} + H_2O \qquad (5)$$

$$H_{ads} + H_3O^+ + e \Rightarrow H_2 + H_2O^+ \qquad (6)$$

First step is rate determining for some metals whereas the second one is rate determining for others. The transition probability for the reaction of Eq.5 is discussed below.

First quantum mechanical model for the reaction of this type was suggested by Dogonadze, Kuznetsov and Levich[22,23]. It is assumed that the adsorbed hydrogen H_{ads} represents a neutral atom vibrating with a high frequency at the electrode surface. A double Born - Oppenheimer approximation was used. The first one served to separate the electron and nuclear motions. The second one served to separate the proton motion from the motion of heavier species. For entirely nonadiabatic reaction the transition probability was calculated in the first order of the perturbation theory[22,23]

$$W=Z^{-1}\sum_{m,n}\exp(-E_m^i/k_BT)\langle\psi_i(q)\chi_m^i(r,q)|V_i|\chi_n^f(r,q)\psi_f(q)\rangle^2\delta(E_i-E_f) \qquad (7)$$

where Z is the vibrational partition function of the initial state, E^i and E^f are vibrational energies of nuclei, V^i is the electron matrix element $\chi_m^i(r;q)$ and $\chi_n^f(r;q)$ are the wave functions of the proton, $\psi_i(q)$ and $\psi_f(q)$ are the wavefunctions of all the other nuclei. In classical limit for the latters using Condon approximation we obtain[12,22-24]

$$W=(1/\Omega)\int dR\ \exp[-U(R)/k_BT]\ W(R) \qquad (8)$$

$$W(R)=Z_p^{-1}\sum_{m,n}\exp(-E_m^i/k_BT)\ \langle\chi_m^i(r,R)|\chi_n^f(r,R)\rangle^2\ W_s(\Delta F+E_n^f-E_m^i) \qquad (9)$$

where U(R) is the intermolecular interaction potential between the proton donor and proton acceptor, Ω is the volume of the system, W(R) is the probability of proton transfer at fixed distance R between the donor and the acceptor, E_m^i and E_n^f are the vibrational excitation energies for the proton, W_s is the probability of the transition for all other degrees of freedom which in classical limit has the form of Eq.1.

For reaction which are close to thermoneutral ($\Delta F=0$) the major contribution to the sum in Eq.9 is due to the term corresponding to the tunnelling of the proton between unexcited initial and final vibrational states (m=n=0) and the expression for the transition probability is similar to that for the electron transfer. However in general the transitions between excited vibrational states of the proton also give contribution to the transition probability. The physical mechanism of the transition may be explained in terms of generalized Franck - Condon principle. Due to a fluctuation of molecular surroundings, the matching of vibrational proton energy levels between which the proton tunnelling occurs takes place.

Different overlapping of the wave functions of various isotopes of hydrogen explains in particular the proton/deutron kinetic isotope effect[25-27]. It should be emphasized that the distance dependence of the transition probability W(R) for reactions of proton transfer is much stronger than for the electron transfer. Therefore the approach of the reactants may play very important role in these reactions. It explains in particular a temperature dependence of kinetic isotope effect[25-27] and may also lead to a distortion of Arrhenius temperature dependence of the transition probability.

A simple parabolic model for the intermolecular potential U(R) used in ref.24 gave the following temperature dependence for the transition probability

$$W \sim \exp\left[-d/(a+bT)\right] \quad (10)$$

where a, b and d are quantities determined by the parameters of physical model.

It may be seen that if a<<bT, we obtain usual Arrhenius temperature dependence. In the opposite limit a>>bT the temperature dependence becomes exponential[28]

$$W \sim \exp(+\gamma T) \quad (11)$$

More complicated models for intermolecular interaction (in particular, Morse-like potentials) were used in ref.27 in the discussion of kinetic isotope effect. It was shown, in particular, that the approach of reactants lead to a decrease of the kinetic isotope effect (see also [29,30]).

The effects of this type are called distance dependent tunnelling or fluctuational preparation of the the potential barrier[11].

Similar to electron transfer reactions a simple model for proton transfer between two unexcited vibrational states gives a quadratic dependence of the exponent on the electrode potential in the expression for the electric current (see Eq.2). However corresponding experimental dependence is linear in a broad interval of potentials. The incorporation of the transitions between the excited vibrational states of the proton improves agreement between the theory and experiment. Still better results may be obtained in the charge variation model suggested in ref.31. This model takes into account possible variation of the charge of the adsorbed hydrogen due to the fluctuations in the solvent. The Newns - Anderson approach was used to calculate the energy of the electron and the charge of the adsorbed hydrogen. In fact it was the first application of Anderson model to electrochemical kinetics. Variation of charge of the adsorbed hydrogen in the activation-deactivation process leads to a change of its interaction with the solvent polarization which becomes non-linear in the polarization. It produces a distortion of the shape of the diabatic free energy surface of the final state which becomes nonparabolic. This may result in an exponential dependence of the current on the electrode potential in a broad range of potential variation.

4. Concluding remarks

The above discussion demonstrates that the charge transfer reactions at electrochemical interfaces represent quantum mechanical phenomena. The quantum mechanical effects are due to both electrode and solution sides of the reaction system. The electrochemical reactions are a part of a broad class of charge transfer process. The theory of these processes is developing in various directions. In particular, a great attention during recent years is paid to adiabatic reactions[32-37]. A stochastic approach is used for the calculation of the rate constant. On the other hand similar results may be obtained in the framework of the effective Hamiltonian method developed earlier[10,12]. For adiabatic reactions electron transmission coefficient is equal to unity and the preexponential factor is proportional to an effective frequency ω_{eff} which depends on the frequency and relaxational characteristics of nuclear modes.

For electrochemical reactions theoretical treatment of adiabatic transitions is complicated due to multielectron character of the process. The results obtained in various models[38-41] are rather close to those which follow from the model suggested in ref.42. Quite new field of the theory of electrochemical kinetics is related to electron transfer reactions at superconductor electrodes which is discussed in one of the lectures included in this volume.

References

1. R.Gurney, *Proc.Roy.Soc.* **A134** (1931) 1371.
2. R.A.Marcus, *J.Phys.Chem.* **24** (1956) 966, 979.
3. R.A.Marcus, *J.Phys.Chem.* **26** (1957) 867.
4. V.G.Levich and R.R.Dogonadze, *Dokl.Akad.Nauk SSSR, ser.Fiz.Khim.* **124** (1959) 123.
5. R.R.Dogonadze and Yu.A.Chizmadzhev, *Dokl.Akad.Nauk SSSR, ser.Fiz.Khim.* **145** (1962) 849.
6. R.R.Dogonadze, A.M.Kuznetsov and Yu.A.Chizmadzhev, *Zh.Fiz.Khim.* **38** (1964) 1195.

7. R.A.Marcus, *Can.J.Chem.* 37 (1959) 155.
8. R.A.Marcus, *J.Chem.Phys.* 38 (1963) 1335, 1965.
9. H.Gerischer, *Z.Phys.Chem. N.F.* 26 (1960) 223, 325.
10. A.M.Kuznetsov, J.Ulstrup and M.A.Vorotyntsev, in *Chemical Physics of Solvation*, eds. R.R.Dogonadze, E.Kalman, A.A.Kornyshev and J.Ulstrup, (Elsevier, Amsterdam - Oxford - New York - Tokyo, 1988), Part C, p.163
11. A.M.Kuznetsov, *Modern Aspects of Electrochemistry* 20 (1989) 95
12. R.R.Dogonadze and A.M.Kuznetsov, *Comprehensive Treatise of Electrochemistry* 7 (1983) 1
13. J.Ulstrup,*Charge Transfer Processes in Condensed Media*, (Springer-Verlag, Berlin - Heidelberg - New York, 1979)
14. A.M.Kuznetsov, *Nouv.J.Chemie* 5 (1981) 427
15. H.J.Kim and J.T.Hynes, *J.Phys.Chem.* 94 (1990) 2736
16. R.R.Dogonadze and A.M.Kuznetsov, *Kinetics of Heterogeneous Chemical Reactions in Solutions, in Itogi Nauki i Tekhniki* (VINITI, Moscow, 1978), Ser.Kinetics and Catalysis, Vol.5.
17. A.M.Kuznetsov and Yu.I.Kharkats, *in The interface structure and electrochemical processes at the boundary between the two immiscible liquids, ed. V.E.Kazarinov* (Springer-Verlag, Berlin - Heidelberg, 1987) p.11.
18. E.D.German and A.M.Kuznetsov, *Khim.Fiz.* 7 (1988) 305
19. E.D.german and A.M.Kuznetsov, *Electrochim.Acta* 26 (1981) 1595
20. A.M.Kuznetsov, *Elektrokhimiya* 7 (1971) 1067
21. Yu.I.Dakhnovskii and A.A.Ovchinnikov, *Dokl.Akad.Nauk.SSSR* 270 (1983) 119
22. R.R.Dogonadze, A.M.Kuznetsov and V.G.Levich, *Electrochim.Acta* 13 (1968) 1025
23. R.R.Dogonadze, A.M.Kuznetsov and V.G.Levich, *Elektrokhimiya* 3 (1967) 739
24. E.D.German, R.R.Dogonadze, A.M.Kuznetsov, V.G.Levich and Yu.I.Kharkats, *Elektrokhimiya* 6 (1970) 350
25. E.D.German, R.R.Dogonadze and A.M.Kuznetsov, *J.C.S.*Faraday Trans.II 76 (1980) 1128
26. E.D.German and A.M.Kuznetsov, *J.C.S.Faraday Trans.I* 77 (1981) 397
27. E.D.German and A.M.Kuznetsov, *J.C.S.Faraday Trans.II* 77 (1981) 2203
28. L.I.Trakhtenberg, *Zh.Fiz.Khim.* 54 (1980) 1324
29. J.Suhnel and K.Gustav, *Chem.Phys.* 70 (1982) 109
30. D.Borgis and J.T.Hynes, *in The Enzyme Catalysis Process*, ed.A.Cooper

(NATO ASI SERIES, Plenum, 1989) p.96
31. A.M.Kuznetsov, *J.Electroanal.Chem.* 159 (1983) 241
32. H.Kramers, *Physica* 7 11 (1940) 284
33. L.D.Zusman, *Teor.Eksperim.Khim.* 15 (1979) 227
34. J.T.Hynes, *J.Phys.Chem.* 90 (1986) 3701
35. H.Sumi and R.A.Marcus, *J.Chem.Phys.* 84 (1986)) 4272
36. A.M.Kuznetsov, *Electrochim.Acta* 32 (1987) 1271
37. J.N.Onuchic, *J.Chem.Phys.* 86 (1987) 3924
38. W.Schmickler, *J.Electroanal.Chem.* 204 (1986) 31
39. L.D.Zusman, *Elektrokhimiya* 21 (1985) 621
40. A.M.Kuznetsov, *J.Electroanal.Chem.* 241 (1988) 45
41. K.L.Sebastian, *J.Chem.Phys.* 90 (1989) 5056
42. R.R.Dogonadze, A.M.Kuznetsov and M.A.Vorotyntsev, *J.Electroanal.Chem.* 25 (1970) App.17

EXPERIMENTAL STUDY OF THE ELEMENTARY ACT OF PROTON TRANSFER

LEV I. KRISHTALIK
A.N.Frumkin Institute of Electrochemistry, Academy of Sciences of the USSR, Moscow,117071, Leninskii prosp.31, USSR

ABSTRACT

A review has been given on the problems of the barrierless proton donor discharge, proton tunneling and the dependence of the transfer coefficient α on the process parameters.New experimental data and new problems connected with them have been considered.

This paper is aimed at the discussion of some problems of the elementary act of proton transfer,mainly in electrode processes but,in some cases, also for homogeneous reactions. A special attention will be paid to the recently obtained data.

1.The process phenomenology.Barrierless reactions

For the most charge transfer processes,the typical potential energy scheme is represented by Curves 1 and 2 of Fig.1 (so called ordinary or normal process).Curves with a minimum reflect the existence of some equilibrium state characteristic of any charged particle in a condensed phase;the coordinate X represents symbolically all the coordinates determining the system energy.The potential barrier is formed by the intersection of the curves corresponding to the initial and final states.We see that quite general phenomenological considerations are the reason for the existence of the potential barrier and for the main tendencies in its variation with the change of the system parameters.These results are independent of the physical nature of the variable X,i.e. of the particular mechanism of reaction.

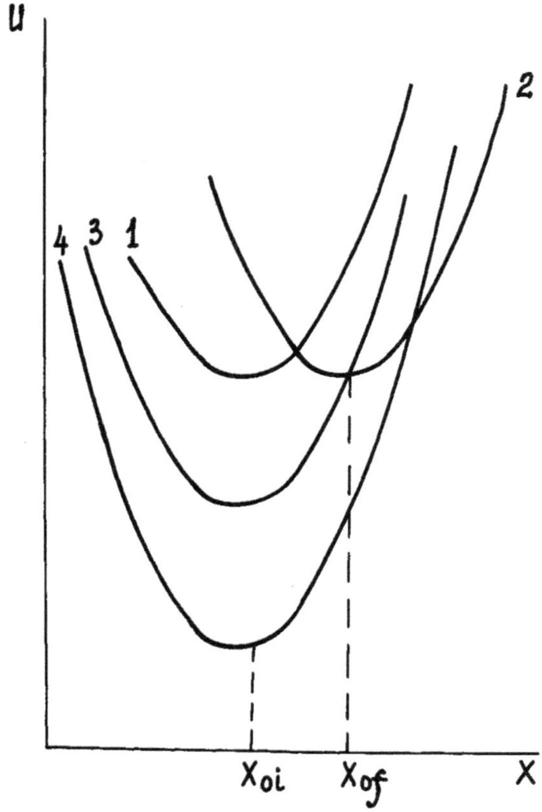

Fig.1.Scheme of potential curves

For the process being endothermic enough, the curve of initial state crosses the curve of the final state at its minimum (Curves 2 and 3, Fig.1). In this case, the activation energy equals to the energy of the elementary act of reaction without any additional barrier over this energy gap. Such kind of reactions was called barrierless ones[1]. The reverse process (2 → 3) is activationless. With the initial state energy level decreasing further, the barrier reappears (Curve 4; so called inverted region[2]). In the case of electrochemical reactions, however, the inverted region does not manifest itself. The reason is that in any electrode reaction one of the reactants is electron. In a metal electrode, electron can occupy any possible energy level, hence we have a wide spectrum of initial (or final) states. As it was shown first by Gerischer[3], in the normal region the optimal reaction path involves an electron in vicinity of Fermi level (±kT). In the inverted region, however, the situation changes substantially[4]. Let Curve 4 correspond to an electron at the Fermi level at a definite electrode potential. It is evident that the activation energy of this reaction is substantially higher than the activation energy of reaction involving an exited electron from the level corresponding to Curve 3 plus the excitation energy (the difference of minima 3 and 4). Therefore, for any initial equilibrium level lying under the Curve 3, the optimal reaction path is the electronic excitation to the level 3 with a subsequent thermal activation along the Curve 3. The total effective activation energy is equal to the difference of the energy minima 4 and 2, i.e. is equal to the energy of the elementary act of reaction; that means, it is a barrierless process. The reverse activationless reaction involves an electron transfer onto an exited level.

A change of the electrode potential, i.e. the change of the elementary act energy, gives rise to the relative shift of potential curves resulting in the change of activation energy. The latter makes a fraction α of the total change of the equilibrium energy. The quantity α is called usually the transfer coefficient or the symmetry factor. Its value in the normal region is determined by the ratio of the slopes of potential curves and is usually close to 0.5. In the barrierless region $\alpha = 1$.

The discharge (e.g. cathodic reaction) kinetics is described by the equation (see, e.g.[5]):

$$i = kc^{1-\alpha} \exp\{-(z-\alpha)\psi_1 F/RT\} \exp(\alpha\eta F/RT) \qquad (1)$$

where i is current density (i.e. the reaction rate per cm^2 expressed in electric units); k is rate constant; c - concentration of the discharging particle bearing the charge z ; ψ_1- potential at the point of localization of the center of charge of this particle related to the potential in the bulk of solution;η - overpotential,i.e. the absolute value of the difference of potentials under current and in equilibrium (the absence of any concentration gradients is supposed). In the case of an ordinary discharge, Eq.1 predicts,in a quantitative agreement with the experiment, a substantial effect of the reactant concentration and the double layer structure (i.e. of ψ_1-potential) on the reaction rate.For the barrierless discharge $\alpha = 1$ and both the reactant concentration and the double layer structure (at $z = 1$) cannot influence the polarization curve.This fact,together with the value of the polarization curve slope in semi logarithmic coordinates ($2.3\ RT / F = 59\ mV$ instead of $2.3\ RT/ 0.5\ F = 118\ mV$ at $\alpha = 0.5$) are the criteria of the experimental observation of the barrierless discharge.

The barrierless discharge can take place only for the processes endoergonic enough (in the terms of the elementary act theory,under conditions when the equilibrium energy of the elementary act exceeds the reorganization energy).In particular,for the hydrogen ion discharge this situation becomes possible at a low enough overpotential,i.e. at a low energy level of the initial state, combined with the high level of the final state (adsorbed hydrogen atom),i.e. with an endothermic hydrogen adsorption.Such kind of relationship one could expect for the hydrogen evolution on mercury in the region of low overvoltages.

To reach low hydrogen overpotentials at not too low current densities, i.e. current densities accessible to a reliable experiment, we have used the acidified concentrated solutions of salts of strongly adsorbing anions (iodide, bromide)[6]. As it is shown on Fig.2, under such conditions at the mercury cathode a process takes place which meets the criteria of a barrierless discharge: the slope of polarization curve equals to 0.06 V and the overpotential is independent of H^+ ions concentration and the double layer structure (I^-,Br^-concentration); at the same time, there are strong effects of these factors for the ordinary discharge in the same solutions.

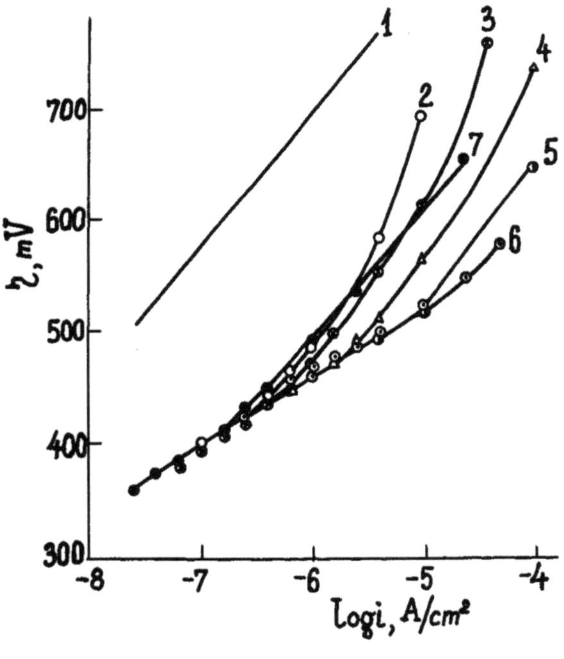

Fig.2.Polarization curves of cathodic hydrogen evolution on mercury at 25^{0}C in solutions: 1 - diluted acids solutions in absence of surface active anions;2 - 0.8 KI + 0.8 HCl;3 - 3.2 KI + 0.15 HCl;4 - 3.2 KI + 0.45 HCl;5 - 6.0 KI + 0.55 HCl;6 - 3.0 KI + 0.9 HCl + 1.3*10^{-3} $N(C_4H_9)_4Br$;7 - 6.6 NaBr + 0.85 HCl.

Some objections were put forward against the evidences given above.The first is the question why such an effect was not observed in other laboratories working at so low (or even somewhat lower) current densities. The answer is that not the current density by itself but the overpotential value is of importance here. Without surface active anions the overpotential at ambient temperatures is to high to enable the experimental observation of barrierless discharge (see the upper curve on Fig 2).By the way, at silver cathode the hydrogen overpotential is sufficiently low and at the same time the hydrogen adsorption is endothermic.In this system,a section of the polarization curve with the slope of 0.06 V was observed in many papers, but it wasn't realized that here a barrierless process is present. Only a special study which has demonstrated that for this section, in contrast to the section with the slope 0.12V,overpotential is independent of solution composition,revealed the existence of barrierless process (more detailed review on this system as well as on other cases of barrierless electrode processes see in[7]).

The second objection is that the slope of 0.06 V might be due not to the change of the mechanism of the elementary act but to the change of the rate determining step in this multistage reaction.Such a slope is to be expected for the reversible first step (discharge — ionization) with a subsequent slow, potential-independent step which rate is proportional to the surface concentration of adsorbed hydrogen.It was supposed, for the silver cathode, that this slow step is the process of surface diffusion of H atoms from the discharge sites to the sites of desorption.Possibly, such a hypothesis could be somehow substantiated for an inhomogeneous surface of solid electrode,but for homogeneous liquid mercury electrode it is hardly plausible.Another conceivable candidate for such a rate determining step is the activationless electrochemical desorption. The photoemission experiments has shown unambiguously that in acid enough solutions (corresponding to the experimental conditions of Fig.2) formation of H_2 molecule proceeds not via reaction atom plus molecule but by atom plus ion reaction ($H+H^+$;the experiment and its discussion see[8-10]).The analysis of the energetics of the elementary act also leads to conclusion that, at low overpotentials,activationless electrochemical desorption is practicable only for the atom plus ion reaction.However, in case of rate determining H + H^+ desorption the overpotential must depend on H^+ concentration,and this contradicts the experiment.

And the last objection was that in concentrated solutions of surface active salts the structure of the double layer may be rather complex, and the observed independence of the overpotential from the solution composition might be caused by some accidental compensation of different effects of some unknown nature.This reasoning,which from the very beginning did not seem to be really convincing,now has been disproved experimentally.Dr.Tsionskii has developed recently a new technique for study the electrode kinetics up to very high temperatures(300°C) which enables one to obtain quite reliable

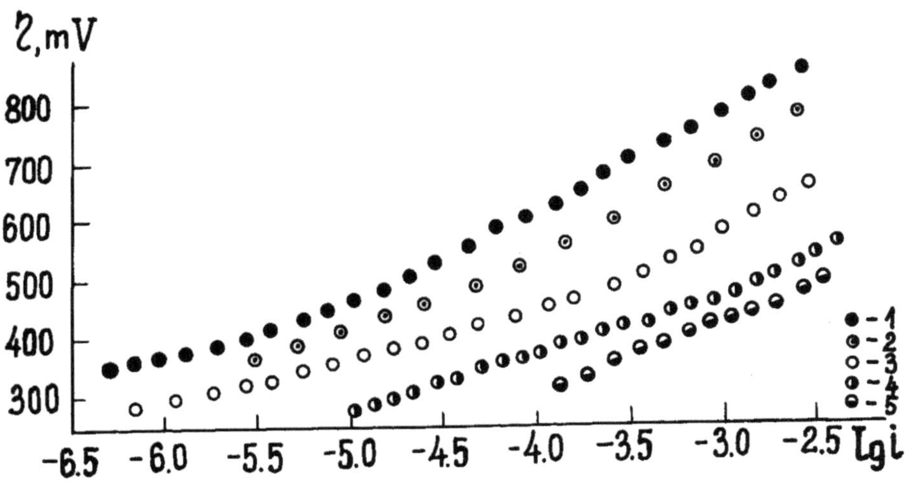

Fig.3. Polarization curves for hydrogen evolution on mercury in 1M HCl at temperatures:1-123°C;2-145°C;3-188°C;4-240°C;5-302°C.

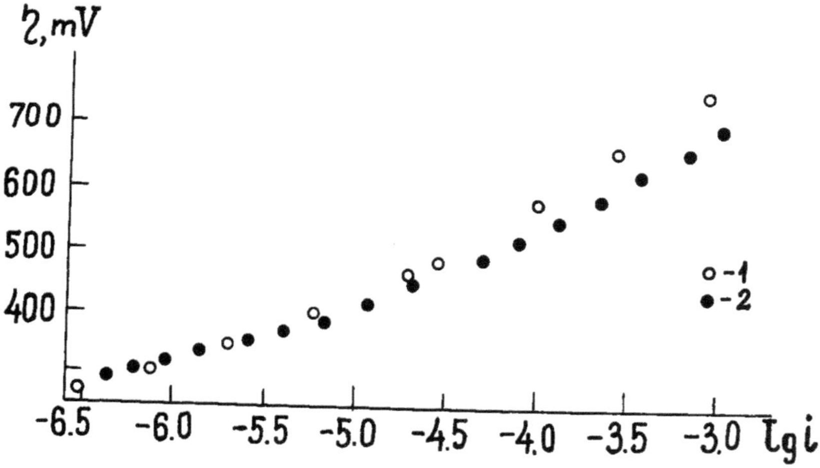

Fig.4.Polarization curves for hydrogen evolution on mercury at 145°C in solutions: 1-0.5 M HCl+0.005 M $N(C_4H_9)_4Br$; 2-0.5 M HCl.

data in dilute electrolyte solutions. A strong increase in temperature results in a drastic decrease in overpotential,and the sections of polarization curves with a slope corresponding to $\alpha \cong 1$ become observable in surface inactive solutions at rather high current densities (Fig.3).

In full accordance with theory, it was observed that the addition of a surface active cation causes a marked effect in a region with $\alpha=1/2$ but no effect in barrierless region (Fig.4).The data for high temperatures and the data for moderate ones (the latter - in presence of iodide) form a single temperature dependence (Fig.5).Therefore, there are in principle no differences in behavior of the system in presence and in absence of strongly adsorbing anions.Hence, the specific features of the region with $\alpha = 1$ cannot be ascribed to some unidentified effects due to the complex structure of the double layer.

The final evidence of the existence of barrierless discharge seems to be of principal importance. First, the conclusion on its existence was derived from rather general phenomenological considerations, and, hence, it was substantial to test this conclusion to prove the picture on the whole. Second, as a matter of fact, the purely phenomenological approach does not give yet an exhaustive substantiation of barrierless processes: the further analysis shows that to that end it is necessary to take into account the existence of two principally different degrees of freedom, namely quantum and classical ones. Hence, barrierless processes appear to be important for the deeper understanding of the mechanism of the elementary act of charge transfer.

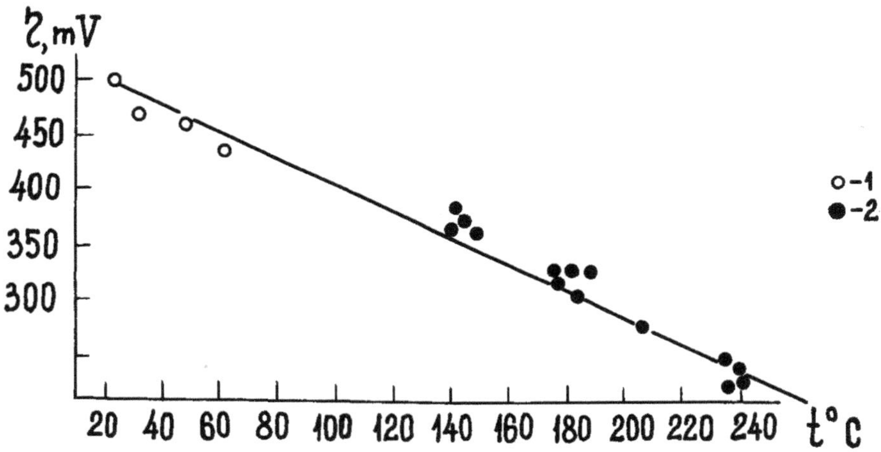

Fig.5. Temperature dependence of hydrogen overpotential in the barrierless discharge region at current density $10^{-5.4}$ A cm^{-2}; 1- in presence of iodide; 2- in absence of surface active additives.

2. Proton transfer mechanism: gradual bond stretching or tunneling ?

One of specific features of the modern quantum mechanical theory of charge transfer reactions, in particular proton transfer reactions is a division of all degrees of freedom of the system in two main types: into classical and quantum ones. The criteria determining the type of the mode are the inequalities $\hbar\omega \ll kT$ and $\hbar\omega \gg kT$ where ω is the frequency characteristic for a given degree of freedom[12]. For the classical degrees of freedom, an overbarrier transition is typical, its probability depending on the probability of thermal fluctuations having the given energy. The quantum degrees undergo an underbarrier tunnel transition taking place after the energy levels of the quantum particle in the initial and final states have been equalized as a result of thermal fluctuations of the classical subsystem. The scheme of such a process, viz. a proton discharge is shown on Fig.6. For proton, characteristic frequencies of vibrations are ≥ 15 kT, hence its behavior must be substantially quantum one[4].

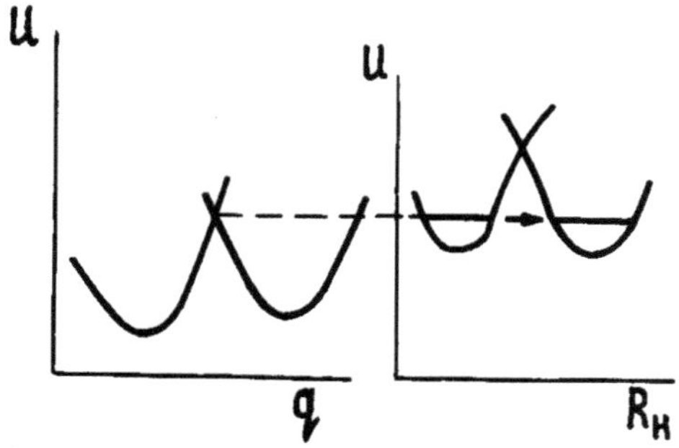

Fig.6. Scheme of potential curves for the ordinary discharge region in function of the classical coordinate (the medium polarization q) at the left side and in function of the quantum coordinate (the proton position R_H) at the right side.

According to Dogonadze - Kuznetsov - Levich model, proton tunnels predominantly from the ground level of the initial state to the ground level of the final state. Proton motion affects the pre-exponential factor in the reaction rate equation while the activation energy is determined by the movement of classical subsystem, in our case by the solvent reorganization. Here lies the principal distinction between this model and the model by Horiuti and Polanyi which was widely accepted before that. In the latter, the energy expenditure necessary to stretch the hydrogen covalent bond was considered as the principal reason for the emergence of the activation barrier. In subsequent modifications of the model, the possibility of proton tunneling through the top of the barrier was taken into account but the main idea on the bond stretching energy as the main (if not the single) component of activation energy was kept intact.

It is worth to stress that, from phenomenological point of view, both models are similar (the generalized coordinate X at Fig.1 can be considered as a coordinate q describing the medium polarization as well as the proton coordinate R_H). Hence, the main general regularities of the process may be interpreted successfully in the framework of the both models. Therefore, to choose between these two alternatives one has to study such phenomena for which the models predict qualitatively different results.

One of phenomena of the kind is barrierless discharge. If X (Fig.1) is identical to R_H (or involves both R_H and q) then the state immediately after discharge, i.e. an adsorbed hydrogen atom, is labile. It is just one of points on the single potential curve having only one minimum. The absence of some discrete, separate state for adsorbed hydrogen means in fact the impossibility of discharge.

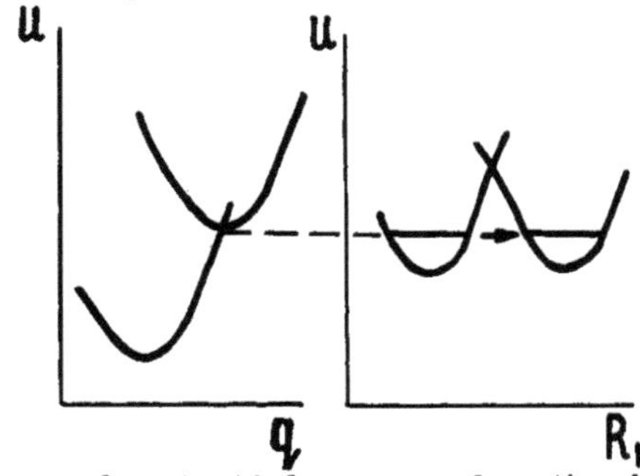

Fig.7. Scheme of potential curves for the barrierless discharge region in function of the classical coordinate (the medium polarization q) at the left side and in function of the quantum coordinate (the proton position R_H) at the right side.

The process could occur if the newly formed H atom could combine immediately with the second hydrogen ion (an activationless electrochemical desorption which, as was discussed above, involves just H^+ ions).However, in that case reaction will be of the second order in respect to hydrogen ions whereas the first order was found experimentally.

On the other hand, in the framework of quantum mechanical model the situation looks quite different (see Fig.7).The barrier is absent only along the medium coordinate q but doesn't disappear along the proton coordinate R_H. In other words, in this model two discrete different states remain both for ordinary and barrierless processes.Adsorbed hydrogen possesses some finite life time determined by the probability of the tunneling in the reverse direction (this time increases due to repulsion between two products of the H_3O^+ discharge,viz. water molecule and H atom; see e.g. chapter 5 in[7]). During this time span, a second hydrogen ion can approach the reaction zone and enter the activationless electrochemical desorption process; this second step doesn't affect the total reaction rate.Therefore,the very existence of the barrierless discharge is the consequence of the qualitatively different behavior of quantum and classical degrees of freedom and evidences in favor of the main concept of quantum mechanical model of the elementary act of charge transfer.

The other consequence of this model which also agrees with the experiment is the practical equality of pre-exponential factors for the barrierless and ordinary discharge of hydrogen ions. Indeed,as it is evident from comparison of Figs.6 and 7,at quite different shape of barriers along the medium coordinate the barriers along the proton coordinate are similar,but it is precisely that barrier which determines the tunneling probability (some difference in pre-exponentials might arise as a result of different contributions of vibrationally-excited states in normal and barrierless regions).On the other hand, at the activation barrier depending substantially on the proton coordinate,the tunneling (at the top of the barrier) could take place only in ordinary but not in barrierless discharge,that means the pre-exponentials had to be substantially different.

There are several other evidences in favor of the quantum

mechanical model of proton transfer (for review see [7,14]).Let us just list them briefly:

a) decrease in pre-exponential factor with increasing energy of hydrogen adsorption on metal: a larger localization of proton renders its underbarrier transition to be more difficult;

b) increase in H/T isotope separation factor with increasing hydrogen adsorption energy: the same reason as for point "a";

c) a different dependence of the isotope separation factor on electrode potential in acid and alkaline solutions in spite of virtually the same activation barrier height (more detailed discussion see below);

d) at constant overpotential, the activation energy of the proton donor discharge does not depend on the energy of the covalent bond to be ruptured but depends substantially on the solvent nature.

In all the cases listed above, the predictions of the quantum mechanical model were confirmed experimentally. At the same time, the model of activation via gradual bond stretching disagreed qualitatively with the experiment (or it was necessary,in framework of this model, to resort to several additional assumptions different for each of relations studied).

3. Dependence of α on the process parameters

Last years, an assertion was many times spoken out that α increases markedly with temperature rising (for review see [15]).This result was,however, obtained in the systems where the experimentally measurable value b can be affected substantially by different side effects and hence does not reflect the true value of α characteristic of the elementary act.These systems are,first, catalytically active metals for which changes of surface coverage with temperature and/or potential as well as interference of more than one reaction step are possible.The second type of systems used are concentrated aqueous acid solutions or non aqueous (alcoholic) solutions liquid at low temperatures.However, in these solutions two kinds of proton donors are present,namely hydrogen ions and undissociated molecules of acid (or contact ion pairs).The relative contribution of these donors depends on potential and temperature.

The technique of a pure electrochemical experiment at high temperatures developed by Dr. Tsionskii enabled us to study the temperature dependence of b for the process of hydrogen evolution on

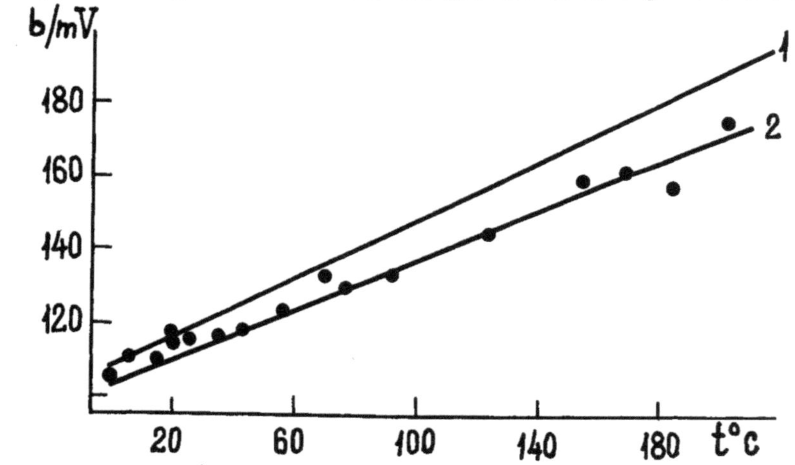

Fig.8. The dependence of the Tafel slope b on temperature: 1 - calculated at $\alpha = 1/2$ and ψ_1 = const; 2 - calculated at $\alpha = 1/2$ and ψ_1 depending on potential (inter- or extrapolated using the literature data on the double layer structure); the points are experimental data.

mercury in dilute aqueous acid solutions, i.e. without the interference of the side effects discussed above[16]. As can be seen from Fig.8, b rises with temperature somewhat slowlier than it corresponds to the simple relation $b = 2.3\ RT\ /\ \alpha F$ .However, the temperature increase causes an increase in correction to b value due to the dependence of ψ_1- potential on the electrode potential (the main contribution here originates from a substantial decrease of the overpotential, i.e. the range of potentials under study becomes closer to the potential of zero charge, and the nearer to p.z.c. the stronger is the dependence of ψ_1 on electrode potential). Taking into account this correction, the experimental data, within of error limits, conform with the constancy of α in the whole temperature interval studied.

The most complex problem which has not been solved till now is the problem of the dependence of α on potential. Two experimental findings are to be explained here: the constancy of $\alpha = 0.5$ in a very wide range of potentials (about 1.5 V) and a very narrow interval of energies (about 3 kT) in which a transition from $\alpha = 0.5$ to $\alpha = 1$ takes place. One could doubt in the last assertion because our previous data (Fig.2) were obtained in systems with a rather complex double layer structure, and hence one could suppose some distortion of the α – potential dependence in the region of the ordinary discharge. However, the new data (see Figs.3 and 4) are devoid of this drawback, and they show quite clear that the change of α proceeds really in a very narrow potential range. This fact has not find a reasonable explanation in framework of any model of the process because for that purpose one has to substantiate a very unusual shape of potential curves having a slope changing rather sharply from a large value to zero (at the curve minimum).

In the harmonic approximation, the true activation energy is expressed as follows

$$E^{\neq} = (E_s + \Delta I)^2 / 4 E_s \qquad (2)$$

where E_s is reorganization energy, ΔI is the energy of the elementary act (it may be both positive and negative). Assuming the dependence of the reaction rate on ΔI being determined by the corresponding dependence of $E^{\neq}$ we obtain the expression for the true transfer coefficient

$$\alpha = d E^{\neq} / d \Delta I = 0.5 (1 + \Delta I / E_s) \qquad (3)$$

The linear dependence of α on ΔI was observed for several homogeneous reactions of proton transfer between OH- (or NH-) acids and O- (or N-) bases[17] but in electrode reactions, as it was stressed above, α is constant. An attempt to explain this constancy by the assumption that $\Delta I \ll E_s$ contradicts the experiment. As it was discussed above, the barrierless process starts at the moment when $E^{\neq} = E_s$. This circumstance allows one to find the value of E_s. For proton discharge, as well for already mentioned homogeneous reactions, it was evaluated as $\cong$ 1 eV [18]. A similar value results also from theoretical calculations of the reorganization energy for the discharge of an aqueous ion $H_9O_4^+$ (here an account was taken of the phenomenon that ,with a real metal, the image forces don't contribute significantly into reorganization energy)[19,20].

There were some attempts to explain the constancy of α by some effects specific for electrode reactions only.One of explanations is based on the idea on a partial charge transfer between adsorbed hydrogen and the metal[21]. In this case, the degree of charge transfer depends on potential,and the reorganization energy depends, in its turn, upon the charge of the hydrogen atom.At the definite choice of the electron repulsion parameter (in the Anderson model of adsorption),one can obtain a very good constancy of α in a wide range of potentials. The result, however, is sensitive to the choice of the repulsion parameter and there is no independent substantiation of this parameter.

The other way to decrease the curvature of the polarization curve (i.e. to make α more constant) consists in the account of some approachment of hydronium ion to the electrode under the action of electric field.This results in facilitating of proton tunneling, so in parallel to the change of activation energy with changing ΔI the pre-exponential factor changes too.The existence of this effect seems to be doubtless (more detailed discussion see below) but it can hardly be responsible for the whole picture because a constant value of α was observed also for discharge of a neutral proton donor – water molecule – which approachment to the electrode is much less influenced by the electric field.

Last years, the experimental data were obtained showing the α constancy in the wide energy range not only for electrochemical but for homogeneous proton transfer reactions too.Dissimilarly to the processes discussed before, in that case one deals with reactions of CH-acids and not of OH- or NH- ones. The most striking example is the proton transfer between cation-radicals of polymethylbenzenes and substituted pyridines having constant $\alpha \cong 0.3$ in the energy range of 1.4 eV[22].In this case, it is impossible to consider a specific effect of an electric field and one has to find some other effects. One of possible reasons here is the involvement in the process of some vibrationally-excited states, both initial (mainly for endothermic reactions) and final ones (the latter for exothermic processes)[23,24]. The change of ΔI results in the change of the excited states contribution, and the $ln\ k - \Delta I$ appears to be closer to a straight line than it would follow from Eq.2.

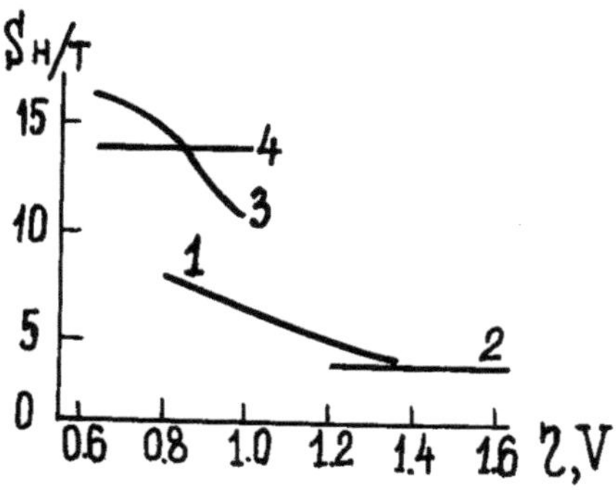

Fig.9. The dependence of the isotope separation factor on the hydrogen overpotential at mercury (curves 1 and 2) and at gallium (curves 3 and 4) in acid (1 and 3) and alkaline (2 and 4) solutions. The reaction path in alkaline solutions

$H_2O + e = H_{ads} + OH^-$ $\quad$ $H_{ads} + H_2O + e = H_2 + OH^-$

in acid solutions

$H_3O^+ + e = H_{ads} + H_2O$ $\quad$ $H_{ads} + H_3O^+ + e = H_2 + H_2O$

As it was mentioned before, the field influencing the approach of an ion to the electrode affects the probability of proton tunneling. The data proving this effect are presented at Fig.9 showing the dependence of the isotope separation factor (H/T) on the electrode potential in acid and alkaline solutions on mercury and gallium electrodes[25]. For cathodic hydrogen evolution proceeding according to the discharge - electrochemical desorption mechanism, the separation factor measured experimentally is determined by corresponding factors for the each of the steps

$$2/S = 1/S_d + 1/S_{e.d.} \qquad (4)$$

In acid solutions, where the discharging particle is H_3O^+ ion (most probably, hydronium ion takes part in electrochemical desorption too) S decreases markedly with increasing potential whereas in alkaline solutions, where the only proton donor is the water molecule, S is constant. This result can be explained only by the action of the electric field attracting the cation to the electrode but not influencing practically the position of a neutral molecule. A closer approach of the ion, i.e. a decrease in the proton transfer distance facilitates proton tunneling. The easier the tunneling is the smaller is the difference in the rate of transfer of light and heavy isotopes, hence the S value decreases.

Reactants come closer before proton tunneling in the course of thermal motions (the process is facilitated by electric field if present). At short intermolecular distances which are especially favorable for tunneling a further approachment is hindered by strong repulsion forces. The tunneling of heavy isotope is less probable than the light one. Hence, for heavy-hydrogen "proton donors", it is preferable to come to a shorter distance in spite of an additional energy expenditure against repulsion forces. This can explain the higher effective activation energy which was many times observed for the deuteron transfer reaction as compared to the proton transfer. Earlier, one explained this effect in the framework of the Horiuti - Polanyi model as due to the difference of zero energies of H- and D- donors. However, from the point of view of the Dogonadze - Kuznetsov - Levich model this difference does not play any substantial role because the tunneling of both proton and deuteron takes place at the equalized energy levels in initial and final states (see the scheme on Fig.6; some small effect remains connected with a slightly different positions of the $U(q)$ minima at the left part of the figure; the corresponding difference equals to the difference of differences of H and D zero energies in initial and final states).

We have explained the difference in activation energies for H- and D- donors by the effect of nonequal distances of approach of reactants. Hence we could expect this difference to disappear if some reason will make the reactants to transfer H^+ and D^+ at the same distance. Such a situation seems to have place at the catalytic hydrolysis by serine proteases. For the esters hydrolysis, the rate determining step is the nucleophillic attack on carbon atom of acylenzyme with a simultaneous detachment of proton by imidazol group of His-57 [26]:

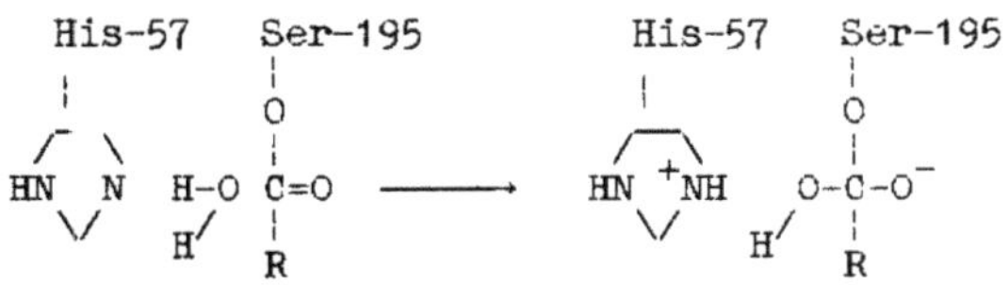

(5)

A mutual disposition of two external reactants is fixed rather rigidly, and H_2O (or D_2O) molecule cannot approach nitrogen at any arbitrary distance because its oxygen attacks simultaneously carbonyl oxygen. It was shown that for two serine proteases -α-chymotrypsin (native and modified) and β-trypsin - activation energies of ester hydrolysis reactions in light and heavy water coincide, only pre-exponentials being different reflecting different

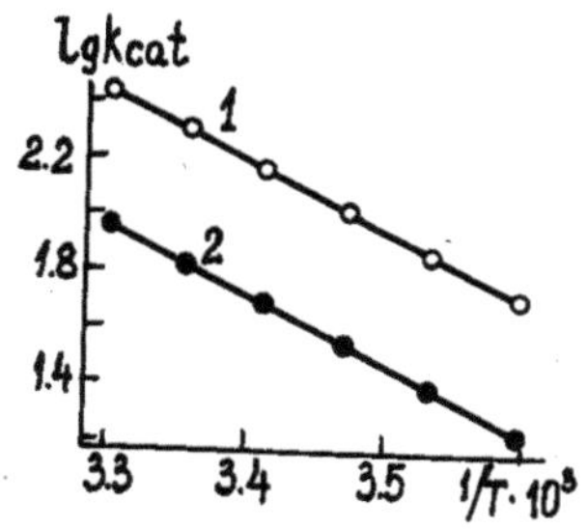

Fig.10. The temperature dependence of the rate of hydrolysis of N-acetyltyrosine ethyl ester by α-chymotrypsine in H_2O (1) and D_2O (2).

tunneling probabilities (for an example see Fig.10[27], for a review [7]).A similar result was obtained recently for a nonenzymatic hydrolysis of imidazolbenzoic acid ethyl ester, a substance mimicking the structure of the serine proteases active site[28].

As it was mentioned above the approachment under the action of electric field cannot explain neither the linearity of polarization curves in alkaline solutions nor the linearity of corresponding dependencies for homogeneous reactions of type of studied in[22].It was already pointed out that one of the possible explanations of the effect might be an account of the contribution of excited states.It should be noted that the rectilinear dependencies were observed for C-H acids while the curvilinear ones for O-H (N-H) acids. In the second case, proton transfer is preceded by formation of a strong hydrogen bond characteristic of a rather low barrier between two minima.In this case, already the first excited state is a single state in a two-well potential,and hence one hardly can speak here on the transition from one well to another with partaking of initial or final excited states.In the case of C-H acids, however, hydrogen bond is weak (if does exist at all) and hence the existence of excited states related to one of two possible proton localizations is quite probable.

The model of potential - dependent approach of proton donor to electrode proves to be insufficient also on the basis of new data on the kinetic isotope effect at barrierless and ordinary discharge presented at Fig.11[11]. In contrast to previously discussed data of Fig.9,here the other value is given, namely the ratio of discharge currents of hydronium ions in light and heavy water and not the separation factor S.The latter depends upon both reaction steps (see Eq.4), the first one only on kinetic parameters of the discharge step.It is evident that kinetic isotope effect decreases with rising potential for barrierless process but increases (however slowlier) for ordinary discharge.The last effect was previously observed , at lower temperature,by Post and Hiskey[29].

Starting from the balance of repulsion and attraction forces and taking into account an increase in tunneling probability with decreasing distance of H_3O^+ ion from the electrode, one can find the distance corresponding to the maximum reaction probability. It is

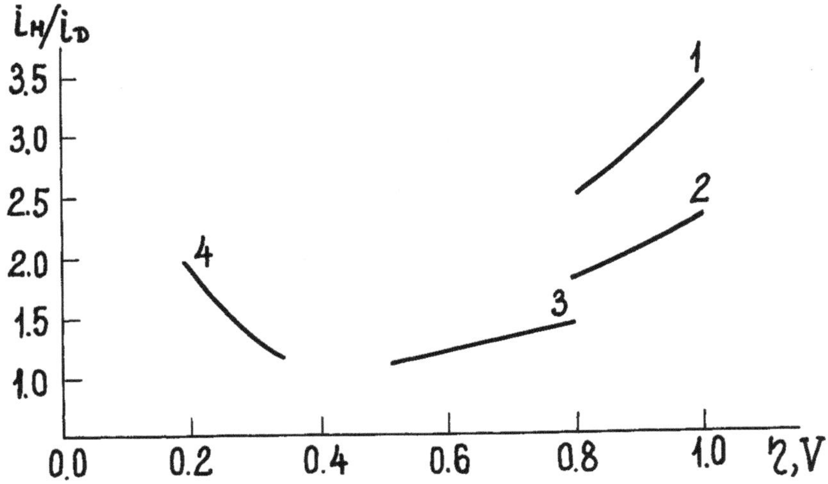

Fig.11. The dependence of the isotope kinetic effect on overpotential. 1,2,3 – ordinary discharge, 4 – barrierless discharge: 1 – 30°C, 2 – 70°C, 3,4 – 140°C.

determined by the following condition[30]:

$$\frac{\partial \sigma}{\partial R} + (1-\alpha)\left(\frac{\partial U_i}{\partial R} - \frac{\partial \gamma}{\partial R}\Phi\right) + \alpha \frac{\partial U_f}{\partial R} \qquad (6)$$

Here σ is tunneling factor, U_i and U_f are repulsion energies for H_3O^+ and H_2O, Φ - potential drop in double layer, γ - fraction of Φ corresponding to the potential drop between metal and the point at distance R from the latter. To find the optimal value of R one should express explicitly the dependence of $\sigma, U_{i,f}$ and γ on R. It is clear, however, that at α = 1 the term depending on Φ disappears from Eq.6, i.e. at barrierless process the optimal discharge distance does not depend on potential. Correspondingly, the isotope effect is to be independent of potential. From the other hand, it is possible to show that at α = 1/2 the isotope effect should decrease with rising potential. The experimental data contradict qualitatively these conclusions.

For the region of an ordinary activated proton transfer, an analysis of tunneling probability with an account of contribution of vibrationally-excited states was done in[23,24].It shows that for endothermic processes the isotope effect increases when the energy difference decreases (in our case, when overpotential rises).This is due to a gradual decrease of contribution of excited states for which tunneling is more facile;in general,any hindrance for tunneling gives some relative preference to the isotope which tunnels easier,i.e. increases discharge rate of H-donor in comparison to D.In our case,this effect of excited states appears to be more important than the opposite influence of electric field on the proton approachment, and hence we observe, in total, a slow increase of isotope effect with potential.

In the case of barrierless discharge, the situation becomes qualitatively different.Due to a much larger value of vibrational quantum for the light isotope,already its first excited state lies in the energy range corresponding to ordinary discharge.Therefore activation energy for the proton transfer from the excited level becomes higher than from the ground one,their difference increasing with the rise of overpotential resulting in an increase of contribution of ground state.On the contrary,for the heavy isotope, in the same overpotential range, the process with participation of the first excited level remains barrierless, and its effective activation energy is the same as for the ground level (the situation here is similar to the problem with involvement of excited electronic levels of metal,see Fig.1).As a result, with potential increasing the contribution of vibrationally excited levels decreases for H and does not change for D, and hence the relative rate of H discharge diminishes.

The question may arise why the shape of curves for isotope effect presented on Figs.9 and 11 is qualitatively different (we can compare only the curves in the region of ordinary discharge).As it was mentioned above, the data of Fig.9, in contrast to Fig.11, reflect the isotope effect not for one but for two steps of the total process,namely discharge and electrochemical desorption steps;

as it is clear from Eq.4, the prevailing contribution is due to the step having the smaller isotope effect. It seems to be quite probable that this step is the electrochemical desorption which is activationless in the range of high overpotentials[31]. For the activationless process, the involvement of initial excited states is practically excluded; as to final vibrationally excited states, the corresponding frequency for H_2 (or D_2) is very high and one could expect their rather low contribution(and, probably, virtually constant in a definite potential range) Therefore, the potential dependence of the separation factor for electrochemical desorption will be determined not by the excited states contribution but by the influence of the electrode electric field upon the optimal discharge distance. This results in a decrease of separation factor with increasing potential in acid and in its constancy in alkaline solutions.

We see that taking into account the contribution of vibrationally - excited states together with the influence of the field of electrode on the proton donor localization enables us to explain qualitatively the general picture observed. However, we lack till now a full clarity based on the quantitative description of all details.

References

1. L.I.Krishtalik, *Zhurn. fiz. khimii (Russ.)* 34 (1960) 117.
2. R.A.Marcus, *Faraday Discuss. Chem. Soc.* 74 (1982) 7.
3. H.Gerischer, *Z. phys. Chem. N.F.* 26 (1960) 223.
4. R.R.Dogonadze, A.M.Kuznetsov and V.G.Levich, *Elektrokhimiya (Russ.)* 3 (1967) 739; *Electrochimica Acta* 13 (1968) 1025.
5. L.I.Krishtalik, in *Comprehensive Treatise of Electrochemistry*, eds. B.E.Conway, J.O'M.Bockris, E.Yeager, S.U.M.Khan and R.E.White, 7 (Plenum, New York etc.,1983) p.87.
6. L.I.Krishtalik, *Zhurn. fiz. khimii (Russ)* 39 (1965) 642; *Electrochimica Acta* 13 (1968) 1045.
7. L.I.Krishtalik, *Charge Transfer Reactions in Electrochemical and Chemical Processes*,(Consultants Bureau, New York etc.,1986).

8. Z.A.Rotenberg, V.A.Lakomov and Yu.V.Pleskov, *Elektrokhimiya (Russ.)* 6 (1970) 515; *J.electroanalyt. Chem.* 27 (1970) 403.
9. V.A.Benderskii, A.G.Krivenko and A.A.Ovchinnikov, *J. electroanalyt. Chem.* 111 (1980) 19.
10.L.I.Krishtalik, *J. electroanalyt. Chem.* 130 (1981) 9.
11.L.B.Kriksunov, L.I.Krishtalik and V.M.Tsionskii, *Elektrokhimiya (Russ.)* 25 (1989) 692; *Electrochimica Acta* in press.
12.R.R.Dogonadze and A.M.Kuznetsov, *Elektrokhimiya (Russ.)* 3 (1967) 1324.
13.J.Horiuti and M.Polanyi, *Acta physicochimica URSS* 2 (1935) 505.
14.L.I.Krishtalik, *Faraday Discuss.Chem.Soc.* 74 (1982) 205.
15.B.E.Conway, in *Modern Aspects of Electrochemistry*, eds. B.E.Conway,R.E.White and J.O'M.Bockris 16 (Plenum,New York,1985).
16.V.M.Tsionskii, L.I.Krishtalik and L.B.Kriksunov, *Electrochimica Acta* 33 (1988) 623.
17.M.Eigen, *Angew.Chemie (Internat. Edition)* 3 (1964) 1.
18.L.I.Krishtalik, *J.electroanalyt.Chem.* 136 (1982) 7.
19.P.G.Dzhavakhidze, A.A.Kornyshev and L.I.Krishtalik, *J.electroanalyt.Chem.* 228 (1987) 329.
20.L.I.Krishtalik, N.M.Alpatova and E.V.Ovsyannikova, *Electrochimica Acta* in press.
21.A.M.Kuznetsov, *J.electroanalyt.Chem.* 159 (1983) 241.
22.C.I.Schlesener, C.Amotore and J.K.Kochi, *J.Amer.Chem.Soc.* 106 (1984) 7472.
23.Yu.I.Kharkats and J.Ulstrup, *J.electroanalyt.Chem.* 65 (1975) 555.
24.A.M.Kuznetsov, *J.electroanalyt.Chem.* 151 (1983) 227.
25.V.M.Tsionskii and L.I.Krishtalik *Elektrokhimiya (Russ.)* 6 (1970) 265; *J.electroanalyt.Chem.* 31 (1971) 363.
26.A.Fersht, *Enzyme Structure and Mechanism*,(W.H.Freeman,Reading and San Francisco,1977).
27.D.E.Khoshtariya, V.V.Topolev and L.I.Krishtalik, *Bioorganicheskaya Khimiya (Russ.)* 4 (1978) 1341.
28.N.G.Goguadze, D.E.Khoshtariya, L.I.Krishtalik, A.K.Yatsimirskii, S.A.Deiko and A.A.Neverov, *Zhurn. fiz. khimii (Russ.)* 63 (1989) 248.
29.B.Post and C.F.Hiskey, *J. Amer. Chem. Soc.* 72 (1950) 4203; 73 (1951) 161.

30.L.I.Krishtalik, V.M.Tsionskii and L.B.Kriksunov, *Elektrokhimiya (Russ.)* in press.
31.L.I.Krishtalik, *Elektrokhimiya (Russ.)* 27 (1991) N° 3.

ELECTRON TRANSFER REACTIONS IN SOLUTION: THE ROLE OF SOLVATION DYNAMICS

BARTON B. SMITH, AND JAMES T. HYNES
Department of Chemistry and Biochemistry, University of Colorado
Boulder, CO 80309-0215, USA

ABSTRACT

A brief account is given of some highlights of recent studies of the influence, for electron transfer (ET) reactions, of solvation dynamics. The general necessity of a generalized Langevin equation (GLE) description for the solvation dynamics is illustrated, for the case of rapidly relaxing dipolar aprotic solvents. The GLE description allows both for solvent inertia and memory friction. Such a description goes well beyond that of most currently popular solvation dynamical theories. It is also shown that, for such solvents, the solvent dynamically induced activation barrier recrossings leading to departures from a Marcus Transition State Theory description are quite different in character from those described in many recent theories. Instead a Grote-Hynes theoretical approach is appropriate. In this theory, attention is focussed on events occurring at the barrier top, and not in the wells, as in other theories.

1. Introduction

Outer sphere electron transfer (ET) reactions occupy a special place in the catalogue of solution reaction classes. The reaction coordinate is a solvent coordinate, and the activation barrier owes its origin to the solvent. These key aspects have been recognized and described some time ago by Marcus[1] and Hush.[2] In particular, the formulation by Marcus for the ET reaction rate constant has seen widespread use and gained considerable acceptance.[3] In recent years, new aspects of the influence of the solvent on ET reaction rates have come under scrutiny. These include the role of the solvent dynamics in affecting the

ET rate (in addition to the solvent free energetic barrier effects included in Marcus Theory). In particular, the role of solvent dynamics in reducing the ET rate below its Transition State Theory value is of interest. Here we give a brief account of some of our recent efforts[4-6] to elucidate this question.

The outline of this contribution is as follows. We first give a brief account of the characterization of the solvent dynamics *per se*, focussing attention on a generalized Langevin description, including a time-dependent friction on the solvent coordinate. We then recount the impact, examined via Molecular Dynamics computer simulation, of the solvent friction on the ET rate constant. This impact can be understood in terms of recrossings of the transition state, such that the rate constant falls below its Transition State Theory value. In our survey, we restrict our focus to the highlights of all these aspects; the reader is referred to the original articles[4-6] for more extensive discussion and further details.

We consider here only the homogeneous ET case. Nonetheless, we expect that many of our conclusions will also apply, albeit in somewhat modified form, for solvation dynamics and ET reactions occurring at electrodes. Indeed, this is an active area of investigation.[5]

2. Solvation Dynamics

Before considering the ET reaction problem, it is useful to address the issue of solvation dynamics in the absence of a real reaction. For this purpose, we specialize for convenience to a model finite dipolar pair $A^{-1/2}A^{1/2}$, with fractional charges. Even though we do not address ET reactions just yet, it is useful to label that pair as "reactant" R and the symmetrically related pair $A^{1/2}A^{-1/2}$ as the "product" P. The solute is immersed in a solvent of finite dipolar molecules, which is[4] a caricature of methyl chloride, and is representative of several dipolar aprotic solvents currently under experimental investigation.[8] The detailed description of the solute and the solvent can be found in Ref. 4.

The relevant microscopic solvent coordinate is

$$\Delta E = H_R - H_P \tag{1}$$

which is the difference in Coulomb interaction potential energy, for a given configuration of all the solvent molecules, between the solvent and the solute in the P and R charge states. (In a dielectric continuum representation of the solvent, this same energy difference would be expressed in terms of the orientational polarization of the solvent; see Sec. 3.)

A convenient measure of the dynamics of ΔE is given by the normalized equilibrium time correlation function (tcf)

$$\Delta(t) = \langle(\delta\Delta E)^2\rangle^{-1} \langle\delta\Delta E \delta\Delta E(t)\rangle_R \ , \tag{2}$$

where $\delta\Delta E$ is the fluctuation

$$\delta\Delta E = \Delta E - \langle\Delta E\rangle_R \tag{3}$$

about the average when the solute is in the reactant state. MD results for $\Delta(t)$ for three different mass cases are shown in Fig. 1.

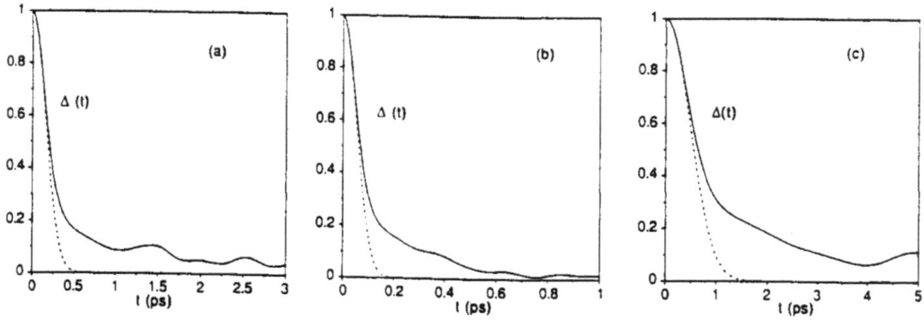

Figure 1. MD results[5] for Δ(t) for three different mass values. Case (a), all atomic masses = 40 amu; (b) 40 amu/9; (c) 9 x 40 amu. Dashed lines indicate the Gaussian approximation.

The most direct and appealing, although approximate, connection of Δ(t) to the solvation dynamics is as follows. If one imagines as in Fig. 2 that in an electronic transition the dipole shift $A^{-1/2}A^{1/2} \rightarrow A^{1/2}A^{-1/2}$ is induced by laser radiation at t=0, and the "product" state $A^{1/2}A^{-1/2}$ radiates, then there will be a resulting time dependent fluorescence (TDF) shift in emission frequency as the solvent relaxes to equilibrium with the new solute charge distribution. This shift, normalized to unity at t=0, is proportional to Δ(t).

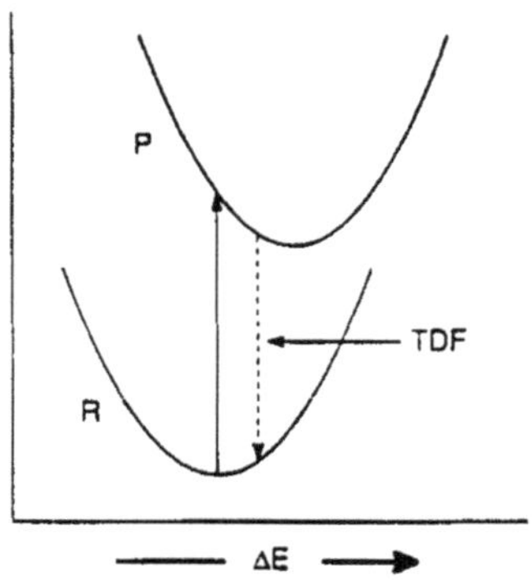

Figure 2. Schematic illustration of TDF. The "product" state $A^{1/2}A^{-1/2}$ has been assigned (arbitrarily) an additional electronic energy to mimic an excited electronic state.

This approximate relation is a linear response theory result, and has been exploited in many studies.[8] An important point is that the MD simulated time dependence of Δ(t) shown in Fig. 1 is rather far from the simple exponential dependence which is predicted by dielectric continuum theories.[9] A convenient measure of the relaxation speed is the solvation time τ defined by the area

$$\tau = \int_0^\infty dt\Delta(t) \quad , \tag{4}$$

which is tabulated in Table 1.

Table 1. Dynamical Solvent Properties

System	τ(ps)	$\omega(ps^{-1})$	$\zeta(t{=}0)(ps^{-2})$	$\zeta(ps^{-1})$
Light[a]	0.12	19	1680	45
Standard[b]	0.49	6.6	180	21
Heavy[c]	1.2	2.2	20	5.8

[a]m = (40/9) amu.[5]
[b]m = 40 amu.[4]
[c]m = (9 x 40) amu.[5]

The next point of note is that a significant fraction of the initial decay of $\Delta(t)$ is well described by the simple Gaussian behavior

$$\Delta_G(t) = \exp(-\omega^2 t^2/2) \quad , \tag{5}$$

where the solvent frequency ω is given by[4,6]

$$\omega^2 = \langle(\delta\Delta E)^2\rangle^{-1}\langle(\delta\dot{\Delta E})^2\rangle \tag{6}$$

This important feature, which is included in almost none of the current dynamical theories of solvation dynamics, was first pointed out by Carter and Hynes.[6] The Gaussian decay arises physically from the initial simple inertial motion of the solvent molecules: At time t=0, those molecules will move with their translational and rotational velocities, governed by Maxwellian distributions, and change thereby the value of $\delta\Delta E$. Note that no intermolecular forces or torques govern that initial motion (although $\delta\Delta E$ itself depends on the intermolecular interactions). From Eq. (6) and the equipartition theorem, the solvent frequency should scale as $m^{-1/2}$, which is corroborated in Table 1. This accounts, for example, for the much faster Gaussian decay for the lighter solvent in Fig. 1.

The influence of the intermolecular forces and torques are responsible for the longer time departures from the Gaussian behavior evident in Fig. 1 They, as well as the shorter time dynamics, can be described via a Generalized Langevin equation (GLE):

$$\delta\ddot{\Delta E}(t) = -\omega^2\delta\Delta E(t) - \int_0^t d\tau\zeta(t-\tau)\,\delta\dot{\Delta E}(\tau) \quad . \tag{7}$$

(We suppress the "random force" term in this equation throughout, for simplicity). This depicts a generalized Brownian motion of the solvent coordinate in, e.g., the reactant well in Fig. 2, whose frequency is ω, and $\zeta(t)$ is the time dependent friction on the solvent coordinate. In the simplest picture, each solvent molecule can be imagined to rotate and reorient its dipole moment and thus change $\delta\Delta E$; since such reorientational motion will be impeded by e.g., other solvent molecules, there will result a net damping on $\delta\Delta E$.

The GLE was first introduced to describe solvation dynamics in Ref. 10, in a dielectric continuum context (and has found a number of interesting applications and extensions since[11]). However, Eq. (7) is in fact exact for the $\delta\Delta E$ dynamics in $\Delta(t)$.[12]

Two more restricted approximations to Eq. (7) are of interest as simpler descriptions of the solvation dynamics. The first is a simple Langevin equation (LE),

$$\delta\ddot{\Delta}E(t) = -\omega^2\delta\Delta E(t) - \zeta\delta\dot{\Delta}E(t) . \tag{8}$$

Here the friction constant

$$\zeta = \int_0^\infty dt\zeta(t) , \tag{9}$$

measures the overall, long-time influence of the damping of the $\delta\Delta E$ motion. The LE completely ignores the time dependent aspects of $\zeta(t)$ in the GLE, assuming that they are rapid compared to the time scale of the velocity $\delta\dot{\Delta}E$, so that the whole friction acts instantly. A further and more drastic approximation would be to assume that $\delta\dot{\Delta}E$ changes rapidly compared to $\delta\Delta E$ itself in such a way that the "potential" and frictional forces cancel, so that the net acceleration is negligible and can be ignored. This is a diffusive limit in which $\Delta(t)$ decays exponentially:

$$\Delta(t) \approx \exp(-\omega^2 t/\zeta) . \tag{10}$$

Though not usually stated in this language, this overdamped description amounts to the most commonly employed characterization of solvent dynamics; in it, the solvation time is

$$\tau = \zeta/\omega^2 . \tag{11}$$

In a dielectric continuum treatment,[9] τ is proportional to the longitudinal dielectric relaxation time of the solvent. Figure 3 shows that neither the LE nor the diffusive approximations--which are fairly close to each other--well portray the solvation dynamics. A full GLE description, including the time dependent friction, is necessary. Unfortunately there currently exists no analytic theory which tells us what the time dependent friction $\zeta(t)$ looks like. We therefore have to fall back on using the MD results to find this out. Two aspects are of most interest--the amplitude and lifetime.

The friction $\zeta(t)$ can be extracted[4-6,12] from $\Delta(t)$ and the GLE via Fourier transform techniques, and is displayed in Fig. 4 for several mass cases. For the light solvent, the initial amplitude $\zeta(t=0)$ of the friction is very large, while it is very small for the massive solvent: $\zeta(t=0)$ scales[5] as $\zeta(t=0) \propto m^{-1}$. On the other hand, the "lifetimes" of $\zeta(t)$ vary in the opposite direction,e.g., dynamics governing the fluctuating "force" on the solvent coordinate are the most rapid for the lighter solvent. As a result, the friction constant changes less rapidly than m^{-1} (cf. Table 1); since ω^2 scales as m^{-1}, this means that the solvation time $\tau=\zeta/\omega^2$ lengthens as the solvent becomes more massive. [It should be emphasized here that Eq.(11) holds[6,10] even with a GLE, using Eqs. (4) and (9), even though $\Delta(t)$ is nonexponential.] At a more detailed level, it turns out[5,6,12] that the longer

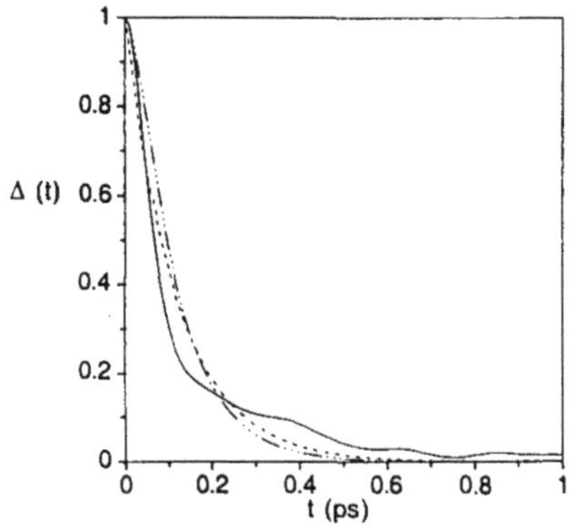

Figure 3. The Langevin equation (—•••—) and diffusive (- - -)approximations to $\Delta(t)$ (——) for the light mass case.

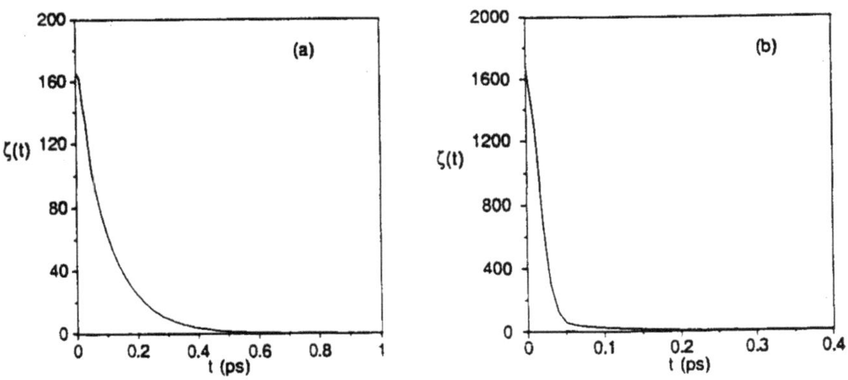

Figure 4. The time dependent friction coefficient $\zeta(t)$ for (a) the 40 amu mass case[4] and (b) the 40 amu/9 mass case.[5]

time "tail" in the tcf $\Delta(t)$ apparent in Fig. 1 arises from the tail in $\zeta(t)$, which is therefore essential in accounting for the bimodal time structure of $\Delta(t)$. A goal for the future is to clarify the microscopic motions responsible for the bimodal aspects of $\zeta(t)$--short time Gaussian plus longer time tail--to provide a predictive theory for the friction.

The conclusion to be drawn from these (and other[13]) results is that, for rapidly relaxing polar aprotic solvents, the relaxation dynamics of the solvent differs significantly from the exponential behavior expected from a dielectric continuum perspective, and a full GLE treatment is required. While further work is clearly needed to understand the molecular origins of the important features of the time dependent friction, even at the current stage this description can be usefully employed to examine solvent dynamical effects for ET rates, which is our next topic.

3. Electron Transfer Reactions

We now turn to the ET reaction problem. There has been a veritable spate of theoretical papers[10,14,15] dealing with the possible role of solvent dynamics in causing departures from the standard Marcus TST rate theory.[1,3] The first of these is due to Zusman[14] and for the most part, further works are variations on this. Zusman focussed on the solvent dynamics within each of the R and P wells, in the diffusive approximation so that the exponential decay Eq. (10) is assumed to hold in each well. The simplest form of the Zusman Theory is for a cusped, i.e., sharply peaked, barrier reaction, for which the transmission coefficient $\kappa = k/k^{TST}$ measuring the departure of the rate constant k from its TST approximation is given by

$$\kappa_Z = (\omega\tau)^{-1}(\pi\Delta G_{na}^{\pm}/k_BT)^{1/2} \ . \tag{12}$$

Here $\Delta G_{na}^{\pm}$ is the nonadiabatic barrier height, i.e., the free energy at the intersection of nonadiabatic R and P curves minus the free energy at the minimum of the R well, which has frequency ω. The solvent relaxation time τ is that appearing in Eqs. (4) and (11). The picture here is that the rate limiting steps for the reaction are the diffusive climbing of the R well wall and the diffusive slide down the P well wall. The passage over the barrier top itself is supposed to be rapid and direct, i.e., there is no solvent-induced recrossing associated with the barrier top *per se*; instead, recrossings of the barrier supposedly arise from "reapproach" from the region of the wells before the bottom of either of the R or P wells is reached. Fig. 5 illustrates the concept.

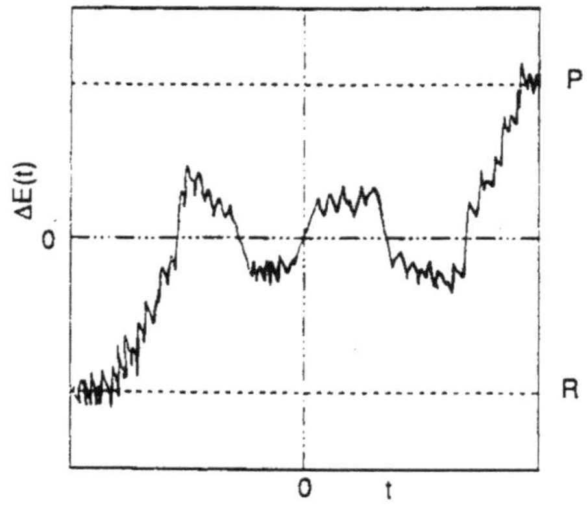

Figure 5. Schematic illustration of the trajectories envisioned in the Zusman and allied theories.

The ET reaction considered is a simplified model,

$$A^{-1/2}A^{1/2} \rightarrow A^{1/2}A^{-1/2} \tag{13}$$

in the solvent described above. The technical and computational rationales for this somewhat artificial fractional charge model can be found in Ref. 4; however, the model is sufficiently realistic to allow us to explicitly address the key dynamical issues.

The MD simulation of the reaction can be effected via the electronically adiabatic Hamiltonian[4]

$$H_{ad} = \frac{H_R + H_P}{2} - \frac{1}{2}\sqrt{(\Delta E)^2 + 4\beta^2} \quad , \tag{14}$$

where $H_{R(P)}$ is the system Hamiltonian when the solute has the R(P) charge distribution and β is the electronic coupling. This is appropriate as representative for many adiabatic ET reactions; in addition, solvent dynamical effects are expected to be most pronounced in the electronically adiabatic limit.[10] With this Hamiltonian, the solute is always in its ground electronic state whatever the solvent configuration may be. The barrier is traversed as ΔE progresses from values appropriate to the neighborhood of equilibrium with R to those similarly appropriate to P. In particular, the electronic charge character proceeds smoothly from that of R, through the transition state distribution A^0A^0 (which is a neutral pair), on to the P charge distribution. Reaction free energy profiles are illustrated in Fig. 6.

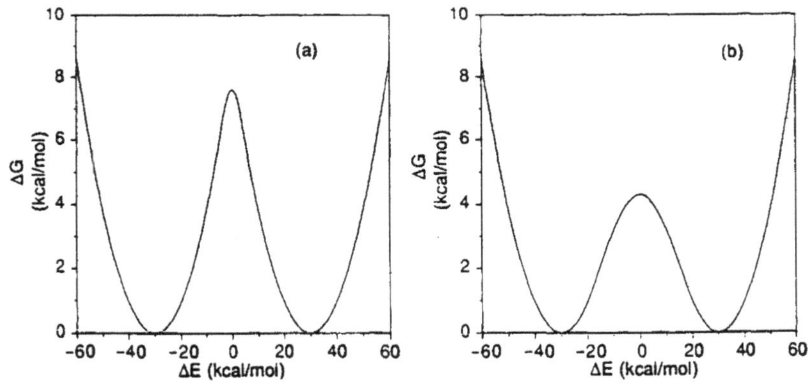

Figure 6. Reaction free energy profiles[4] for two electronic coupling values. (a) $\beta = 1$ kcal/mol; (b) $\beta = 5$ kcal/mol.

The transmission coefficient κ can be directly calculated in an MD simulation for the ET reaction Eq. (14) by procedures described elsewhere involving special constraint dynamics techniques,[4] as can the various quantities in κ_Z. The results are shown in Table 2 for several choices of the electronic coupling β. The first point of interest is that for $\beta = 1$ kcal/mol, κ is quite close to unity; there are few recrossings of the barrier and the Marcus TST Theory is thus an excellent approximation.

Table 2 also indicates that κ_Z gives very poor results for the $\beta = 1$ kcal/mol case. This can be improved by taking into account the fact that the rate of unidirectional crossing the barrier top itself is not infinitely fast, but rather can be estimated from the Marcus TST

Table 2. ET Transmission Coefficients

	β = 1 kcal/mol			β = 5 kcal/mol
System[a]	κ_{MD}	κ_Z	κ_Z'	κ_{MD}
Light[5]	0.87 ± 0.09	3.0	0.75	0.64 ± 0.15
Standard[4]	0.95 ± 0.04	2.2	0.68	0.59 ± 0.11
Heavy[5]	0.86 ± 0.10	2.6	0.72	0.65 ± 0.15

[a]see Table 1 for explanation.

formula. This gives[4,10]

$$\kappa_Z' = (1 + \kappa_Z)^{-1}\,\kappa_Z \tag{15}$$

and, as Table 2 shows, predicts transmission coefficients much closer to the MD values. But, despite this apparent success, there is still a serious problem. For in either Eqs. (14) or (15), the barrier recrossings responsible for the transmission coefficient are supposed to arise from trajectories such as those in Fig. 5, i.e., arising from individual single recrossings subsequent to erratic motion in the R and P wells before the trajectory settles down into the bottom of either well. Fig. 7 shows that the MD computed trajectories are nothing like this. Instead, the recrossings that occur are exclusively associated with the barrier top region. Any trajectory that makes it into the P (or R) well proper in fact continues on to the bottom of that well without returning to the transition state and recrossing it.

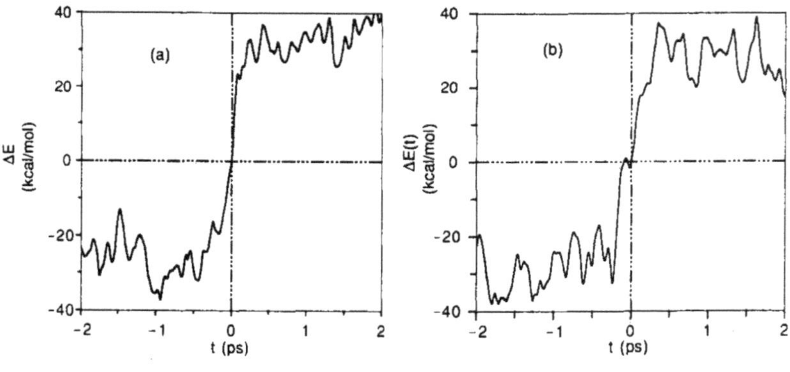

Figure 7. Examples of MD computed trajectories[5] for the light mass case for β = 1 kcal/mol.

When the coupling is increased to β = 5 kcal/mol and the barrier becomes more rounded, Table 2 shows that the transmission coefficient is smaller and there are noticeable, though not enormous, departures from the Marcus TST theory. The character of the barrier recrossings is illustrated in Fig. 8. Again, these recrossings occur in the immediate neighborhood of the transition state, and not in the wells. It is not even appropriate to attempt to apply the Zusman approach or its variants (which assume cusped barriers) to this smooth barrier situation.

We have seen for each of the β coupling value ET reactions that the barrier recrossings occur exclusively in the barrier top region. A theory for just such dynamical rate effects--which has proved highly successful for a wide variety of MD-simulated chemical reaction types[4,16] and in the interpretation of experimental rates[17]--is Grote-Hynes Theory.[2,18] In this theory, the transmission coefficient is given by

$$\kappa_{GH} = \left[\kappa_{GH} + \omega_b^{-1} \int_0^{\infty} dt e^{-\omega_b \kappa_{GH} t} \zeta^{\ddagger}(t) \right]^{-1} , \qquad (16)$$

where ω_b is the barrier frequency and $\zeta^{\ddagger}(t)$ is the time dependent friction coefficient at the barrier top. In the present context, GH Theory is based on the assumption that the GLE

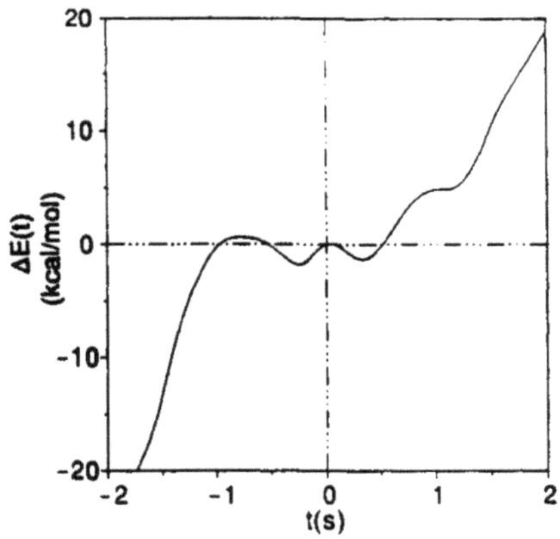

Figure 8. Typical barrier recrossing for the β = 5 kcal/mol, heavy mass case.[5]

$$\ddot{\Delta E}(t) = \omega_b^2 \Delta E(t) - \int_0^t d\tau \zeta^{\ddagger}(t-\tau) \dot{\Delta E}(\tau) \qquad (17)$$

holds in the vicinity of the barrier top $\Delta E = 0$.[4,10] The barrier frequency can be estimated[4] from the formula $\omega_b = \omega[(2\Delta G^{\ddagger}/\beta) - 1]^{1/2}$. The friction $\zeta^{\ddagger}(t)$ is a more difficult quantity to obtain than the well friction, but Zichi et al.[4] have argued that it can be approximated by the time dependent friction appropriate to the reference situation of a neutral pair in the same solvent. This approximate identification derives from the observation, noted above, that in the ET reaction, the transition state charge distribution is that of a neutral pair. The neutral pair friction $\zeta_{NP}(t)$ can be extracted[4,5,13] from the TDF studies of Carter and Hynes,[6] and with the approximation $\zeta^{\ddagger}(t) = \zeta_{NP}(t)$, Zichi et al.[4] showed that κ_{GH} could be estimated for the ET reaction. For example, the results[4] for the standard mass case are $\kappa_{GH} = 0.94$ and $\kappa_{GH} = 0.74$ for $\beta = 1$ kcal/mol and $\beta = 5$ kcal/mol respectively. These agree within the error bars with the MD simulation values in Table 2.

All these results suggest that the conventional dynamic approaches to ET reactions in dipolar aprotic solvents based on relaxation in the wells do not have the correct picture of the reaction, even for fairly cusped barrier reactions. Instead, it is the solvent dynamics occurring near the barrier top, and the associated time dependent friction, that are the critical features in the reaction. It is however possible that, for cusped barrier ET reactions in much more slowly relaxing solvents, the well dynamics could begin to play an important role. We intend to explore this in future MD studies.

Acknowledgments

It is a pleasure to thank E. A. Carter (UCLA), G. Ciccotti (Rome), M. Ferrario (Messina), R. Kapral (Toronto) and D. A. Zichi (Agouron) for their collaboration on some of the work described here, as indicated in the references. This work was supported by NSF grant CHE 88-07852 to JTH. We also thank Professor C. Koval (Colorado) for assistance provided through Department of Energy Grant No. DE-FG02-84ER 13247. This work was also supported by a CRCW Fellowship from the University of Colorado (JTH), and by a grant from the Pittsburgh Supercomputing Center

REFERENCES

1. R. A. Marcus, *J. Chem. Phys.* **24** (1956) 966, 979.
2. N. S. Hush, Trans. Faraday Soc. **57** (1961) 557.
3. See, e.g., the reviews by M. D. Newton and M. D. Sutin, *Annu. Rev. Phys. Chem.* **35** (1984) 437 and R. D. Cannon and J. F. Endicott, in *Mechanisms of Inorganic and Organometallic Reactions*, ed. M. V. Twigg, (Plenum, New York, 1989), Vol. 6, p. 3.
4. D. A. Zichi, G. Ciccotti, J. T. Hynes and M. Ferrario, *J. Phys. Chem.* **93** (1989) 6261. Reference may be found here to other simulation work focussed on different issues in electron transfer.
5. B. B. Smith and J. T. Hynes, unpublished.
6. E. A. Carter and J. T. Hynes, *J. Chem. Phys.*, (1991) in press; J. T. Hynes, E. A. Carter, G. Ciccotti, H. J. Kim, D. A. Zichi, M. Ferrario and R. Kapral, in *Perspectives in Photosynthesis*, eds. J. Jortner and B. Pullman, (Kluwer, Dordrecht, 1990), p. 133.
7. E. W. Castner, Jr., M. Maroncelli and G. R. Fleming, *J. Chem. Phys.* **86** (1987) 1090; M. A. Kahlow, W. Jarzeba, T. J. Kang and P. F. Barbara, ibid. **90** (1989) 151 and earlier papers of the Barbara group.
8. For recent reviews, see, e.g., J. D. Simon, Acc. Chem. Res. **21** (1988) 128; P. F. Barbara and W. Jarzeba, *Adv. Photochem.* **15** (1990) 1; M. Maroncelli, J. MacInnes and G. R. Fleming, *Science*, **243** (1989) 1674. These reviews summarize some of the major experimental contributions of the Simon, Barbara, Fleming and other groups.
9. See, e.g., G. van der Zwan and J. T. Hynes, *J. Phys. Chem.* **89** (1985) 4181 and references therein. For the new features that appear in electrolyte solutions, see G. van der Zwan and J. T. Hynes, *Chem. Phys.* (1991) in press.
10. J. T. Hynes, *J. Phys. Chem.* **90** (1986) 3701.
11. See, e.g., P. F. Barbara, T. J. Kang, W. Jarzeba and T. Fonseca in *Perspectives in Photosynthesis*, eds. J. Jortner and B. Pullman, (Kluwer, Dordrecht, 1990), p. 273; T. J. Kang, W. Jarzeba, P. F. Barbara and T. Fonseca, *Chem. Phys.* **149** (1990) 81.
12. D. A. Zichi, H. J. Kim, E. A. Carter and J. T. Hynes, unpublished.
13. See, e.g., M. Maroncelli, *J. Chem. Phys.* **94** (1991) 2084.
14. L. D. Zusman, *Chem. Phys.* **49** (1980) 295.
15. The literature is very extensive here. Some references are: I. V. Alexandrov and R. G. Gabrielyan, *Mol. Phys.* **37** (1979) 1963; H. L. Friedman and M. D. Newton, *Faraday Discuss. Chem. Soc.* **74** (1982) 73; D. F. Calef and P. G. Wolynes, *J. Phys. Chem.* **87** (1983) 3387; H. Sumi and R. A. Marcus, *J. Chem. Phys.* **84** (1986) 4272; W. Nadler and R. A. Marcus, *J. Chem. Phys.* **86** (1987) 3906;M. Sparpaglione and S. Mukamel, *J. Chem. Phys.* **88** (1988) 3263; I. Rips and J. Jortner, *J. Chem. Phys.* **87** (1987) 2090; Yu. I. Dakhnovskii and A. A. Ovchinikov, *Mol. Phys.* **58** (1986) 237; L. D. Zusman, *Chem. Phys.* **51** (1988) 119; M. Murillo and R. I. Cukier, *J. Chem. Phys.* **89**(1988) 6736 ; T. Fonseca, *J. Chem. Phys.* **91** (1989) 2869.
16. B. J. Gertner, K. R. Wilson and J. T. Hynes, *J. Chem. Phys.* **90** (1989) 3537; J. P. Bergsma, B. J. Gertner, K. R. Wilson and J. T. Hynes, *J. Chem. Phys.* **86** (1987) 1356; B. J. Gertner, J. P. Bergsma, K. R. Wilson, S. Lee and J. T.

Hynes, *J. Chem. Phys.* **86** (1987) 1377; J. P. Bergsma, J. R.Reimers, K. R. Wilson and J. T. Hynes, *J. Chem. Phys.* **85** (1986) 5625; W. Kierstead, K. R. Wilson and J. T. Hynes, *J. Chem. Phys.* submitted; G. Ciccotti, M. Ferrario, J. T. Hynes and R. Kapral, *J. Chem. Phys.* **93** (1990) 7137; S. B. Zhu, J. Lee and G. W.Robinson, *J. Phys. Chem.* **92** (1988) 2401; B. J. Berne, M. Borkovec and J. E. Straub, *J. Phys. Chem.* **92** (1988) 3711 ; B. Roux and M. Karplus (personal communication).

17. B. Bagchi and D. W. Oxtoby, *J. Chem. Phys.* **78** (1983) 2735; J. Ashcroft, M. Besnard, V. Aquada and J. Jonas, *Chem. Phys. Lett.* **110** (1984), 430; D. M. Zeglinski and D. H. Waldeck, *J. Phys. Chem.* **92** (1988) 692; N. Sivakumar, E. A. Hoburg and D. H. Waldeck, *J. Chem. Phys.*, **90** (1989) 2305; N. S. Park and D. H. Waldeck, *J. Phys. Chem.*, in press and G. E. McManis and M. J. Weaver, *J. Chem. Phys.* **90** (1989) 1720. (This last reference can be consulted for extensive references to experimental work on ET reaction rates.)
18. R. F. Grote and J. T. Hynes, *J. Chem. Phys.* **73**(1980) 2715; J. T. Hynes, in *The Theory of Chemical Reaction Dynamics*, Vol. IV, ed. M. Baer, (CRC Press, Boca Raton, Fl, 1985), p. 171.

LOW TEMPERATURE ELECTROCHEMISTRY AT NORMAL CONDUCTOR/FROZEN ELECTROLYTE INTERFACE

Z.Borkowska*, M.Cappadonia** and U.Stimming***

*Institute of Physical Chemistry, Polish Academy of Sciences, Warsaw, Poland
**Fritz-Haber-Institut der Max-Planck-Gesellschaft, D-1000 Berlin 33, Germany[1]
***Department of Chemical Engineering and Applied Chemistry,
Columbia University, New York, NY 10027, U.S.A.[2]

Introduction

The frozen electrolyte technique (**FREECE**=**FR**ozen **E**lectrolyte **E**lectro**C**h**E**mistry) is based on the experimental result that frozen electrolytes are suitable for electrochemical studies [1]. This technique has been used in our laboratory [2] and also by others [3][4][5] [6][7][8] to investigate interfacial electrochemical behavior. An argument will be given as to why the FREECE technique is advantageous in a number of respects and what kind of electrolyte systems can be used. Reference is made to electrochemical results such as interfacial reactions and double layer properties.

Why frozen electrolytes (FREECE)?

The frozen electrolyte technique is an approach in which the electrolyte is used in its liquid as well as its solid state. Strictly speaking the requirement is that also in the solid state one has a one phase system only. In principle, this one phase requirement would be fulfilled also for glasses. The one phase condition usually requires, however, that the electrolyte is a stoichiometric compound, i.e. in the case of an aqueous electrolyte the compounds are hydrates.

There are several aspects which set the FREECE technique apart from common electrochemical investigations in aqueous solutions:

(i) The usual treatment of the electrolyte is that of a solvent and a solute. A description in terms of thermodynamics treats the solute as a perturbation of the solvent and tries to work under the limitation of ideal conditions. The use of a correction factor (activity coefficient) underlines this approach.

(ii) In dilute solutions which fulfill the above requirement considerable concentration gradients of ions at the electrode-electrolyte interface can occur due to charging of the electrode. For a 10^{-3} M $HClO_4$ the surface concentration of anions, c_s, can approach 1 mole for a charge of $\sigma \approx +10\mu C/cm^2$ (approx. 300 mV above U_{pzc}, the potential of zero charge). Such concentration gradients which are, on the other hand, much smaller in concentrated solutions can make the evaluation of interfacial properties with respect to reactions and the double layer difficult.

---

[1] New address: European Patent Office, D-8000 München 2, Germany

[2] Present Address: Department of Electrochemistry, University of Ulm, D-7900 Ulm, Germany

(iii) The validity of a model is thus restricted to low electrolyte concentrations and small charges on the electrode. Most electrochemical measurements, however, are performed in 1 molar electrolytes and at charge densities including values $\sigma > 20\mu C/cm^2$. This requires the development of models of the electrode-concentrated electrolyte interface.
(iv) A stoichiometric compound can have a defined melting point with a one phase condition in the liquid and the solid state, except for the temperature of any phase transition. The influence of the liquid-solid phase transition of the electrolyte on electrochemical behavior can be investigated.
(v) Electrolytes with a stoichiometric composition can exhibit defined structures which can be investigated with common techniques such as X-ray diffraction etc..
(vi) Since the use of a solid electrolyte requires a sufficiently high ionic conductivity the temperature range can be quite extensive in which electrochemical behavior is investigated in the same electrolyte (liquid and solid).

In summary, electrolytes used so far are acid and alkali hydroxide hydrates which possess the following properties:

- stoichiometric composition,
- aqueous electrolytes,
- high ionic concentration,
- defined structure,
- high conductivity in the solid state.

Frozen electrolyte properties

The electrolyte first used in FREECE studies was $HClO_4 \cdot 5.5H_2O$ [9] which is the highest hydrate of the perchloric acid according to the phase diagram $HClO_4$ - H_2O [10]. It crystallizes in a type I clathrate structure [11][12] which possesses a pronounced cage structure with the perchlorate anions occupying the cages. The protons are part of the water network and largely delocalized. The electrical conductivity and NMR properties were investigated in the liquid and solid state [13]. Recently, the impedance spectroscopic behavior of $HClO_4 \cdot 5.5H_2O$ was investigated in the temperature range 4.2 to 300 K [14]. The electrical conductivity which is due to the proton mobility does not change discontinuously at the freezing point of the electrolyte and seems to become temperature independent below approx. 80 K. The limiting dielectric constant which is reached below that temperature is $\epsilon_\infty \approx 11$, a value considerably above the value of water which is $\epsilon_\infty \approx 3$.

Other acid hydrates such as $HBF_4 \cdot 5.75H_2O$ and $HPF_6 \cdot 7.67H_2O$ which crystallize also in a clathrate structure [11][12] show a behavior that is very similar to that of $HClO_4 \cdot 5.5H_2O$ [15]. In addition to acid hydrates also hydroxide hydrates are suitable for FREECE measurements. Investigating the highest hydrates of $LiOH$, $NaOH$, KOH and $CsOH$ it was found [16] that only the KOH and $CsOH$ hydrates show a sufficiently high electrical conductivity in the solid state. The KOH system is currently under investigation also with respect to NMR properties [17]. Mootz and coworkers have shown quite recently that hydroxides with large cations such as tetramethylammonium can form a number of clathrate type structured stoichiometric hydrates [18]. Summarizing, one can say that a number of stoichiometric hydroxide and acid hydrates exist that have a high conductivity in the solid state and can thus be used in FREECE measurements or exhibit properties which are promising for such measurements. The common aspects appears to be a suitable configuration of the water network which allows a relatively high mobility of protons due to a favorable tunneling from and/or to neighboring water molecules.

Interfacial properties reactions

The most important electrochemical reactions that were investigated so far were the Fe^{2+}/Fe^{3+} reaction on platinum under equilibrium conditions, hydrogen evolution on Pt [1][7], Ag and Cu [19] and Au [5][19]. The electrolyte has been $HClO_4 \cdot 5.5H_2O$. The two most important aspects were the evaluation of the influence of the liquid-solid transition of the electrolyte and the investigation of electrochemical kinetics in a wide range of temperatures. The influence of the phase transition of the electrolyte is relatively small. Activation energies increase slightly with the liquid-solid transition but transfer coefficients are not influenced. The latter, however, show a pronounced temperature dependence for the hydrogen evolution on Au, Ag and Cu but not for Pt. At temperatures $T < 140$ K the current appears to become potential independent for Ag and Cu. This may be an indication that the formation of an adsorbed hydrogen atom from a proton becomes frozen as it was found from surface science experiments for the reverse reaction on Pt{111} at about the same temperature [20]. In a recent publication [7] it was claimed that hydrogen evolution on Pt was investigated down to 50 K. The results of the kinetics regarding the temperature dependence of the transfer coefficient appear to be different [7] as compared to earlier results [9].

Capacity measurements were performed on Au [3][4][21] and on Ag and Cu [21] [22]. In all systems and in the potential range investigated the temperature dependence of the capacity exhibits a positive value dC/dT. The most remarkable result, however, is that in all cases the double layer capacity does not show any discontinuity at the freezing point of the electrolyte. This indicates a structure of the double layer that is very similar in the liquid and the frozen electrolyte.

Structure of the electrochemical double layer

The capacity measurements performed in liquid and frozen $HClO_4 \cdot 5.5H_2O$ indicate that the structure of the electrolyte that exists in the solid state is obviously maintained in the liquid electrolyte. In order to further investigate this aspect in liquid electrolytes cyclic voltammetry was performed on gold electrodes in aqueous perchloric acid of various concentrations in the range 10^{-3} M $HClO_4$ to $HClO_4 \cdot 5.5H_2O$ (approx. 7 M $HClO_4$) [23]. Only pH related effects and changes in the cyclic voltammogram in the double layer region due to diffuse double layer effects are observed. The acid concentration has no further influence on the cyclic voltammogram between hydrogen and oxygen evolution, and especially the oxygen adsorption peak which is otherwise very sensitive to adsorption and the interfacial structure is not shifted.

This can be understood considering the fact that the oxygen adsorption peak on gold is observed at rather high positive charges exceeding 30 $\mu C/cm^2$. At such a high surface charge the concentration of counter charges, i.e. ClO_4^- anions, can reach a concentration that is several molar and the structure can become similar to that in the perchloric acid hydrate, $HClO_4 \cdot 5.5H_2O$. As shown above, the clathrate structure which is observed in the solid state of this hydrate in the bulk remains present to some extent on the electrode surface in liquid $HClO_4 \cdot 5.5H_2O$. At lower perchloric acid concentration but at high positive charge densities a clathrate-like ion-water structure is conceivably be formed at the electrode surface.

In a step further one may speculate on a possible connection between the anion-water structure at the interface and the tendency of anions to show specific interaction with a metal electrode [24]. Comparing anions such as perchlorate, tetrafluoroborate and hexafluorophosphate with ions such as sulphate and chloride it is known that the former group shows a small degree of interaction with most metals but the latter exhibit strong interaction (chloride even stronger than sulphate, i.e. specific adsorption. It is interesting to note that the first three anions form acid hydrates which crystallizes in the type I clathrate structure [12] and all of them show a good conductivity in the solid state [15]. Hydrochloric acid and sulfuric acid do not form such cage structures, and it is known of the sulfuric acid hydrates that they form, depending on the amount of water, zero, one and two dimensional open structures [25]. This possible correlation between the structure of the acid hydrate and the adsorption behavior of the anions should be investigated further to possibly better understand interfacial behavior and its relevance to the structure of the double layer. It was Grahame who stated in his seminal paper on the capacity behavior of mercury in fluoride solutions as a function of temperature that the double layer appears to be a "'pseudo-crystalline ice-like' layer of solvent" [26].

Conclusions

There are several aspects which make FREECE measurements beneficial for the investigation of electrochemical behavior. The main aspects are a wide range of temperatures accessible, the solid-liquid transition of the electrolyte and the comparison with more dilute aqueous solutions. The FREECE technique has possibly the ability to bridge different areas in electrochemistry, such as solid electrolytes, aqueous solutions and molten salts which, so far, exist rather independent of each other.

References

[1] U.Stimming and W.Schmickler, J.Electroanal.Chem. 150 (1983) 125

[2] see some summarizing description: U.Stimming in M.P.Soriaga (ed.) "Electrochemical Surface Science", ACS Symposium Series, Vol.378, American Chemical Society, Washington, DC 1988, p.275

[3] T.Iwasita, S.Röttgermann and W.Schmickler, J.Electroanal.Chem. 196 (1985) 203

[4] A.Hamelin, S.Röttgermann and W.Schmickler, J.Electroanal.Chem. 230 (1987) 281

[5] U.Frese and W.Schmickler, Ber.Bunsenges.phys.Chem. 92 (1988) 1412

[6] A.Matsunaga, K.Itoh, A.Fujishima and K.Honda, J.Electroanal.Chem. 205 (1986) 343

[7] J.O'M.Bockris and J.Wass, J.Electroanal.Chem. 267 (1989) 325

[8] J.O'M.Bockris and J.Wass, J.Electroanal.Chem. 267 (1989) 329

[9] U.Frese, T.Iwasita, W.Schmickler and U.Stimming, J.Phys.Chem. 89 (1985) 1059

[10] G.Mascherpa, Revue de Chimie minerale 2 (1965) 379

[11] D.Mootz and M.Wiebke, Z.Kristallogr. 170 (1985) 194

[12] D.Mootz, E.-F.Oellers and M.Wiebke, J.Am.Chem.Soc. 109 (1987) 1200

[13] T.-H.Huang, E.Davis, U.Frese and U.Stimming, J.Phys.Chem. 92 (1988) 6874

[14] M.Cappadonia and U.Stimming, J.Electroanal.Chem. 300 (1991) in press

[15] Z.Borkowska, M.Cappadonia and U.Stimming, Electrochim.Acta, in press

[16] E.Kriegsmann, Z.Borkowska, U.Stimming and U.Frese, paper 586, Extended Abstracts of the 175th Meeting of the Electrochemical Society, Los Angeles, California, U.S.A., 1989

[17] S.Greenbaum, U.Stimming and coworkers, in preparation

[18] D.Mootz and R.Seidel, J.Inclus.Phenom.Mol.Recogn.Chem. 8 (1990) 139

[19] U.Frese und U.Stimming, J.Electroanal.Chem. 198 (1986) 409

[20] F.T.Wagner and T.E.Moylan, Surf.Sci. 206 (1988) 187

[21] T.Dinan and U.Stimming, J.Electrochem.Soc. 133 (1986) 1662

[22] U.Frese and U.Stimming, in J.W.Schultze (ed.) "Principles of Electrode Reactions", Dechema Monographs, Vol. 102, 1986, p.247

[23] Z.Borkowska and U.Stimming, J.Electroanal.Chem., in press

[24] Z.Borkowska and U.Stimming, in preparation

[25] D.Mootz and A.Merschenz-Quack, Z.Naturforsch. 42B (1987) 1231

[26] D.C.Grahame, J.Am.Chem.Soc. 79 (1957) 2093

Low-Temperature Electrochemistry at Superconducting Electrodes

W. J. Lorenz
Institute of Physical Chemistry and Electrochemistry
University of Karlsruhe, Karlsruhe, FRG

G. Saemann-Ischenko
Institute of Physics
University of Erlangen-Nürnberg, Erlangen, FRG

M. W. Breiter
Institute of Technical Electrochemistry
University of Vienna, Vienna, Austria

Abstract

The charge transfer across the interface superconductor / ionic conductor represents a relatively unknown field at present. This paper deals with the first electrochemical studies in this area. Both n-type classical and p- and n-type high T_c superconductors (HTSC) are considered. Different solid electrolytes (Ag ß″-alumina, Ag^+ion conducting glass, and $RbAg_4I_5$) are used in experiments covering a temperature range down to 10 K. A transient technique in the time domain and Electrochemical Impedance Spectroscopy (EIS) in the frequency domain are used to study the silver deposition as the Faradaic charge transfer process at that interface. The results show an admittance peak in transient measurements or a corresponding negative peak of the polarization resistance in EIS-measurements at $T \approx T_c$. These findings are attributed to an increase of the exchange current density around T_c. The effect is much less on an A15 compound than on HTSCs. The effect can only be interpreted as a quantumelectrochemical phenomenon caused by the contribution of Cooper pairs crossing the electrochemical double layer as correlated charge carriers. A proximity-like phenomenon can be excluded.

1. Introduction

Phenomena and processes occuring on superconductor (SC)/ionic conductor (IC) interfaces were absolutely unknown up to the discovery of high T_c superconductors (HTSC) as indicated in Table1. The reasons for this situation are the following:

i) high ohmic resistances of liquid and solid electrolytes at low temperatures,

ii) insufficient sensitivity of electronic equipment for measurements,

iii) contact problems with solid electrolytes,

iv) neither theoretical nor practical interest in classical superconductor / ionic conductor interfaces due to very low critical temperatures: $T_c \leq 23$ K.

The situation changed principally after the discovery of HTSCs with 23 K $\leq T_c \leq$ 125 K. Electrochemical studies involving HTSCs are of interest for different reasons in two temperature ranges:

1. $T \gg T_c$: surface stability, ageing effects, phase boundary reactions as metal deposition, corrosion processes and redox processes, electrocatalysis, photoelectrochemistry.

2. $T \approx T_c$: double layer structure, proximity-like effects, kinetics and mechanisms of charge-transfer processes of electrons, defectelectrons, correlated charge carriers and ions above and below T_c.

2. Experimental

Low-temperature electrochemistry at superconding electrodes, i.e. topic 2, is dealt with in this paper. The systems studied by electrochemical methods in the temperature range 10 K $\leq T \leq$ 298 K were of the type (1-10):

Table 1

Possible types of charge carriers crossing interfaces between condensed matter (symmetrical matrix)

	Electronic conduction			Ionic conduction	
	M	SC	SeC	IC	In
M	e,h	e,h,*)	e,h	e,h,I	—
SC		e,h 2e,2h **)	e,h	?	—
SeC			e,h	e,h,I	—
IC				I	***)
In					—

*) Proximity effect under special conditions
**) Josephson effect under special conditions
***) I for ionic crystals

Symbols:

M	=	material with metal-like conductivity
SC	=	superconducting material
SeC	=	semiconducting material
IC	=	ionic conductor
In	=	insulator
e	=	electron
h	=	defectelectron
2e, 2h	=	Cooper pairs
I	=	ion

Ag/HTSC or SC/Solid Electrolyte (SE)/Ag

with pressure contacts between polycrystalline HTSC and SE samples (Table 2) and silver conducting epoxy resin as silver contact material. The same results were obtained with sputtered silver as contacts on top of the SC, HTSC and SE surfaces.

Non-superconducting material such as $YBa_2Cu_3O_{6.5}$ or Ag served as reference substances instead of using SC or HTSC.

The methods used are the four-probe technique for the determination of the specific conducitivity of SEs (Fig. 1c) and two- and three-probe measurements for the determination of the SC/SE and HTSC/SE interfacial behaviour (Fig. 1a,b).

A special transient technique in the time domain (Fig. 2) and EIS in the frequency domain were applied potentiostatically at the rest potential. Bias polarization was not performed in order to avoid steady state polarization effects within the SEs.

3. Results

i) Electrochemical measurements using SEs under exact temperature control could be performed (1-10).

ii) The intrinsic ionic conductivities of the used SEs were found to be rather small at low temperatures, however, high enough to start electrochemical studies on the interfacial behaviour of SC/SE and HTSC/SE heterojunctions at temperatures above and below T_c (1-10).

iii) The intrinsic conduction mechanism of the SEs was found to change at low temperature (Fig.3): Above $T \approx 115 \pm 10$ K the conductance of silver ions in the SEs is a thermally activated process with an activation energy of $Ea = 0.25 \pm 0.05$ eV. Below this temperature, an apparent zero activation energy for the motion of charge carriers is observed. This change of the mechanism of ionic conductivity might be caused by a tunneling process of Ag^+-ions between energetically nearly equivalent configurations within clusters or so-called "associated regions" within the

Table 2

High T_c superconductors (HTSC), classical superconductors (SC), and solid electrolytes (SE) used in the experiments

Type	Composition	Conduction-Mechanism ρ, χ (T)	T_c/K
		HTSCs	
$YBa_2Cu_3O_{7-\delta}$	$0 \leq \delta \leq 0.5$	p	92
$YBa_2Cu_{3(1-x)}Zn_{3x}O_{7-\delta}$	$0 \leq x \leq 0.06$	p	30 ... 92
$Tl_2Ba_2CaCu_2O_8$		p	102 ...105
$Tl_2Ba_2Ca_2Cu_3O_{10}$		p	113 ...115
$Tl_{0.7}Pb_{0.3}Sr_2CaCu_2O_{7-\delta}$		p	44
$Tl_{0.5}Pb_{0.5}Sr_2CaCu_2O_{7-\delta}$		p	74
$Nd_{2-x}Ce_xCuO_{4-\delta}$	$0 \leq x \leq 0.2$ $0 \leq \delta \leq 0.04$	n	23
		SC	
$Nb_{76.1}Al_{17.7}Ge_{6.2}$ (A15 Compound)		n	19.3
			Glass transition-like temperature T_g/K
		SEs	
Ag ß''- alumina (Ag_2O . 5 Al_2O_3)		Ag^+	110 ± 10
Ag^+ ion conducting glass		Ag^+	110 ± 10
Rb Ag_4J_5		Ag^+	110 ± 10

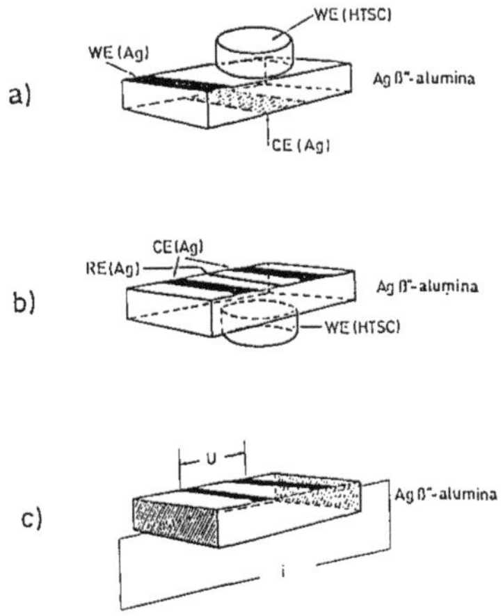

Fig. 1: Schematic representation of the electrochemi-al cell. Two-probe technique (a), three-probe technique (b), four-probe technique (c).

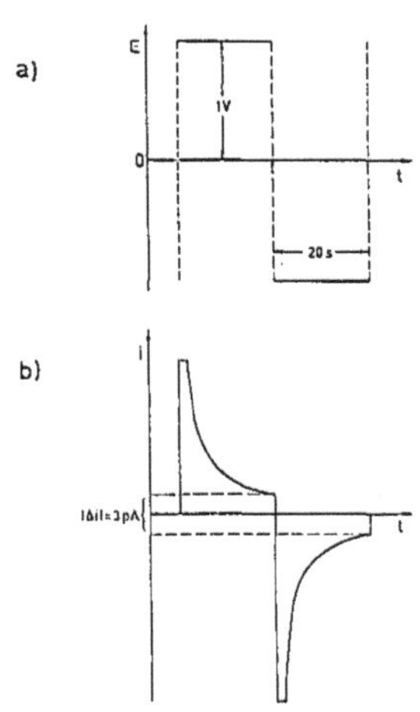

Fig. 2: Transient technique.

Potentiostatic system perturbation with ΔE = 1 V.(a) and system response (b) at the interface $YBa_2Cu_3O_{7-\delta}$/Ag β"-alumina at T = 93 K using two-probe technique.

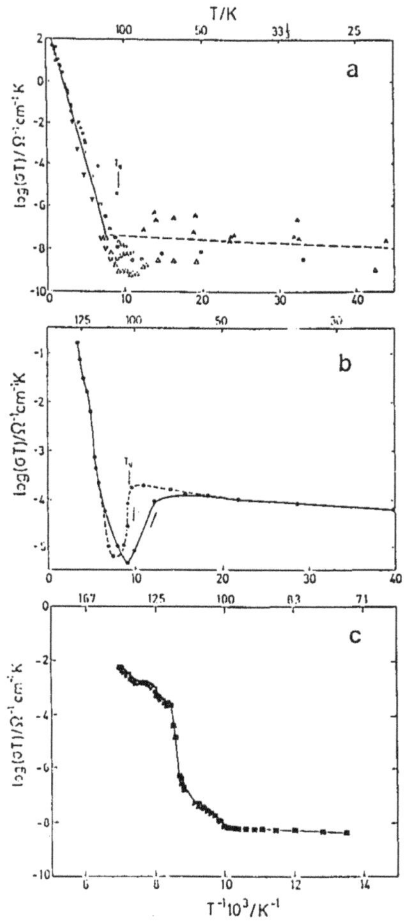

Fig. 3: Temperature dependence of the specific conductivity of different solid electrolytes measured by
Electrochemical Impedance Spectroscopy (EIS) (•, Δ, ∇, ■) or
transient technique (o, Δ, ∇)
using four-probe technique.

(a) Ag ß''-alumina, (b) Ag^{+}ion conducting glass, (c) $RbAg_4J_5$

SEs, corresponding to a two-level tunneling system (TLS-states). A "glass transition" at T ≈ 110 K was previously found for Me ß-alumina with Me = Na, Ag by other authors. Below this temperature, a barrierless ionic motion of Me-cations has been proposed (1,2,9,10).

iv) Electrochemical measurements in various systems HTSC(polycrystalline)/SE and classical SC/SE showed: The quasi-dc currents of transients, which corresponds to the quasi-steady state admittance, $Y_{s\to 0}$, of the system under test, displayed a significant maximum $|\Delta i|_{max}$ at $T \approx T_c$ in their temperature dependence (Figs. 4, 5 and Tab. 3). This admittance peak corresponds to the respective minimum of the polarization resistance, R_p, in EIS measurements at $T \approx T_c$ (Fig. 6-8). The polarization resistance is defined as $R_p = \lim_{(s \to 0)} \{Z\}$, where Z denotes the impedance of the system under test, $Z = Y^{-1}$. The maximum of $|\Delta i|$ and the corresponding minimum of R_p are strongly correlated to T_c but independent of the SE used in the heterojunctions SC/SE or HTSC/SE (1-8).

4. Discussion

In principle, there are two interpretations for the observed enhancement effect, i.e. the increase of the quasi-steady state admittance or the decrease of the polarization resistance of the systems under test at $T \approx T_c$:

i) An injection of Cooper pairs as correlated charge carriers at $T \leq T_c$ into the SE might change the conduction mechanism in the interfacial region of the SE resulting in a decrease of the ohmic resistance, R_Ω, of the SE. This interpretation corresponds to a proximity-like effect (1-8).

ii) Both, the increase of the quasi-steady state admittance or the decrease of the polarization resistance might also correspond to an enhancement of the exchange current density of the Faradaic phase boundary reaction caused by a transfer

Table 3

Comparison between physical*) and electrochemical**) determination of T_c

HTSC	T_c ($\rho(T)$, $\chi(T)$)	$T_c(\lvert\Delta i\rvert_{max}, R_p)$	$\lvert\Delta i\rvert_{max}/\lvert\Delta i\rvert$
$YBa_2Cu_3O_{7-\delta}$	92	93 ± 2	2.0 ± 0.2
$Tl_2Ba_2CaCu_2O_8$	105	100 ± 5	1.3 ± 0.2
$Tl_2Ba_2Ca_2Cu_3O_{10}$	113	124 ± 6	1.6 ± 0.2
$YBa_2Cu_{3(1-x)}Zn_{3x}O_7$			
x = 0	92	93 ± 2	2.0 ± 0.2
x = 0.02	75	75 ± 2	2.0 ± 0.2
x = 0.04	50	50 ± 5	1.8 ± 0.2
x = 0.06	40	32 ± 6	1.6 ± 0.2
$Nd_{1.85}Ce_{0.15}CuO_{4-\delta}$	23	23 ± 1	1.6 ± 0.2
A15 Superconductor $Nb_{76.1}Al_{17.7}Ge_{6.2}$	19.3	20 ± 1	1.15 ± 0.05

*) T_c was determined measuring either the specific resistivity $\rho(T)$ using dc polarization or the magnetic susceptibility $\chi(T)$ using ac technique.

**) T_c was determined from $|\Delta i|$ vs. vs. T^{-1} plots where $|\Delta i| = i_a(\tau) - i_c(\tau)$ denotes the difference of the anodic (a) and cathodic (c) quasi-dc currents i of potentiostatic transients after a transient time of $\tau = 20$ s. The value of $|\Delta i|$ corresponds to the admittance of the system measured.
T_c was also determined from a R_p vs. T^{-1} plot in the case of the $YBa_2Cu_3O_{7-\delta}/RbAg_4J_5$ interface where R_p represents the polarization resistance of EIS measurements.

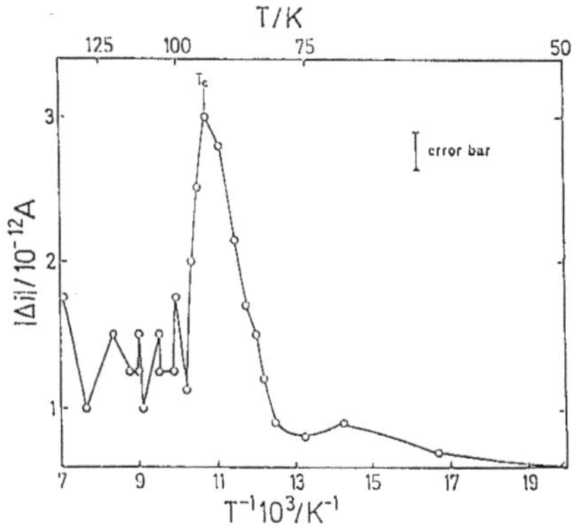

Fig. 4: $|\Delta i|$ vs. T^{-1} Arrhenius plot for the $YBa_2Cu_3O_{7-\delta}$/Ag ß''-alumina interface measured by the transient technique with $\Delta E = 1$ V using two-probe technique.

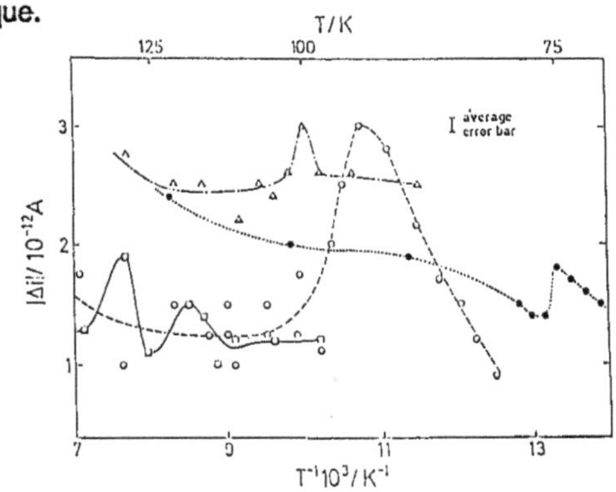

Fig. 5: $|\Delta i|$ vs. T^{-1} Arrhenius plots for different HTSC/SE interfaces measured by the transient technique with $\Delta E = 1$ V using two-probe technique (curves are only eye guides).

Interfaces:

$YBa_2Cu_{3(1-x)}Zn_{3x}O_7$ (x=0.2)/Ag ß''-alumina (•)

Y $Ba_2Cu_3O_7$/Ag ß''-alumina (o)

$Tl_2Ba_2CaCu_2O_8$/Ag ß''-alumina (Δ)

$Tl_2Ba_2Ca_2Cu_3O_{10}$/Ag ß''-alumina (□)

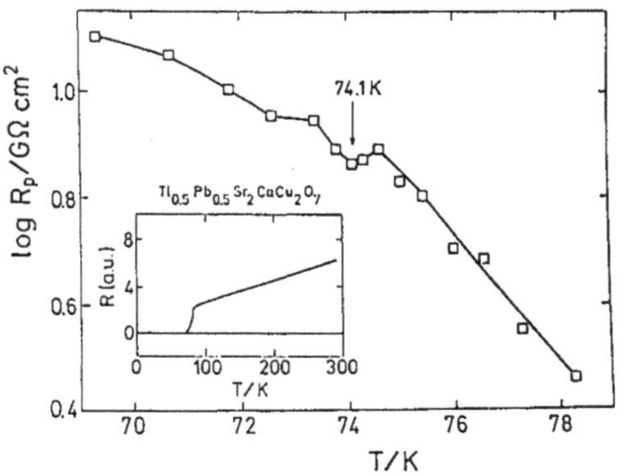

Fig. 6: R_p vs. T plot for the interface $Tl_{0.5}Pb_{0.5}Sr_2CaCu_2O_{7-\delta}$ (δ = 0)/$RbAg_4J_5$ measured by EIS using two-probe technique.

Insert:

resistivity R of the HTSC used measured by EIS using four-probe technique (T_c = 74 K).

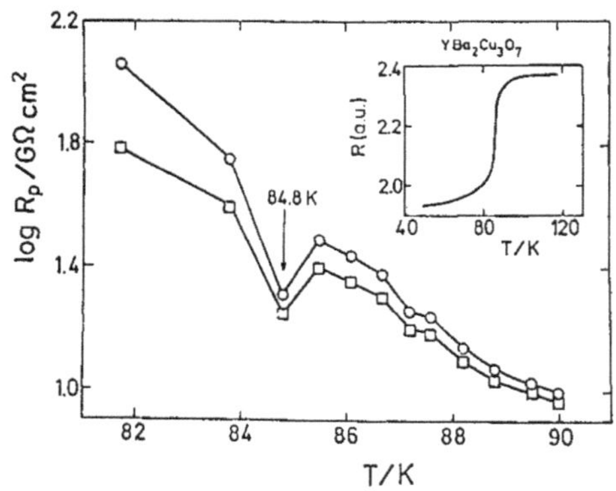

Fig. 7: R_p vs. T plot for the interface Y $Ba_2Cu_3O_{7-\delta}$/$RbAg_4J_5$ measured by EIS using two-probe technique.

(□) uncorrected data, (o) data corrected for instrumental impedance.

Insert:

resistivity R of the HTSC used measured by EIS using two-probe technique (T_c = 85 K).

contribution of correlated charge carriers at $T \leq T_c$. The Faradaic process occuring at the studied interfaces SC/SE and HTSC/SE is:

[1a] n-type HTSC or SC: $Ag+(SE) + e^{-}(HTSC) = Ag$ (Interface)

[1b] p-type HTSC: $Ag+(SE) = Ag$ (Interface) $+ p^{+}(HTSC)$

As long as the kinetics of this Faradaic process is purely activation controlled, the polarization resistance, R_p, is equal to the charge transfer resistance, R_t, which is expected to be temperature-dependent and inversely related to the exchange current density, i_o, of the Ag/Ag^+ system:

[2] $R_t = (kT)/(z.e_o.i_o)$

where k is the Boltzmann constant, and z and e_o denote the ionic charge and the elementary charge, respectively. In the case of a thermally activated classical charge transfer process of quasiparticels or ions, i_o is strongly temperature-dependent according to an Arrhenius-like equation $i_o = i_o'.exp(-E_a/kT)$, where E_a denotes the activation energy of the charge transfer process of quasiparticels. Indeed, a strong increase of R_p with decreasing temperature was found for $T > T_c$. However, $|\Delta i|$ and R_p approach nearly constant values at $T << T_c$ excepting their non-monotoneous behaviour at $T \approx T_c$. The constant values at $T << T_c$ indicate a temperature-independent tunneling process of charge carriers rather than a thermally activated one.

The observed non-monotoneous temperature dependence of $|\Delta i|_{max}$ and R_p exactly at $T = T_c$ (Table 3) can also not be explained only on the basis of a classical charge transfer process of quasiparticles (electrones or holes) across the interfaces SC/SE and HRSC/SE. Additionally, it must be taken into consideration that Cooper pairs as correlated charge carriers are formed at $T \leq T_c$ and may also influence the electrode kinetics and possibly the mechanism of the charge transfer process. Therefore, it can be assumed that Cooper pairs will not only change the structure of the electrochemical double layer at the SC/SE and HTSC/SE interfaces, but that they also will penetrate through the double layer and participate in the charge transfer process. However, Cooper pairs are assumed to decay in quasiparticles before filling up unoccupied and spatially separated but close energy levels within the SE.

The direction of the charge transfer of Cooper pairs across the SC/SE and HTSC/SE interfaces clearly depends on the conduction type of the SC or HTSC and on the sign of the polarization (cf. eq. [1]).

The above interpretation corresponds to a kinetic effect (1-8).

Transfer function analysis of EIS measurements indicates that the experimental results cannot be explained on the basis of a proximity-like effect: The ohmic resistance, R_Ω, of the SE was estimated to be more than two orders of magnitude lower than the overall impedance of the systems Ag/SC or HTSC/SE/Ag (Fig. 9). Therefore, a proximity-like effect may have only minor importance if it occurs.

The measured high impedances in the systems Ag/SC/SE/Ag and Ag/HTSC/SE/Ag are obviously caused by the Faradaic phase boundary reactions at the SC/ SE, HTSC/SE and Ag/SE interfaces (Fig. 9). Uneqivocally, the extraordinary temperature effect of $|\Delta i|$ and R_p, exactly at $T \approx T_c$, is correlated to the influence of Cooper pairs, formed at $T \leq T_c$, on the kinetics and mechanism of the charge transfer process at the SC/SE or HTSC/SE interfaces because no maximum of $|\Delta i|$ or the corresponding minimum of R_p could be observed around T_c in the corresponding reference systems Ag/SE/Ag or Ag/non-superconducting ceramics/SE/Ag in which only quasiparticles can penetrate across the electrochemical double layers. In agreement with this, the classical energy representation of Fermiparticles at the SC/SE or HTSC/SE interfaces at $T > T_c$ (Fig. 10a) and at $T < T_c$ (Fg. 10b) can not

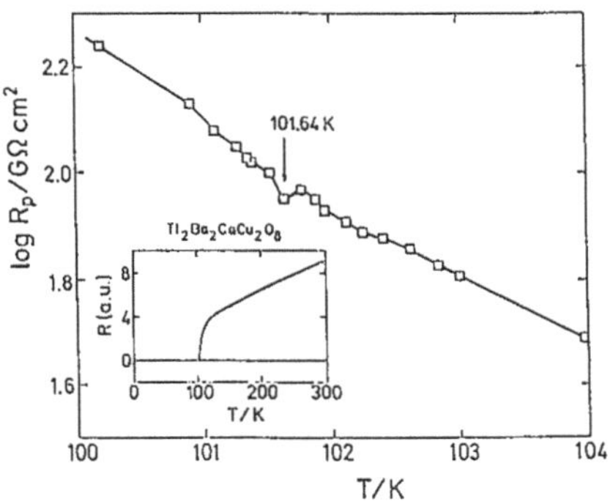

Fig. 8: R_p vs. T plot for the interface $Tl_2Ba_2CaCu_2O_8/RbAg_4J_5$ measured by EIS using two-probe technique.

Insert:
resistivity R of the HTSC used measured by EIS using four-probe technique (T_c = 102 K).

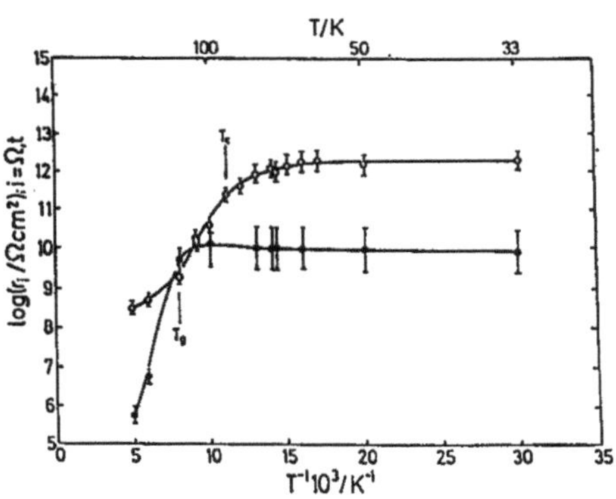

Fig. 9: Comparison between the specific ohmic resistance, $r_\Omega = R_\Omega.A$ (•), and the specific charge transfer resistance $r_t = R_t.A$ (o) as a function of the reciprocal temperature in the system $YBa_2Cu_3O_{7-\delta}$/Ag β''-alumina/Ag, where A denotes the geometrical area.

R_Ω was obtained from transient and EIS measurements of the SE using a four-probe technique with ΔE = 1 V, and R_t from transient measurements of the electrochemical cell using a two-probe technique with ΔE = 1 V.

explain the observed kinetic enhancement effect in the excahnge current density. Therefore, the observed low-temperature effect on SC/SE and HTSC/SE interfaces is currently interpreted as a real quantumelectrochemical kinetic phenomenon caused by the formation of Cooper pairs (Bosons) as correlated charge carriers within the SC or HTSC at $T \leq T_c$ and their ability to cross the interfaces SC/SE and HTSC/SE, to contribute to the charge transfer process and to stimulate the interfacial Faradaic reaction.

The following problem, however, has not been solved: Why does a hump but not a jump of $|\Delta i|$ or R_p occur at $T \approx T_c$ in the transient or EIS measurements? Different possible explanations exist at present:

i) R_p might include more contributions than that of an electrode reaction with pure charge transfer controlled electrode process characterized by R_t, i.e. impedance contributions of mass transport, chemical reactions, sorption phenomena, and electrocrystallization steps which need not be affected by the formation of Cooper pairs.

ii) A self-inhibition of the charge transfer process of Cooper pairs through the electrochemical double layer can be produced by the electrodeposition of relatively thick layers of non-superconducting metal in deposited metal clusters in the interphase.

iii) The decrease of $|\Delta i|$ or the increase of R_t at $T < T_c$ might also be attributed to the energy gap, 2Δ, which is opened at T_c and increases with decreasing temperature at $T < T_c$. This gap may inhibit the participation of uncoupled charge carriers in the transfer proccess across the HTSC or SC/SE interfaces at $T < T_c$.

In very recent EIS two-probe measurements in the system $Ag/HTSC/RbAg_4I_5/Ag$ a jump of R_p instead of a hump has been sometimes observed. At present we are looking for the experimental conditions responsible for the appearance of the hump

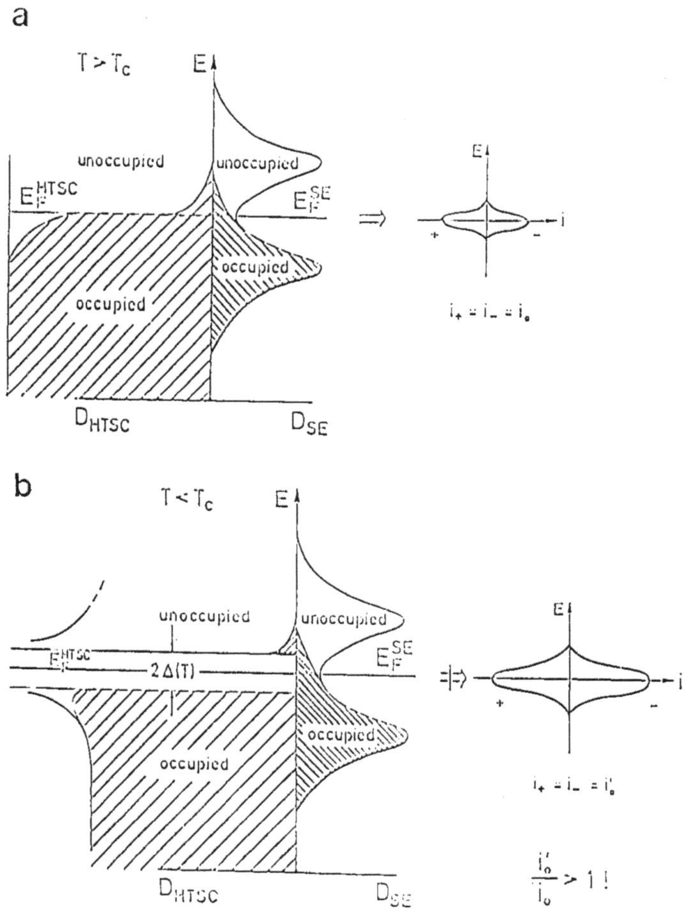

Fig. 10: Schematic energy diagram of Fermi-particels at the superconductor/ionic conductor interface at electrochemical equilibrium ($E = E_{rev}$).

(a) : $T > T_c$ (metallic state)

(b) : $T \leq T_c$ (superconducting state)

allone and its reproducibilty.

More experimental and theoretical investigations are necessary for a better understanding of the extraordinary low-temperature behaviour of the electrode kinetics just around T_c. Very recently, some experimental results and theoretical interpretations described in this paper could be confirmed by other authors. Bockris and Wass observed a temperature-independent proton tunneling process in frozen perchloric acid, below a glass-like transition temperature of about 150 K (11) and the appearance of a hump of the exchange current density i_0 at $T \approx T_c$ in the system $YBa_2Cu_3O_{7-\delta}$/frozen $HClO_4.5.5\ H_2O$ (12). Stimming confirmed the existence of proton tunneling processes at low temperatures in different systems: $HClO_4.5.5\ H_2O$, $HBF_4.5.75\ H_2O$, and $HPF_6.7.6\ H_2O$ (13). Murray reported a non-monotoneous temperature- dependence of the double layer capacitance around T_c in the system $Tl_2Ba_2Ca_2Cu_3O_{10}$/mixture of liquid organic electrolytes with low melting temperature (14). Kuznetsov (15) and Shapiro et al. (16) considered theoretically the possible charge transfer mechanisms at SC/IC interfaces on the basis of a quantummechanical treatment. Kuznetsov (15) demonstrated that Cooper pairs may contribute to the charge transfer process but that their influence on the electrode kinetics seems to be rather small for liquid electrolytes due to the relatively high reorganisation energy of the solvent. This restriction obviously does not remain by using SEs in the heterojunction SC/IC (14). In the paper of Shapiro et al. (16), only the hindered charge transfer of quasiparticles on a SC/IC interface at $T \leq T_c$ is considered but not the additional contribution of Cooper pairs in this temperature range.

Another open question is if the appearance of the quantumelectro-chemical kinetic effect is strongly connected with correlated energy states in both phases, i.e. Cooper pairs as correlated charge carriers within the SC or HTSC at $T \leq T_c$ and TLS states within the SEs below a glass-like transition temperatures $T \leq T_g$.

In the future, the low-temperature electrochemistry on HTSC/SE interfaces will not only be very interesting from the theoretical point of view but may also lead to very important practical applications. Surface modification of HTSCs including ion exchange, electrodeposition of metals, microstructuring, in-situ formation of HTSC/semiconductor surface contacts, etc. may be offered by low-temperature electrochemistry as a soft-method for varying the chemical, physical or electronic surface properties of HTSCs without affecting the superconducting state.

5. Acknowledgement

The authors gratefully acknowledge the financial support by the German Bundesministerium für Forschung und Technologie (BMFT R & D project 423-4013 - 13N55425) and the succesful scientific collaboration with A. Marok, A. Froese, R. Pötzschke, M. Ebert, J. Doneit, I. Engelhardt, A. Pinkowski, K. Jüttner (University of Karlsruhe, FRG), J. Markl, J. Ströbel, B. Roas (University of Erlangen, FRG), and T. Zetterer and K. F. Renk (University of Regensburg, FRG).

6. References

(1) A. Pinkowski, J. Doneit, K. Jüttner, W. J. Lorenz, G. Saemann-Ischenko, and M. W. Breiter, Europhys. Lett. 9 (1989) 269.

(2) A. Pinkowski, J. Doneit, K. Jüttner, W. J. Lorenz, G. Saemann-Ischenko, T. Zetterer, and M. W. Breiter, Electrochim. Acta 34 (1989) 1113.

(3) A. Pinkowski, J. Doneit, K. Jüttner, W. J. Lorenz, G. Saemann-Ischenko, T. Zetterer, and M. W. Breiter, Physica C 162 - 164 (1989) 1039.

(4) M. W. Breiter, W. J. Lorenz, and G. Saemann-Ischenko, Surface Science 230 (1990) 213.

(5) A. Pinkowski, K. Jüttner, and W. J. Lorenz, J. Electroanal. Chem. 287 (1990) 203.

(6) R. Pötzschke, A. Froese, S. Häckel, G. Herbst, J. Doneit, A. Pinkowski, K. Jüttner, W. J. Lorenz, P. Fickenscher, G. Saemann-Ischenko, T. Zetterer, K. F. Renk, and M. W. Breiter, UNESCO Expert Workshop on "Contribution of Electrochemistry for Conservation and Saving Energy and Environmental Protection", Gaussig Castle, Oct. 30 - Nov. 3, 1989 Proceedings Part 1, (Dresden University of Technology, 1990), p. 87.

(7) A. Pinkowski, J. Doneit, K. Jüttner, W. J. Lorenz, G. Saemann-Ischenko, T. Zetterer, and M. W. Breiter, DECHEMA-Monographs, Vol. 121, (VCH Verlagsgesellschaft Weinheim, 1990), p. 247.

(8) A. Pinkowski, K. Jüttner, W. J. Lorenz, G. Saemann-Ischenko, and M. W. Breiter, J. Electroanal. Chem. 286 (1990) 253.

(9) A. Pinkowski, T. Chierchie, and W. J. Lorenz, J. Electroanal. Chem. 285 (1990) 241.

(10) A. Pinkowski, W. J. Lorenz, G. Saemann-Ischenko, and M. W. Breiter, Solid State Ionics 40/41 (1990) 822.

(11) J. O'M. Bockris and J. Wass, J. Electroanal. Chem. 267 (1990) 325.

(12) J. O'M. Bockris and J. Wass, J. Electroanal. Chem. 267 (1990) 329 .

(13) U. Stimming, private communication.

(14) R. W. Murray, private communication.

(15) A. M. Kuznetsov, J. Electroanal. Chem. 278 (1990) 1.

(16) L. D. Zusman, A. B. Gelman, and B. Ya. Shapiro, Chem. Phys. Lett. 167 (1990) 457.

THEORY OF CHARGE TRANSFER REACTIONS AT SUPERCONDUCTING ELECTRODES

ALEXANDER M. KUZNETSOV
The A.N. Frumkin Institute of Electrochemistry of the Academy of sciences of the USSR. Leninskii prospect 31, Moscow 117071, USSR

ABSTRACT

A theory of electron transfer at superconductor/electrolyte interface is presented. Electric current due to transitions of normal electrons is analyzed in various limits. An effect of tunnel transitions of Cooper pairs to the reactants in the electrolyte is predicted. A maximum of the electric current near T_c is explained.

1. Introduction

Metals and semiconductors were traditionally used as electrodes in electrochemical systems. Later the metal electrodes covered by isolating, semiconducting or polymer film were applied. So the theory of electrochemical reactions was first developed for these electrodes[1-5]. The superconductors were also attractive materials for the applications of the theory of charge transfer. However since the electrochemical experiments were not possible due to extremely low temperature T_c of the transition to the superconducting state, they did not attract the attention of theorists and experimentalists until recent time.

The discovery of high-temperature superconductors (HTSC) with T_c higher than 77K creates the interest of researchers to these systems as possible electrode materials[5-12]. Recent fascinating successes in the high-temperature superconductivity give hope for further increase of the temperature of superconducting transition T_c.

At present there are mainly two kinds of electrochemical investigations involving the superconductors. One of them is the investigation of the properties of superconductors at high temperatures when they are in the normal state. In particular, their stability and corrosion processes are investigated in various electrolytes. The oxides of the type Y-Ba_2-Cu_3O_7, Bi-Ca-Sr-Cu-O, Tl-Ca-Ba-Cu-O, $La_{1.8}Sr_{0.2}CuO_4$ and $Bi_2Sr_{2.2}Ca_{0.8}Cu_2O_8$ are mainly investigated.

Another type of experiments is presented by the study of the kinetics of electron and proton transfer reactions in a broad temperature range including the temperature of superconducting transition T_c. Most of them are performed at the contact at superconductors of the type $YBa_2Cu_3O_{7-\delta}$, $YBa_2Cu_{3(1-x)}Zn_{3x}O_{7-\delta}$, $Tl_2Ba_2CaCuO_8$, $Tl_2Ba_2Ca_2Cu_3O_{10}$ with solid lectrolyteAg β'-alumina[5-11]. Only few experiments using liquid electrolyte mixtures have been recently reported[12].

Rather common observation of most experiments is a hump of the current near the temperature T_c. The increase of the current at the maximum point does not exceed usually 2 or 3 times.

First theory of elementary processes at a superconductor/electrolyte interface was presented in ref.13 where quantum mechanical theory of the elementary act of charge transfer[1-4,14,15] was extended to the systems under discussion. A similar approach was used a little later in ref.16.

2. Some electronic properties of superconductors

The conduction mechanism of HTSC is still not well understood. The main problem is the unknown nature of coupling between the electrons forming Cooper pairs which are responsible for superconductivity. However one may expect that electron pairing as a mechanism of superconductivity is common for low-temperature superconductors and for HTSC.

So it is useful to consider as a first step an idealized model to see what specific features of the superconductor are of importance for phenomena under discussion and which of their peculiarities may be expected to be of not great importance.

Let us consider a uniform superconductor with a smooth surface at the contact with a medium having ionic conductivity. There is an energy gap $2\Delta(T)$ in the excitation spectrum near the Fermi energy ε_F. The density of states for the "normal" electrons may have singularities of the type

$$\rho(\varepsilon)=(\varepsilon-\varepsilon_F)\rho_0(\varepsilon)/[(\varepsilon-\varepsilon_F)^2-\Delta^2]^{1/2} \qquad (1)$$

near the energies $\varepsilon-\varepsilon_F=\pm\Delta$ where ε_F is the Fermi energy and $\rho_0(\varepsilon)$ is the density of states for normal metal which is a smooth function of the electron energy ε. In what follows the latter will be considered as a constant for the sake of simplicity.

The quantity Δ depends on the temperature. Often it may be approximated by a power function. For model calculations we shall use simplest approximation in the form

$$\Delta(T)=\Delta_0(1-T/T_c)^{1/2} \tag{2}$$

The normal state of the superconductor is the state where pairs of the electron one-particle states are occupied (super-fluid states or Cooper pairs). The size of a Cooper pair may vary from $\xi_0\sim10\text{Å}$ to $\xi\sim1000\text{Å}$.

3. Transitions of normal electrons

Transition probability per unit time for the transfer of normal electron from electrode to an acceptor was calculated in various limiting cases[17].

3.1. High-temperature limit for medium modes

This limit assumes that vibrational frequencies for most of the nuclear modes satisfy the condition $\hbar\omega \ll k_BT$. Two limiting cases are to be distinguished.

Strong coupling with the medium modes. If the reorganization energy of the medium E_r is big ($E_r \gg k_BT$), the transition probability has a form

$$W(\varepsilon)=V^2(\pi/\hbar^2k_BTE_r)^{1/2}\ \exp\left[-(\Delta F(\varepsilon)+E_r)^2/4E_rk_BT\right] \tag{3}$$

where V is the electron matrix element and $\Delta F(\varepsilon)$ is the free energy of transition.

Weak coupling with the medium modes. If $\lambda \ll (\hbar\Omega^{(m)}/k_BT) \ll 1$ (where $\lambda \sim 2E_r/\hbar\Omega^{(m)}$ and $\Omega^{(m)}$ is a characteristic frequency of medium modes) the transition probability has a resonance form

$$W=(2\pi/\hbar)V^2(1/\pi)\lambda k_BT/[(\Delta F)^2+(\lambda k_BT)^2] \tag{4}$$

3.2. Low-temperature limit for medium modes

It corresponds to the condition $\hbar\Omega^{(m)}\gg k_BT$. If the vibrational spectrum of the medium is continuous, at strong coupling with medium modes the transition probability depends weakly on $\Delta F(\varepsilon)$ at $\Delta F(\varepsilon)\leq 0$ and goes fast to zero at $\Delta F(\varepsilon)>0$. So it may be approximated as

$$W(\Delta F)\cong \theta(-\Delta F)W_0(\Delta F) \tag{5}$$

However if the vibrational spectrum of the medium is discrete, then for not too great reorganization energy transition probability from a given energy level of the electron in the electrode to the acceptor has maxima as a function of the free energy of the transition, the one which corresponds to participation of different vibrational levels. Similar maxima appear in the dependence of the transition probability as a function of electron energy at fixed position of the electron energy level of the acceptor.

At weak coupling ($\lambda<<1$) transition probability is described be Eq.4.

4. Electric current due to normal electrons

General expression for cathodic electric current has the form[13]

$$i_e=e\int dx\ c(x)\ \left[\int_{-\infty}^{-\Delta} d\varepsilon + \int_{\Delta}^{\infty} d\varepsilon\ \right]\rho(\varepsilon)f(\varepsilon)W(\varepsilon) \tag{6}$$

where $c(x)$ is the concentration of acceptor at the distance x from the electrode, $f(\varepsilon)$ is the Fermi distribution function. We consider several limiting cases.

4.1. Strong coupling limit

High-temperature approximation. Approximate calculation of the integrals in Eq.6 with $W(\varepsilon)$ described by Eq.3 gives[13]

$$i_c\sim i_0(T)\exp\left[-\frac{\Delta(T)-e\eta}{2k_BT}\right] \tag{7}$$

where $i_0(T)$ is the exchange current for normal metal and $\eta=\phi-\phi_0$ is the overpotential, i.e. the shift of the electrode potential from the equilibrium potential ϕ_0.

Low-temperature approximation. In the case of continuous vibrational spectrum of the medium the expression for i_c has the form (see Eq.5 for W)

$$i_c \sim \int_{\Delta}^{\infty} d\varepsilon\ \rho(\varepsilon) f(\varepsilon) W_0(\varepsilon) + \theta(-e\eta-\Delta) \int_{-e\eta-\Delta}^{\Delta} d\varepsilon\ \rho(\varepsilon) W_0(\varepsilon) \sim$$
$$a\ \exp[-\Delta(T)/k_B T] + \theta(-e\eta-\Delta) b(e\eta) e\eta \qquad (8)$$

where a and $b(e\eta)$ are weak functions of η. The first term is small and plays a role only in the region $e\eta+\Delta>0$. The second term at small $e\eta$ is approximately linear in η, however at greater values of η it tends to level-off due to a decrease of $W(\varepsilon)$ with the increase of ε (Fig.1).

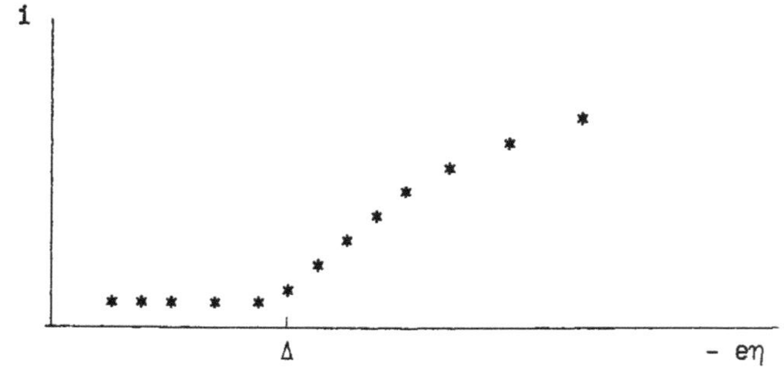

Fig.1. Dependence of the electric current on the overpotential at strong coupling with the medium in low-temperature limit.

If the vibrational spectrum of the medium is discrete, the dependence of the current on the overpotential may have steps corresponding to various vibrational levels of the accepting mode.

Weak coupling limit. At weak coupling with the medium modes the expression for W has the form of Eq.4 for both high-temperature and low-temperature limits. Qualitative plot of i_c vs. $e\eta$ is shown in Fig.2.

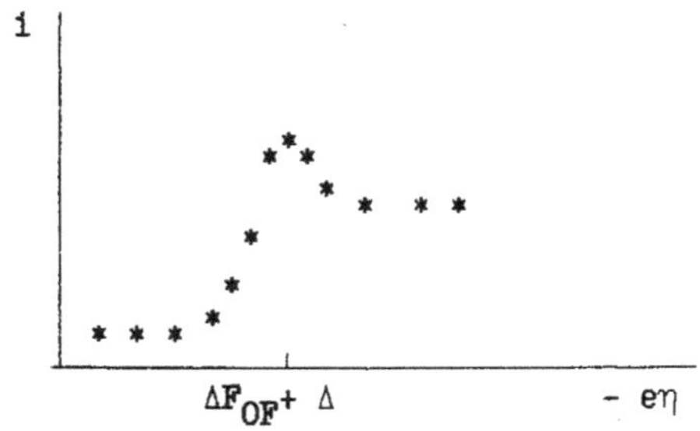

Fig.2. Dependence of the electric current on the overpotential at weak coupling with the medium.

For narrow resonance ($\lambda << 1$) the value of the current at the point of maximum is of the order $1/\lambda^{1/2}$. The right-hand wing of the curve is described by Eq.9

$$i_c \sim V^2\rho(\Delta F_{OF}-e\eta+\varepsilon_F)f(\Delta F_{OF}-e\eta) \tag{9}$$

where ΔF_{OF} is the free energy of the transition for electron transfer from the Fermi level at equilibrium potential.
The left-hand wing is described by Eq.10

$$i_c \sim V^2W(\Delta F_{OF}-e\eta+\Delta)\int\rho(\varepsilon)d\varepsilon \quad (\sim\lambda << 1) \tag{10}$$

5. Temperature dependence of the current near T_c

The analysis of Eq.6 with $W(\varepsilon)$ described by Eq.3 shows that in this case electric current decreases monotonously with decrease of temperature. In low-temperature approximation the temperature behaviour of the current depends on the form of $W(\varepsilon)$. If the dependence of W on ε is neglected, the electric current is also monotonous function of temperature. It is approximately valid for continuous vibrational spectrum of the medium. If the vibrational spectrum of the medium is discrete, the electric current in principle may exhibit a maximum in the neighbourhood of T_c. Since $W(\varepsilon)$ here has maxima as a function of ε, the qualitative situation is similar to that for weak coupling limit (which is discussed in detail below).

The temperature dependence of the current in the case of weak coupling with the medium depends on the overpotential $e\eta$. If at a given value of $e\eta$ the resonance of Eq.4 is centered at the Fermi level, viz.

$$W(\varepsilon)=(2\pi/\hbar)V^2(1/\pi)\lambda k_BT/\left[(\varepsilon-\varepsilon_F)^2+(\lambda k_BT)^2\right] \qquad (11)$$

or at the energy levels $\varepsilon>\varepsilon_F$, the electric current is a monotonous function of temperature decreasing with the decrease of T.

If however it is centered at an energy $\varepsilon_F-\delta$ $(\delta>0)$, the electric current may exhibit a hump in the neighbourhood of T_c. The ratio of the electric current to the electric current i_T at the temperature T_c is described in this case by Eq.(12)

$$i_c/i_{T_c}=(1/\pi)\int_{\Delta}^{\infty} d(\varepsilon-\varepsilon_F)\frac{\varepsilon-\varepsilon_F}{\left[(\varepsilon-\varepsilon_F)^2-\Delta^2\right]^{1/2}}\frac{1}{1+e^{(\varepsilon-\varepsilon_F)/k_BT}}\frac{\lambda k_BT}{\left[(\varepsilon-\varepsilon_F+\delta)^2+(\lambda k_BT)^2\right]}+$$

$$(1/\pi)\int_{\Delta}^{\infty} d(\varepsilon-\varepsilon_F)\frac{\varepsilon-\varepsilon_F}{\left[(\varepsilon-\varepsilon_F)^2-\Delta^2\right]^{1/2}}\frac{1}{1+e^{(\varepsilon-\varepsilon_F)/k_BT}}\frac{\lambda k_BT}{\left[(\varepsilon-\varepsilon_F-\delta)^2+(\lambda k_BT)^2\right]} \qquad (12)$$

Here the first term corresponds to the integration over the electron energy above the Fermi level, and the second one describes the region below the Fermi level.

Qualitatively the possibility of the appearance of the maximum in $i_c$vs.T plot may be seen in the limit of extremely narrow resonance (λ 0). If at a given T the gap $\Delta(T)$ is not too close to δ $(\Delta(T)<\delta)$ and $\delta>k_BT$, the current is equal to

$$\frac{i_c}{i_{T_c}} \approx \frac{\delta}{\left[\delta^2-\Delta^2(T)\right]^{1/2}} \quad \frac{1}{1+e^{-\delta/k_BT}} \sim \frac{\delta}{\left[\delta^2-\Delta^2(T)\right]^{1/2}} \tag{13}$$

since the first term is negligibly small.

At $T=T_c$ we have $\Delta=0$ and $i_c=i_T$. As T decreases, $\Delta(T)$ increases and so does the ratio i_c/i_T. When $\Delta(T)$ becomes greater than δ the integral in the second term of Eq.12 also decreases due to the shift of the lower integration limit to the right of the resonance maximum.

The results of calculations according to Eq.12 with $\Delta(T)=\Delta_0(1-T/T_c)^{1/2}$ are shown in Fig.3.

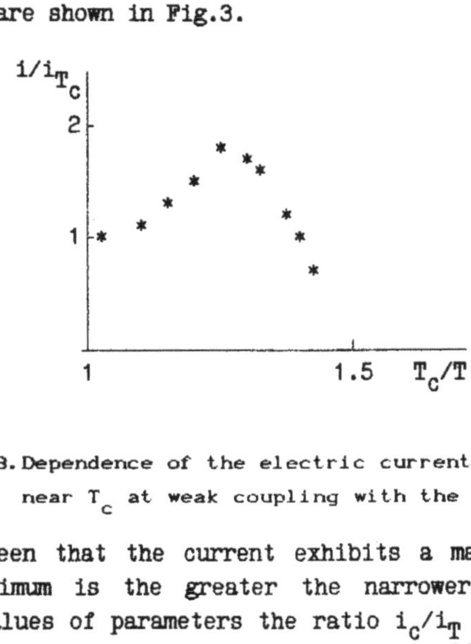

Fig.3. Dependence of the electric current on the temperature near T_c at weak coupling with the medium.

It may be seen that the current exhibits a maximum near $T=T_c$.The height of the maximum is the greater the narrower is the resonance. Depending on the values of parameters the ratio i_c/i_T may amount to 2 - 3 or even greater. In principle the qualitative results of this kind may exist also for other types of W vs. ε dependencies for which the major contribution to the current comes from the electron energies except for a small neighbourhood of the Fermi level. However the maximum is pronounced only for rather narrow volcano type dependence of W on ε. The height of the maximum increases also with the increase of δ, i.e. the existence of the maximum depends also on the overpotential.

6. Electric current due to the transitions involving superfluid electrons

Along with normal electrons superfluid electrons also may participate in the electrode process. Such a possibility was first predicted in ref.13. Two-electron processes due to the transitions of Cooper pairs from the electrode to the acceptors were discussed in the high temperature limit for nuclear modes for the case of strong coupling with the medium modes[13].

Different mechanisms may be distinguished:

(1) A direct transfer of two electrons to one acceptor in one elementary act. This process is characterized by a rather small matrix element and a big reorganization energy of the medium which is 4 times as great as the reorganization energy for one-electron process.

(2) A superexchange (through virtual intermediate state) transition of two electrons to two different acceptors located at some distance from each other. This distance must be comparable with the size of Cooper pair and if it is sufficiently large, the reorganization energy of the medium is only twice as large as that for one-electron process.

(3) A transition through real intermediate state without the relaxation of the energy in it. In this case the activation barrier is determined by the highest crossing point of the potential energy surfaces and the reorganization energy is the same as for one-electron process.

The expressions for the transition probability and more detailed discussion may be found in ref.13.

Different limiting cases and mechanisms are discussed qualitatively below.

6.1 One-electron reactions resulting from the decay of Cooper pairs at the interface

This mechanism corresponds to the transfer of one electron of Cooper pair to an acceptor with the simultaneous transition of another one to the states of normal electrons. Since there is an energy gap in excitation spectrum, in high-temperature limit for medium modes it leads to the appearance of the factor of the type $\exp(-\Delta/2k_BT)$ in the transition probability and hence in the electric current (in the region near the equilibrium potential). In the low-temperature limit a similar factor has the form of $\exp(-\Delta/k_BT)$ which determines the probability for corresponding electron energy level in the metal to be unoccupied. Thus in this mechanism the current is lower than one-electron current at normal metal.

Illumination of the electrode with the light of the energy $\hbar\omega \geq 2\Delta$ also leads to decay of Cooper pairs and hence to an increase of the concentration of normal electrons. Therefore it must result in an increase of electric current.. At small light intensities one may expect that an excess current will be proportional to the intensity of the light.

6.2 The superexchange transfer of Cooper pair in the low-temperature limit. The low-temperature limit for superexchange transfer is very similar to low-temperature limit for one-electron processes[17,18]. The difference consists only in the form of electron matrix element. In the case of superexchange mechanism it is the matrix element of the second order of the perturbation theory. Therefore with due account of this fact all the results of the theory of low-temperature reactions may be applied to the transfer of Cooper pairs (see Section II and refs.17,18). The main result of this limit is that the transition probability is nonzero practically only at $\Delta F_0 \leq 0$ (see Eq.5) and weakly depends on temperature and ΔF_0.

7. Concluding remarks

The main results are briefly summarized below.

7.1. The current due to normal electrons

(1) In the high-temperature region for vibrational medium modes the dependence of the current on overpotential η is Tafel-like (exponential) for the case of strong coupling with the medium modes. The temperature dependence of the current is activational. The current depends exponentially on the energy gap as $\exp(-\alpha\Delta/k_BT)$ (see Eq.7) where α is the symmetry factor.

In the low-temperature region the temperature dependence of the current is weak and at $|e\eta|<\Delta$ it involves a factor $\exp(-\Delta/k_BT)$ (see Eq.8).Thus in both cases the current is sensitive to the energy gap. Therefore a decrease of the gap at constant temperature (e.g.due to the application of the magnetic field) must lead to an exponential increase of the current.

At $|e\eta|>\Delta$ the dependence of the current on overpotential in low-temperature limit becomes non-exponential (see Eq.8). Therefore varying Δ at a constant η we can switch from the region of the exponential behaviour to the region of almost linear behaviour of i vs.η dependence. Both in high-temperature and low-temperature regions the current monotonously decreases with temperature for continuous vibrational spectrum of the medium.

(2) At weak coupling with the medium modes the dependence of the current on the overpotential has a maximum near some value of η. Temperature dependence of the current in general is rather weak both in high-temperature and low-temperature regions except a neighbourhood of the transition temperature T_c. In some region of overpotentials the temperature dependence of the current may exhibit a maximum near T_c. A similar behaviour may be expected for the case of discrete vibrational spectrum of the medium in the limit of strong coupling. Thus the experimental temperature behaviour of the current may be explained if we assume that around T_c either the coupling with the medium modes is weak or the coupling is strong but the vibrational spectrum is quantum and discrete. However some other (still unknown) reasons of such behaviour may exist.

7.2. Superfluid electrons. The superfluid electrons may also participate in the elementary act of electrochemical reaction. Two main mechanisms are predicted: one-electron reactions with the decay of Cooper pair and two-electron reactions.

References

1. R.R.Dogonadze and Yu.A.Chizmadzhev, *Dokl.Akad.Nauk SSSR,ser. Fiz.Khim.* 145 (1962) 849
2. R.R.Dogonadze and A.M.Kuznetsov, *Progr.Surf.Sci.* 6 (1975) 1
3. R.R.Dogonadze and A.M.Kuznetsov, *Comprehensive Treatise of Electrochemistry* (Plenum Publishing Corp., New York) 7 (1983) 1
4. J.Ulstrup, *Charge Transfer Processes in Condensed Media* (Springer-Verlag, Berlin - Heidelberg - New York, 1979)
5. S.Hackel, G.Herbst, J.Doneit, M.Bachtler, A.Pinkowski, W.J.Lorenz, E.Grabner, M.W.Breiter, U.Stimming and G.Staikov, *J.Electrochem.Soc.* 135 (1988) 456C
6. A.Pinkowski, J.Doneit, K.Juttner, W.J.Lorenz, G.Saemann-Ischenko, T.Zetterer and M.W.Breiter, *J.Electrochem.Soc.* 136 (1989) 234C
7. A.Pinkowski, J.Doneit, K.Juttner, W.J.Lorenz, G.Saemann-Ischenko and M.W.Breiter, *Europhys.Lett.* 9 (1989) 269
8. A.Pinkowski, J.Doneit, K.Juttner, W.J.Lorenz, G.Saemann-Ischenko, T.Zetterer and M.W.Breiter, *Electrochim.Aacta* 34 (1989) 1113

9. A.Pinkowski, J.Doneit, K.Juttner, W.J.Lorenz, G.Saemann-Ischenko, T.Zetterer and M.W.Breiter, *Physica* C162-164 (1989) 1039

10. M.W.Breiter, W.J.Lorenz and G.Saemann-Ischenko, *Surf.Sci.* 230 (1990) 213

11. J.O'M.Bockris and J.Wass, *J.Electroanal.Chem.* 267 (1989) 329

12. J.T.McDevitt, S.Ching, M.Sullivan and R.W.Murray, *J.Am.Chem.Soc.* 111 (1989) 4528

13. A.M.Kuznetsov, *J.Electroanal.Chem.* 278 (1990) 1

14. A.M.Kuznetsov, *Modern Aspects of Electrochemistry* 20 (1989) 95

15. A.M.Kuznetsov, J.Ulstrup and M.A.Vorotyntsev, *in Chemical Physics of Solvation*, eds.R.R.Dogonadze, E.Kalman, A.A.Kornyshev and J.Ulstrup (Elsevier, Amsterdam - oxford - New York - Tokyo, 1980) p.163

16. L.D.Zusman, A.B.Gelman and B.Ya.Shapiro, *Chem.Phys.Lett.* 167 (1990) 457

17. R.R.Dogonadze and A.M.Kuznetsov, *Kinetics of Heterogeneous Chemical Reactions in Solutions, in Itogi Nauki i Tekhniki* (VINITI, Moscow, 1978), ser.Kinetics and Catalysis, vol.5.

18. R.R.Dogonadze, A.M.Kuznetsov, M.A.Vorotyntsev and M.G.Zaqaraya, *J.Electroanal.Chem.* 75 (1977) 315

APPLIED RESEARCH

ELECTROCHEMICAL SOLAR ENERGY CONVERSION

HEINZ GERISCHER
Fritz-Haber-Institut der Max-Planck-Gesellschaft,
Faradayweg 4-6, D-W-1000 Berlin 33, F. R. G.

Abstract

The principles of solar energy conversion in photoelectrochemical cells are briefly reviewed. Cells for the generation of elektric power and for energy storage in form of electrochemical energy are described. These systems are compared with solid state photovoltaic devices, and the inherent difficulties for the operation of the electrochemical systems are analyzed.

1. Introduction

The driving force of photosynthesis is a photoelectrochemical reaction in which excited chlorophyl molecules transfer electrons to an acceptor complex which can reduce CO_2 in a complex series of catalytic reactions to carbohydrates. The oxidized chlorophyl molecule transfers its positive charge to another complex which can oxidize water to oxygen [1,2]. This occurs at two sides of a membrane with a highly organized, complex structure which can hardly be synthesized artificially. The principle of light energy conversion in photosynthesis is, however, the same as in most other photoreactions. Light absorption creates an excited electronic state which can simultaneously generate carriers with negative and positive charge. If recombination of these charge carriers can be prevented, the excitation energy can be converted into electrical or electrochemical energy.

Some photoredox reactions in homogeneous solution are known to produce species with different redox potentials which can be used for the construction of photogalvanic cells [3]. The efficiency of energy conversion by such processes is, however, very low because most of the energy is lost by the reverse redox reaction between the primary products. Much more efficient are photoreactions at interfaces of semiconductors [4-7]. These processes are closely related to solid state photovoltaic devices and will be described and evaluated.

2. Photoelectrochemical Cells with Light Absorption in the Solid

Photocurrents and photovoltages can be generated at semiconductor/electrolyte contacts if a depletion layer exists in the semiconductor. This is the case for n-type semiconductors in the presence of a redox system with a positive enough redox potential to extract electrons from the semicnductor. For p-type semiconductors one needs a respectively negative redox system which extracts holes. Under illumination, electron hole pairs are separated by the electric field in the depletion layer. At open circuit, the majority carriers accumulate in the bulk and a counter voltage is created until charge separation is compensated by recombination. If the circuit is closed vs. a counter electrode, a photocurrent is generated. This is outlined in **figure 1** for an n-type semiconductor.

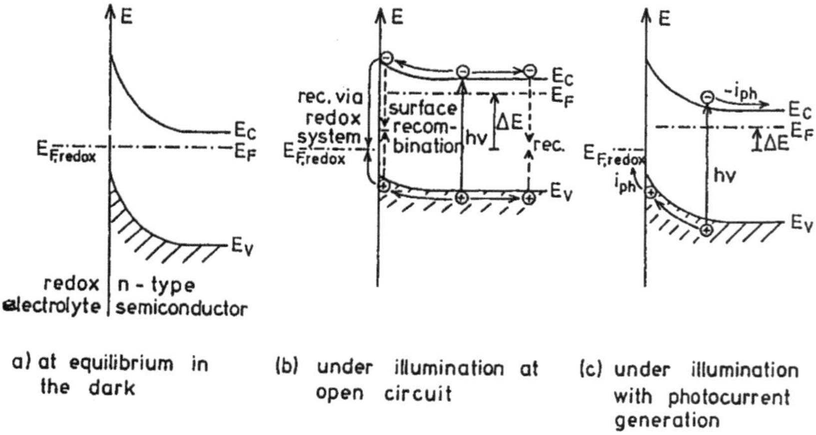

Fig. 1: Generation of photovoltages and photocurrents at an n-type semiconductor electrode.

In a closed cell the photovoltage acts as a driving force for the photocurrent, compensating at the steady state the ohmic and interfacial voltage losses. Such a closed cell and its function is depicted in **figure 2** in terms of an electronic energy diagram. This picture also indicates the unavoidable losses of free energy connected with such cells [4,8].

The power produced by such a cell depends, as in any photovoltaic system, on the external resistance of the consumer [9]. **Figure 3** shows some power characteristics of photoelectrochemical cells and the maximal efficiencies for solar energy conversion at the optimal adjustment of the consumer system.

The power generation at such contacts corresponds completely, with regard to mechanism, to the processes at a solid state Schottky barrier cell [10,11], the only dif-

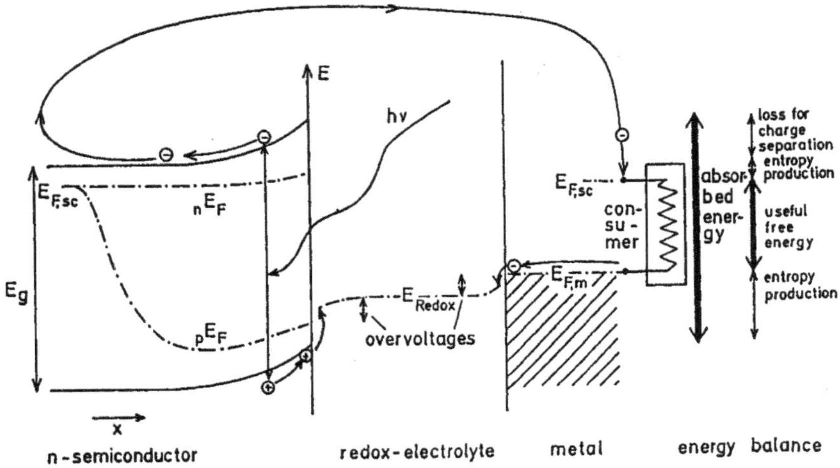

Fig. 2: Energy diagram of a photoelectrochemical cell for the generation of electric power.

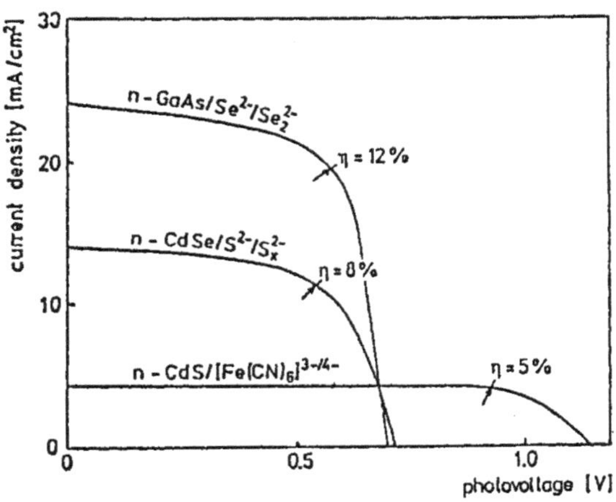

Fig. 3: Power characteristics of some photoelectrochemical cells.

ference being that the metal is replaced by the redox electrolyte and a counter electrode to close the electronic circuit. This opens new possibilities for the use of the photovoltage. Instead of immediately consuming the electric power of the cell like in photovoltaic systems, one can directly store the excess energy in the form of electrochemical redox energy [12]. For this purpose, only another redox reaction at the counterelectrode is needed. The two redox systems with different redox potentials, which have to be kept separated by a suitable membrane or similar means, form a redox battery. The stored energy of this battery can be used at any time when it is needed. The principle design of such a system is shown in **figure 4**.

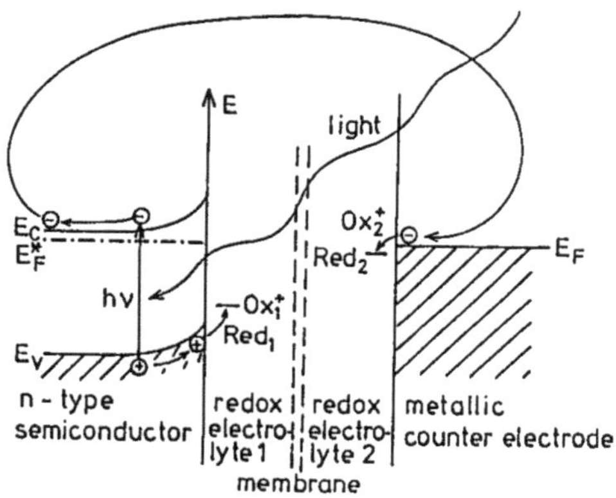

Fig. 4: Principle of a photoelectrolytic storage cell in terms of electronic energies.

The photoelectrolytic decomposition of water, in which the energy can be stored as H_2 and O_2, is an attractive process for this purpose [13-16]. Such a system would not necessarily need a separator between the electrode compartments since the losses by the reverse reactions at the electrode remain small due to the limited solubility of the gases. The necessary photovoltage, however, must exceed the decomposition voltage plus the overvoltages for the electrode reactions [4]. Therefore, the band gap of the semiconductor has to be larger than 2.1 or even 2.4 eV [4,8,16]. The ideal conditions shown in **figure 5** have, unfortunately, not been met by available semiconductors.

One reason is that it is not sufficient to have a semiconductor with the right band gap. The band edge positions must also match the redox energy levels of the electrode reactions [4]. Furthermore, the semiconductors must be stable against photodecomposition [17,18]. The latter condition seems to be fulfilled only by oxides. The valence band edge of stable oxides is, however, at such deep energies that the driving

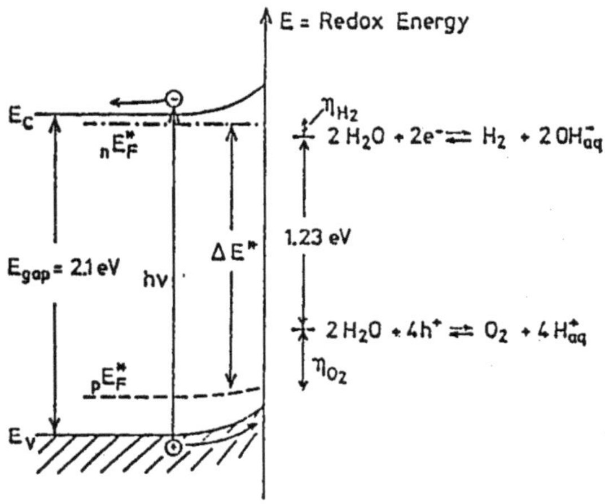

Fig. 5: Ideal conditions for a semiconductor to perform water photoelectrolysis.

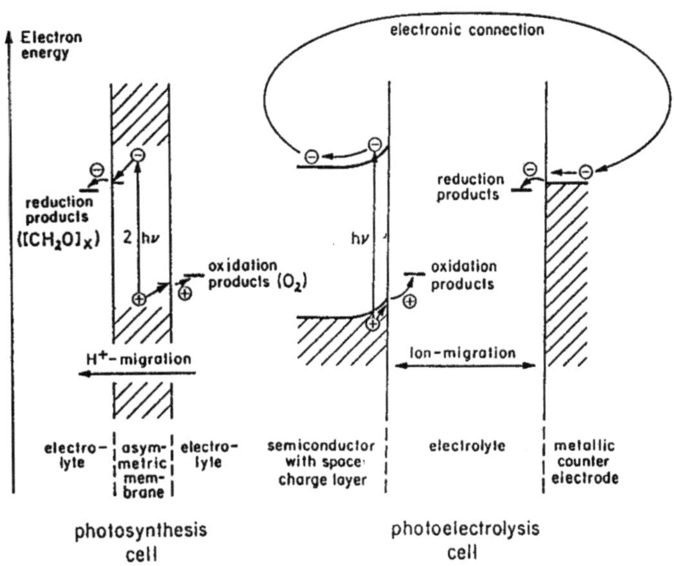

Fig. 6: Comparison between photosynthesis and photoelectrolysis.

force for water oxidation by holes becomes unnecessarily high and much excess energy is wasted. In order to give the electrons in the conduction band enough energy for water reduction, i.e., hydrogen evolution, a band gap of 3 eV or more is needed to reach a high enough driving force [4,8,16]. The result is that too little solar light is absorbed in materials with such a band gap which can split water, as the correlation between theoretical efficiency and band gap indicates [19,20].

In the course of evolution, nature developed an enormously sophisticated photoelectrochemical system for water decomposition in the photosynthesis apparatus of the plants. In this case, the excess energy of the electrons is immediately used for the reduction of CO_2, in which the protons left over from the H_2O oxidation are consumed, and the gained energy is stored in the form of carbohydrates. **Figure 6** shows the analogy between the photosynthesis process and a photoelectrolysis cell in terms of the electronic energy states involved in the respective electron transfer steps.

3. System with Charge Injection from Excited Molecules into the Semiconductor

Semiconductors with a band gap below 2 e V absorb a large fraction of the solar light; they are, however, unstable against photodecomposition. Some semiconductors with a wide band gap like TiO_2 or SnO_2 are very stable against photodecomposition but absorb too little solar light. They can, however, be spectrally sensitized by the adsorption of dyes at the interface which can inject electrons or holes after light absorption [21,22]. The energetic conditions for electron or hole injection from an excited molecule are depicted in **figure 7**. If the oxidized dye molecules are reduced by a redox system in the electrolyte, a photovoltaic cell can be constructed of which an energy scheme is shown in **figure 8**.

The limitations of such devices are in the limited light absorption of monolayers of dyes. Thicker layers cannot be used because the dye layer itself would form a high resistance for electron transfer and transport. However, Grätzel found rather high yields with highly dispersed surfaces of TiO_2, sensitized by dyes forming chelate complexes on the surface. He recently constructed a cell following the outline of figure 8 with the redox reaction $2Br^- \rightleftharpoons Br_2 + 2e^-$ for the regeneration of the dye and a conversion efficiency for monochromatic light in the maximum of the absorption spectrum of 72 % and a solar energy conversion efficiency of 12 % [23]. The open question is the long time stability of the device, since dyes are usually slowly decomposing.

4. Evaluation and Prospects

Solar light conversion into other forms of free energy is limited by a general restriction due to the wide spectrum of solar radiation [19,20]. Semiconductors have a threshold for light absorption given by their band gap. The energy of light quanta in excess of the band gap energy is, after absorption, quickly dissipated, since electrons

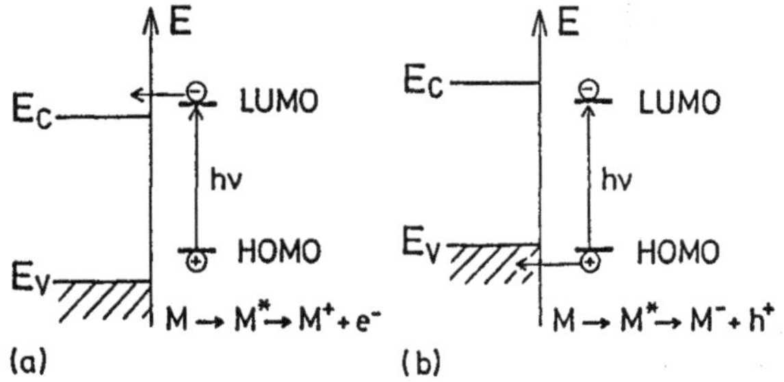

Fig. 7: Charge injection from excited molecules into solids.

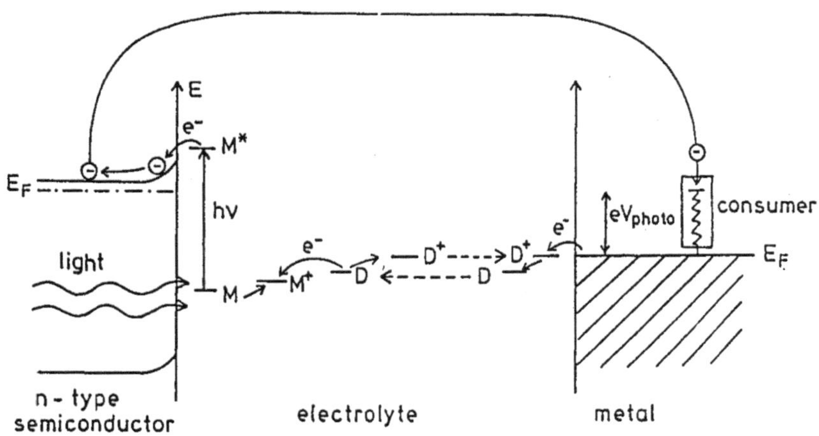

Fig. 8: Photoelectrochemical cell with sensitized semiconductor electrodes.

and holes are stable only at the energy levels around the band edges. As a result of these two effects, the theoretical conversion efficiency depends on the band gap width and has a maximum in the range of 1.4 to 1.6 eV. This is shown in **figure 9** for an ideal threshold absorber exposed to the solar radiation in space.

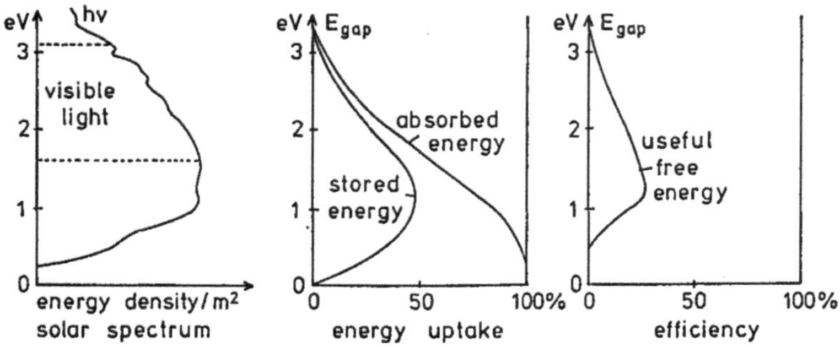

Fig. 9: Energy conversion by a threshold absorber (E_{gap} = energy threshold for absorption).

This restriction is valid for all photovoltaic and photosynthetic systems. The question is therefore whether photoelectrochemical cells (PEC) have some advantages over photovoltaic cells. In the mode of electric power generation, such an advantage could only be cheaper production costs. Even in its simplest construction with a transparent redox electrolyte and a transparent counter electrode, as shown in figure 10, the PEC has, however, the disadvantage of needing a second interface in the form of the counter electrode, where additional energy losses (overvoltages) in the electron transfer reaction are unavoidable. In addition to this, the transport of electric charge through the electrolyte has to occur by diffusion of ions. This is a much slower process, even in electrolytes of high concentration, than the charge transport in an electronic conductor. An additional problem is the sealing of the electrolyte and the photocorrosion of the semiconductor. Although the corrosion can be kept very small in some cases, the lifetime suffers even from very low corrision rates. Commercial solar cells require lifetimes of many years for an economic application. Such problems can be overcome much more easily in solid state devices.

The next question is whether the possibility of immediate energy storage can give advantages to the photoelectrochemical systems. We have seen that nature created an effective system in photosynthesis. The success of this system is, however, not in the energy conversion efficiency, which is theoretically about 10 % and practically in the best plants 4 %. The very high catalytic activity for H_2O oxidation and CO_2 reduction keeps, however, the additional losses by overvoltages in these reactions extremely low. With electronic devices we can easily reach higher energy conversion efficiencies. But where nature is still much superior, this is the electrocatalytic efficiency for water splitting. Unfortunately, the semiconductor electrodes which generate enough energy for H_2O decomposition are poor catalysts for these reactions. They

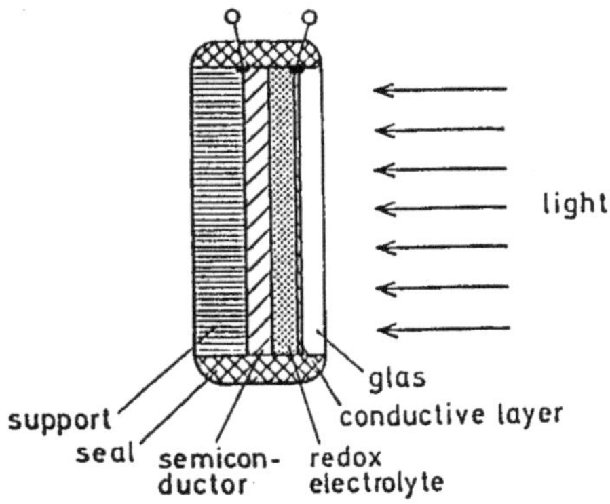

Fig. 10: Model of a photovoltaic cell with a semiconductor-redox electrolyte junction.

have to be activated by the deposition of metallic catalysts, which can be done but introduces additional problems [24,25]. Semiconductors with band gaps in the range of the optimum for solar energy conversion do not generate high enough photovoltages for water decomposition. One needs either two photoactive junctions in series [26] or an auxiliary external voltage in a hybrid system [27].

Texas Instruments developed an integrated device with two p-n junctions of silicon in series as voltage generators for a redox battery, consisting of a compartment with a $Br_2/2Br^-$ electrode as the anode and a compartment with a $H_2/2H^+$ electrode as the cathode, as storage system [28]. This system has not found practical application up to now. It has to compete with solid state photovoltaic systems as power generators for normal H_2O electrolysis. Separation of the function of power generation and energy storage has the great advantage that all the restrictions, imposed on the construction by the electrolyte and the counter electrode, are avoided. The electrolysis can be performed separately at higher current densities. The optimal adjustment between photovoltaic systems and electrolysis cells as well as the improvement of electrocatalysis remain important tasks for the electrochemist. It appears unlikely that the direct conversion of light energy by electrochemical processes, except in photosynthesis, will play a considerable role in the energy production of the future.

In the near future, photoelectrochemical studies will probably be dominated by fundamental research, enriching our knowledge of material properties and extending our understanding of nature.

5. References

1. H. T. Witt, E. Schlodder, K. Brettel and Ö. Saygin, *Ber. Bunsenges. Phys. Chem.* **90** (1986) 1015.
2. H. T. Witt, *New Journal of Chemistry* 11 (1987) 91.
3. M. D. Archer and M. I. C. Ferreira, in *Photochemical Conversion and Storage of Solar Energy*, ed. J. S. Connolly, (Academic Press, NewYork, 1981), p. 201.
4. H. Gerischer, in *Topics in Applied Physics*, Vol. 31: *Solar Energy Conversion*, ed. B. O. Seraphin, (Springer, Berlin, 1979), p. 115.
5. R. Memming, *Electrochim. Acta* **25** (1980) 77.
6. A. Heller, *Acc. Chem. Res.* 14 (1981) 154.
7. H. Tributsch, in *Modern Aspects of Electrochemistry*, Vol. 17, eds. J. O'M. Bockris, B. E. Conway and R. E. White, (Plenum Press, New York, 1986), p. 303.
8. H. Gerischer, in *Photoelectrochemistry, Photocatalysis and Photoreactors*, ed. M. Schiavello, NATO ASI Series, C 146, (D. Reidel, Dordrecht, 1985), p. 39.
9. S. M. Sze, *Physics of Semiconductor Devices*, (Wiley, New York, 1969).
10. H. Gerischer, *J. Electroanal. Chem.* **58** (1975) 263.
11. D. L. Feucht, *J. Vac. Sci. Technol.* 14 (1977) 57.
12. H. Gerischer, in *Photoelectrochemistry, Photocatalysis and Photoreactors*, ed. M. Schiavello, NATO ASI Series, C 146, (D. Reidel, Dordrecht, 1985), p. 93.
13. A. Fujishima and K. Honda, *Nature* **238** (1972) 37.
14. M. Tomkiewicz and H. Fay, *Appl. Phys.* **18** (1979) 1.
15. A. Heller, in *Photochemical Conversion and Storage of Energy*, ed. J. Rabani, (Weizmann Science Press, Jerusalem, 1982), p. 63.
16. J. B. Goodenough, in *Photoelectrochemistry, Photocatalysis and Photoreactors*, ed. M. Schiavello, NATO ASI Series, C 146, (D. Reidel, Dordrecht, 1985), p. 153.
17. H. Gerischer, *J. Electroanal. Chem.* **82** (1977) 133; *J. Vac. Sci. Technol.* **15** (1978) 1422.
18. A. J. Bard and M. S. Wrighton, *J. Electrochem. Soc.* **124** (1977) 1706.
19. R. T. Ross and Ta-Lee Hsiao, *J. Appl. Phys.* **48** (1977) 4783.
20. R. T. Ross, *J. Chem. Phys.* **45** (1966) 1.
21. H. Gerischer and F. Willig, in *Topics in Current Chemistry*: Vol. 61: *Physical and Chemical Applications of Dyestuffs*, (Springer, Berlin, 1976), p. 31.
22. R. Memming, in *Progress in Surface Science*, Vol. 17, ed. S. G. Davison, (Pergamon Press, New York, Oxford, 1986), p. 7.
23. N. Vlachopoulos, P. Liska, J. Augustynski and M. Grätzel, *J. Am. Chem. Soc.* **110** (1988) 1216.
24. A. Heller and B. Miller, *Electrochim. Acta* **25** (1980) 29.
25. M. P. Dare-Edwards, A. Hamnett and J. B. Goodenough, *J. Electroanal. Chem.* **119** (1981) 109.
26. A. J. Nozik, *Appl. Phys. Lett.* **28** (1976) 156.
27. E. Aharon and A. Heller, *J. Electrochem. Soc.* **129** (1982) 2865.
28. J. D. Luttmer, D. Konrad, and I. Trachtenberg, *J. Electrochem. Soc.* **132** (1985) 1054.

POWER SOURCES

MANFRED W. BREITER

Institut für Technische Elektrochemie, TU Wien, 9 Getreidemarkt, A-1060 Wien, Austria

ABSTRACT

Trends in the development and production of power sources are described. Systems for the production of electrical energy by the conversion of chemical energy are considered first, distinguishing between high and low temperature systems. Rechargeable batteries for the storage of electrical energy are dealt with next, differentiating again between high and low temperature systems.

1. INTRODUCTION

In general, high temperature systems are considered for both energy production and energy storage in stationary installations. The so-called sodium batteries are an exception because of their planned use for vehicle propulsion.

The development of a new system proceeds in the following way. Once the idea has been conceived, the production of single cells for testing and improving is started. This stage of the development involves studies of fundamental questions in electrode kinetics, corrosion and electrochemical engineering on the one hand and in materials science on the other hand. After single cells with satisfactory performance can be made the emphasis shifts to other questions like system reliability, overall efficiency, costs, life and pilot plants for production. This second stage involves mainly engineers and materials scientists. The electrochemists continue to study specific fundamental questions to a smaller extent than in the first stage. The second stage is very expensive. The expectations of quick profitability and poor management were detrimental to the development of many promising systems, especially in USA.

2. ENERGY PRODUCTION

2.1. HIGH TEMPERATURE SYSTEMS

High temperature systems consist of fuel cells (conversion of chemical energy into electrical energy) and the alkali metal thermo-electric converter (direct conversion of thermal energy into electrical energy). The latter converter was originally called [1] "Sodium Heat Engine".

2.1.1. FUEL CELLS

There are two different systems under consideration:

- Molten carbonate fuel cell (MCFC)

- Solid Oxide fuel cell (SOFC)

Both these systems have the advantage over the low temperature fuel cell that

a) the heat from the exhaust gases can be used in turbines and generators to improve the overall efficiency.

b) the anode is not so sensitive to carbon monoxide.

Both systems have reached the engineering stage. The MCFC is more advanced[2]. Stacks in the 100 kW class are tested. For the SOFC a planar design, somewhat similar to that of the MCFC, is evaluated[3]. Answers to a number of fundamental questions are looked for. These questions are for MCFC:

c) Will it be possible to keep the present cathode (lithiated nickel oxide) which dissolves slowly, leading to the shortening of the cells by nickel dendrites which are formed in the vicinity of the anode ?

d) Can the long term loss of electrolyte by evaporation be compensated sufficiently by the addition of carbonates from reservoirs or by the use of different electrolytes ?

e) To what extent can sulfur be present in the fuel because of the formation of sulphides in the carbonates ?

f) Will internal reforming of the fuel be feasible ?

For OSFC the fundamental questions are:

g) Can other oxide ion conductors which possess a satisfactory conductivity at lower temperatures than stabilized zirconia be used ? The so-called BIMEVOX materials[4] are promising candidates.

h) Is chemical cogeneration like the oxidation of methanol to formaldehyde economically feasible ?

2.1.2. ALKALI METAL THERMO-ELECTRIC CONVERTER

This system was originally conceived with Na beta" alumina. Essentially, it consists of a concentration cell with different sodium activities. A diagram of the alkali metal thermo-electric converter is shown in Fig.1. The projected efficiency is compared with the efficiencies of other systems in Fig.2. The reason of the attractiveness of the converter becomes evident from Fig.2.

The main problem is the performance[5] of the porous electrode, for instance Mo, because of alloy formation between Na and Mo. Developmental work is going on in several laboratories in USA and Europe.

2.2. LOW TEMPERATURE SYSTEMS

2.2.1. FUEL CELLS

Four types of fuel cells are considered:

- Phopshoric acid fuel cell (PAFC)
- Alkaline fuel cell (AFC)
- Fuel cells with membranes (earlier: SPEFC, now: IEMFC)
- Direct methanol fuel cell (DMFC)

The upper three cells require an external reforming of the fuel and to a different extent the reduction of carbon monoxide content in the reformed gas. The latter condition depends upon the electrocatalyst of the anode and the electrolyte. Additionally the alkaline fuel cell requires the scrubbing of CO_2 from the air,fed to the cathode.

Applications for PAFC which represents[3,6] the most advanced system are sought in power plants. The feasibility of a power plant in the MWh class was demonstrated. However, environmental concerns or military use must be present to pay the higher price of electricity production in such a

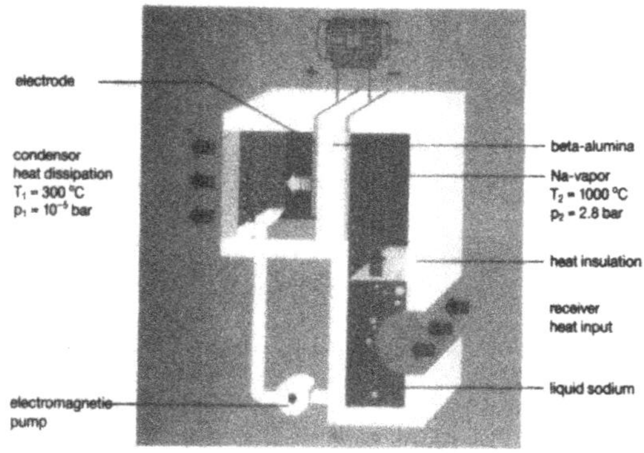

Fig.1. Schematic diagram of AMTEC Cell

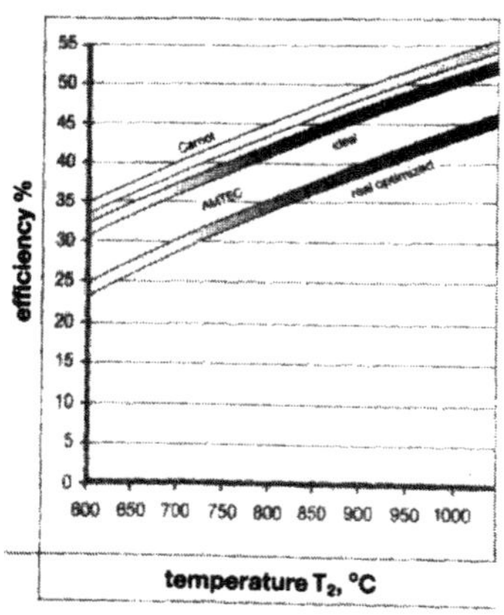

Fig.2. Efficiencies of AMTEC Systems and according to Carnot Cycle

plant. Improvements from the use of ternary Pt alloys in the oxygen electrode are expected[6] for the efficieny and the life of PAFCs.

Application for the alkaline fuel cells[3,7] are considered for vehicle propulsion and other special purposes like military and space-related ones.

Great progress has been achieved with the ion exchange membrane fuel cells in the last five years[3,8-10]. The electrocatalyst partcicles are directly pressed onto the membrane. Current densities up to 2 A/cm^2 can be drawn at reasonable cell voltages (0.7 V). The price of the membranes is coming down as a result of increased competition. Applications in vehicle propulsion appear likely.

Methanol is an attractive alternate fuel from the economic and environmental point of view. Renewed effort has been put[3] into the development of the direct methanol fuel cell since the need for external reforming is absent. The most successful approach is not yet known.

2.2.2. NON-RECHARGEABLE BATTERIES

Considerable progress has been accomplished with respect to shelf-life and capacity of primary batteries (compare Table 1). The performance data for several systems of interest are shown in Fig.3. The market for this type of batteries expands continuously (compare Table 2) although their production is not desirable from an environmemtal point of view (removal of used batteries).

Table 1: Comparison of the capacity of D-size cells from 1960 to 1985

1960	1979	1965	1975	1985
NH_4Cl	$ZnCl_2$	KOH	KOH	KOH
0.12 Wh/cm^3	0.15 Wh/cm^3	0.21 Wh/cm^3	0.25 Wh/cm^3	0.3 Wh/cm^3

Table 2: Percentage of alkaline cells in markets

	USA	Japan	Europpe
Alk. cells with electrol. MnO_2	60%	20%	20%
Heavy duty Leclanche´	10%	80%	70%
Cells with natural MnO_2	30%	0%	10%

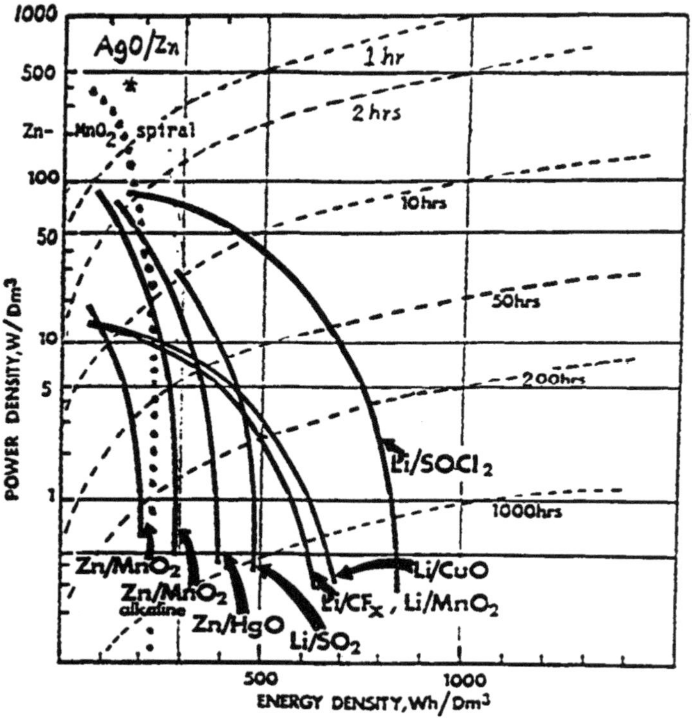

Fig.2. Performance Data for different Primary Batteries

The replacement of primary cells by rechargeable cells is highly desirable. Three examples for the advancement of primary batteries are considered in more detail:

- Manganes dioxide/ Zinc (aqueous electrolytes)
- Lithium/ Manganese dioxide (organic electrolytes)
- Lithium/ Sulfuryl chlorid (organic electrolyte)

The manganese dioxide operates as an insertion electrode for H^+ or Li^+ resp. in the first two systems. A passivating layer which prevents the reaction of lithium with the organic solvent forms on the Li electrode in the second and third system.

Great progress has been made in the reduction of the amount of Hg, added to the Zn electrode in order to reduce the side reaction of H_2 production in the MnO_2/Zn cells. Safety concerns exist for the $Li/LiAlCl_4/SO_2Cl_2/C$ battery which involves a complicated chemistry in the cathode department. A rough estimate shows that about 33000 MWh are yearly produced by throw-way batteries. Since the energy for the manufactoring of these primary batteries is 5 to 10 times larger than 33000 MWh, these batteries should be replaced for reasons of energy savings by rechargeable batteries in the long run.

3. ENERGY STORAGE

3.2. HIGH TEMPERATURE RECHARGEABLE BATTERIES

Three systems are considered here:

- $Li-Al/ FeS_2$
- Na/ S (Beta battery)
- Na/ $MeCl_2$ in $NaAlCl_4$ (Zebra battery, Me = Fe, Ni,Cr,Co)

It appears that the sodium batteries have been developed to a greater extent than the $Li-Al/ FeS_2$ system. The company ABB expects the introduction of the Na/ S battery for the propulsion of vehicles into the market between 1990 and 1992. It is the author's opinion that commercial vehicles the maintenance of which is kept by trained workers or golf carts are to be considered as application in the first place. The personal car, powered by a beta battery, looks problematic with respect to overnight maintenance by the owner. The main problems of the beta battery are:

a) Uneven charging and dischraging of individual cells in the battery

b) Cracking during thermal cycling

To overcome the problem (b) the sulfur electrode was replaced[11] by metal chloride in $NaAlCl_4$. This zebra battery has other advantages. Cell operation becomes feasible at temperatures below 300°C, easing the requirements for cell construction. Safety problems are reduced because the reaction of Na with the molten electrolyte leads to the formation of solid products which slow down the rate of the reaction. Modules of about 30 kWh (appr. 90 Wh/kg, larger than 110 W/kg) were cycled with an efficiency greater than 90% over 100 times. Recently an attempt was reported[12] to use tetracyanoethylene (TCNE) in $NaAlCl_4$.

3.2 LOW TEMPERATURE RECHARGEABLE BATTERIES

The lead acid battery still represents the most important rechargeable battery from the commercial point of view. Sales of this battery accounted for 70% of the battery market in 1986. Small improvements have a great impact because of the large volume of sales. The best laboratory- type Pb/PbO_2 cells have energy densities of about 50 Wh/kg and can be cycled up to 800 times to 80% depth of discharge. Peak power densities can be maintained at the large rate of about 400 W/kg for 1 sec at discharge depths of up to 50%. Today's largest load leveling unit (40 MWh, 40 MW at a discharge rate of 4 h) was completed in the summer of 1988 in Chino, about 50 miles East of Los Angeles.

Ni/Cd batteries were the next important type of rechargeable batteries in 1985. It is to be expected that these batteries will be replaced by other types during the next decade because the Ni/Cd batteries are becoming too expensive and they contain poisonous cadmium.

New developments which are promising occur in the following areas:

- Nickel/ Hydrogen (Ni/ H_2)
- Zinc/ Manganes dioxide (Zn/ MnO_2)
- Lithium/ Intercalating compound (Zn/ MeO)

- Zinc/ Bromine (Zn/ Br_2)

Ni/H_2 batteries are, besides Ni/Cd batteries, the most widely used system in space applications where the price does not matter much. Efforts are under way to produce less expensive batteries of this type. In another approach the hydrogen is stored in nickel alloys.

The rechargeability of MnO_2 in alkaline solutions can be affected by certain additives (Ca and Ba advantageously, Bi and Pb questionably, Al detrimentally). 100 to 200 cycles which may represent years of service for many applications are achievable[13] at low cost and small Hg content. The decrease in rechargeability is due to the capacity loss at the Zn electrode. Energy densities equal to or larger than 80 Wh/kg appear feasible.

A large number of systems with Li is under development (compare Fig.3). The practical energy densities are still far below the theoretical ones for most cells. Modeling studies for an assessment of the energy densities for packaged batteries with an electrolyte, based on complexes between polyethylene oxide and lithium salt, and a composite anode of V_6O_{13} were recently carried out[14] and led to relatively large values of up to 450 Wh/kg for low density metallic or metallized plastic current collectors. For vehicle propulsion designs of 200 Wh/kg and about 700 W/kg were considered possible. Experimental verification of the latter figures is lacking..

Another promising candidate is the Zn/Br_2 battery. In contrast to the other systems, briefly described under 3.2, a flow design is necessary since the electrolyte has to be circulated. The conceptual design, developed by EXXON, was improved[15] by the company SEA in Austria. The following goals are achievable at present: 65 - 80 Wh/kg, 60 - 80 W/kg, 60 - 75 % energy efficiency, cycle life 500 - 1000, 2 - 5 h discharge.

4. REFERENCES

1. N.Weber, Energy Conversion, 14 (1974) 1

2. K.Kishida, Ber.Bunsenges.Phys.Chem., 94 (1190) 94

3. See K.Kordesch, Ber.Bunsenges.Phys.Chem., 94 (1990) 902

4. F.Abraham, J.C.Boivin, G.Mairesse and G.Nowogrocki, J.Solid State Chem.,40/41 (1990) 934

5. R.M.Williams, M.E.Loveland, B.Jeffries-Nakamura, M. L.Underwood, C.P.Bankston, H.Leduc and J.T.Kummer, J.Electrochem. Soc. 137 (1990) 1709

6. P.Stonehart, Ber.Bunsenges.Phys.Chem.,94 (1990) 913

7. A.Winsel, O.Führer, K.Rühling and Ch.Fischer, Ber. Bunsenges.Phys.Chem., 94 (1990) 926

8. K.Strasser, Ber.Bunsenges.Phys.Chem.,94 (1990) 1000

9. K.Ledjeff, J.Ahn, D.Zylka and A.Heinzel, Ber.Bunsenges.Phys.Chem., 94 (1990) 1005

10. G.G.Scherer, Ber.Bunsenges.Phys.Chem., 94 (1990) 1008

11. J.Coetzer, J.Power Sources, 18 (1986) 377

12. B.V.Ratnakuman, S.di Stefano, R.M.Williams, G.Nagasubramanian and C.P.Bankston, J.Appl.Electrochem. 20 (1990) 357

13. K.Kordesch, W.Harer, Y.Sharma, R.Ujj, K.Tomantschger and D.Freeman, 33 rd Internat. Power Sources Symp., Cherry Hill, USA, June 13 - 16, 1988

14. M.Z.A.Munshi and B.B.Owens, Solid State Ionics, 38 (1990) 103

15. G.Tomazić, Österr.Chemie-Zeitschrift, 87 (1986) 104, Österr.Zeitschrift Elektrizitätswirtschaft, 49 (1987) 13

TURBULENT MASS TRANSFER IN ELECTROCHEMICAL SYSTEMS
Turbulence for electrochemistry. Electrochemistry for turbulence

M. A. VOROTYNTSEV
A.N.Frumkin Institute of Electrochemistry, Academy of Sciences of the USSR
Moscow 117071, Leninsky prospekt 31, USSR

ABSTRACT

Key problems of turbulent mass transfer at a solid wall are overviewed: closure problem for the concentration field, information on wall turbulence, applications of microelectrodes to study the structure of turbulence, correlation properties of current fluctuations.

1. Introduction

If sweeping the electrode potential in presence of some electroactive species, the current through the electrode is determined both by kinetics of electrode reaction and reactive species transport to the electrode. The latter phenomenon is usually a rate-determining one at sufficiently high overpotentials. The value of this limiting diffusion current, I_d, can be found if solving the convection-diffusion equation,

$$\partial c/\partial t + v\nabla c - D\nabla^2 c = 0, \qquad (1)$$

for the concentration field, c(r,t), of a reagent, D being its molecular diffusivity, v(r,t), the fluid velocity field.

The velocity field has been found theoretically or experimentally for numerous simple geometries of the system if the reagent concentration is low enough to neglect the effects of its variation in the hydrodynamic flow. Then, the corresponding distributions of the concentration field, c(r,t), have been found after solving equation (1) so that the diffusion flux density at the boundary of the flow, $J=-D\partial c/\partial \nu$, has been expressed through the hydrodynamic characteristics of the flow, the molecular diffusivity, D, and parameters of the boundary conditions for the concentration field.

Much more complicated problem arises in the case of a turbulent regime of the flow. Mass transport is governed by an irregular fluctuative velocity field, $v(r,t)$. As a result, both the concentration field and the boundary mass fluxes are also fluctuative so that only some averaged characteristics of these random distributions are worth of consideration. The simplest quantities of this kind are the average concentration field, $\bar{c}(r,t)$, and the average surface mass flux, J. These are the characteristics which have been mostly considered.

2. Average concentration distribution

If splitting all quantities into the average and the pulsation parts, e.g. $v=u+v'$, $c=\bar{c}+c'$, the average of the latters being zero, $\langle v'\rangle=0$, $\langle c'\rangle=0$, then equation (1) gives an equation for $\bar{c}=\langle c\rangle$ after carrying out an averaging operation,

$$\partial\bar{c}/\partial t+u\nabla\bar{c}-D\nabla^2\bar{c}=-J_t,\quad J_t(r,t)=\langle c'(r,t)v'(r,t)\rangle \tag{2}$$

Unlike eqn.(1) the equation for the average concentration is not closed since it contains another unknown quantity, the turbulent mass flux, J_t. To solve this closure problem means to express J_t through the average concentration field as well as the molecular diffusivity, D, and some average hydrodynamic quantities.

Numerous semi-empirical and phenomenological approaches have been proposed for its solution. Most of them included popular assumptions like locality hypothesis on the relation between J_t and the average concentration gradient,

$$J_t(r,t)\cong-D_t(r,t)\nabla\bar{c}(r,t), \tag{3}$$

the eddy diffusivity, D_t, being related to a similar hydrodynamic quantity, the eddy viscosity, ν_t, by Reynolds analogy,

$$D_t\cong\nu_t \tag{4}$$

Since an intensity of velocity pulsations is diminishing at approaching a solid boundary, then both these eddy characteristics are to be having a similar behaviour. A power law has usually been postulated for them,

$$D_t\sim\nu_t\sim y^k, \tag{5}$$

y being the distance from the solid wall. If taking into account a relation for ν_t, $\nu_t\sim\langle v_x'v_y'\rangle$, the value of the exponent, k, is to be 3 or more. The minimal value, k=3, has been accepted by many authors. But their numerous opponents were giving arguments in favour of a larger value, k=4, assuming the coefficient at y^3 term being zero. Various experimental data for average mass transport have not been precise enough to distinguish between these two variants.

Most theoretical activity has been paid to the simplest geometry of the system: stationary mass transfer across the diffusion boundary layer of a constant thickness, δ_D. All one-point characteristics like the $\bar{c}$ and u depend on the normal coordinate, y, only, the average velocity is directed along the x-axis. Since molecular Prandtl-Schmidt number is very large for liquids, e.g. $Pr \sim 10^3$ for water, the whole region of concentration variation is located within a thin viscous sublayer of the turbulent flow near the solid wall so that simple power expansions can be used for hydrodynamic quantities, e.g. $u \cong \alpha y$.

Above hypotheses (3)-(5) result in estimates both of the average mass flux at the solid wall, $\langle J \rangle$, and of its dependence on the molecular diffusivity, D, which are in reasonable agreement with experimental data for both assumed k values, 3 or 4. But this success do not open a way how similar problems for a more complicated geometry can be treated. In addition, mass transfer experiments cannot be used to study characteristics of wall turbulence.

The boundary layer of a turbulent flow includes three regions, a logarithmic sublayer, buffer zone and viscous sublayer. The latter two have a very small total thickness. Therefore there are limitations in employing usual techniques like thermoanemometric, stroboscopic or laser interferometric ones to study local turbulence characteristics within these two internal regions, especially within the viscous sublayer. Possibilities to deduce this information from theoretical reasoning are restricted since a similar closure problem for hydrodynamic characteristics has not been overcome as yet. The situation concerning pulsations of the normal component of the fluid velocity, v_y', is especially complicated. At the same time this is the characteristics which determines the turbulent mass flux to the flow boundary and its knowledge is of crucial importance for theoretical chemical engineering.

The closure problem for the average mass transport, equation (2), has been treated within a statistical hydrodynamics theory[1-7] based on an equation for the concentrational fluctuations, $c'(r,t)$, obtained by subtracting eqn (2) from eqn (1),

$$\partial c'/\partial t + u\nabla c' - D\nabla^2 c' = -v'\nabla\bar{c} - \nabla(v'c' - \langle v'c'\rangle), \qquad (6)$$

Its solution was based on a linear approach, i.e. on neglecting the second-order terms, $v'c'-\langle v'c'\rangle$. As a result, a unified relation for the turbulent mass flux has been found[1-7],

$$J_t(r,t) \cong -\int \Delta(r,t;r',t')\nabla\bar{c}(r',t')dr'dt' \qquad (7)$$

though expressions for the turbulent diffusivity density, Δ, are somewhat different in these publications depending on an approximation used. The locality hypothesis, eqn (3), has also been approximately justified for the case of a stationary one-dimensional mass transport, the turbulent diffusivity being expressed through characteristics of the normal velocity pulsations[1,5-7]. This relation gives a power behaviour, $D_t \sim y^3$, in the close vicinity of the solid wall, in agreement with the Reynolds analogy (4), but the turbulent Prandtl number, $Pr_t \equiv \nu_t/D_t$, turns out to exceed considerably unity and depends significantly on its molecular analog, Pr. Moreover, this law for $D_t(y)$ is valid within a very thin layer which plays a minor role in mass transport characteristics. Quite a different relation for $D_t(y)$ was obtained at larger distances from the wall,

$$D_t(y) \cong \int_{-\infty}^{t} \langle v_y'(r,t)v_y'(r,t')\rangle\, dt' \sim \langle v_y'^2\rangle\, T_\nu, \qquad (8)$$

T_ν being the hydrodynamical correlation time. This law corresponds to a different power law behaviour, $D_t \sim y^4$, so that the Reynolds analogy is not valid in this region, $Pr_t \sim y^{-1}$. The total mass transport is governed mostly by the latter region where equation (8) can be approximately used.

Relations (3) and (8) for the turbulent mass flux result in a closed equation for the average concentration, $\bar{c}(y)$, so that the mass flux across the diffusion boundary layer can be expressed through the product, $\langle v_y'^2\rangle T_\nu$, at the outer boundary of the viscous sublayer. Since the corresponding experimental data are not especially exact as yet it is probably more reasonable to use mass transport data, the Nu(Pr) dependence, to find the above combination of these hydrodynamical characteristics. This idea has been employed[8] for the first evaluation of the correlation time of the normal velocity pulsations, $T_\nu \sim (2\text{-}7)\bar{\alpha}$, $\bar{\alpha} = (du/dy)_{y=0}$.

3. Electrochemical methods in turbulence

Though such consideration of the average mass transport has given quite a new information on the normal velocity fluctuations it is evident that the whole variety of space-time correlation properties of the velocity fluctuations, $\langle v_\alpha'(r,t)v_\beta'(r',t')\rangle$, within the viscous sublayer cannot be determined from this experimental characteristics. Much more extensive information is covered in fluctuative properties of the current density through the boundary, $J'(s,t) \equiv J(s,t) - \langle J\rangle$, since these fluctuations are produced by the fluid velocity pulsations.

This idea to study the wall turbulence by an electrochemical method proposed for the first time by Hanratty[9] has attracted numerous scientists. Most attention has been paid to the fluctuative properties of an individual microelectrode[9-20]. The average current through such an electrode is directly connected with the average friction, i.e. $\bar{\alpha} = (du/dy)_{y=0}$, whereas its fluctuations, $I'(t)$, are determined by the local velocity pulsations so that their spectral density,

$$W_I(\omega) \equiv |I(\omega)|^2, \quad I(\omega) = \int I'(t)e^{i\omega t}dt \tag{9}$$

is related to the velocity one, W_v. If the size of the electrode is small enough the dominant role is played by the longitudinal velocity component, $v_x' \cong \alpha'(x,z,t)y$. If the second-order terms of eqn (6) are neglected a linear relation is valid between the corresponding spectral densities,

$$W_I(\omega) \cong |H(\omega)|^2 W_\alpha(\omega) \tag{10}$$

Numerous analytical and numerical calculations of the transfer function, $H(\omega)$, are now available[9-20] though there is still some divergence between some of those results.

It has been recently found[19,20] that current fluctuations of a microelectrode are markedly influenced by the normal velocity fluctuations, v_y', as well if its size is not too small. The corresponding transfer functions of the relation between $W_I(\omega)$ and the spectral functions of both velocity components, $v_x(r,t)$ and $v_y(r,t)$, are now available[19,20] in an analytical and numerical forms.

If a circular microelectrode is composed of two isolated ones then the measurement of their currents, $I_1(t)$ and $I_2(t)$, is enabling one to find the spectral function of the transverse velocity component, v_z', which is related to the difference of those currents[21].

All the above measurements are giving an information on the temporal autocorrelation of the velocity components at the same spatial point. The cross-correlation, i.e. correlation of the velocity components at different spatial points, $\langle v_\alpha'(r,t)v_\beta'(r',t')\rangle$, can only be studied with the use of a set of microelectrodes. This idea has been realized[22] with respect to the x-component in both directions along the wall, x and z. Similar information on the v_z' correlation might probably be obtained with the use of a set of composed microelectrodes.

4. Normal component of velocity pulsations

The most challenging problem is to measure characteristics of the normal velocity component, v_y'. An estimate of its integral intensity, $\langle v_y'^2\rangle$, within the viscous sublayer has been obtained[22] with use of the continuity equation for the velocity components, after simultaneous measurement of two other components in different spatial points, with subsequent carrying out an operation of numerical differentiation. The value found turned out to be much smaller than that obtained by an extrapolation of the results within the buffer zone. So, this is an important achievement of the electrochemical method. At the same time it is hardly possible to hope for determination of the whole correlation function, $\langle v_y'(r,t)v_y'(r',t')\rangle$, along this line which would be highly important for better understanding of the wall turbulence.

An alternative way to solve this problem has been proposed[7,23-26]. Its basic idea is to carry out simultaneous measurements of current density fluctuations at a set of wall points within a stabilized diffusion boundary layer of a constant thickness. Such experimental data might be expected to give an immediate information on the normal velocity pulsations, v_y', since this is the only component which is efficient in interacting with the average concentration gradient, $\nabla\bar{c}(y)$, eqn (6), in such a system, unlike the case of an individual microelectrode. The technical realization of this idea means to incorporate a set of microelectrodes inside an extensive electrode, all those having the same electrode potential, i.e. equal average current densities. Then, instantaneous currents of these microelectrodes, $I'(x_i,z_i,t)$, could be measured, x_i,z_i being coordinates of their centers. This would give us auto- and cross-correlation functions of the current densities, $\langle j'(x,z,y)j'(x',z',t')\rangle$, at varying differences of the variables, x-x', z-z' and t-t'. One could expect that functions are directly related to the spatial and temporal correlation of the normal velocity pulsations, $\langle v_y'(r,t)v_y'(r',t')\rangle$, near the solid wall.

Our theoretical studies have confirmed this intuitive idea but the above relation between the current and velocity fluctuative properties turned out to be much more complicated than that for an individual microelectrode.

It is easy to derive this relation within a linear theory, if neglecting the second-order fluctuative terms in eqn (6). Then,

$$c'(r,t) \cong -\int G(r,t;r',t')\, v_y'(r',t')\, d\bar{c}/dy\, dr'dt' \qquad (11)$$

G being the Green function of the differential operator in the left-hand side of eqn (6), and the transfer function analogous to $H(\omega)$, eqn (10), would be expressed through G and $d\bar{c}/dy$. But the linear theory has been found to be not valid in the case of mass transport across the diffusion boundary layer of a constant thickness[24-26].

Though each individual concentration fluctuation, $c'(r,t)$, is small compared with the average concentration difference, $|c'| \ll c_0$, this fluctuation is living during an extremely long time interval, $t_D \sim \delta_D^2/D$. As a result, the overall intensity of concentration fluctuations, $\langle c'^2 \rangle$, is about c_0^2. Moreover, each concentration cloud, $c'(r,t)$, is considerably shifted by the average flow, $u(y)$, during its long life-time, t_D. In addition, though the intensity of each individual v' pulsation is rather weak their combined action on a c' fluctuation is quite strong and results in appearance of an effective turbulent diffusion which is smearing out this concentration field.

These conclusions to be drawn[24-26] called for an analysis of the infinite chains of coupled equations for various correlation functions, $\langle c'c'v'\ldots v' \rangle$, $\langle c'v'\ldots v' \rangle$ and $\langle v'\ldots v' \rangle$. As a result, closed equations have been derived for binary correlators,

$$\tilde{L}\, \langle c'(r,t)v'(r',t') \rangle \cong -\langle v'(r,t)v'(r',t') \rangle \nabla\bar{c}(r,t), \qquad (12)$$

$$\tilde{L}\,\tilde{L}\, \langle c'(r,t)c'(r',t') \rangle \cong \langle v'(r,t)v'(r',t') \rangle \nabla\bar{c}(r,t)\nabla\bar{c}(r',t') +$$
$$+ \int dr''dt'' \langle v'(r,t)v'(r'',t'') \rangle\, \langle c'(r'',t'')c'(r',t') \rangle, \qquad (13)$$

$$\tilde{L} \equiv \partial/\partial t + u\nabla - \nabla_\alpha (D\delta_{\alpha\beta} + D_{\alpha\beta})\nabla_\beta, \quad D_{\alpha\beta}(r) = \int_{-\infty}^{t} \langle v_\alpha'(r,t)v_\beta'(r,t') \rangle\, dt',$$

(tensor indices are mostly omitted). An approximate solution of eqn (13) has enabled us[25] to derive an expression for the transfer function, Q,

$$\tilde{I}(æ,\zeta,\omega)=|Q(æ,\zeta,\omega)|^2\ \tilde{B}(æ,\zeta,\omega), \tag{14}$$

$$\tilde{B}(æ,\zeta,\omega)=\int_{-\infty}^{\infty}dxdzdt\ e^{ixæ+iz\zeta+it\omega}\ \langle\beta(0,0,0)\beta(x,z,t)\rangle,\ \beta(x,z,t)\cong v_y(r,t)/y^2$$

The similar 3-dimensional Fourier transformation relates $\tilde{I}(æ,\zeta,\omega)$ to the correlation function of the current density, $\langle j'(x,z,t)j'(x',z',t')\rangle$. The latter could be measured within an experiment with a set of microelectrodes inside a large electrode, as discussed above.

The influence of nonlinear terms in eqn (6), i.e. higher-order terms in the chain of coupled equations for the correlation functions turned out to be considerable. For example, their account results in a threefold decrease of the transfer function, $|Q(æ,\zeta,\omega)|^2$, eqn (14), in comparison with that within the linear theory.

Thus, this direction of studies might be expected to produce the whole picture of space-time correlation properties of the most important hydrodynamic characteristic, the normal component of velocity pulsations, $v_y'(r,t)$, within the viscous sublayer of the flow.

REFERENCES

1. P.I.Geshev, Sov.J.Applied Mech.Techn.Phys. N 2(1974)61.
2. C.A.Petty, Chem.Eng.Sci. 30(1975)413.
3. C.A.Petty, Chem.Eng.Commun. 7(1980)261,275.
4. C.A.Petty, Physicochem.Hydrodyn. 8(1987)31.
5. S.A.Martemyanov,M.A.Vorotyntsev and B.M.Grafov, Sov.Electrochemistry 15(1979)913,916.
6. M.A.Vorotyntsev,S.A.Martemyanov and B.M.Grafov, JETP-Sov.Physics 79(1980)1797.
7. M.A.Vorotyntsev,S.A.Martemyanov and B.M.Grafov, J.Electroanal.Chem. 179(1984)1.
8. S.A.Martemyanov,M.A.Vorotyntsev and B.M.Grafov, Vestn.Moscow State Univ. (Math. and Mech.) N 3(1980)67.
9. J.E.Mitchell and T.J.Hanratty, J.Fluid Mech. 26(1966)199.
10. K.K.Sirkar and T.J.Hanratty, J.Fluid Mech. 44(1970)605.
11. G.Fortuna and T.J.Hanratty, Int.J.Heat Mass Transfer 14(1971)1499.
12. T.J.Hanratty and J.A.Campbell, in Fluid Mechanics Measurements,

ed. R.J.Goldstein (Hemisphere,New York,1983),p.559.
13. V.E.Nakoryakov,A.P.Burdukov,O.N.Kashinsky and P.I.Geshev, Electrodiffusion Method to Study a Local Structure of Turbulent Flows, ed.V.G.Gazenko (IT AN,Novosibirsk,1986).
14. Yu.Ye.Bogolyubov,P.I.Geshev,V.E.Nakoryakov and I.A.Ogorodnikov, Sov J.Applied Mech.Techn.Phys. 4 (1972)112.
15. C.Deslouis,B.Tribollet and L.Viet, in 4th Intern.Conference on Physicochemical Hydrodynamics, Ann.NY Acad.Sci. 404 (1983)471.
16. A.Ambari,C.Deslouis and B.Tribollet, Int.J.Heat Mass Transfer 29 (1986)35.
17. C.Deslouis,O.Gil and B.Tribollet, J.Fluid Mech.(in press).
18. T.J.Hanratty and E.Vassiliadou, in Transport Phenomena in Turbulent Flows, eds. M.Hirata & N.Kasagi (Hemisphere,New York,1988),p.255.
19. S.A.Martemyanov, Sov.Electrochemistry 21 (1985)842,942.
20. E.F.Skurygin,S.A.Martemyanov,M.A.Vorotyntsev and B.M.Grafov, Sov.Electrochemistry 25 (1989)772.
21. C.Deslouis,O.Gil and V.Sobolik, Int.J.Heat Mass Transfer 33 (1990)1363.
22. D.S.Finnicum and T.J.Hanratty, Phys.Fluids 28 (1985)1654.
23. M.A.Vorotyntsev,S.A.Martemyanov and B.M.Grafov, Sov.Electrochemistry 16 (1980)915,919.
24. E.F.Skurygin and M.A.Vorotyntsev, Sov.Electrochemistry 23 (1987)1001.
25. E.F.Skurygin,M.A.Vorotyntsev and S.A.Martemyanov, Sov.Electrochemistry 25 (1989)663,668.
26. E.F.Skurygin,M.A.Vorotyntsev and S.A.Martemyanov, J.Electroanal.Chem. 259 (1989)285.

LIST OF CONTRIBUTORS

G. Aloisi	Dipartimento di Chimica Università di Firenze Via G. Capponi, 9 50121 Firenze, Italy
S. Amokrane	Structure et réactivité aux interfaces Université P. et M. Curie F(74) 4 Place Jussieu 75230 Paris Cedex 05, France
J.-P. Badiali	Structure et réactivité aux interfaces Université P. et M. Curie F(74) 4 Place Jussieu 75230 Paris Cedex 05, France
Ph. Bopp	Institut für Physikalische Chemie Rheinisch-Westfälische Technische Hochschule Templergraben 59 D-5100 Aachen, Germany
Z. Borkowska	Institute of Physical Chemistry Polish Academy of Sciences Warsaw 01-224, Poland
M.W. Breiter	Institut für Technische Elektrochemie Technische Universität Wien 9 Getreidenmarkt A-1060 Wien, Austria
P. Brito	Instituto di Fisica Facultad de Ciencias Exactas y Tecnologia Universidad Nacional de Tucuman Av. Independencia 1800 4000 Tucuman, Argentina
M. Cappadonia	European Patent Office D-800 München 2, Germany
H.E. Christensen	Chemistry Department A Building 207 The Technical University of Denmark 2800 Lyngby, Denmark

L.S. Conrad	Chemistry Department A Building 207 The Technical University of Denmark 2800 Lyngby, Denmark
T. Fonseca	Department of Chemistry Colorado State University Fort Collins, CO 80523, USA
A. Friedrich	Fritz-Haber-Institut der Max-Planck-Gesellschaft Faradayweg 4-6 D-1000 Berlin 33, Germany
A.M. Funtikov	The A.N. Frumkin Institute of Electrochemistry Academy of Sciences of the USSR Leninskii prosp. 31 117071 Moscow, USSR
H. Gerischer	Fritz-Haber-Institut der Max-Planck-Gesellschaft Faradayweg 4-6 D-1000 Berlin 33, Germany
J.K. Gimzewski	IBM Zürich Research Laboratory CH-8803 Rüschlikon, Switzerland
R. Guidelli	Dipartimento di Chimica Università di Firenze Via G. Capponi, 9 50121 Firenze, Italy
J.M. Hammerstad-Pedersen	Chemistry Department A Building 207 The Technical University of Denmark 2800 Lyngby, Denmark
A.T. Hubbard	Department of Chemistry University of Cincinnati Cincinnati, OH 45221-0172, USA
J.T. Hynes	Department of Chemistry and Biochemistry University of Colorado Boulder, CO 80309-0215, USA

D.E. Irish	Department of Chemistry University of Waterloo Ontario N1L 3G1, Canada
G. Jarzabek	Institute of Physical Chemistry Polish Academy of Sciences Warsaw 01-224, Poland
V.E. Kazarinov	The A.N. Frumkin Institute of Electrochemistry Academy of Sciences of the USSR Leninskii prosp. 31 117071 Moscow, USSR
H.J. Kim	Department of Chemistry and Biochemistry University of Colorado Boulder, CO 80309-0215, USA
D.M. Kolb	Institut für Elektrochemie Universität Ulm D-7900 Ulm, Germany
A.A. Kornyshev	The A.N. Frumkin Institute of Electrochemistry Academy of Sciences of the USSR Leninskii prosp. 31 117071 Moscow, USSR
D.A. Kossakowski	The A.N. Frumkin Institute of Electrochemistry Academy of Sciences of the USSR Leninskii prosp. 31 117071 Moscow, USSR and Department of Physical and Chemical Biology Moscow Physico-Technical Institute Institutskii per., 9 141700 Dolgoprudny, USSR
L.I. Krishtalik	The A.N. Frumkin Institute of Electrochemistry Academy of Sciences of the USSR Leninskii prosp. 31 117071 Moscow, USSR
A.M. Kuznetsov	The A.N. Frumkin Institute of Electrochemistry Academy of Sciences of the USSR Leninskii prosp. 31 117071 Moscow, USSR

B.M. Ladanyi	Department of Chemistry Colorado State University Fort Collins, CO 80523, USA
A. Liebsch	Institut für Festkörperforschung der Forschungscentrum Jülich Postfach 1913 D-5170 Jülich, Germany
J. Lipkowski	Department of Chemistry and Biochemistry University of Guelph Guelph, Ontario N1G 2W1, Canada
S.H. Liu	Solid State Division Oak Ridge National Labortory Oak Ridge, TN 37831-6032, USA
W.J. Lorenz	Institut für Physikalische Chemie und Elektrochemie Universität Karlsruhe Karlsruhe, Germany
B. Pettinger	Fritz-Haber-Institut der Max-Planck-Gesellschaft Faradayweg 4-6 D-1000 Berlin 33, Germany
Ping Gao	Department of Chemistry University of Cincinnati Cincinnati, OH 45221-0172, USA
G. Saemann-Ischenko	Institut für Physik Universität Erlangen-Nürnberg Erlangen, Germany
J.K. Sass	Fritz-Haber-Institut der Max-Planck-Gesellschaft Faradayweg 4-6 D-1000 Berlin 33, Germany
B.B. Smith	Department of Chemistry and Biochemistry University of Colorado Boulder, CO 80309-0215, USA
U. Stimming	Institut für Elektrochemie Universität Ulm D-7900 Ulm, Germany

L. Stolberg	Department of Chemistry and Biochemistry University of Guelph Guelph, Ontario N1G 2W1, Canada
M.P. Tosi	Dipartimento di Fisica Teorica Università di Trieste Strada Costiera 11 I-34100 Trieste, Italy
J. Ulstrup	Chemistry Department A Building 207 The Technical University of Denmark 2800 Lyngby, Denmark
M.I. Urbakh	Faculty of Exact Sciences School of Chemistry Tel-Aviv University Ramat Aviv 69 978 Tel-Aviv, Israel
M.A. Vorotyntsev	The A.N. Frumkin Institute of Electrochemistry Academy of Sciences of the USSR Leninskii prosp. 31 117071 Moscow, USSR